Macmillan Building and Surveying Series
Series Editor: Ivor H. Seeley
Emeritus Professor, The Nottingham Trent University

Advanced Building Measurement, second edition Ivor H. Seeley
Advanced Valuation Diane Butler and David Richmond
Applied Valuation Diane Butler
Asset Valuation Michael Rayner
Building Economics, fourth edition Ivor H. Seeley
Building Maintenance, second edition Ivor H. Seeley
Building Maintenance Technology Lee How Son and George C. S. Yuen
Building Procurement Alan E. Turner
Building Project Appraisal Keith Hutchinson
Building Quantities Explained, fourth edition Ivor H. Seeley
Building Surveys, Reports and Dilapidations Ivor H. Seeley
Building Technology, fifth edition Ivor H. Seeley
Civil Engineering Contract Administration and Control, second edition Ivor H. Seeley
Civil Engineering Quantities, fifth edition Ivor H. Seeley
Commercial Lease Renewals Philip Freedman and Eric F. Shapiro
Computers and Quantity Surveyors Adrian Smith
Conflicts in Construction – Avoiding, Managing, Resolving Jeff Whitfield
Constructability in Building and Engineering Projects Alan Griffith and Tony Sidwell
Construction Contract Claims Reg Thomas
Construction Law Michael F. James
Contract Planning and Contractual Procedures, third edition B. Cooke
Contract Planning Case Studies B. Cooke
Cost Estimation of Structures in Commercial Buildings Surinder Singh
Design–Build Explained David E. L. Janssens
Development Site Evaluation N. P. Taylor
Environmental Management in Construction Alan Griffith
Environmental Science in Building, third edition R. McMullan
European Construction – Procedures and Techniques B. Cooke and G. Walker
Facilities Management – An Explanation J. Alan Park
Greener Buildings – Environmental Impact of Property Stuart Johnson
Housing Associations Helen Cope
Housing Management: Changing Practice Christine Davies (Editor)
Information and Technology Applications in Commercial Property Rosemary Feenan and Tim Dixon (Editors)
Introduction to Building Services, second edition Christopher A. Howard and Eric C. Curd
Introduction to Valuation, third edition David Richmond
Marketing and Property People Owen Bevan
Principles of Property Investment and Pricing, second edition W. D. Fraser
Project Management and Control D. W. J. Day
Property Finance David Isaac
Property Valuation Techniques David Isaac and Terry Steley
Public Works Engineering Ivor H. Seeley
Quality Assurance in Building Alan Griffith
Quantity Surveying Practice Ivor H. Seeley
Recreation Planning and Development Neil Ravenscroft
Resource Management for Construction M. R. Canter
Small Building Works Management Alan Griffith

(continued overleaf)

Structural Detailing, second edition P. Newton
Sub-Contracting under the JCT Standard Forms of Building Contract Jennie Price
Urban Land Economics and Public Policy, fifth edition P. N. Balchin, J. L. Kieve and G. H. Bull
Urban Renewal – Theory and Practice Chris Couch
1980 JCT Standard Form of Building Contract, second edition R. F. Fellows

BUILDING ECONOMICS

Appraisal and control of building design cost and efficiency

IVOR H. SEELEY

BSc (Est Man), MA, PhD, FRICS,
CEng, FICE, FCIOB, FCIH
Chartered Quantity Surveyor
Emeritus Professor of The Nottingham Trent University

Fourth Edition

First edition 1972
Reprinted three times
Second edition 1976
Reprinted four times
Third edition 1983
Reprinted seven times
Fourth edition 1996

Published by
MACMILLAN PRESS LTD
Houndmills, Basingstoke, Hampshire RG21 6XS
and London
Companies and representatives
throughout the world

ISBN 0–333–63835–2

A catalogue record for this book is available from the British Library

10 9 8 7 6 5 4 3 2
05 04 03 02 01 00 99

Printed in Malaysia

CONTENTS

Preface to Fourth Edition ix

Acknowledgements xi

List of Figures xiii

List of Tables xv

1 **The Concept of Cost Control** 1
Introduction; historical development of cost control processes; European context; need for cost control; main aims of cost control; cost, price and value; the importance of building; building output and costs; tendering arrangements; building procurement; public procurement in the European Community; comparison of cost planning and approximate estimating; cost control terminology; BCIS; BMI; CEEC; other sources of European information; cost implications of dimensional co-ordination; introduction to value management/engineering; computer aids to cost planning; expert systems; examiners' reports.

2 **Cost Implications of Design Variables and Quality Assurance** 31
Plan shape; size of building; perimeter/floor area ratios; circulation space; storey heights; total height of buildings; grouping of buildings; relative costs of flats and houses; implications of variations in the number of storeys of buildings; column spacings; floor spans; floor loadings; cost estimation of structures of commercial buildings; design cost criteria; constructability; quality assurance.

3 **Functional Requirements and Cost Implications of Constructional Methods** 56
Low and high rise buildings; substructures; structural components; walling; roofing; flooring; doors and windows; finishings; service installations; automation; fire protection; external works.

4 **Influence of Site and Market Conditions and Economics of Prefabrication, Industrialised and System Building** 90
Effect of site conditions on building costs; use of plant; site productivity; market considerations; cost implications of prefabrication and standardisation; industrialised/system building methods; industrialised and system built housing; problems with industrialised/system building; economics of industrialised/system building; the future in industrialised/system building and their main defects; refurbishment of tower blocks; CLASP.

5 **Economics of Residential Development** 113
Background to public housing provision; later housing developments in the UK; use of land for housing purposes; assessment of housing need; some financial aspects of housing provision; dwelling types; housing requirements of occupants; patterns of development to meet varying density requirements; economics of housing layouts; some overseas developments; car parking provision; rehabilitation and modernisation of older dwellings; latest trends in housing.

6 Approximate Estimating 154
Purpose and form of approximate estimating techniques; unit method; cube method; superficial or floor area method; storey-enclosure method; approximate quantities; elemental cost analyses; comparative estimates; interpolation method.

7 Cost Planning Theories and Techniques 174
Plan of work; cost control procedure; information required by architect and building client; role of the quantity surveyor during the design stage; cost planning techniques; building industry code; C1/SfB classification system; cost planning of mechanical and electrical services; cost control during execution of contract; cost control by the contractor.

8 Cost Modelling 202
Introduction; accuracy; historical development of cost modelling; BCIS on-line approximate estimating package; Bucknall Austin building cost model; Davis Langdon and Everest cost model; purposes of cost modelling; approaches to cost modelling; types of model; risk analysis/Monte Carlo simulation; statistical models and regression analysis; other cost modelling applications; simulated modelling; network cost modelling systems; expert systems; conclusions.

9 Cost Analyses, Indices and Data 211
Cost analyses; standard form of cost analysis; cost limits; building cost/price indices; application and use of cost analyses; cost data; cost research.

10 Practical Application of Cost Control Techniques 248
Worked examples 1, 2 and 3 covering the preparation of a preliminary estimate, first cost plan, and cost checks and cost reconciliation during the design process; conclusions; cost control of engineering services.

11 Value Management 277
General principles; value management definitions; reasons for the client commissioning VM studies; alternative approaches: the charette; 40 hour management workshop/study; one–two day workshop/study; other alternatives; VM strategy; comparison of VM and cost management; VM techniques – functional analysis; FAST diagrams; criteria scoring/alternative analysis matrix; VM case studies: Computer Centre, Northern England; Bank Processing Centre, Northern England; conclusions.

12 Valuation Processes 296
The concept of value and investment; methods of valuation; valuation tables; rental value; premiums; service charges.

13 Life Cycle Costing 308
Concept of life cycle costing; volume and impact of building maintenance work; value for money; practical problems which affect life cycle costing; life cycle costing terminology; the technology of maintenance; types of maintenance; current and future payments; maintenance and running costs; occupancy costs; the lives of buildings and components; life cycle cost plans; practical life cycle costing examples; energy conservation; prediction errors; effect of taxation and insurance; maintenance cost records and data; greener buildings; facilities management; intelligent buildings.

14 Land Use and Value Determinants 380
Changing land use requirements; land use planning; land values; factors influencing development; encumbrances and easements; matters determining land use and value; land and building values.

15 Economics of Building Development 403

Nature of property; the essence of development; development properties; budgeting for public and private development; land acquisition procedures and problems; financial considerations; sources of development finance; project finance; developer's budget; feasibility study; choice between building lease or purchase.

16 Environmental Economics and the Construction Industry 434

Environmental management; concept of environmental economics; public and private investment; structure of the construction industry; variations in workload on the construction industry; relationship of output of construction industry to available resources; effect of government action on the construction industry; European Union proposals for growth and competitiveness; urban renewal and town centre redevelopment; new and expanding towns; building conservation and urban regeneration; cost benefit analysis; private finance projects.

Appendix 1: Amount of £1 table 476
Appendix 2: Present value of £1 table 478
Appendix 3: Amount of £1 per annum table 480
Appendix 4: Annual sinking fund table 482
Appendix 5: Present value of £1 per annum or years' purchase table 484
Appendix 6: Metric conversion table 486
Appendix 7: Life cycle cost plan 488

References 493

Index 499

PREFACE TO FOURTH EDITION

The fourth edition has been comprehensively rewritten, updated and extended to cover the latest developments and techniques in this rapidly expanding, changing and vital area of activity. Cost management has become the most important single facet of the work undertaken by the quantity surveyor, with the primary objective of controlling construction costs and obtaining value for money, set against perceived performance expectations. It is imperative to secure projects that satisfy the client's requirements with regard to cost, time, function and quality. There has developed a pressing need to refine the tools of cost prediction and control, and for the quantity surveyor to possess a wide knowledge of the factors influencing construction costs and other related development aspects; this book seeks to address this need.

Hence the scope of the book has been enlarged substantially to take account of the widening and more sophisticated cost management and control service required by the discerning client. It is believed that the new edition will meet the needs of the practising surveyor and student more effectively, by providing the breadth of knowledge needed to make informed judgements on the many diverse matters coming within the remit of this subject. The book should be of value to students on quantity surveying degree, higher diploma, BTEC and NVQ courses, be of considerable use to practising quantity surveyors as a handy means of reference, while architects, building surveyors, property managers and contractors may find much of interest within its pages.

The RICS Report *Quantity Surveying 2000* (QS 2000) – The Future Role of the Chartered Quantity Surveyor, published in 1991, emphasised the need to provide more accurate and robust forecasts of construction costs and prices and particularly the development of the techniques of early cost advice, construction cost and market forecasting and cost control, including the use of cost modelling and the application of latest computer technology. Seen in this context, cost forecasting is a truly professional service, requiring the exercise of a high level of expert discretion and judgement in conditions of considerable uncertainty over client requirements, design and future cost and price movements in the industry and in the economy generally. It is hoped that the revised text will assist the reader in undertaking these activities and having regard to the desirability of incorporating future costs in comparative analyses.

Two new chapters have been added: one on 'Value Management' and an introductory chapter on cost modelling, as these areas have assumed much greater importance and the quantity surveyor needs to be familiar with their main characteristics, uses and applications.

In addition, the text has been widened significantly to incorporate many new aspects including the European context, expert systems, quality assurance, constructability, cost implications of automation and fire protection, latest housing developments in the UK and overseas with their cost consequences, cost control by the contractor, extension of cost planning of M&E services, enlargement of life cycle costing, occupancy costs, facilities management, energy conservation, greener buildings, intelligent buildings, land and building values, property characteristics, sources of development finance, project finance, environmental management, environmental impact assessment, urban regeneration, building conservation and private finance projects.

The number of case studies and supporting tables and diagrams has also been increased considerably to assist in bringing the subject to life and to make the text more easily readable.

Unfortunately, fluctuating building and land prices and interest rates will continue to affect the rates and prices contained in the text, which are mainly those operative in 1992 to 1994. Hence these should be taken as indicative and are used primarily to illustrate the principles and techniques involved. In practice current local rates and prices should be used wherever possible.

Nottingham
Autumn 1994

IVOR H SEELEY

ACKNOWLEDGEMENTS

The author acknowledges with gratitude the willing co-operation and assistance received from many organisations and individuals, so many that it is not possible to mention them all individually.

Valuable information has been obtained from a variety of sources including RICS Journals, the *Architect's Journal*, *Building*, *Building Economist*, *Housing Associations Weekly*, *Housing*, *Inside Housing*, *New Builder*, *Estates Times*, *Building Technology and Management and Specification*, and from the work of the Department of the Environment, the Treasury, Building Research Establishment, British Standards Institution, Energy Efficiency Office (DoE), Royal Institution of Chartered Surveyors, Chartered Institute of Housing, National Federation of Housing Associations, CIRIA and the National Association of Lift Manufacturers, for which the author is most grateful.

The author is also indebted to the following:

The RICS Building Cost Information Service and Building Maintenance Information, and particularly the wealth of help and information which was so readily provided by Douglas Robertson and Joe Martin on cost analyses, cost indices and other cost and construction data, which was of great value in illustrating and amplifying the text.

Macmillan Press Ltd for permission to quote from *Building Maintenance*, and to reproduce figure 13.4 and tables 13.4 and 13.5, to quote from *Public Works Engineering*, and reproduce figures 5.6 to 5.9 inclusive, and to quote from *Civil Engineering Contract Administration and Control*.

Tim Carter MSc FRICS FAPM of Davis Langdon Management, Chester, for supplying a large amount of practical information on value management, including valuable case studies, with consent to reproduce from them. This forms the backbone of chapter 11 and includes figures 11.2 and 11.4 to 11.9 inclusive.

Dr Paul Townsend BSc PhD FRICS for contributing to the cost modelling chapter.

Bruce Watson FRICS FACostE of Engineering Cost Management Ltd, Croydon, for cost information on engineering services (tables 10.6 to 10.10 inclusive).

DoE for data on construction output and employment, housebuilding completions and PUBSEC.

RICS Publications for figure 13.1, reproduced by permission of the RICS which owns the copyright.

The Valuation Office for kind permission to quote from the tables in Property Market Report (tables 14.1 and 14.3 to 14.7 inclusive).

Richard Ellis, Jones Lang Wootton and Hillier Parker for helpful advice and information on property values.

David Hoar FRICS, Quantity Surveying Practice Manager, Nottinghamshire County Council, for information on energy conservation in schools and on CLASP.

Willmott Dixon Ltd for the supply of comparative social housing costs contained in table 2.1.

Brian Drake FRICS of Drake & Kannemeyer, Godalming, for information on CEEC.

Brian Norton BSc ARICS of Currie & Brown Inc for permission to reproduce figures 11.4 and 11.5.

BRE, Garston, Watford WD2 7JR for permission to refer to various digests and information papers.

The Controller of Her Majesty's Stationery Office for permission to reproduce figs 5.9 and 5.10 from *Flats and Houses 1958*.

The *Architect's Journal* for permission to reproduce figure 5.5.

The Building Group for the housing cost index (table 9.8).

The Energy Efficiency Office (DoE) for abundant help on many aspects of energy conservation and the use of tables 13.22, 13.23, 13.27 and 13.28.

J Alan Park FRICS MCIOB ACIArb of Stride Treglown Management, Bristol (figure 13.7).

WT Partnership (tables 13.13, 13.14 and 13.15).

National Federation of Housing Associations (tables 13.24 and 13.25).

HM Treasury (figure 16.1).

The Housing Corporation (table 15.2).

Society of Chief Quantity Surveyors in Local Government (SCQS) (appendix 7: format).

Nationwide Building Society (table 15.1).

Last but by no means least, my most grateful thanks are due to the publisher for abundant help and the utmost consideration throughout the production of the book, with special thanks to Malcolm Stewart. My wife and family displayed their customary patience and understanding as the pressure of the book rewrite intensified.

LIST OF FIGURES

1.1	Cost adjustment opportunities	6
1.2	Building cost trends	9
1.3	Construction output by sector	11
2.1	Higher cost of buildings of irregular shape	32
2.2	Effect of change in size of buildings	35
2.3	Perimeter/floor area ratios	38
2.4	Means of access to flats	40
2.5	Hotel circulation patterns and relationships	42
5.1	Real house prices	121
5.2	Three-storey flats infill development – Harlow New Town	128
5.3	Radburn layout – Cwmbran	130
5.4	Radburn layout – Andover	131
5.5	Residential development – Odham's Walk, Covent Garden	133
5.6	Developments from Tsuen Wan to Tsing Yi Island, Hong Kong	134
5.7	Layout plan, Ang Mo Kio New Town, neighbourhood 3, Singapore	135
5.8	Housing development at Ang Mo Kio New Town, Singapore	136
5.9	Layout plan, Yishun New Town, neighbourhood 2 (part), Singapore	136
5.10	Residential development at 250 habitable rooms per hectare	138
5.11	Residential development at 350 habitable rooms per hectare	140
5.12	Blocks of lock-up garages	142
6.1	Approximate estimating – block of six unit factories	158
6.2	Office block – estimating by storey-enclosure method	163
6.3	Factory check office – approximate quantities	165
7.1	Sequence of design team's work	177
8.1	Development stages for a model	203
8.2	Sample probability distributions for risk analysis	207
10.1	Cost planning example 10.2 – sketch design of four-storey block of flats	252
10.2	Cost planning example 10.2 – three-storey block of flats	257
10.3	Cost planning example 10.3 – social club	264
11.1	Extract from a typical FAST diagram	284
11.2	Compiling a FAST diagram	285
11.3	A criteria scoring matrix	288
11.4	Summary of FAST diagram of Computer Centre project	289
11.5	Detailed FAST diagram of site preparation work to Computer Centre project	289
11.6	Summary of FAST diagram of Bank Processing Centre	290
11.7	Detailed FAST diagram of provision of acceptable working environment to Bank Processing Centre	290
11.8	Detailed FAST diagram of secure operations to Bank Processing Centre	291

11.9	Value management: optimum timing for study	294
13.1	Life cycle cost and the RIBA plan of work	312
13.2	Breakdown of total costs	317
13.3	Types of maintenance	318
13.4	Maintenance feedback	327
13.5	Occupancy expenditure patterns	335
13.6	Average and good practice energy costs for four different office types	365
13.7	Facilities management: operation flowchart	377
15.1	Comparison of rehab and new build costs	419
16.1	Examples of private finance projects	471

LIST OF TABLES

1.1	Labour costs June 1993	10
2.1	Comparative costs of one-, two-, three- and four-bedroom new build social housing in the Midlands, Eastern and Southern England (1993 prices)	36
2.2	Typical relative proportional costs of local authority houses and flats	44
2.3	Design cost criteria	53
3.1	Relative costs of houses and low flats (1993)	57
3.2	Social housing – relative substructure costs	65
3.3	Cost relationships of non-loadbearing partitions	74
3.4	Comparative costs of alternative coverings to pitched roofs	77
3.5	Cost relationships of floor finishings	78
3.6	Cost relationships of internal doors	78
3.7	Capital costs of domestic central heating systems (1993 prices)	81
3.8	Comparative costs of basic mechanical engineering services	83
3.9	Comparative air conditioning installation and running costs for a medium sized office building	84
4.1	Typical CLASP project costs (1992)	112
5.1	Categorisation of housing tenures in England (percentages)	115
5.2	Housebuilding completions; by sector and number of bedrooms	119
5.3	Comparative costs of different dwelling forms	120
5.4	Costs of refurbishment work	148
5.5	Summary of typical recommended energy saving measures for gas heated dwellings	151
5.6	Summary of typical recommended energy saving measures for electrically heated dwellings	151
6.1	Comparison of cube, floor area and storey-enclosure approximate estimating methods	161
6.2	Comparison of cube, floor area and storey-enclosure price rates	161
6.3	Schedule of comparative costs of different constructional methods	171
7.1	Plan of work for design team operation	175
7.2	Cost plan of office block	188
7.3	Detailed cost plan of substructural work to school	189
7.4	Part of cost study using comparative method	192
7.5	Comparison of elemental and comparative cost planning	193
8.1	Example of relationship between construction cost per square metre and number of storeys	208
9.1	Concise cost analysis: thirty sheltered flats	215
9.2	Concise cost analysis: warehouse	216
9.3	Detailed cost analysis: advance factories	217
9.4	Tender price and building cost indices	226
9.5	Building cost indices for different types of buildings	227

9.6 Davis Belfield and Everest indices 228
9.7 Pubsec tender price index of public sector building non-housing 229
9.8 Housing cost index 231
9.9 Analysis of groups of elements in typical school 233
9.10 Weighted analysis of house and siteworks 233
9.11 Regional factors 234
9.12 Average building prices 243
10.1 Cost planning example 10.1 – concise cost analysis of factory 249
10.2 Cost planning example 10.2 – detailed cost analysis of block of flats (3-storey) 254
10.3 Cost planning example 10.2 – initial cost plan: four-storey block of flats, Greater London 263
10.4 Cost planning example 10.3 – initial cost plan and record of cost checks of social club 267
10.5 Suggested procedure for cost control of engineering services in buildings 271
10.6 Elemental cost summary: engineering services 272
10.7 Cold water supply – comparative costs of outlet points 273
10.8 Elemental rates for air conditioning for offices: variable air volume system 273
10.9 Elemental rates for air conditioning for offices: induction system 274
10.10 Elemental rates for air conditioning for offices: brief specification notes 275
13.1 Breakdown of typical total costs for various types of buildings 310
13.2 Construction output including repairs and maintenance 311
13.3 Breakdown of cleaning costs of a higher education building 324
13.4 Maintenance manual materials schedule 325
13.5 Maintenance manual cleaning schedule 325
13.6 Local authority traditional housing: capitalised maintenance costs as a percentage of initial costs 329
13.7 Sources and causes of typical local authority housing maintenance costs 330
13.8 Distribution of maintenance expenditure between different buildings 331
13.9 University ranked average expenditure 1985/86 to 1989/90 331
13.10 Average occupancy costs 333
13.11 Sample property occupancy cost analyses 336
13.12 Percentage breakdown of cleaning expenditure 337
13.13 Characteristics of heating boilers 338
13.14 Relative costs of heat emitters 338
13.15 Typical elemental breakdown of LTHW heating installation with gas fired boiler plant and continuous perimeter natural convectors for office building of 5000 m^2 339
13.16 Typical lives of building components 341
13.17 Comparative heating costs 346
13.18 Comparison of costs of installing and maintaining heating systems 346
13.19 Relative capital costs of air conditioning plant with different glazing combinations 347
13.20 Conversion of billed units to kWh 357
13.21 Total energy usage in typical non-domestic buildings 358
13.22 Typical annual fuel costs for dwellings with different NHER ratings 358
13.23 Typical costs of domestic heat losses, energy conservation work and annual savings 359
13.24 Average additional capital cost versus fuel saving for each size of unit: gas heated 360
13.25 Average additional capital cost versus fuel saving for each size of unit: electric systems 360
13.26 Extra costs and savings (three bedroom house): heating by Economy 7 off-peak electricity 361
13.27 Energy efficiency methods in schools 364
13.28 Predicted disaggregated annual energy use for advance factory unit 366

13.29 Savings and payback periods for energy efficiency measures in refurbished industrial buildings 367
13.30 Energy efficiency in refurbished public houses 368
13.31 Effect of errors in predicting lives of buildings 369
13.32 Typical BMI standard form of property occupancy cost analysis 373
14.1 Residential building land prices (October 1993) 396
14.2 Private sector housing land prices (at constant average density) and housing prices, 1978–91 397
14.3 Offices: rental values (October 1993) 398
14.4 Shops: rental values (October 1993) 399
14.5 Industrial buildings: capital and rental values (October 1993) 401
14.6 Industrial and warehouse land values (October 1993) 401
14.7 Agricultural land and property values (October 1993) 402
15.1 House prices in the United Kingdom (second quarter 1993) 412
15.2 Housing association development: forecast mid-range figures for approvals (1993/97) 416
16.1 Construction output by type of work 441
16.2 Private contractors: number of firms 443
16.3 Private contractors: total employment 443
16.4 Construction manpower: employees in employment 444

'It is unwise to pay too much; but it is worse to pay too little.'
John Ruskin

1 THE CONCEPT OF COST CONTROL

INTRODUCTION

Cost control aims at ensuring that resources are used to the best advantage. With the alternating high costs and acute shortage of funds, the majority of promoters of building work are insisting on projects being designed and executed to give maximum value for money. A good illustration was the Housing Corporation formulating performance expectations in 1989, which prescribed the criteria which housing associations were expected to achieve in the provision and management of housing stock and emphasised the need for rigorous self-monitoring. Quantity surveyors are employed to an increasing extent during the design stage to advise architects on the probable cost implications of their design decisions and to assist in obtaining economical and efficient designs. As buildings become more complex and building clients more exacting in their requirements, so it becomes necessary to improve and refine the cost control tools. Constantly changing prices, restrictions on the use of capital and high interest rates have caused building clients to demand that their professional advisers should accept cost as an element in design, and that they should ensure suitably balanced costs throughout all parts of the building, as well as an accurately forecast overall cost, despite the difficulties entailed.

An RICS Report: QS2000 (1991) rightly emphasised that clients need early and accurate cost advice, more often than not well in advance of site acquisition and of a commitment to build. In helping to define clients' requirements in financial terms, quantity surveyors are exerting considerable inflence on any resulting design. This requires not only a knowledge of construction and construction costs, but also a knowledge of the property market and an ability to anticipate and visualise clients' detailed requirements. Seen in this context, cost estimating is a truly professional service, requiring the exercise of a high level of expert discretion and judgement in conditions of considerable uncertainty. An important challenge is thus to improve the accuracy of quantity surveyors' cost estimates, given uncertainties over client requirements, design and future cost and price movements in the industry and in the economy generally.

An important part of the quantity surveyor's function is concerned with ensuring that the client receives value for money in building work. Advice may be given on the strategic planning of a project which will affect the decision whether or not to build, where to build, how quickly to build and the effect of time on costs or prices and on profitability. During the design stage, advice is needed on the relationship of capital costs to maintenance costs and on the cost implications of design variables and differing constructional techniques. The cost control process should be continued through the construction period to ensure that the cost of the building is kept within the agreed cost limits. Building cost can be considered as a medium relating purpose and design and it certainly forms an important aspect of design. A client is very much concerned with quality, cost and time: he wants the building to be soundly constructed at a reasonable cost and within a specified period of time. The term 'cost' which is used extensively throughout this book signifies cost to the client as distinct from the cost of labour, plant and materials incurred by contracting and subcontracting organisations. Hence it signifies the amount which the client will have to pay the contractor to construct the building, but not the actual cost to the contractor of building it. Some confusion has arisen between the terms 'price' and 'cost' and it has been suggested that 'price' should be used when describing the cost to the client and 'cost' when dealing with contracting costs, and this aspect is dealt with more fully later in the chapter.

HISTORICAL DEVELOPMENT OF COST CONTROL PROCESSES

General Background

Cost planning, as at present operated, is the logical extension of a process which has continued since the days of the eighteenth-century measurers, who were employed to measure and value the cost of work after it was both designed and executed. Thus the measurers were involved after the building was erected, to measure and value, and to argue with the client and architect on behalf of groups of tradesmen, who at that time had not been brought together under a main contractor. The main contractor system, which became fully operative in the early nineteenth century, implied price competition before construction which previously rarely occurred. The measurers soon realised that a new function was required and that they possessed the necessary skills to undertake it. It was in response to this situation that they developed the skills of premeasuring, of taking off quantities from the drawings before construction started and assembling them in a bill of quantities to provide a rational basis for competition. At that time it constituted an extremely valuable contribution to the building process. Hence, with the development of the quantity surveying profession, the work was measured and priced before execution, but after design.

The next development was the introduction of approximate estimating techniques which attempted to give a forecast of the probable tender figure, although the basis of the computations often left much to be desired. It was subsequently realised that by the use of cost planning techniques and the methods of cost analysis on which they depend, probable cost can be reasonably accurately determined early in the design process, and sometimes even before the design is commenced, and can be identified with the client's own required limit of cost in the sure knowledge that the development of the design can be controlled to accord with it. Cost planning began to evolve as a process of precosting which attempted to represent the total picture of anticipated cost in a way which provided a clear statement of the issues and isolated the courses of action and their relative costs, so as to provide a guide to decision making.

Since the last war many building projects have become increasingly complex and many specialist skills have become associated with their design, creating vastly different problems from those of prewar days. Clients require, and are entitled to expect, positive assurance that their money is being spent wisely and well. In the wider sphere the nation needs to use the national resources to their optimum effectiveness. The quantity surveyor has a vital role to play as financial adviser to the building industry in producing economically viable solutions to many kinds of building, engineering and development problems; preparing budget estimates; cost planning; the appraisal of alternative cost solutions, procurement arrangements and tendering procedures; preparing contract and subcontract documents; advising on the most satisfactory financial relationships between contractor and client; price evaluation of building and associated engineering works; cost checking; forecasting anticipated final costs; settlement of contractors' claims; cost control; cost analysis; life cycle costing and cost benefit studies.

Developments in information technology are having a significant effect on the provision of services within the construction industry from computer aided design (CAD) and automated bills of quantities through to the development of expert systems. Furthermore, the emphasis of the quantity surveyor's advice is likely to change increasingly from that of supplying information for others to act upon, to the controlling of information and giving direct advice to the client. A competent and experienced quantity surveyor should be able to provide a building package of consultants to provide the best compilation of design, cost and construction for a client, which should prove to be a very useful approach within the European context. The modern concept of chartered surveyors as 'the property profession' is based on the premise that surveyors can add greater value to property and land than other professions or organisations.

From the architect's viewpoint, the quantity surveyor is required to be an encyclopaedia of information on every aspect of building costs. With new constructional techniques and materials frequently being introduced, the quantity surveyor must be fully cogniscant with them and be able to advise on suitability and comparative long-term costs. He must be able to work harmoniously with the architect and other specialists and determine the likely costs simultaneously with the planning of the building. The quantity surveyor in his role as a building economist needs to know the effects of private and public investment policies and aesthetic and planning factors, all of which play a part in determining the system of economic forces which underpin the building process.

European Context

Barry (1992) asserted that the emergence of the single European market highlighted the need for the effective cost control of projects which, in future, will be increasingly international in character.

In a single scheme, developer, funder, designer and contractor could all come from different countries and they will need more than a common language to bring it to a successful conclusion. As if the complexities of working within the EU were not enough, in 1992 other countries were operating in Europe, notably Japan and South Korea, while the reunification of Germany and the liberalisation of what was once the Eastern Bloc doubled the potential size of the European market.

To enable the property and construction industries of a score of countries to operate successfully together constitutes a major undertaking. Because it involves financial and institutional interests and procedures, the immensity of the task outweighs the purely quantitative one of harmonising technical standards, of which over 2500 had been agreed in 1992.

The concordance document for evaluating capital development costs across national boundaries, devised by the Construction Economics European Committee (CEEC), enables quantity surveyors and allied professionals in one country to exchange meaningful cost information with their counterparts elsewhere on the basis of gross internal floor areas and the elemental analysis of building tender prices. The system uses conversion protocols which identify specific elements and subelements in internationally recognised forms. Furthermore, these are designed to be computer compatible, making it relatively straightforward to convert national analyses to EU standard ones (Barry, 1992).

Barry (1992) also foresees a likely future development of a European cost databank, which could provide analyses almost on demand by searching databases within individual member countries and making automatic conversions to EC concordance format.

In 1993, significant progress had been made on the vexed question of unit rates, historically bedevilled by the misleading nature of cost conversions based on fluctuating exchange rates. The EC agreed on a substantial number of items for which prices can realistically be compared. From these it has been possible to derive an international elemental exchange rate which is independent of monetary exchange rates, which Barry (1992) considers a truly revolutionary development. The process has been aided significantly by the evolution of EU-wide purchasing power relatives (PPRs) initially for office developments, but they need using with discretion by qualified and experienced construction economists. They can, for example, be influenced by disproportionate cost increases in a specific material over and above national levels of inflation which would produce distorted comparisons. Other variables include differences in labour costs and productivity rates; availability of materials, labour and specialists; standards of supervision; and the extent of mechanisation and prefabrication. In essence PPRs provide a potentially very effective cost management tool for international development (Barry, 1992).

The long-term advantages of such innovations as the concordance formula and non-monetary exchange rates have been highlighted by events in Eastern Europe in the early 1990s. While it will take time for these economies to play a full part in a wider community, joint ventures were already being established and construction expertise was being imported. Inflation will remain a serious problem, with devaluation being used as an economic tool to help counter its effects, and deciding which currency to use poses problems (Barry, 1992).

NEED FOR COST CONTROL

It is vital to operate an effective cost control procedure during the design stage of a project to keep the total cost of the scheme within the building client's budget. Where the lowest tender is substantially above the initial estimate, the design may have to be modified considerably or, even worse, the project may have to be abandoned. Pressures from ten main sources have combined to stress the importance of effectively controlling building costs at the design stage. The need to continue the monitoring of costs throughout the construction stage is described in chapter 7.

(1) There is greater urgency for the completion of projects, to reduce the amount of unproductive capital or borrowed money, and few building clients have sufficient time for the redesign of schemes consequent upon the receipt of excessively high tenders.

(2) Building clients' needs are becoming more complicated, more consultants are being engaged and the estimation of probable costs becomes more difficult. Furthermore, individual developments tend to be block-sized rather than plot-sized in scope.

(3) Employing organisations, both public and private, are themselves adopting more sophisticated techniques for the forecasting and control of expenditure, and they in their turn expect a high level of efficiency and expertise from their professional advisers for building projects and require a broad and comprehensive range of services (RICS, 1991).

(4) The introduction of new constructional techniques, materials and components creates greater problems in assessing the capital and maintenance costs of buildings.

(5) Changing prices, restrictions on the use of capital, variable interest rates and low contractors' profit margins all make effective cost control that more important.

(6) The move towards reduced waste and greater use of scarce resources creates the need for more accurate forecasting and improved cost control.

(7) There is an increasing demand for integrated design to secure an efficient combination of building and services elements in complex developments, such as hospitals, with effective cost planning to optimise the design solution within the budget figure.

(8) Rising energy costs necessitate costing alternative heating and thermal insulation measures.

(9) More attention is being paid to life cycle costing and total cost appraisal.

(10) Demand patterns centre around the elements of value-added, finance, management and investment (RICS, 1991).

The method of approach and form of cost control are often dictated by the type of development and nature of the promoter. Some examples will serve to illustrate this point.

(1) *Single house*. Prospective owners usually have a reasonably clear picture of the amount and form of accommodation required and of the price that they are prepared to pay, but they may be open to persuasion on such matters as fittings, finishings and form of central heating and thermal insulation. The designer's task of equating accommodation and cost is frequently made more difficult by the client having a preconceived idea of cost, often based on information extracted

from house magazines which can easily be misleading. Single houses built to individual clients' requirements are bound to be more expensive than those forming part of an estate and, for this reason, it is incumbent upon an architect, who may or may not be supported by a quantity surveyor, to exercise the greatest care and skill in the design of the project with constant checks on costs, both at the design stage and during construction.

(2) *Housing estate developments*. In speculative housing work the aim is to provide the type of accommodation that the majority of purchasers want at a reasonable price. This entails a considerable amount of market research to establish current demand, and in the late 1980s and early 1990s a substantial fall in demand resulted in many thousands of unsold new houses and extensive surplus land banks, both at reducing values, causing substantial financial problems for developers. The estate developer's task is made difficult by the majority of prospective purchasers seeking value for money – wanting the maximum amount of accommodation and fittings at the lowest possible price and yet, at the same time, expecting a reasonable standard of quality. The developer is often faced with considerable competition from other builders and from existing properties, and should be continually examining the possibilities offered by the introduction of new materials, components and constructional techniques, and the comparative economics of alternative layouts. This latter aspect is considered in more detail in chapter 5.

(3) *Local authority and housing association development*. Local authorities and housing associations are frequently subject to strict control of expenditure on new projects by the sanctioning government departments or organisations. Local authorities require loan sanction from central government for the majority of new building projects and consent will be withheld if the estimated cost does not show value for money. Up until 1981, the Government operated cost yardsticks, which had a severely restricting effect on design and often prevented full consideration of total costs, as investigated in chapter 13. Housing associations secure a substantial proportion of their funds from the Housing Corporation on the basis of annual bids, and they are subject to very strict financial controls which became progressively more severe in the 1990s.

(4) *Commercial and industrial development*. The circumstances vary tremendously with this class of development. Sometimes the owner is anxious to keep the initial constructional costs to an absolute minimum when available funds are restricted, and he may not be unduly concerned with maintenance costs on which some measure of tax relief will be forthcoming. In other cases clients are very much concerned with 'total costs' and wish to see a sensible relationship between initial costs and maintenance costs.

(5) *Development companies*. Many large scale developments, such as commercial buildings, are financed by development companies whose primary objective is to erect the building for investment or sale.

There are two basic procedures. Firstly, the property may be erected for a specific occupier and the development company agrees accommodation requirements and an acceptable rent figure with the occupier. The development company often borrows money from a large financial organisation, such as a bank or insurance company, and needs a reasonable margin of profit. Secondly, the property may be erected as part of a speculative venture, and the developer and his advisers need to exercise great care in selecting and acquiring likely sites following extensive market research, particularly in times of recession, and should aim at putting them to the most profitable permitted use. 'Permitted use' means uses of land permitted under planning regulations. The rent or profit received will be determined largely by the extent and type of accommodation, its location and the state of the property market. Expert advice is needed on the most desirable pattern of accommodation and quality of finish, and on current rent levels. The general practice or valuation surveyor is well fitted to advise on these aspects.

Blocks of offices in central areas of large towns and cities may let reasonably well, whereas small groups of shops on local authority housing estates may not. It is imperative that a realistic assessment of the probable demand for the particular type of building is made before any provision takes place. In all cases the quantity surveyor should aim to supply accurate forecasts of costs based on extensive cost records suitably adapted as necessary, and to ensure that the money allocated to the project is spent wisely.

MAIN AIMS OF COST CONTROL

The implementation of effective cost control procedures enables the architect to be kept fully informed of the cost implications of all his design decisions. It necessitates close collaboration between the architect, quantity surveyor and any other consultants that have been engaged, throughout the design stage. In public offices and integrated practices it is probably easier to secure the desired closeness of working, than with separate architectural and quantity surveying practices possibly located some distance apart. A concerted effort should enable any difficulties of physical separation to be overcome. There is an overwhelming need for close working relationships throughout the design stage and of mutual confidence and understanding between the various professional advisers. From these desiderata and the preference of many clients for a one-stop service, more integrated or multidisciplinary practices and consortia are likely to be established, with the prime aim of offering a better and more comprehensive service to the client.

The main aims of cost control are threefold.

(1) To give the building client good value for money – a building which is soundly constructed, of satisfactory quality and appearance and well suited to perform the functions for which it is required, combined with economical construction and layout, low future maintenance and operating costs, and completed on schedule as lost time is money, and in accordance with the agreed brief.

(2) To achieve a balanced and logical distribution of the available funds between the various parts of the building. Thus the sums allocated to cladding, insulation, finishings, services and other elements of the building will be properly related to the class of building and to each other.

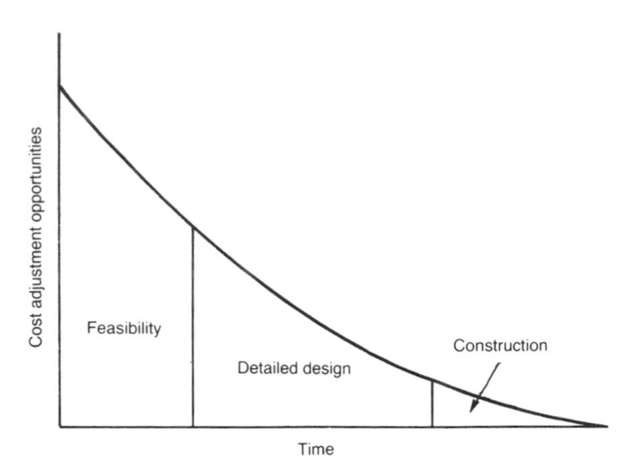

Figure 1.1 Cost adjustment opportunities

(3) To keep total expenditure within the amount agreed by the client, frequently based on an approximate estimate of cost prepared by the quantity surveyor in the early stages of the design process. There is a need for strict cost discipline throughout all stages of design and execution to ensure that the initial estimate, tender figure and final account sum are all closely related. This entails a satisfactory frame of cost reference (estimate and cost plan), ample cost checks and the means of applying remedial action where necessary (cost reconciliation).

Figure 1.1 illustrates diagramatically how the opportunities for making cost adjustments reduce substantially as the project progresses from the feasibility stage through to the maintenance period.

COST, PRICE AND VALUE

Both 'cost' and 'price' are referred to generally within construction and quantity surveying circles. 'Cost' is the resultant of labour, materials, plant and management deployed for a specific activity and is charged according to the accounting system of an organisation. Overheads and/or profit may or may not, by definition, be included in the 'cost'. 'Price' is the amount a purchaser, for example a client will pay for an item, an element or a product, such as a complete building.

The prediction of construction cost and price remains a central part of the provision of quantity surveying services. Together with procurement and contractual advice, the provision and understanding of cost/price information remains crucial, as a management support function, within the overall management of construction projects.

Construction 'cost' and 'price' are apparently simple concepts but the prediction of either is not. Each is generally the result of a complex series of interactions determined by specification quality, design, resources and their management, procurement programme, project location and market conditions. The interaction of a large number of human actions within the building process makes the comparison of any prediction of cost and price with the resources that are eventually used generally difficult and often quite tenuous. A RICS working party on cost information and data services in 1990 concluded that more research was required to refine the cost/price relationship in construction.

'Value' comprises a measure of the relationship between supply and demand. Thus an increase in the value of an object can arise through an increase in demand or a decrease in supply. Indeed, every organisation has a common goal – to provide value to their customers or clients at a profit. Customer value is a function of quality, delivery and price. An important objective of an organisation's internal processes must be to increase quality and delivery performance while improving productivity to enhance profitability and offer competitive rates. Profitability can only be improved if the organisation knows and manages the real cost of its operations. In the severely depressed economic climate of the early 1990s, many organisations turned their attention to controlling and reducing overheads as the key to improved productivity and, sometimes, survival (Smith, 1992).

Smith (1992) has defined effective cost management as a set of techniques and methods for planning, implementing, measuring and reporting, designed to improve the productivity of an organisation's services and related processes; a long-term continuous process of cost improvement. By taking an approach geared to the needs of the longer term, management is better placed to identify and eliminate only those activities and costs which add little or no value to the service delivered to clients over the whole of the economic cycle, and to focus resources on the areas which will ensure the best return when the market expands.

THE IMPORTANCE OF BUILDING

A substantial amount of the value of fixed capital produced in this country each year emanates from the construction industry, and in 1992 this accounted for about 6 per cent of the GDP and amounted to over £40b. In excess of 40 per cent of building work consisted of repairs and maintenance. The industry employed almost one and a half million people in 1988 but this figure was reduced by over 400 000 in the depths of the recession in 1992/93. Furthermore, if employees in allied industries, such as manufacturers and suppliers of building materials and components, and professional advisers and administrative personnel on building projects were to be included, the total number of persons involved would be increased by about 400 000. These statistics give an indication of the immensity of the work undertaken by the construction industry, which embraces both building and civil engineering work, and its relative importance in the national economy, even in the middle of the worst recession in the UK in living memory.

Employers in the construction industry embrace a wide range of interests from local authorities, government departments and housing associations to industrialists, development companies and private individuals. Public authorities accounted for about 35 per cent of the work undertaken by the construction industry in 1992 and any reduction in the volume of work in the public sector has serious effects for the industry in times of financial crisis as it depresses work on the country's infrastructure. To put the brake on large projects after inception leads to inefficient use of resources because these are the buildings which require the longest time for planning. Hence any curtailing or restriction of building work should have regard to the preplanning periods required for different types and sizes of project, with a view to mitigating, or at least minimising, the harmful effects of such restrictions.

Building work embraces a wide range of activities in both the types and sizes of project undertaken. There are surprisingly large variations in the size of projects within any particular use class. Residential buildings, for instance, can range from small, comparatively simple single-storey elderly persons' dwellings to complex schemes of high rise development to secure residential densities in the order of 400 persons per hectare, although these were far less popular in the 1980s and 1990s than in the 1960s to 1970s. Similarly, industrial buildings can range from small workshops of about 200 m^2 floor area to large factories occupying many hundreds of thousands of square metres. Buildings also vary in the form of the benefits that they generate. A residential building has a direct value to the occupier in the satisfaction that it gives him, whereas the value of an industrial building is more related to the products which can be manufactured within it.

BUILDING OUTPUT AND COSTS

Significance of Building Costs

Over the years many clients have been concerned about the rising cost/price of building work, but the position was reversed in the period 1990/92 when, because of a substantial fall in demand, building prices fell significantly in many parts of the country and many building contractors experienced considerable financial difficulties, with large numbers being forced to call in receivers and this included some of the very large firms. Although it was evident that this situation could not continue as contractors had absorbed virtually all their overheads and profit in order to survive. This shows very clearly the direct relationship between supply and demand and the significant effect of market conditions on building prices.

Figure 1.2 shows average estimated building costs compared with average tender prices over the period 1984 to 1993 based on a datum of 100 in 1985 (average), with forecasts up to 1995. There was a steadying of tender prices in the early 1980s as building costs rose significantly faster than tender prices, following the building boom of 1978–81. Subsequently, in the period 1987 to 1988 tender prices rose more sharply as contractors were no longer able to absorb such a large proportion of the increased costs of labour, materials and plant, and the demand for building work rose substantially. In the period 1990 to 1992 tender prices generally fell sharply because of the industry's reduced workload. This caused activities which were highly labour intensive, such as painting and most repair and maintenance work to be proportionately more expensive.

In the early 1990s the construction industry suffered extreme difficulties mainly stemming from severe cuts in public building programmes, a reduction in the private sector and high interest rates. The small and medium-sized building firms appear to fare the worst, having less financial resilience to meet fluctuations of demand, and their larger dependence on private house-building makes them particularly susceptible, as it is here that the variable financial climate tends to make its most immediate and disruptive impact. However, a number of the largest firms also went into receivership and firms concentrating on house building suffered considerably. The number of new dwellings completed annually has progressively declined since 1988.

There are many contributory factors to the high level of building costs, many of which are outside the control of the contractors themselves. Table 1.1 gives an assessment of 'all-in' weekly rates for craft operatives and labourers employed in the London area in June 1993. Some of the allowances will vary from one firm to another but the number and extent of the labour oncosts are disturbingly high. Labour oncosts include non-productive overtime, sick pay, public and annual holidays with pay, working rule agreement payments, insurance, CITB levy, tool money and trade supervision. The payment of training levy to the Construction Industry Training Board has

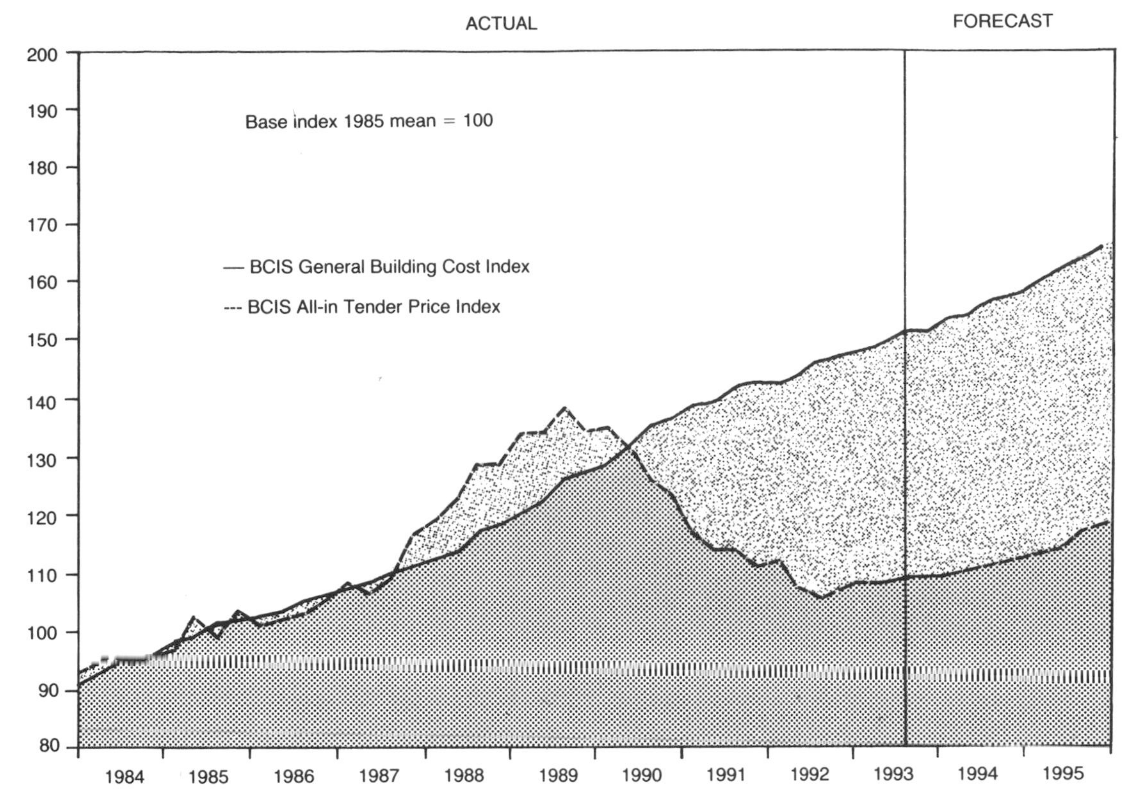

Figure 1.2 Building cost trends (Source: BCIS, Quarterly Review of Building Prices)

caused considerable resentment, particularly among the smaller building firms who have for the most part received little in the way of grants. It has been estimated that well in excess of £1b was outstanding to the construction industry in the early 1990s as a result of the operation of periods for honouring of certificates, retention provisions and periods of final measurements. In view of variable interest rates paid by many contractors, a review of the financial arrangements of building contracts is very desirable, although it is important that building clients are not faced with unjustifiable risks. The widespread liquidity problem within the building industry is accentuated by the following factors:

(1) The sometimes unnecessarily high level of retention monies held against work in progress under the conditions of contract.

(2) The further amount of 'hidden retention' represented by the usual timelag between the execution of work and the receipt of the corresponding payment by the contractor or subcontractor concerned.

(3) Delays in the issue of certificates and payment of final accounts.

(4) Delays caused by claims.

Contractors' costs can be classified broadly into four categories:

(1) Quantity related: costs which have a direct relationship to the permanent work, such as most materials.

(2) Occurrence related: costs related to a specific event or occurrence such as bringing plant, like excavation plant, to a site, assembling and transporting it around the site as necessary, and dismantling and removing it when it is no longer required.

Table 1.1 Labour costs June 1993

Basic costs – labour rates
All-in hourly wage rate for building craft operative & labourer

Rates effective from 28th June 1993	Craft operative	Labourer
Total hours worked per year	1898	1898
Hours non-productive overtime per year	80	80
Days sick paid per year (days sick per year)	5 (8)	5 (8)
Days public holiday per year	8	8
Sick pay: per day	£12.10	£12.10
Tool money: per week	£0.97	–
Annual holiday with pay contribution	£18.50	£18.50
BASIC HOURLY RATE	£4.090	£3.485
27% plus rate	£1.105	£0.940
TOTAL HOURLY RATE	£5.195	£4.425
	£	£
Basic wages	9 860.11	8 398.65
Non-productive overtime	415.60	354.00
Sick pay (days sick × sick pay)	60.50	60.50
Public holidays with pay	332.48	283.20
Sub-total	10 668.69	9 096.35
National Insurance, 10.4% 8.6%	1 109.54	782.29
Tool Money (44.6 weeks)	43.26	
Annual holiday with pay (47 weeks)	869.50	869.50
Training	26.67	22.74
Sub-total	12 717.67	10 770.88
Severance pay – add 2%	254.35	215.42
Employers insurance – add 2%	259.44	219.73
TOTAL COST	13 231.46	11 206.02
COST PER HOUR (total cost/number of productive hours)	£6.97	£5.90
INDIRECT OVERHEADS AND PROFIT add 20%	£8.36	£7.08

Source: *BMI Price Book 1994*.

(3) Time related: costs related to the length of time on site such as hired plant.

(4) Value related: charges such as fire insurance and probably establishment charges which are related to project value.

As has been illustrated, the severe recession in the early 1990s had a significant effect on the ratio between building costs and tender prices. For example, building costs rose by about 5 per cent in line with inflation, while tender prices fell by 8 per cent as contractors continued to cut their profit margins and oncosts in order to survive, but this practice could not continue as there was virtually nothing left to cut. Hence in 1992 tender prices remained almost static.

Building Output

The changes in the nature and value of building work between 1981 and 1994 are shown in figure

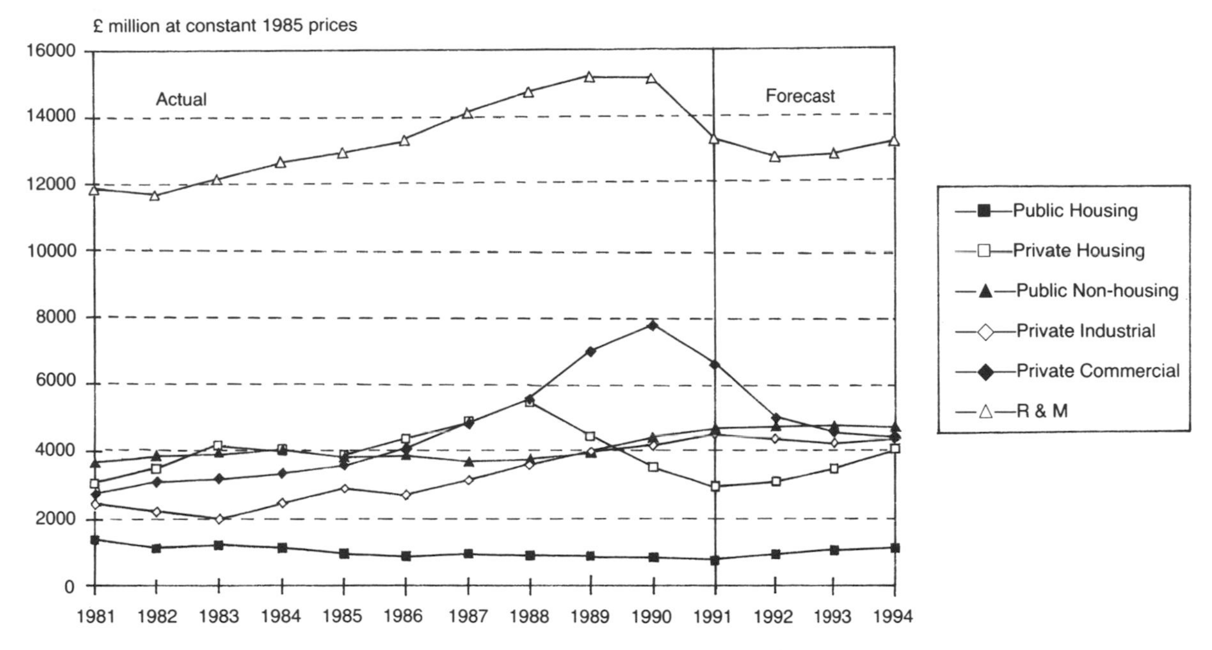

Figure 1.3 Construction output by sector (Source: RCIS)

1.3. By the early 1990s public sector housing had reached an all-time low because of drastic cuts in public expenditure, while private housing was proceeding on a restricted scale because of the frequent difficulty in selling the dwellings. The valiant efforts of housing associations to provide good quality affordable social housing was being undermined by progressively reducing levels of grant from the Housing Corporation, who provided much of their finance. Both the private commercial and industrial sectors also showed a considerable downturn as commercial and industrial activity continued to decline throughout 1990 to 1992 and there was extensive surplus accommodation available in most parts of the country.

In the early 1990s the situation in the construction industry had become so serious that the Group of Eight, prior to being abandoned, representing the contractors and allied professions, was pressing the Government to adopt reflationary measures and to introduce zero rating on maintenance and repair work, which was also in a depressed state. Under-investment in the construction industry had reached crisis point, resulting in substantial unemployment and loss of capacity, which would create serious difficulties in the event of a significant upturn and probable overheating of the industry, accompanied by escalating prices. In 1992/93 there was considerable pressure from the Building Employers Council (BEC) and the Construction Industry Council (CIC) for increased investment on much needed work on the infrastructure, housing, educational buildings frequently in a very poor state of repair, energy conservation, hospitals and urban renewal, which collectively formed the life blood of the nation. The Government's response in the Autumn 1992 Statement was very restricted largely because of very high public sector borrowing rate (PSBR) and the emphasis on privatisation.

By 1993 one in four construction industry workers were unemployed (450 000), over 3000 building firms became bankrupt in 1992 and there were 300 000 homeless people. In 1992 (the depth of the recession) the value of building output was about two-thirds of that achieved in 1988 (the peak of the building boom). The largest decline was in the commercial sector where there was a large over-supply of accommodation. Unfortunately the number of insolvencies in the construction industry is likely to increase as the country comes out of recession, as many contractors and subcontractors are so starved of resources that when they obtain new contracts they will be unable to secure the necessary funding to enable them to undertake the work.

Building Productivity

Productivity has been defined as a measure of the quantity of output of goods and services that can be produced for a given input of factors of production, such as land, labour, capital, energy and entrepreneurial skills. Because of the world wide financial recession in the early 1990s and relatively high interest rates, clients were increasingly requiring better value for money. This created the need for substantial increases in output and the better use of all resources. A review of post-war policies show the main steps that have been taken. The Ministry of Housing and Local Government produced some disturbing reports after the Second World War which showed that in 1947, forty-five per cent more man-hours were required to perform a given quantity of house-building work as compared with 1938–39, and that this represented a decline in productivity of 31 per cent. By 1949 there had been a considerable improvement, although the man-hours were still twenty-six per cent above the 1938–39 level, with productivity 20 per cent below pre-war level, and in 1951 the position was still much the same. The Ministry (MHLG, 1952) made recommendations to local authorities with a view to reducing house-building costs. These recommendations included reducing room heights by 6 inches (150 mm) making possible a saving of £15–£20 per house; reducing the floor areas of three-bedroom houses by 100 sq ft (9.3 m^2), saving about £90 per house and omitting the second WC. Towards the end of 1950 a number of local authorities were considering ways of building houses to let at lower rentals, as it was becoming apparent in some areas that many persons on the housing lists were unable to afford the rents resulting from the current house plans.

Schemes for improved productivity are as relevant to the design team as to the contractor. Designers of individual buildings, by considering the way each stage of construction may be handled, can ensure designs that are relatively easy to build, resulting in improved buildability or constructability. Designers who concentrate on a particular building type, or always employ similar construction techniques, have an opportunity to encourage, and sometimes promote, the introduction of new techniques and new components. In the 1960s it was believed that industrialised building offered scope for innovation and for the integration of design and production. Bishop (1975) postulated that detailed design could simplify work, or at least eliminate work which could be tackled by a gang without interruption from other gangs, in the following ways:

(1) *Simplification*. For example, arranging brickwork in long runs between quoins and reducing waste.

(2) *Ensuring continuity*. Such as by using joist hangers so that bricklayers' work is not interrupted while carpenters set joists; or arranging precast concrete stair flights to be self-supporting from one floor to another rather than by building in half-landings.

(3) *Separating the work of gangs*. For instance, detailing claddings so that joints and flashings can be completed independently of the erection gang.

(4) *Reducing the number of separate operations*. For example,detailing reinforcement and construction joints so that concrete floors and walls may be cast in one operation.

(5) *Making mechanisation feasible*. Such as by specifying flooring of uniform thickness so that the screeds may be finished at one level, thus making the use of power floats more practicable.

(6) *Improved site management*. For instance, using recognised planning techniques and effective incentive bonus schemes.

It is vital that bills of quantities are so formulated that it is immediately apparent to an estimator that the building has been designed for the use of economical and easy construction processes, in order that the contractor can submit a more advantageous price for the project.

Since the 1952 report of the Ministry of Housing and Local Government there has been a steady reduction in labour requirements for house building, with the average labour commitment per house dropping from about 2650 to around 1600 manhours. This improved efficiency is the cumulative effect of a variety of factors: including increased mechanisation; improved management techniques; increased prefabrication and industrialisation; greater utilisation of incentive bonus schemes, and improved training methods. Early surveys of the Building Research Establishment (BRE) showed that on the least efficient sites the manhours per house were three times as great as on the most efficient. It was also found that there is no strong relationship between the plan area of a dwelling and the manhour requirements. Productivity was found to be high in firms paying a target bonus; where subcontractors were employed on the work rather than the main contractor's own labour; in small firms in which the builder himself worked with the tools; and in firms working regularly on house building.

If manhours of each trade for the whole project are considered sufficient for measuring productivity, then the information can usually be obtained from weekly timesheets. It is however likely that it will be necessary to apportion the men's time between specific parts or elements of the work and/or to different activities, when special arrangements will have to be made to collect the information by direct observation.

The output of work performed is normally measured by a functional unit varying with the level of detail at which the study is being made, for example, number of houses built or number of bricks laid. The great variety of building types and techniques makes it difficult to define the work content in a way to give one physical measurement of the work done. At the coarser levels of detail the dwelling or bed-space can serve as a suitable unit in house building. At the other end of the scale the output must be considered in terms of the work done, for example, the output of two gangs of bricklayers erecting similar buildings could be compared on the manhour requirements per thousand bricks laid. On the other hand the work content of, for instance, carpenters' second fixings is less easy to compare as so much depends upon design detail and specification. At an even finer level of detail this unit of output would be broken down to a list of tasks, such as 'hang door' or 'fix cupboard'. Furthermore, when comparing the effect of different forms of construction on productivity it is always important to determine how far the observed differences arise both from the particular design and specification or from organisational aspects, including the ability and motivation of different operatives.

Many local authorities attempted in the late 1960s and 1970s to reduce building costs and increase efficiency by the use of industrialised building and consortia arrangements. Industrialised building aimed at producing building systems and methods that permitted rapid assembly on site with the minimum use of labour, and this subject will be considered further in chapter 4. The establishment of consortia enabled bulk orders to be placed for components with the advantage of discounts for large quantities, serial tendering, development of designs and feedback of information from sites. Similarly the National Building Agency was set up to help the smaller housing authorities to pool and collate their requirements into phased programmes and to obtain more efficient and more economical working arrangements, although the NBA was subsequently disbanded. In the 1980s and early 1990s building productivity increased with the use of improved building management techniques, including increased cost monitoring, the establishment of performance goals or targets and more extensive comparisons of objectives and achievements.

In the past builders and designers have experienced difficulties in keeping abreast of new materials and components and there has often quite naturally been a reluctance to use products which were neither well-established nor officially approved. This impediment to progress was largely removed by the then Minister of Public Building and Works setting up the Agrément Board in 1966 with the use of the Building Research Establishment as its technical agent. The Board's principal objective is to assist innovation in the building industry in respect of building materials, products, components and processes, by providing an assessment on the basis of examination, testing and other investigations. Products and processes, which, in the opinion of the Board, are likely to give a satisfactory performance will qualify for a statement or certificate identifying the product and its method of use or of installation, together with a summary of appropriate design data. In general, these certificates will cover the period before a British Standard Specification can be devised and issued and will, in appropriate cases, indicate compliance with the relevant part of the Building Regulations.

A comparison of UK and USA procedures indicated that higher productivity was achieved in the USA through less complicated construction details, more rational construction methods and the use of larger plant and more powered hand tools, although there was ample evidence of less controls and lower safety standards.

The Role of the Quantity Surveyor

It is advisable that the quantity surveyor is involved in the planning and design stages of a project. He can clarify and amplify quality decisions by bringing to bear the full spectrum of his professional knowledge of what is available in the market, to reinforce the architect's range of experience, synthesise the professional skills of the various members of the design team, help to incorporate features which the contractor can provide efficiently and competitively in favour of those which create greater difficulties; and, by the careful balancing of alternatives in terms of time involvement, quality and economics, he can provide a sound basis for decision-making by the architect and client. The quantity surveyor's prime aim should always be to assist building clients in obtaining optimum value for money. Quantity surveying activities have expanded significantly in the fields of feasibility studies, cost advice during the early stages of projects and cost control during the design and post-contract stages, aided by computerised methods.

The quantity surveyor can give advice on the strategic planning of a project which may affect the decision whether or not to build, where to build, how quickly to build and the effect of time on costs or prices and on profitability. Cost control throughout the design and construction stages may be extended to the giving of advice on taxation matters and cash flow, while the optimum use of production resources is also important. Building clients are becoming more knowledgeable and are demanding a much wider range of expert cost advice than quantity surveyors were accustomed to giving in the past. There is still a need to develop further contract documents related to performance specifications and coordinated project information (CPI) and to produce bills that can be used during the course of projects for management purposes and to monitor costs. There is a developing market for a quantity surveying contribution in building services, civil and heavy engineering, project management and rehabilitation work, aided by increased computerisation.

In the exercise of cost planning techniques the quantity surveyor is concerned with many issues of building economics, some involving returns as well as costs, and some of these are listed now:

(1) Substitution between capital and running costs to secure the minimum total cost.

(2) Investigation of the different ways of producing the same building at lower cost.

(3) Finding ways of slightly altering a building so that for the marginally greater use of resources, the returns are more than proportionately increased.

(4) Investigating methods of using the same resources to produce a different building which could give greater returns.

The quantity surveyor can make a significant contribution in the field of construction economics internationally and his contribution in the

European Community (EC) will be examined later in the chapter.

TENDERING ARRANGEMENTS

Tendering Procedure

Conventional tendering procedures have been criticised on the grounds that they fail to take full advantage of modern techniques and inhibit the optimum use of the contractor's expertise. All tendering procedures aim at selecting a suitable contractor and obtaining from him at an appropriate time an acceptable offer, or tender, upon which a contract can be let. Negotiation may be the best approach in certain circumstances, such as additions to existing contracts and specialist work. Tendering is a complex process whereby the tendering contractor who is anxious to secure a particular contract needs to pitch his tender at the right level, having regard to the resources and method of execution without reducing his profit margin to an excessively low level. Readers requiring more information on bidding strategy are referred to Skitmore (1989) and Smith (1995).

Governmental Committee Recommendations

As long ago as 1944 the Simon Committee drew attention to the fact that 'low prices resulting from indiscriminate tendering result in bad building' and that 'resources are wasted when many firms tender for the same job'. The Banwell Report (1964) suggested that invitations to tender should be limited to a realistic number of firms, all of whom were capable of executing the work to a recognised standard of competence. The Banwell Committee appeared to favour the general use of standing approved lists of contractors and that *ad hoc* lists should be used mainly when the work was of a specialist or one-off nature. The Committee further recommended that the period allowed for tendering should be adequate for the type of project and welcomed 'firm price' contracts. The Ministry of Housing and Local Government issued revised model standing orders to local authorities in 1964 to facilitate the wider use of selective tendering procedures, and in 1965 the Ministry gave guidance to local authorities on the operation of selective tendering.

In 1965 a working party was established by the Economic Development Committee for Building to examine the Banwell Report and its implementation and its report was submitted in 1967. This report considered that insufficient attention was paid to the importance of time and its proper use and that clients seldom define their requirements in sufficient detail at the start of negotiations. It favoured the main contractor joining the design team at an early stage. The report further urged the wider adoption of the practices detailed in *Selective Tendering for Local Authorities* (1965) and *Early Selection of Contractors and Serial Tendering* (1966), although the principal guide is the NJCC *Code of Procedure for Single-Stage Selective Tendering* (1994). They recognised that in the public sector this would require a more flexible approach to satisfy standards of accountability. Although the working party saw the merit in 'firm price' contracts, they stressed the difficulties involved in producing firm tenders in a market where materials prices tend to fluctuate widely and contractors are often invited to tender on incomplete documentation.

Selective tendering is now widely adopted and 'firm price' contracts are in general use for contracts not exceeding two years' duration. Further codes were subsequently issued for two-stage tendering (1994) and design and build (1985).

BUILDING PROCUREMENT

There are a variety of client/contractor relationships and the choice will be influenced considerably by the particular circumstances. The building procurement methods range from *cost reimbursement* or *cost plus* contracts at one end of the scale to truly lump sums, such as package deals, at the other. The essential difference between these two extremes devolves upon which party to the contract is to carry the risk of making a loss (or profit) and the incentives which are built into the contract to encourage the contractor to provide an efficient and economic service to the client. A useful introduction to the nature and choice of the different building contractual arrangements, now more commonly referred to as building procurement methods, supported by a good range of carefully selected case studies is provided by Turner (1990). A brief examination and comparison of the main categories of building procurement method follow.

Cost plus contracts. These contracts are sometimes referred to as cost *reimbursement* contracts or *prime cost* contracts. In practice they can take three quite different forms.

(1) Cost plus percentage contracts are those in which the contractor is paid the actual cost of the work plus an agreed percentage of the actual or allowable cost to cover overheads and profit. They are useful in an emergency, when there is insufficient time available to prepare detailed schemes prior to commencement of the work, but it will be apparent that an unscrupulous contractor could increase his profit by delaying the completion of the works. No incentive exists for the contractor to complete the works as quickly as possible or try to reduce costs.

(2) Cost plus fixed fee contracts are those in which the sum paid to the contractor will be the actual cost incurred in the execution of the work plus a fixed lump sum, which has been previously agreed upon and does not fluctuate with the final cost of the project. No real incentive exists for the contractor to secure efficient working, although it is to his advantage to earn the fixed fee as quickly as possible and so release his resources for other work. This type of contract is superior to the cost plus percentage type of contract.

(3) Cost plus fluctuating fee contracts are those in which the contractor is paid the actual cost of the work plus a fee, with the amount of the fee being determined by reference to the allowable cost by some form of sliding scale. Thus the lower the actual cost of the works, the greater will be the value of the fee that the contractor receives. An incentive then exists for the contractor to carry out the work as quickly and as cheaply as possible, and it does constitute the best form of cost plus contract from the employer's viewpoint (Seeley, 1984).

Target cost contracts. These contracts were introduced in recent years to encourage contractors to execute the work as cheaply and efficiently as possible. A basic fee is generally quoted as a percentage of an agreed target estimate often obtained from a priced bill of quantities. The target estimate may be adjusted for variations in quantity and design and fluctuations in the cost of labour and materials. The actual fee paid to the contractor is obtained by increasing or reducing the basic fee by an agreed percentage of the saving or excess between the actual cost and the adjusted target estimate. In some cases a bonus or penalty based on the time of completion may also be applied. Target cost contracts can be useful when dealing with unusual or particularly difficult situations, but then the real difficulty lies in the agreement of a realistic target. They should not however be entered into lightly as they are expensive to manage, requiring accurate measurement and careful costing on the employer's behalf (Aqua Group, 1990a).

Trench (1991) has described in detail a refined system of target cost building procurement, whereby:

(1) a reasonably equitable balance is set between client's risk and contractor's risk;

(2) the quality of design is improved because trade contractors contribute to the design and detailing early in the project and possibly before the main contractor is appointed; and

(3) the contractor appointed to manage the final design and the work of subcontractors has an incentive to complete the project to an agreed target cost.

Savings or overruns, within a band, are shared with the client according to an agreed formula which reflects market conditions and the level of risk perceived by the contractors. In addition the client can obtain a guaranteed maximum price so he knows before building work is started what the ceiling of the construction costs will be.

The system comprises the following eight stages:

(1) feasibility and setting up of the system;
(2) planning consent, general arrangement and configuration;
(3) detailed design with specialist trade input;
(4) appointment of the design and manage contractor;
(5) preparation of the client's requirements;
(6) agreement of target cost and notice to proceed;
(7) design co-ordination and execution of works by design and manage contractor;
(8) adjustment of fees and final accounts.

Measure and value contracts. These contracts include those based on schedules of rates, approximate quantities and bills of quantities. The great merit of these contracts lies in the predetermined nature of the mechanism for financial control provided by the precontract agreed rates. The risk of making a profit or loss is with the contractor. Another adaption of the orthodox bill of quantities contract is serial or two-stage tendering whereby a series of contracts are let to a single contractor. Serial contracting is based on a standing offer, made by a contractor in competition, to enter into a number of lump sum contracts in accordance with the terms and conditions set out in a master or notional bill of quantities. This procedure is particularly suitable for a known programme of building work over a period of time in a specific locality, and where there is a high degree of standardisation of construction. The advantages claimed for the system include improved relationships with the contractor, more effective cost control and faster and more economical planning and execution of projects. It is also argued that the most usual organisation of design/construction relationships is perpetuated by the competitive tendering system and inhibits communication between the different parties involved in a building project. A survey carried out by Davis Belfield and Everest (1994) for the RICS showed that projects based on bills of quantities (firm or approximate) comprised 44.06 per cent of contracts by value in 1993, compared with 64.70 per cent in 1985 (the survey covered about 12 per cent of new construction orders).

Contracts based on drawings and specification. These are often described as 'lump sum' contracts although they may be subject to adjustment in certain instances. They form a useful type of contract where the work is limited in extent and reasonably certain in its scope. They have on occasions been used where the works are uncertain in character and extent, and by entering into a lump sum contract the employer hoped to place the onus on the contractor for deciding the full extent of the works and the responsibility for covering any additional costs which could not be foreseen before the works were commenced. The client would then pay a fixed sum for the works, regardless of their actual cost, and this constitutes an undesirable practice from the contractor's point of view. A survey carried out by Davis Belfield and Everest (1994) for the RICS showed that 9.98 per cent of contracts valued used this method in 1993, compared with 13.13 per cent in 1985.

Design and build contracts. These constitute a specialised form of contractual relationship in which responsibility for design as well as construction is entrusted to the contractor. The less developed the design, the less detailed the specification and hence the less precise must be the calculation of the price. Contingencies must be included to provide for the unknown. Design and build contracts have been employed for a wide range of projects including local authority and housing association housing, hospitals, defence installations and factories. In housing schemes about six contractors may be invited by a local authority to submit a complete development scheme for a large site. The contractors may use their own design teams or private architects to prepare schemes within the specified requirements of densities and costs, and the successful contractor will subsequently be required to collaborate with the authority's architect. This form of approach is particularly favoured where special factors operate, such as the use of building systems.

Contractor-sponsored systems make it essential for the contractor to be brought into the design team. It has also become necessary to evolve procedures which will allow competition between contractors offering different systems. This has involved two separate stages: first, a competition to find the contractor who can best satisfy the functional, aesthetic and economic needs; and second, a period of negotiation with the selected contractor using data, especially prices, derived from the first stage. Quantity surveyors can offer various services in connection with contractor-sponsored contracts at the precontract, tender and contract stages.

More detailed information on the operation of design and build contracts is supplied by Janssens (1991). In 1993 it was estimated that about 25 per cent of building contracts were let on this basis, although a survey by Davis Belfield and Everest (1994) for the RICS showed 35.70 per cent of contracts by value using this method.

In order to maintain sound financial control of contractor-designed projects, one typical approach is to contact not more than three or four contractors, all of whom must have adequate resources and experience to undertake the project. The brief submitted to contractors should not be so detailed as to preclude contractors from using their initiative and know-how. The compiling of a cost benefit analysis against the client's requirements puts them in true perspective; an example being the inclusion of balconies – are they for instance being introduced as an amenity feature or to assist in maintenance? The brief is accompanied by all relevant documents including surveys, consents, statutory requirements, 1:100 plans, sections and elevations, and larger typical detail drawings. Contractors would be required to submit lump figures to cover site management costs, head office overheads, design costs and profit, and superstructure work. Substructures, external works and services will be evaluated in a provisional bill of quantities, and provisional sums will be listed by the client. In the final assessment regard has to be paid to aesthetic as well as financial aspects. To control costs on the project, lump sum figures for building work will have to be broken down and checked.

Design and manage contracts. Turner (1990) has aptly described how the 'design and management' system combines some of the characteristics of 'design and build' with those of 'management'. A single firm is appointed, following a selection process that may include some degree of competition on price, although this is not usually the main selection criterion. The client will need adequate in-house skills or obtain appropriate professional services in order to formulate his requirements and carry out his responsibilities under the contract.

The main components of the system are:

(1) establishing the need to build;
(2) determining the client's requirements;
(3) selecting and inviting tenderers to bid;
(4) the contractor or contractors preparing their proposals for management, design, time and cost;
(5) evaluation and acceptance of a tender which becomes a contract;
(6) management, design and construction of the works.

In this system the client enters into a contract with a design and manage contractor and possibly a scope designer, while the design and manage contractor, in addition to his contract with the client, will be in contract with consultants for design and/or cost consultancy services and with works contractors who may number as many as 60 to 100 organisations. Alternatively, the client may enter into a contract with a design and manage consultant who will be selected from one of the building professions and who will enter into similar contractual arrangements with a design and/or cost consultancy and numerous works contractors, as the design and manage contractor. These systems offer advantages when factors such as timing, controllable variation, complexity, quality level and competition are significant.

Develop and construct contracts. In this type of contract consultants design the building to a partial stage, often called 'scope design', the competitive tenders are obtained from contractors that develop and complete the design and then construct the building. The system offers maximum flexibility as the 'develop' part can begin from anywhere between RIBA Plan of Work stage C up to any point starting before the end of stage E. Develop and construct requires analysis/creativity of a design team, which may be chosen in competition, and then acceptance/implementation by a construction team which may also be chosen in competition. This system offers advantages where timing, good quality level, price certainty, competition, professional responsibility and risk avoidance are considered important (Turner, 1990).

Management contracting. In management contracting the contractor is paid a fee to manage the construction of a project on behalf of the client, and it is therefore a contract to manage, procure and supervise as opposed to a contract to build. Thus the management contractor is a member of the client's team and works closely with the professional consultants. In the various forms of contract the client is referred to as the employer, but client is probably the better term to use in the context of this book as most parties to a contract are employers.

The building work is subdivided into subcontracts, usually referred to as packages. They are normally let in competition although some may be negotiated or let on a cost reimbursement basis, with all discounts normally reverting to the client. The management contractor provides the site management team and such preliminary items as site offices, canteen, hoardings and other communal services, and these are usually reimbursed at cost although they can form part of the management contractor's tender as a fixed lump sum. The management contractor will also be paid a fee encompassing overheads (head office charges) and profit, and is often expressed as a percentage of the final prime cost of the works (Aqua Group, 1990a).

A cost plan will be prepared before the management contractor is appointed and following his appointment he will liaise with the quantity surveyor to prepare an estimate of the prime cost (EPC) showing the estimated value of all the subcontract packages, site management and other costs, for agreement by the client and architect and establishing the required level of specification. Valuations and certificates for payment are prepared in the usual way and cost reporting is performed on a package by package basis against the EPC (Aqua Group, 1990a).

The management contractor can be selected either on a negotiated basis or more commonly in competition. Negotiation is computed on the basis of the management fee together with the contractor's estimate of site management requirements. Whereas if the contractor is to be selected in competition, documentation is prepared inviting several contractors to submit their proposals and this will include general arrangement drawings, known specification information, expected contract value, details of key dates and of the contract document.

The management contractor's submission would include the management fee, estimate of site management and preliminary costs, any comments on the estimated cost of the work, a method statement giving outline proposals for carrying out the work including a draft list of work packages (subcontracts), draft contract programme, details of proposed management team, and a proposed typical subcontract form. Selection will be made largely on the ability of the contractor to provide the building on time and to budget (Aqua Group, 1990a).

With a management contract the contractor can be appointed at an earlier date than with most other procurement methods, but set against this, some of the packages are not let until a late stage in the project, thereby creating a lack of certainty. The system is best suited to large, complex projects exceeding £20m in contract value, which exhibit particular problems that militate against the use of fixed price contract procedures. Typical examples are:

(1) projects for which complicated machinery and/or computer equipment are to installed concurrently with the building works;

(2) projects for which the design process will of necessity continue throughout most of the construction period;

(3) projects on which construction problems are such that it is necessary or desirable that the

design and management team includes a suitably experienced building contractor appointed on such a basis that his interests are largely synonymous with the client's professional consultants (Seeley, 1984b). The largest contract management project completed in Europe in 1986 was carried out for BAA on the £200m Terminal 4 complex at Heathrow (Seeley, 1992). A survey by Davis Langdon and Everest (1994) for the RICS showed 6.17 per cent of contracts by value using this method in 1993, compared with 14.40 per cent in 1985.

Construction management. The construction work is carried out by 'works contractors' engaged directly by the client himself, and hence the client assumes the contractual position of the main contractor. Since most clients do not have the expertise to manage the works contractors, they employ a construction management firm, on a fee basis, to do this on their behalf. The firm could be a contracting organisation or a professional consultant. In general the organisations using this system are large ones with rolling programmes, considerable experience of similar projects and often some in-house expertise.

Construction management has been a common method of procurement in North America for many years past. In more recent times very large projects have been undertaken in this country using the construction management approach and these include Broadgate, Canary Wharf, Stanstead Airport and Sainsbury's major stores, and BAA's Terminal 5 complex at Heathrow will be executed on this basis. In addition, Continental contracting methods tend to operate on a similar basis, and their influence in the UK is likely to be accentuated by the formation of the single market, although this still has to be validated (Herbert, 1991).

A Construction Management Forum Report (1991) concluded that construction management is likely to replace management contracting for future major contracts, and this has already been evidenced at Heathrow. There are distinct similarities between construction management and management contracting, but one major difference is that with construction management all work packages are treated as direct contracts between the client and the various package contractors. The construction manager is appointed in a similar manner to the professional consultants with similar liability to the client. This procedure avoids some of the drawbacks of management contracting, which can prove to be more confrontational and expensive, and carry a greater degree of risk for the client, works contractors and management contractor. Views differ as to the stage at which the construction manager is to be appointed, either before, simultaneously with or after the appointment of the design team (Herbert, 1991). A survey by Davis Langdon and Everest (1994) for the RICS showed 3.94 per cent of contracts by value using this method in 1993, compared with 6.89 per cent in 1989.

PUBLIC PROCUREMENT IN THE EUROPEAN COMMUNITY

The European Commission has undertaken a programme of legislation designed to ensure an open market for public purchasing throughout the EU. Companies from other member states have the opportunity to introduce foreign competition into existing UK markets for the supply of goods, including roads, bridges and public buildings, put out to tender by central and local government, police and health authorities. It also represents an opportunity for UK companies to bid for these contracts in the rest of the EU. This is a substantial opportunity to enter new markets which account for 15 per cent of all EU business (Melia, 1992).

In 1993 five directives were in force throughout most of the EU; namely the Supplies Directive, the Public Works Directive, the Compliance Directive (to ensure the effective enforcement of the public purchasing Directives), the Utilities Directive and the Services Directive.

The stated aim of the Commission is to harmonise national procurement procedures in each area. There are three different types of procedure for awarding contracts: the 'open procedure' where all tenders received are considered, the 'restricted procedure' where parties are required to respond to an invitation to tender, and finally the 'negotiated procedure' which allows public purchasers to negotiate directly with suppliers of their choice in certain circumstances. Large contracts require advertising in the *Official Journal of the European Economic Community*.

The criteria by which a public purchaser decides to appoint a contractor may be based on either the cheapest bid or the most economically advantageous bid. The latter enables public purchasers to assess considerations such as price, financial standing of the relevant party, quality and delivery (and, in the case of services, professional skill). Public purchasers will be obliged to set out such conditions in the contract notice and must assess tenders received in accordance with those criteria. The contract notice must also provide information on provision for employment and working conditions as these must be taken into account by each party submitting a tender (Melia, 1992).

The Commission wants public purchasers not to discriminate against parties submitting a tender solely on the basis of individual national standards and technical specifications. Public purchasers should in future where it is feasible refer to European or international standards. However, this provision would not apply where imposing such rules would lead to problems of incompatibility with existing equipment, increase costs dramatically or limit innovation.

COMPARISON OF COST PLANNING AND APPROXIMATE ESTIMATING

In the past there was on occasions some confusion over the two processes. Admittedly many quantity surveyors were first introduced to cost planning through approximate estimating. Cost planning aims at ascertaining costs before many of the decisions are made relating to the design of a building. It provides a statement of the main issues, identifies the various courses of action, determines the cost implications of each course and provides a comprehensive economic picture of the whole. The architect and quantity surveyor should be continually questioning whether a specific item of cost is really necessary, whether it is giving value for money or whether there is not a better way of performing the particular function, and the latter two aspects have some similarity with value management which is introduced later in this chapter and examined in more detail in chapter 11.

Cost planning does therefore differ significantly from approximate estimating. Approximate estimating aims at providing a preview of the probable tender figure with the method employed often being influenced by the amount of information available. Approximate estimating methods will be examined and compared in chapter 6. Cost planning, on the other hand, does not merely estimate the tender sum but probes much deeper into the cost implications of each part of the building, whereby each design decision is analysed and costed, and the final decision maintains a sensible relationship between cost, quality, utility and appearance. In addition cost planning permits control of expenditure throughout the design and construction stages, and if required throughout the life of the building through life cycle costing, as described in chapter 13.

Hence approximate estimating often plays a largely passive role after the major design decisions have been made, while cost planning bears upon the decisions themselves and plays an active part in the formulation of the design. Indeed, the cost plan should be regarded as one of the elements of design.

COST CONTROL TERMINOLOGY

A number of terms are used widely in cost control work and it is deemed advisable to define and explain these terms prior to their use.

Cost plan. A statement of the proposed expenditure on each section or element of a new building related to a definite standard of quality. Each item of cost is generally regarded as a 'cost target' and is usually expressed in terms of cost per square metre of gross floor area of the building as well as total cost of the element.

Cost check. The process of checking the estimated cost of each section or element of the building as the detailed designs are developed, against the cost target set against it in the cost plan.

Cost analysis. The systematic breakdown of cost data, often on the basis of elements, to assist in estmating the cost and in the cost planning of future projects. Cost analyses are often supplemented with specification notes, data concerning site and market conditions and various quantity factors, such as wall to floor ratios. The form and method of use of cost analyses are dealt with in chapter 9. Cost analysis aims at examining the cost of buildings already planned or built and for which priced bills of quantities and tenders are available. It has been suggested that it is in the nature of a post-mortem but in practice it is more valuable than this as it can assist materially in the design and cost evaluation of new projects.

Approximate estimating. Computing the probable cost of new building works at some stage before the bill of quantities is produced. It is an essential and integral part of the cost planning process.

Element. A component or part of a building that fulfils a specific function(s) irrespective of its design, specification or construction, such as walls, floors and roofs. Many cost plans and cost analyses are prepared on an elemental basis.

Cost research. All methods of investigating building costs and their interrelationship, including maintenance and running costs, in order to build up a positive body of information which will form basic guidelines in planning and controlling the cost of future projects. This includes the compilation of cost databases and other statistical cost information with the aid of computers.

Costs in use. Investigating the total costs of building projects – initial capital costs and maintenance and running costs throughout the predicted lives of the buildings. It provides the only way of obtaining the overall cost picture but does present a number of difficulties in practice, and these are examined in chapter 13. The term costs in use has now been largely superseded by the term life cycle costing. LCC comprises the total cost of an asset over its operating life, including land acquisition costs and subsequent operating costs, and possibly its disposal at the end of its operational life.

Cost study. Breaking down the total cost of buildings with the following objectives:

(1) to reveal the distribution of costs between the various parts of the building;
(2) to relate the cost of any single part or element to its importance as a necessary part of the whole building;
(3) to compare the costs of the same part or element in different buildings;
(4) to consider whether costs could have been apportioned to secure a better building;
(5) to obtain and use cost data in planning future buildings; and
(6) to ensure a proper balance of quantity and quality within the appropriate cost limit.

Cost control. All methods of controlling the cost of building projects within the limits of a predetermined sum, throughout the design and construction stages.

Cost management. The synthesising of traditional quantity surveying skills with structured cost reduction or substitution procedures using a multi-disciplinary team (the full design team).

Cost planning. This is often interpreted as controlling the cost of a project within a predetermined sum during the design stage, and normally envisages the preparation of a cost plan and the carrying out of cost checks. Cost planning uses the information gained by cost analysis to maintain a surer control over costs of future projects.

The following represents a more comprehensive definition: A systematic application of cost criteria to the design process, so as to maintain in the first place a sensible and economic relation between cost, quality, utility and appearance, and, in the second place, such overall control of proposed expenditure as circumstances may dictate. This definition warrants further consideration.

Emphasis is directed towards the proper consideration of design criteria other than cost to produce a properly balanced design, and it is advisable that the quantity surveyor does not lose sight of this. For instance, the cost of a project could be reduced merely by using cheaper materials, finishings and fittings, regardless of the fact that maintenance and running costs would probably be increased considerably in consequence, Furthermore the lower quality materials and components may be quite out of keeping with the class of building in which they are being incorporated. An economically priced project is required, but not necessarily the cheapest as a certain standard of quality has to be maintained. A building must also be designed so that it can satisfactorily perform its required functions. It might, for example, be possible to cheapen the cost of a factory roof by incorporating more columns and thus reducing the roof spans. This approach would be quite fruitless if the factory needed large unrestricted floor areas for its successful operation. Costs can often be reduced by disregarding the aesthetic quality of the building to be erected. Plain façades devoid of any form of embellishment could reduce costs but provide most uninspiring elevations. The author believes that every building client has a duty to assist in improving the environment and, through it, the well-being and general satisfaction of the community. HRH the Prince of Wales (1989) laid particular emphasis on this important aspect. As environmental aspects assume even greater importance, an environmental impact analysis (EIA) should be carried out on major projects (see chapter 16).

Feature. The aggregate of a number of construction units to provide a feature such as piled foundations, concrete stanchion base foundations, curtain walling systems and air conditioning systems.

Unit costs. Costs applied to a unit of work such as a concrete foundation wall, ground floor construction, a door including its frame, ironmongery, painting, lintel and other related items.

Value management. The utilisation of structured functional analysis focussing on the explicit determination of a client's needs and wants related to both cost and worth. This concept will be examined in some depth in chapter 11.

BUILDING COST INFORMATION SERVICE

The Building Cost Information Service (BCIS) is a collaborative venture, operating under the control of the RICS, for the exchange of building cost information to provide subscribers involved in design and construction with extensive data on building economy. It has been operating since 1962 and for the first ten years disseminated data sheets to a membership restricted to chartered quantity surveyor principals. However, since 1972, the membership of the service has been open to those of any discipline willing and able to contribute in accordance with the reciprocal basis of the service. In 1990 it had approximately 1300 subscribers to the main service and 240 to its on-line service (BCIS, 1992a).

The service centres its operations on the processing and amplification of information submitted by subscribers, the preparation and selection of data from all relevant sources and its continuous distribution to subscribers. All the information is stored on a central computer which forms the basis for the BCIS on-line service and also facilitates the production of the hard copy service. The distribution takes the form of regular mailings of information sheets for members to file in loose leaf binders.

The range of building cost information supplied by BICS is very extensive and includes comprehensive briefing; quarterly review of building prices; indices; labour, hours and wages; materials and equipment; techniques, systems and operations; legislation; statistics and economic indicators; regional trends; cost guidelines; publications digest and photocopying service; cost studies; detailed cost analyses; concise cost analyses; building price schedules, including average building prices for over 300 building types; and reference indexes. Much of this information will be described and illustrated in subsequent chapters.

BUILDING MAINTENANCE INFORMATION

This service also operates under the control of the RICS and was established in 1971 to provide independent cost information for professionals in all aspects of property management, maintenance and refurbishment. Subscribers also receive interpretive information on current technical developments and problems affecting costs, and relevant economic trends. The information is obtained directly from property users with whom BMI maintains regular liaison.

Publications and services provided by BMI comprise the following, which will be examined in more detail in subsequent chapters: BMI building maintenance price book; quarterly cost briefings; BMI cost indices; BMI news bulletins; special reports, which have included hospital occupancy costs and condition surveys; building owners' reports, which included property occupancy costs and performance/design data; publications digests; and BMI seminars.

CONSTRUCTION ECONOMICS EUROPEAN COMMITTEE (CEEC)

CEEC constitutes the Construction Economics European Committee/Comité Européen des Economistes de la Construction, and was founded in 1979 by the Royal Institution of Chartered Surveyors, the Society of Chartered Surveyors in the Republic of Ireland and Union Nationale des Techniciens Economistes de la Construction. With the establishment of the single market and the need for close working arrangements between construction economists in all the EC member states, the Committee performs a most valuable role. Each member body has a maximum of three representatives on the Committee with an elected president, two vice presidents and an honorary secretary. In 1992/93 Brian Drake of the UK and a former chairman of BCIS was president.

The aim of the Committee is:

To facilitate the exchange of experience and information between professionally qualified persons who are responsible for construction economics in the EU member states and to initiate studies with a view to:

(1) promoting the training and qualification of persons who are responsible for construction economics and drawing up proposals for the harmonisation and acceptance of standards of training and qualification;
(2) establishing guidelines for the definition, content, control and practice of construction economics;
(3) ensuring adequate representation of qualified persons who are responsible for construction economics in the EEC Commission and in other European institutions;
(4) studying existing and proposed legislation and regulations relating to construction economics, with a view to their harmonisation;
(5) co-ordinating working methods;
(6) establishing European statistics relating to costs and types of construction, procedures and materials.

One of the great difficulties in Pan-European cooperation are the changes in definitions, procedures, and traditions from country to country. In order to facilitate the exchange of experience and information, co-ordinate working methods, promote training and qualification, establish guidelines for the definition content and practice of construction economics, ensure adequate representation to the EC Commission and other European institutions, establish European statistics relating to costs and types of construction, procedures and materials CEEC set up four work groups dealing respectively with: methods and cost information; the use of information technology and green issues; EU matters; and construction economics in business services.

Supervised by the work groups the following desk studies had been launched in 1992 dealing with: national education and post-grade training programmes; sample case study; a CEEC code of good practice in construction economics; roles and liability in construction; procurement of services; methods of measurement; comparison of fee scales; and obstacles to cross border trade (CEEC/UNTEC Conference in Paris, 1993). The work programme devised for each of the four work groups in 1992/93 was most comprehensive and challenging and covered every important facet of construction economics.

The first European Conference on Construction Economy on the theme of 'Added Value in Construction', organised by the CEEC and the Union Nationale des Techniciens Economistes de la Construction (UNTEC) was held in Paris in May 1993. It was followed by the General Meetings of the CEEC and the AEEBC (Association of the European Experts for Building and Construction). There was also an exhibition of materials and equipment for the construction industry with 200 exhibitors. Promotional work for the conference was undertaken by the Union Nationale des Techniciens Economistes de la Construction (UNTEC) who publish a quarterly magazine entitled *Economie & Construction*.

OTHER SOURCES OF EUROPEAN INFORMATION

The Franco-British Construction Industry Group of the French Chamber of Commerce

This group was established with the aim of assisting its members by:

(1) promoting a deeper understanding of the differing requirements and approaches used in the French and UK construction markets;
(2) enabling business opportunities to be more easily identified and pursued;
(3) learning from experiences of other members and invited guests;
(4) providing an information source for members and non-members.

Members include architects, designers, contractors, engineers, developers, building material producers and distributors, quantity surveyors and engineers as well as research, PR and marketing specialists, legal advisers, commercial property advisers and land surveyors. The group meets bi-monthly to provide members with the opportunity for business discussions and social conversation. A feature of each meeting is a talk or presentation on a specific aspect of the construction industry from a French, British or joint venture viewpoint, and members are encouraged to share their experiences. Future plans included trade missions, seminars and publications.

Other Sources of Information

Readers requiring more information on European construction costs are referred to *Spon's European Construction Costs Handbook*, while those seeking more general information on the construction industries in other European countries could find the guides issued by the Construction Industry Research and Information Association (CIRIA) helpful. The RICS has set up an office in Brussels to maintain closer contact with other EC member states, with particular regard to building and surveying matters and reports of its activities appear in *Chartered Surveyor Monthly (CSM)*.

COST IMPLICATIONS OF DIMENSIONAL CO-ORDINATION

It can be argued that economic benefits, both to individual clients and in the total use of resources, follow from dimensional standardisation and consequent variety reduction of components and, in particular, from producing components of a high level of accuracy which will readily fit into place on site with a minimum of on-site labour.

The number of critical dimensions and the specification of low tolerances in several directions on one component should be kept to a minimum, since they increase the probability of misfit on the site and also the checking and quality control necessary in production. The designer may need to make joints capable of absorbing all likely inaccuracies, while the manufacturer has often had to make assumptions about the locations in which they will be used. It may be possible to reduce the need for small tolerances in components by ensuring adequate flexibility on their fixings on other components, such as slot-holes for bolts in precast concrete units. Costs must be considered for both manufacture and site erection. For example, low manufacturing costs may result in extra costs in jointing and in other components. Rejection of components on site will lead to delay in construction and extra costs to the contractor, and delay in completion with a cost penalty to the client.

INTRODUCTION TO VALUE MANAGEMENT/ENGINEERING

This technique will be examined in more detail in chapter 11 but it was considered beneficial to provide a short introductory section at this stage because of its growing importance and close interrelationship with cost management. The process was first introduced in the United States in the 1940s with the objective of securing increased efficiency and productivity in manufacturing industry in the supply of goods from scarce materials and components, but its use in the UK came much later. The Americans use the term 'value engineering' using the Society of American Value Engineers (SAVE) jobplan, whereas in the UK the term 'value management' seems to be preferred.

Value management/engineering is a creative, organised approach aiming at optimising cost and/or performance of a facility or system. It achieves savings or increases product value by identifying and eliminating or modifying components that add cost to the product without contributing to its required function. While developed originally to study manufacturing industry problems, the technique is readily adaptable to construction projects where the product will be the constructed building or facility (Norton, 1992a).

Value management/engineering is more than a cost reduction exercise of the type used by quantity surveyors in cost planning, and illustrated in this book. While there are similarities between the two processes, the techniques and methodology employed are quite distinct and tend to produce different results.

The basic requirements for a value management/engineering study are:

(1) specific allocation of resources for a study distinct from other processes occurring during the design stage;

(2) use of a management plan following a functional analysis approach to problem solving as described in chapter 11;

(3) use of a multi-disciplinary team experienced in value management/engineering methodology to review projects, preferably with members who were not involved in the original design;

(4) recording results to establish the degree of success of the value management/engineering study (Norton, 1992a).

COMPUTER AIDS TO COST PLANNING

Introduction

With computer software becoming cheaper, more accessible and user friendly, even small organisations are encouraged to use computers as valuable tools in the control and management of projects, in addition to their own business arrangements. Rapid developments in the computer industry have provided a cheaper, faster, more robust and more powerful product, enabling the user to operate more efficiently, and, at the same time, clients are more knowledgeable and computer oriented, and they want to be sure that quantity surveyors and contractors have the resources to assist in ensuring that projects will be delivered on time and on budget (Haksever, 1992).

The choice of an appropriate computer system is a problem because of the diversity and non-standardisation of systems. It is important to appoint the right person to analyse the organisation's computer needs, choosing the most appropriate computer system, providing adequate training and obtaining sufficient feedback from the users, and securing a high level of commitment to the system by all involved.

A quarterly news and reviews magazine (*Construction Computing*) is published by the CIOB, which contains an extensive software guide, and a wealth of information is available from the Construction Industry Computing Association (CICA) of which the RICS is a forum member along with the RIBA, CIOB and other construction oriented bodies.

The Construction Industry Computing Association

The Association (CICA) maintains information about computer programs for all aspects of construction projects, from the early design and conceptual stage through to the management and maintenance of properties. In association with RIBA Services, CICA publishes a directory of software, *The Construction Industry Software Selector*, and it also has its own in-house directory and offers a telephone information service to members.

Reports are available from CICA on an extensive range of computing and information technology (IT) topics, which included computer aided design (CAD), communication, project management, knowledge-based systems, software directories, and surveys on the impact, use and future needs of computing in the industry. Both the contracting and surveying groups meet three times a year and topics in 1992 included effective early cost advice – estimating and cost planning, and using databases in surveying – costs, administration and information (Wager, 1992).

Examples of Building Economics Software

A wide range of appropriate software is detailed in *Construction Computing* and a short selection of widely adopted systems is briefly described:

(1) *Techsonix*. Since 1983 this firm became one of the leading suppliers of computer systems for operational processes in the construction industry. The systems covered both pre and post contract activities, with software for cost modelling, cost planning, cost analysis, digitised measurement, bill of quantities production, scanned BoQ input, trade enquiries, estimating, quote comparisons, valuations, variations, final account, cost/value reconciliations, cash flow, cost analysis and cost management, supported by the necessary staff training and development.

(2) *Turner & Townsend Group: Project Cost Database*. This is a sophisticated computer program that stores, sorts and analyses a firm's cost and project data. It is easy to use, allows multiple levels of cost detail, provides multiple search criteria and statistical analysis of project costs.

(3) *E C Harris: Expenditure Forecast*. This program is designed to quickly establish the flow of expenditure over the duration of a construction project. It can be calculated using four different standard curve formats. The budget can be easily updated and, once set, a separate forecast of cost to completion can be used. After receipt of tenders, a revised forecast can be made based upon the contract sum. As the project progresses each valuation can be entered and a projection can be made to show future expenditure. Full reporting facilities are available including the production of a graph or histogram for plotter or laser printer.

(4) *E C Harris: Post Contract Cost Control & Reporting*. This program enables an organisation to keep on computer an up to date record of all data affecting the final cost of a project; supporting the automatic production of full cost reports, account progress and final accounts. The status of all cost information can be set to one of three levels.

EXPERT SYSTEMS

Introduction

As the reader will see frequent references in technical and professional journals to expert systems for use in building economics and related subjects, it was felt advisable to briefly explore the technique at this stage. Expert systems, which are more correctly termed intelligent knowledge based systems, are sophisticated computer programs which manipulate a stored body of knowledge in order to mimic the reasoning experience and judgement of human experts. An expert system may also be able to deal with the end-user inputting incomplete information, by making reasonable assumptions on the basis of other supplied information. In addition, an expert system can often cope with uncertainty in the knowledge it manipulates to find solutions to the end-users' problems (Brandon *et al.*, 1990).

However, expert systems do have limitations, as, for example, they can only cope with well-defined areas of knowledge, which are often of a technical nature. They cannot handle everyday knowledge, are incapable of intuitive thinking and cannot give inspired solutions to problems. They must be focused on a relatively small set of knowledge, are not often proficient at quickly processing numerical information, and cannot handle sensory or pictorial information.

Expert systems differ from conventional computer programs in the following ways:

(1) they involve the representation and manipulation of knowledge rather than data;
(2) heuristic as opposed to algorithmic procedures are used to solve problems;
(3) problems are solved by using inferential methods rather than the repetitive processes of conventional programs.

A survey of quantity surveying organisations carried out by the University of Salford for the RICS identified the areas which commonly generated major problems, requiring specialist knowledge for their solution and which it was felt could be resolved by expert systems. The following areas related to the general field of building economics: feasibility studies; investment appraisal; funding of projects; cash flow forecasting and monitoring; life cycle forecasting and control; cost estimating; and risk analysis and management.

Imaginor Systems

Imaginor Systems was formed 1n 1989 as a partnership between a subsidiary limited company of the Royal Institution of Chartered Surveyors (RICS) and Salford University Business Services Ltd (SUBSL). It was established to develop and market expert knowledge based software (KBS) in the property and construction industry. Its first product called ELSIE-Commercial, evolved from a UK Government Alvey initiative awarded to the two parties in 1986. In 1992 the installed base was in excess of 350 expert packages comprising a mix of ELSIE-Commercial and its later sister product called ELSIE-Industrial. The software operates on any IBM PC and can achieve a high standard of feasibility estimating using KBS at the beginning and early stages of a building project, and packages are mainly purchased by UK quantity surveyors, although the market has extended to architectural practices, design and build contractors, property developers and corporate companies.

The commercial buildings included offices, retail stores/shops, public buildings, hotels, schools, flats and houses, hospitals/clinics, libraries, and sports halls. The industrial buildings included hi-tech units, mid-tech units, start-up units, factories, warehouses, large retail outlets, business parks, and mixed developments. The sales of packages did not achieve the planned targets and many users expressed the view that the system should be applied to a wider range of projects. In 1993, the system was extended to cover mechanical and electrical estimating.

ELSIE-Commercial generates a detailed high quality cost model, expressed in BCIS format (cost models are examined in chapter 8); gives advice on the five main procurement routes; computes likely project duration in three phases; and derives and tests five residual values.

ELSIE-Commercial quickly and accurately produces detailed feasibility reports at the very early concept stages of a commercial building project, even before any drawings are prepared. This makes it a valuable tool for pre-design consultants and developers. It will cater for most building qualities and complexities, and will accommodate up to 80 floors. Detailed analysis can be applied to any internal and external elements in the building using a high capacity knowledge database (Imaginor Systems, 1991).

In operation ELSIE-Commercial displays a series of question-sets on the computer's monitor which are answered by the user. The nature of the series used depends on the module invoked, the project in hand and the answers given to previous question-sets. ELSIE is continually using its expert knowledge to deduce the next question and is simultaneously constructing a virtual model of the desired building and its key attributes. This model represents the design ideas of the user or client. On completion of this stage, an ELSIE model is used to obtain its own unique analysis of the model or its parameters. Each project is stored permanently in a limitless projects database structure managed by ELSIE. A detailed final report can be printed, or the user may choose to query and adjust some of the results from current or past projects.

This system gives users the flexibility to develop various approaches to the same project and can explore, query, refine and optimise different ideas and can be an important tool in meeting a precise brief. The user can also adjust cost assumptions made by ELSIE for local conditions, different building types and changing project requirements. The result is an accurate report containing all computed figures, costs, assumptions made, and advice on how to proceed.

ELSIE Industrial also provides four essential estimating functions in one fully integrated expert package. An ELSIE industrial project can handle up to 100 three-storey buildings with external works analysed in a single process.

The main advantages claimed for both systems are the production of estimates without the existence of a design; save feasibility costs; give very fast results; are easy and intuitive; provide advice from an expert quantity surveying knowledge bank; provide high quality detailed printed reports; are very adaptable using a 'what-if?' feature; and they complement most building software. Imaginor Systems offer a major step forward in the refinement of cost planning techniques, within certain limitations, and further reference will be made to their application in later chapters.

Brandon *et al.* (1990) concluded that practices without a substantial cost/price database of their own and with a project portfolio containing many office-type developments are more likely to use ELSIE directly for producing information for clients. Practices at the other extreme, with an established cost/price database, are less likely to use ELSIE and more inclined to rely on their own computer-based systems, which in some cases have been developed over many years.

EXAMINERS' REPORTS

A brief summary of the reports made by examiners in this subject follows, from which students will find useful hints and descriptions of some of the pitfalls to be avoided.

Many candidates showed a limited knowledge of the meaning of 'building economics' by discussing only its influence on the construction of a building, without making any reference to other factors such as market conditions, choice of site, expected use and life of buildings, together with the influence of running and maintenance costs.

A number of students see the need to bring the contractor into the construction team at an early stage in the design process by means of a suitable contractual procedure, but much fewer recognise the advantage stemming from the use of the contractor's expertise on how to best carry out the project.

Few also recognise the need to achieve financial control during the construction of the building, including the need for an early meeting of those concerned with the execution of the project in order to decide how best to proceed with the work as quickly as possible, while attempting to adhere to the financial limit imposed by the cost plan.

When comparing cost analyses for similar buildings attention must be paid to varying methods of pricing preliminaries and different site conditions. The ratios of circulation space to usable floor space must be taken into account when computing expected returns. With industrial projects account must be taken of the cost of disruption to production of building works. With contractor-sponsored arrangements, a check is needed to ensure compliance with the client's performance specification.

Most answers to a question requiring a definition of 'value for money' were too restricting and were mainly confined to office buildings. A question relating to a developer's budget was also poorly answered and one relating productivity to the nature and scope of design stages was not properly answered as most candidates failed to analyse the problem and answered in generalities.

It was not generally appreciated that where an architect has produced a design that is easy to build and is inherently economical through repetition and standardisation, then the quantity surveyor has a clear responsibility to his client to communicate these economies to the contractor in the bills of quantities.

It is essential that candidates answer the questions set to secure good marks, as so often in practice candidates fail to keep to the question and sometimes misinterpret it. A question requiring candidates' views on how to construct an index for updating cost data was frequently answered by giving lists of indices without making any reference to the problems posed.

Candidates did not always appreciate that the use of cost indices involves problems which stem mainly from the peculiarity of jobs and the difficulty of forecasting future costs from past trends. Candidates also lacked knowledge in the cost implications of alternative constructional techniques. There was also a general failure to grasp the main difficulties associated with the collection and analysis of costs in use data.

2 COST IMPLICATIONS OF DESIGN VARIABLES AND QUALITY ASSURANCE

The costs of buildings are influenced by a variety of factors, some of which are interrelated. It is essential that quantity surveyors should be fully aware of the cost consequences resulting from changes in shape, size, storey heights, total height, fenestration and other building characteristics. The cost effect of the main design variables will be examined and compared in this chapter. The final design of a building will be influenced by a variety of factors, including user's needs, planning and Building Regulations requirements, site conditions and aesthetic requirements. Other factors affecting cost include the form of contract, period for completion, structural form, extent of prefabrication and standardisation and consideration of maintenance and running costs, all of which will be considered in subsequent chapters.

It is important that the method used for expressing building costs permits alternative designs and forms of construction to be examined on a comparable basis. The most convenient unit is the square metre of gross floor area (GFA) measured between the main enclosing walls (inner faces), and making no deductions for internal walls, staircases, lift shafts or other circulation space. The cost of the whole building or any part of it can be equally well expressed in this way. For instance, a six-storey office block in the West Midlands might have cost about £760 per m^2 of floor area in 1993, with internal finishings accounting for approximately £65 per m^2. Considerable care has, however, to be taken in using costs expressed in this way to make allowance for widely differing conditions on different projects. For example, soil conditions can cause quite different foundation costs for otherwise similar buildings, the cost of finishings will be influenced appreciably by the requirements of the specification and lift costs will be affected by the height of the building and the area of each floor, as well as the type of lift.

PLAN SHAPE

The shape of a building has an important effect on cost. As a general rule the simpler the shape of the building the lower will be its unit cost. As a building becomes longer and narrower or its outline is made more complicated and irregular, so the perimeter/floor area ratio will increase, accompanied by a higher unit cost. The significance of perimeter/floor area relationships will be considered in more detail later in the chapter. An irregular outline will also result in increased costs for other reasons; setting out, siteworks and drainage work are all likely to be more complicated and more expensive and this will be apparent by comparing the two buildings shown in figure 2.1, each of which have the same floor area. In building B where there is six per cent more external wall to enclose the same floor area, setting out costs are increased by about fifty per cent, excavation costs by about twenty per cent and drainage costs by approximately twenty-five per cent (two metres of additional 100 mm drain and two extra manholes). The additional costs do not finish there as brickwork and roofing will also be more costly due to the work being more complicated. It is important that both architect and client are aware of the probable additional costs arising from comparatively small changes in the shape of a building. They are then able to consider the advantages to be gained from variations in shape in the full knowledge of the additional costs involved and can adopt a rather rudimentary

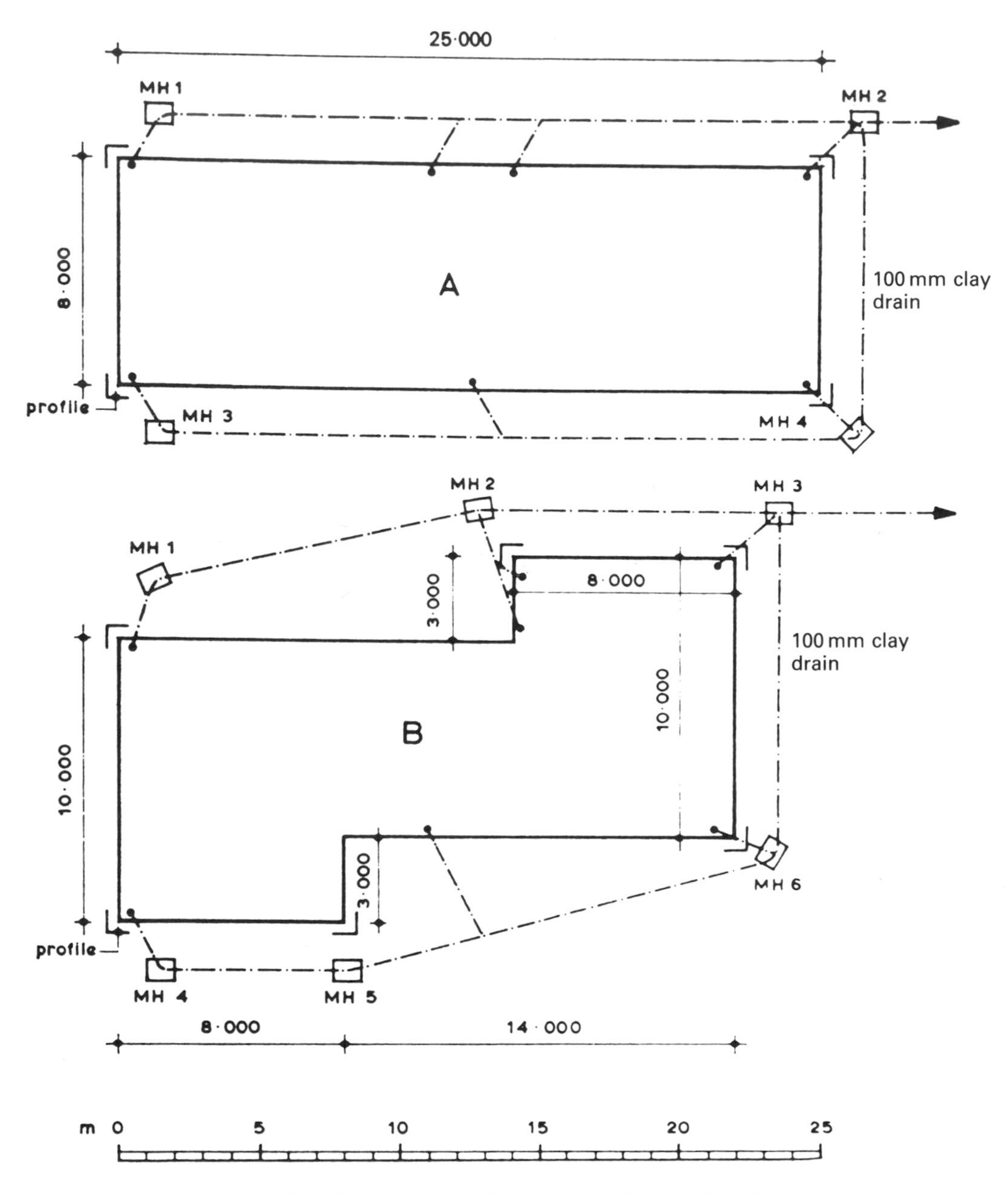

Figure 2.1 Higher cost of buildings of irregular shape

cost benefit approach. It does however involve a subjective judgement as far as the aesthetics are concerned. The running costs of the completed building may also be affected by such factors as higher heat losses (walls and windows), window cleaning and painting. The plan shape of a building also affects the provision of the internal divisions and hence their cost.

Although the simplest plan shape, that is a square building, will be the most economical to construct it would not always be a practicable proposition. In dwellings, smaller offices, schools and hospitals considerable importance is attached to the desirability of securing adequate natural daylighting to most parts of the buildings. A large, square structure would contain areas in the centre of the building which would be deficient in natural lighting. Difficulties could also arise in the planning and internal layout of the accommodation. Hence, although a rectangular shaped building would be more expensive than a square one with the same floor area because of the smaller perimeter/floor area ratio, nevertheless practical or functional aspects, and possibly aesthetic ones in addition, may dictate the provision of a rectangular building. The following example of hotels illustrates the need to maintain a balance between various design criteria, in this case cost, function and appearance. It is pointless for the quantity surveyor to submit cost-saving alternatives which could not function satisfactorily or which would be aesthetically undesirable. Circular buildings, although enclosing the greatest floor area for the smallest perimeter, are uneconomic and result in major internal planning problems.

Certain types of building present their own peculiar problems which in their turn may dictate the form and shape of the building. For instance hotels, for visual reasons, to provide guests with good views and the advertising effect of a prominent building on the skyline, need to be tall. The shape and floor area are closely related to the most economic bedroom per floor ratio and this is generally in the range of forty to fifty. This dictates a tall slab rather than a tower. Slender towers can be aesthetically desirable but their relatively poor ratio of usable to gross floor area often renders them prohibitively expensive.

Taking another example, there are also functional limits on the depth of office blocks. Office buildings with depths of up to 18 m are acceptable in America and Australia, but the rental value obtainable per square metre of floor area reduces with the larger depth of building. Buildings of small depth, up to say 12 m, can be more readily split into small office units and will accommodate a greater variety of tenancies. Admittedly, an office building occupied by a large industrial concern will probably contain mainly large general offices and depths of up to 15 m may well be acceptable. Furthermore, an industrialist normally requires about 900 m^2 of office floor area per floor.

There are occasions when the site itself will dictate the form or shape of the building. In some cases the designer may feel obliged to advise the building client to purchase additional land, where this is practicable, to make the development a more economical proposition. It may be worthwhile to under-utilise an awkwardly shaped site in order to secure a regularly shaped and more economical building. Where a strip form of development of possibly eight to ten storeys in height is involved, means of escape considerations alone may dictate the optimum length of building to secure the maximum net to gross floor area relationship.

The shape of a building may also be influenced by the manner in which it is going to be used. For instance, in factory buildings the determining factors may be co-ordination of manufacturing processes and the form of the machines and finished products. In schools, dwellings and hospitals, and to a more limited extent in offices, shape is influenced considerably by the need to obtain natural lighting. Where the majority of rooms are to rely on natural lighting in daylight hours, the depth of the building is thereby restricted. Otherwise it is necessary to compensate for the increase in depth of building by installing taller windows which may compel increased storey heights. The aim in these circumstances should be to secure an ideal balanced solution which takes into account both the lighting factor and the constructional costs. Deeper rooms result in reduced perimeter/floor area ratios with a subsequent reduction in construction, maintenance and heating costs, but these savings may be offset by increased lighting costs. With taller rooms the conditions are reversed. Where a high density of rooms is required, the use of an external wall as a boundary to a room may compensate for the amount of partitioning that would otherwise be required. It is, therefore sometimes preferable to elongate the building, so that rooms can be entered from either side of a spinal corridor rather than having a deep building with a complex network of corridors to give access to all rooms, together with the possible need for artificial ventilation to internal rooms.

A number of attempts have been made at assessing the efficiency of plan shapes mainly by reference to the wall to floor area ratio and these are examined later in this chapter.

SIZE OF BUILDING

Increases in the size of buildings usually produce reductions in unit cost, such as the cost per square metre of gross floor area. The prime reason for this is that oncosts are likely to account for a smaller proportion of total costs with a larger project, or expressed in another way, they do not rise proportionately with increases in the plan size of a building. Certain fixed costs such as the transportation, erection and dismantling of site buildings and compounds for storage of materials and components, temporary water supply arrangements and the provision of temporary roads, may not vary appreciably with an extension of the size of building and will accordingly constitute a reduced proportion of total costs on a larger project. A larger project is often less costly to build as the wall/floor ratio reduces, rooms tend to be larger with a proportional reduction in the quantity of internal partitons, decorations, skirtings, etc., and there may also be a proportional reduction in the extra cost of windows and doors over walls. With high rise buildings a cost advantage may accrue due to lifts serving a larger floor area and greater number of occupants with an increased plan area.

Figure 2.2 shows the effect of doubling the length of a rectangular building on the ratio of enclosing wall to floor area. The length of external wall per square metre of floor area (one floor only) is reduced from 383 mm to 317 mm, a reduction of 17.25 per cent. The wall thicknesses are ignored to simplify the calculations.

An example will serve to illustrate the cost advantage in lift provision by increasing the area on each floor of multistorey blocks of flats and offices. A six-storey block of offices built in the East Midlands in 1993 has 360 m^2 of gross floor area on each floor and the six floors are served by two passenger lifts. The total cost of the project of £1 609 200 is equivalent to £745/m^2 of gross floor area and the lifts cost £94 000 and are equivalent to £44/m^2 of gross floor area. If the floor area was doubled on each storey the lift provision could remain the same and the cost of lifts would then be reduced to about £22/m^2 of floor area, giving a saving of about 3.0 per cent on total building costs.

Another interesting illustration of the effect of size on building costs is obtained from a comparison of the costs of one, two, three and four bedroom new build social housing provided in table 2.1. Willmott Dixon kindly supplied the 1993 costs covering developments in the Midlands and Eastern and Southern England together with the gross floor areas of the various houses. The following schedule gives average 1993 costs for each of the four different sized houses and their respective costs/m^2 of floor area.

House size (nr of bedrms)	*Average cost* £	*Percentage increase*	*Average cost/m^2* £	*Percentage reduction*
1	22 803		447.5	
2	28 050	22.8	416.0	12.88
3	34 240	50.2	401.2	15.98
4	43 246	89.7	393.5	17.60

This schedule shows significant increases in overall cost, excluding land costs in all cases, as the amount of accommodation is increased, but these are accompanied by reducing unit costs as the wall/floor ratio improves and many of the same fixtures and fittings and siteworks are spread over larger floor areas, thereby producing economies of scale. The high cost ratio of the smaller houses shows up even more forcibly when related to persons being accommodated as against units of floor area. The average cost per person in a one bedroom house amounted to £11 402, as compared with £7012/person in a 2 bedroom house, £6848/person in a 3 bedroom house and £7208/person in a 4 bedroom house.

An analysis of the main constructional groupings shows significant cost variations over the different sites and between the different sized houses, which highlights the dangers of attaching too much significance to a limited number of buildings, although they do illustrate the broad

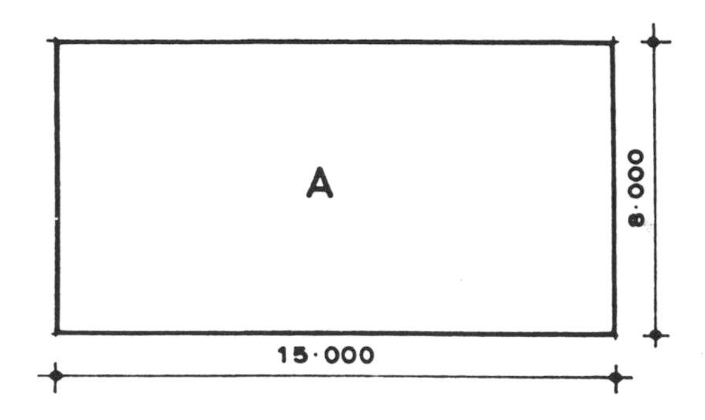

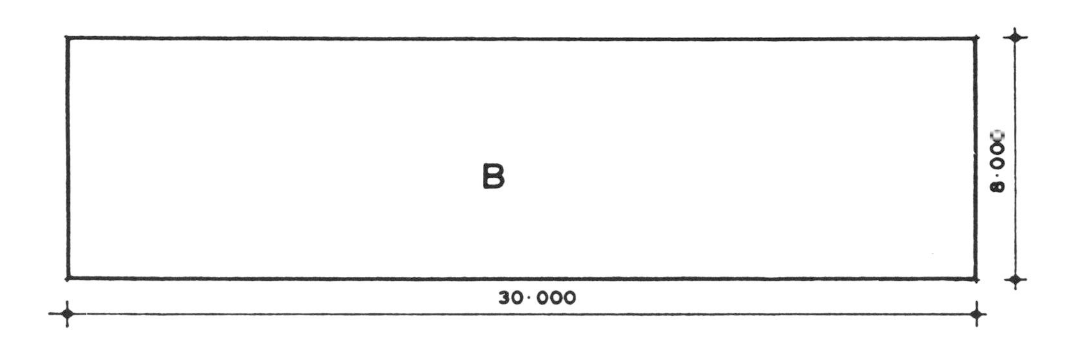

	BUILDING A	BUILDING B
FLOOR AREA	120 m^2	240 m^2
LENGTH OF ENCLOSING WALL	46 m	76 m
LENGTH OF WALL/ M^2 OF FLOOR AREA	383 mm	317 mm

Figure 2.2 Effect of change in size of buildings

pattern. Thus substructure costs range from 5.28 to 9.30 per cent, superstructure costs from 63.20 to 73.47 per cent and siteworks costs from 18.37 to 30.95 per cent. The large variation in siteworks costs is to be anticipated as the site conditions can vary so much from one site to another.

As to locational cost differences, the unit costs of the houses in Eastern England were 12 to 19 per cent higher than those in the Midlands, but they were smaller in floor area and there could well be other design and constructional differences. Whereas the houses in Southern England

Table 2.1 Comparative costs of one-, two-, three- and four-bedroom new build social housing in the Midlands, Eastern and Southern England (1993 prices)

	Total cost		Cost/m²
Midlands			
1 Bedroom (2 Person)			
Substructure	1 638		31.50
Superstructure	13 780		265.00
Siteworks costs	6 032		116.00
Gross floor m²	£21 450	52 m²	412.50
2 Bedroom (4 Person)			
Substructure	1 943		29.00
Superstructure	16 683		249.00
Siteworks costs	7 772		116.00
Gross floor m²	£26 398	67 m²	394.00
3 Bedroom (5 Person)			
Substructure	2 255		27.50
Superstructure	20 254		247.00
Siteworks costs	9 348		114.00
Gross floor m²	£31 857	82 m²	388.50
4 Bedroom (6 Person)			
Substructure	3 048		26.50
Superstructure	28 175		245.00
Siteworks costs	12 995		113.00
Gross floor m²	£44 218	115 m²	384.50
Eastern England			
1 Bedroom (2 Person)			
Substructure	1 800		40.00
Superstructure	16 200		360.00
Siteworks costs	4 050		90.00
Gross floor m²	£22 050	45 m²	490.00
2 Bedroom (4 Person)			
Substructure	2 440		40.00
Superstructure	19 520		320.00
Siteworks costs	5 490		90.00
Gross floor m²	£27 450	61 m²	450.00
3 Bedroom (5 Person)			
Substructure	3 200		40.00
Superstructure	24 800		310.00
Siteworks costs	7 200		90.00
Gross floor m²	£35 200	80 m²	440.00
4 Bedroom (6 Person)			
Substructure	4 040		40.00
Superstructure	30 300		300.00
Siteworks costs	9 090		90.00
Gross floor m²	£43 430	101 m²	430.00
Southern England			
1 Bedroom (2 Person)			
Substructure	1 316		28.00
Superstructure	15 886		338.00
Siteworks costs	7 708		164.00
Gross floor m²	£24 910	47 m²	530.00
2 Bedroom (4 Person)			
Substructure	1 800		24.00
Superstructure	19 575		261.00
Siteworks costs	8 925		119.00
Gross floor m²	£30 300	75 m²	404.00
3 Bedroom (5 Person)			
Substructure	2 090		22.00
Superstructure	22 705		239.00
Siteworks costs	10 830		114.00
Gross floor m²	£35 625	95 m²	375.00
4 Bedroom (6 Person)			
Substructure	2 530		22.00
Superstructure	27 600		240.00
Siteworks costs	11 960		104.00
Gross floor m²	£42 090	115 m²	366.00

Table 2.1 Continued

Source: Willmott Dixon Ltd.

displayed higher unit costs than those in the Midlands for 2 and 3 bedroom houses and lower unit costs for the 3 and 4 bedroom houses, hence there appeared to be no logical pattern and a more detailed breakdown of the component costs would be needed to arrive at meaningful conclusions.

Tender prices when expressed per square metre of gross floor area, tend to fall as the area increases. For example, an increase in the area of flats of x per cent may be expected to reduce the price per square metre by $x/2$ per cent. A decrease in area of x per cent causes a corresponding increase of about $x/2$ per cent in the price per square metre. This form of relationship is however limited to differences in area of up to about fifteen per cent, as over this limit rather more complex relationships operate.

PERIMETER/FLOOR AREA RATIOS

We have already seen that the plan shape directly conditions the external walls, windows and external doors which together constitute a composite element – the enclosing walls or envelope. Different plans can be compared by examining the ratio of the areas of enclosing walls to gross floor area in square metres (known as the wall/floor ratio). The lower the wall/floor ratio, the more economical will be the proposal. The best wall/floor ratio is produced by a circular building, but the saving in quantity of wall is usually more than offset by the much bigger cost of circular work over straight, the increased cost varying between twenty and thirty per cent. The wall to floor ratio is a means of expressing the planning efficiency of a building and is influenced by the plan shape, plan size and storey heights. The ratio is calculated by dividing the external wall area (inclusive of windows and doors) by the gross floor area.

Figure 2.3 shows the outline of two buildings, one of which (building A) is L shaped and the other (building B) has a very irregular outline. Both buildings have an identical gross floor area on each floor of 244 m^2, and assuming that the buildings are each of two storeys, this gives a total floor area of 488 m^2 for each building. Wall thicknesses have been ignored in this example to simplify the calculations. The length of enclosing wall in building A amounts to 70 m while that in building B totals 100 m – an increase of forty-three per cent. Assuming that the height of the walling is 6 m, the areas of enclosing walls are 420 m^2 for building A and 600 m^2 for building B and the wall/floor ratios are

$$\text{building A} = \frac{420}{488} = 0.86$$

$$\text{building B} = \frac{600}{488} = 1.23$$

Building B is very uneconomical with a much greater area of enclosing walls than A. It should be borne in mind that the perimeter cost of a building can be in the order of twenty to thirty per cent of total cost and an external wall can be two to four times as expensive as an internal partition. In this example building B is likely to be at least ten per cent more expensive than building A on account of the much increased perimeter costs.

It has been found that with an economically designed five-person house, the cost of enclosing walls, including windows and doors, is likely to be twenty-five to thirty per cent of the total cost of the superstructure, services should account for about twenty per cent, and internal partitions and associated doors fourteen to fifteen per cent. With traditional construction, the cost of the structure – enclosing walls, windows, roof and floors – increases steadily with an extension of the frontage. As a result wider frontage houses are more expensive than those of narrower frontage with similar floor area. The length of frontage also affects the costs of external services and external works. Consideration also needs to be given to the cost effects of varying the number of dwellings in a terrace where schemes include terraced blocks. The additional cost of an end-of-terrace-house superstructure over that of an intermediate house was likely to be in the order of £1170 to £1930 in 1993 because of the higher cost of gable-end walls over party walls, but this will be partly offset by the increased cost of making provision for rear access to the intermediate dwellings.

Narrow frontage houses become relatively cheaper as the length of terrace increases, with costs falling more sharply, than with the medium and wide frontage houses. Where both narrow and wide frontage houses and short and long terraces are required, it is more economical to use wide frontage houses in the short terraces and the narrow frontage houses in the long ones.

Another aspect which has to be considered when investigating perimeter/floor area ratios is the adequacy of the natural lighting to the interior of the building and the practicability of the internal layout. By reducing the frontage and increasing the depth of a building the amount of natural light reaching the innermost parts will be reduced and may result in increased operating costs through higher artificial lighting charges. A deeper building may also result in wasteful and inconveniently shaped rooms such as long cubicles housing WCs. Thinner walls will also provide greater floor area for the same length of enclosing wall.

Many blocks of flats are rectangular in plan shape, although some are U, L, Y or T shaped. Block shape and depth influence the proportion of external wall area to floor area and the wall/floor ratio varies from about 1.40 for the most complex plans to 0.60 for the most simple and deepest rectagular plans.

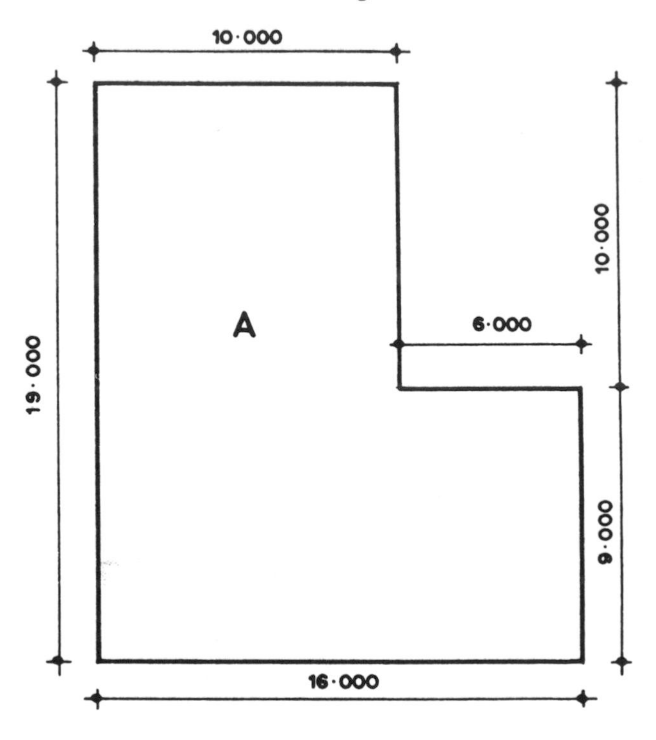

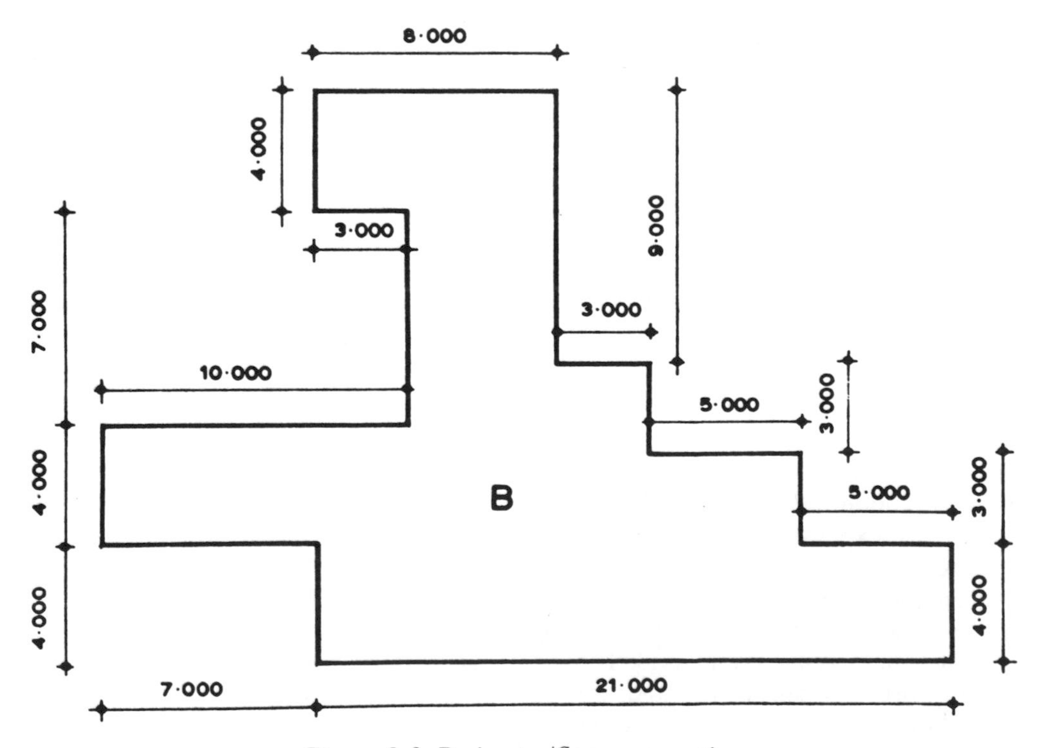

Figure 2.3 Perimeter/floor area ratios

Some analytical work has been undertaken to assist in obtaining optimum plan shape efficiency and these are now detailed, but these must be viewed against the background of the other design factors previously identified.

(1) *plan shape index*: $\frac{g + \sqrt{(g^2 - 16r)}}{g - \sqrt{(g^2 - 16r)}}$ where g is sum of perimeters of each floor divided by number of floors and r is gross floor area divided by number of floors.

(2) *optimum envelope area*: $N\sqrt{N} = \frac{x\sqrt{f}}{2S}$ where N = optimum number of floors; x = roof unit divided by wall unit cost; f = total floor area in m^2; and S = storey height in m (Ashworth, 1994).

CIRCULATION SPACE

An economic layout for a building will have as one of its main aims the reduction of circulation space to an acceptable minimum, having regard to the building type. Circulation space in entrance halls, passages, corridors, stairways and lift wells, can all be regarded as 'dead space' which cannot be used for a profitable purpose and yet involves considerable cost in heating, lighting, cleaning, decorating and in other ways. Almost every type of building requires some circulation space to provide means of access between its constituent parts and in prestige buildings spacious entrance halls and corridors add to the impressiveness and dignity of the buildings.

In the majority of buildings, however, there is a definite need to reduce circulation space to a minimum compatible with the satisfactory functioning of the building. Elimination of lengths of corridor which result in communication through rooms or an entire 'open plan' may not prove to be the most economical proposition if all the costs and benefits of each set of proposals are quantified and evaluated. Reducing the width of corridors to an extent that persons using the building suffer actual inconvenience could not really be justified; corridors may also have to serve as escape routes in case of fire. As with other parts of buildings, cost is not the only criterion which has to be examined – aesthetic and functional qualities are also very important. Circulation space requirements tend to rise with increases in the height of buildings and it is accordingly well worthwhile to give special consideration to circulation aspects when designing high rise buildings.

The proportion of floor space allocated to circulation purposes will vary considerably between different types of building. The following circulation ratios (proportion of circulation space to gross floor area) provide a useful guide: office blocks: nineteen per cent: laboratories: thirteen per cent; flats (four-storey): twenty-one per cent. The reader may find these ratios surprisingly high and their significance will be apparent when the published cost of a building calculated per square metre of gross floor area is converted to the cost of a square metre of usable floor space. For instance an office block costing £720 per m^2 of gross floor area with twenty per cent circulation space is equivalent to £865 per m^2 of usable floor area. This is particularly important in buildings, such as offices and factories, which may be erected for letting where the rent is usually calculated on usable floor area only.

Circulation provision in blocks of flats may take one of the forms illustrated in figure 2.4. One common approach is to provide four flats on each floor, frequently two one-bedroom and two two-bedroom, with access from a common hall. The shape of the block can be rectangular (A) or cruciform (B) and it will be immediately apparent that the cruciform variety has a higher circulation ratio and a much increased perimeter/floor ratio, although it provides a considerably improved elevation and much better living conditions for the occupants. When it is planned to provide more flats on each floor they are usually accommodated in slab blocks, when access may be obtained by internal corridors (C) or external balconies (D). Layouts incorporating external balcony access have a higher circulation ratio but involve less artificial lighting to common parts of the block. Typical circulation ratios follow: the circulation space includes that in the flats themselves as well as communal space.

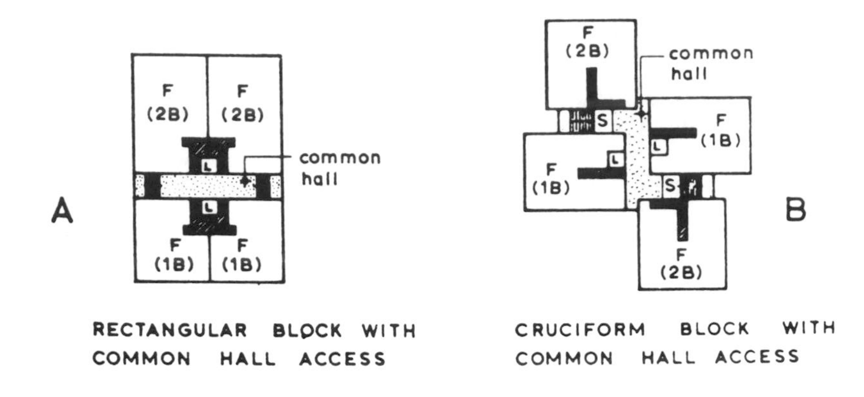

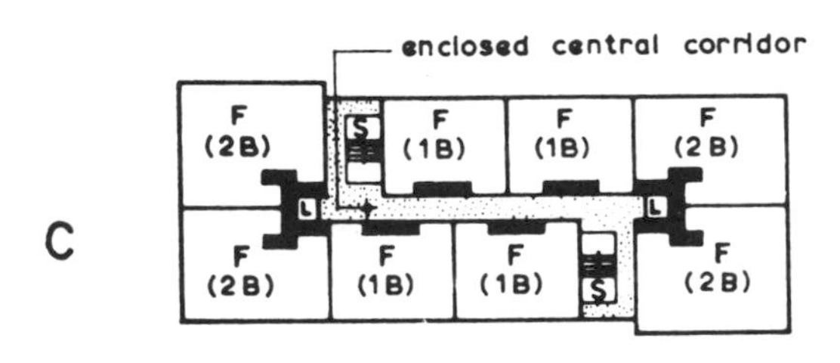

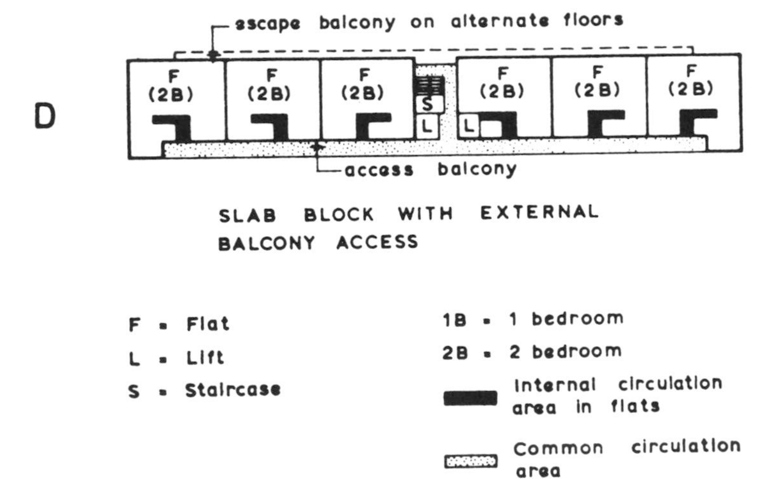

Figure 2.4 Means of access to flats

Plan arrangement	*Circulation ratio*
A. Rectangular block with common landing access	twenty per cent
B. Cruciform block with common landing access	thirty per cent
C. Slab block with internal corridor access	twenty-two per cent
D. Slab block with external balcony access	thirty-two per cent

With certain types of building, planning suitable circulation arrangements can be a complex task. Hotels provide a good example where the routes taken by resident guests, non-resident diners, and staff follow distinct patterns and these establish clear relationships between the hotel's various parts. The layout and planning of hotels must facilitate movements of people and, as far as possible, provide for the separation of guests, staff and maintenance personnel. This is important to prevent annoyance and disturbance of guests and also to enable service facilities to be designed for efficient use. Secondary circulation may also be desirable to separate resident and non-resident guests, as for example by providing direct access to restaurants and banqueting halls. Figure 2.5 illustrates in diagrammatic form typical hotel circulation patterns and relationships. Crown Courts require three completely separate restricted or private circulation systems for judges, juries and defendants, and a fourth system for the public at large.

STOREY HEIGHTS

The storey heights of buildings are mainly determined by the requirements of the users of the building. Variations in storey heights cause changes in the cost of the building without altering the floor area, and this is one of the factors that makes the cube method of approximate estimating so difficult to operate when there are wide variations in the storey height between the buildings being compared. The main constructional items which would be affected by a variation in storey height are walls and partitions, together with their associated finishings and decorations. There will also be a number of subsidiary items which could be affected by an increase in storey height, as follows.

(1) Increased volume to be heated which could necessitate a larger heat source and longer lengths of pipes or cables.

(2) Longer service and waste pipes to supply sanitary appliances.

(3) Possibility of higher roof costs due to increased hoisting.

(4) Increased cost of constructing staircases, and lifts where provided.

(5) Possibility of additional cost in applying finishings and decorations to ceilings, sometimes involving additional scaffolding.

(6) If the impact of the increase in storey height and the number of storeys was considerable, it could result in the need for more costly foundations to support the increased load.

One method of making a rough assessment of the additional cost resulting from an increase in the storey height of a building is to work on the basis that the vertical components of a building in the form of walls, partitions and stanchions account for about thirty per cent of total costs. An example will serve to illustrate the approach.

Estimated cost of building	£1 200 000
Estimated cost of vertical components thirty per cent of £1 200 000	£360 000
Proposal to increase storey heights from 2.60 m to 2.80 m: increased cost would be $\frac{0.20}{2.60} \times 100 \times £360\,000$	£27 692

It would, however, be necessary to consider the possible effect of some or all of the subsidiary items previously listed if the increase in storey height was substantial.

The average clearance height for single-storey factory buildings in a pre-war sample taken by the Building Research Establishment was about 4.5 m, with an average building height of about 5.8 m as few of the buildings had flat roofs. The only trade in which the average height of 4.5 m was much

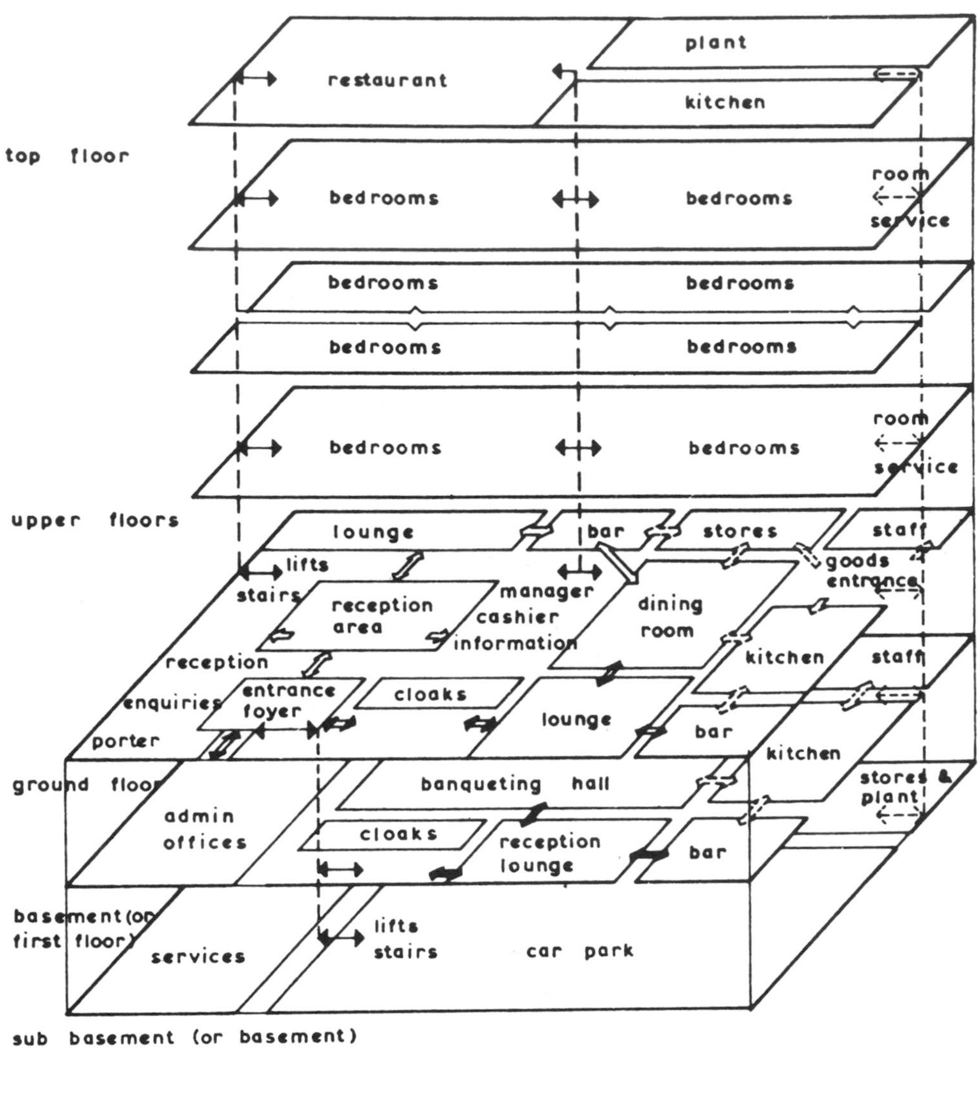

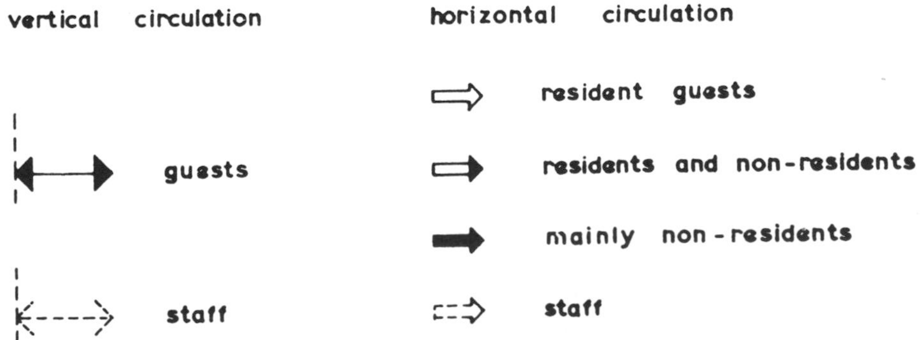

Figure 2.5 Hotel circulation patterns and relationships

exceeded was the engineering group, where in some cases clearance heights were 6 m and 9 m. Such clearances were usually required for overhead cranes and did not often extend over the whole building. In fact, the average height for the engineering industry was 4.9 m, and in light industries few machines exceeded 3 m in height.

In modern commercial buildings it may be necessary to provide space above false ceilings to accommodate service ducts for cables or pipes or air conditioning ducts. In other types of building such as theatres, sports halls, conference centres and churches, increased storey heights are required.

TOTAL HEIGHT OF BUILDINGS

Constructional costs of buildings rise with increases in their height, but these additional costs can be partly offset by the better utilisation of highly priced land and the reduced cost of external circulation works. Private blocks of flats are generally best kept low, for reasons of economy, except in very high cost site locations where luxury rents are obtainable. In similar manner, office developments in tower form are more expensive in cost than low rise, but providing the tower has around 1000 m^2 of gross area per floor the rent obtainable may offset the additional cost.

There are some general principles relating to increases in the number of storeys to a building which ought to be taken into account when high rise buildings are under consideration.

(1) It is sometimes desirable to erect a tall building on a particular site to obtain a large floor area with good daylighting and possibly improved composition of buildings.

(2) The effect of the number of storeys on cost varies with the type, form and construction of the building.

(3) Where the addition of an extra storey will not affect the structural form of the building, then, depending upon the relationship between the cost of walls, floor and roof, construction costs may fall per unit of floor area.

(4) Beyond a certain number of storeys the form of construction changes and unit costs usually rise. The change from load-bearing walls to framed construction is often introduced when buildings exceed four storeys in height.

(5) Foundations costs/m^2 of floor area will fall with increases in the number of storeys provided the form of the foundations remains unchanged. This will be largely dependent upon the soil conditions and the building loads. A large rise in costs will occur where pile foundations have to be substituted for strip or pad foundations; beyond this point it is likely that the same foundations would serve a building with an increased number of storeys.

(6) More expensive plant, such as tower cranes and concrete pumps, is required for the construction of high rise buildings.

(7) Means of vertical circulation in the form of lifts and staircases tend to be increasingly expensive with higher buildings, although fairly sharp increases in costs are likely to occur at the storey heights at which the first and subsequent lifts become necessary.

(8) As a general rule maintenance costs rise with an increasing number of storeys, as maintenance work becomes more costly at higher levels.

(9) Heating costs are likely to fall as the number of storeys increases and the proportion of roof area to walls reduces. Heating costs are influenced considerably by the relationship between the areas of roof and walls, as roofs are points of major heat loss. However the services and associated equipment become more sophisticated and costly with high rise buildings, and their ducting can also increase costs.

(10) Fire protection requirements increase with height as fire fighting equipment becomes more sophisticated, involving the use of wet or dry risers and possibly sprinklers.

(11) Fees of specialist engineers will probably be incurred for the design of foundations and frame, mechanical and electrical services and fire fighting equipment.

(12) As the number of storeys increases, both the structural components and circulation areas tend to occupy more space and the net floor area assumes a smaller proportion of the gross floor

area, thus resulting in a higher cost per sq m of usable floor area. One study showed the cost of office blocks rising fairly uniformly by about two per cent per floor when increasing the height above four storeys.

GROUPING OF BUILDINGS

The grouping and arrangement of buildings on a site can have a significant influence on the total cost of a project. For example, inter-linking buildings often results in savings in costs, usually achieved by a reduction in the quantity of foundations, external walling and other common elements of construction. Furthermore, this approach can also produce lower costs in using and maintaining the buildings (Ashworth, 1994).

The most common examples are dwellings and industrial buildings, but the use of inter-linking is not confined to these building types. Indeed stepped houses, as illustrated in figures 5.2 and 5.3, have become a familiar feature of post war housing developments, primarily on aesthetic grounds to provide breaks in the front elevations. Terraced houses could show savings in the order of 6 per cent compared with semi-detached houses and as much as 12 per cent by comparison with detached dwellings.

RELATIVE COSTS OF FLATS AND HOUSES

Multistorey design generally involves certain features which are not required in two-storey dwellings. These include costs arising from the height of the structure, the provision of common areas for access to the dwellings, additional Building Regulations and fireproofing requirements, and the installation of additional amenities needed to solve problems which only arise in multistorey housing. Such additional amenities include lifts in high blocks, clothes drying facilities, and costly refuse disposal and central heating installations. Table 2.2 shows a summary of typical relative proportions of costs of houses and flats broken down into four basic elements.

Table 2.2 Typical relative proportional costs of local authority houses and flats

Component	*Two-storey house*	*Three-storey flat*	*Eight-storey flat*
	per cent	per cent	per cent
Substructure	11.2	6.7	9.0
Superstructure	52.4	44.6	55.2
Internal finishings	18.9	25.5	13.4
Fittings and services	17.5	23.2	22.4
Total	100.0	100.0	100.0

It is interesting to note that the substructure costs of the three-storey flat account for a smaller proportion of total costs than the two-storey house, as similar foundation costs are spread over a greater floor area, whilst the substructure costs of the eight-storey flat increase relative to the three-storey flat because of the more expensive foundations that are needed. While the superstructure costs of the eight-storey flat are double those of the two-storey house on account of the structural frame and increased constructional costs associated with the taller building, its proportion of total costs does not vary significantly. Balcony facings and balustrades provide another source of additional expenditure with flats as distinct from houses.

Consideration will now be given to a comparison of costs of the various elements of flats and houses.

Substructures

The cost of substructures is influenced by site conditions, such as type of soil, groundwater level and presence of sulphates and other contaminants or obstructions, as well as by the type of building to be erected. Normally, foundations of two-storey dwellings are capable of supporting greater loads, and some economy results from the substructure cost being shared by a larger number of dwelling units in taller blocks. There will, however, be limits on the savings that can be achieved in this way, particularly on poor load-bearing soils. The construction of multistorey blocks in slab form often involves building across the contours to secure satisfactory orientation, and this will result in additional costs in earthworks and provision of foundations. With multistorey blocks expensive piled foundations may be required unless the soil has a high load-bearing capacity, permitting the use of concrete pad foundations, although the higher cost of piled foundations will be offset to some extent by the sharing of the substructure over a larger number of floors. Some typical substructure costs for different building types are given in chapter 10.

Superstructures

Superstructure costs vary considerably between blocks of different designs, and this particular element accounts for much of the extra cost of multistorey development over traditional two-storey housing. Brick load-bearing structures are generally the most economical for three to five-storey blocks, while reinforced concrete columns or walls, or a composite form using both, are frequently used for taller blocks. The need for fire-resisting floors and staircases in multistorey blocks also results in increased costs. As described earlier, increased circulation ratios with multistorey blocks will also produce higher unit usable floor space costs.

Three-storey flats are generally about thirty per cent more expensive than two-storey houses, with costs related to a specific unit of floor area such as the square metre. Increasing the height of blocks of flats from three to five storeys is likely to raise costs by about twelve per cent (six per cent per storey). This trend continues when the total height is further increased to six to eight storeys, with a further rise in costs of about seventeen per cent. The rate of increase in costs appeared surprisingly to flatten above eight storeys in height to about a two per cent addition per floor. There are, however, substantial variations in the cost relationships between blocks of differing heights located in Inner London, Outer London and the provinces. The cost differences are greater than could readily be explained by variations in the cost of labour and materials between different parts of the country, and must presumably have their explanation in part in differences in design and specification and in the materials used. Lower costs can often be achieved by the preparation of simpler designs, keeping different materials to a minimum and incorporating a maximum of standardisation, with closer liaison between designers and producers and more detailed consideration of erection methods at the design stage.

Roofs

Some reduction in the roof cost per dwelling unit is to be expected with multistorey design, where the total roof cost is shared by a larger number of dwellings. In the case of three and four-storey blocks the roof is often of similar construction to that used for two-storey houses. With taller blocks a more expensive flat roof is normally provided which partially offsets the savings from the shared roof.

Internal Finishings

The floor finishings to the upper floors of multistorey flats are often rather more expensive per unit area than the corresponding finishings in two-storey houses. Furthermore, it is necessary to provide additional areas of finishings to the floors, walls and ceilings of common access and circulation areas. The elimination of plaster in flats by the use of fair-faced concrete walls and ceilings combined with 'dry' partitions will help to offset these additional costs.

Joinery Fittings

The cost of joinery fittings per square metre of floor area in multistorey flats is generally higher than the equivalent provision in two-storey houses. There are two main reasons for this: (1) A higher standard of provision of cupboards and storage space is almost invariably expected by flat occupiers. (2) Flats are usually smaller in floor area than houses and so the cost of joinery fittings per unit of floor area is proportionately greater.

Plumbing

Plumbing costs are normally higher in multistorey blocks because of the need for larger pipes and additional valves. It is possible to reduce the extra costs by the careful grouping of sanitary appliances to restrict the number of vertical pipe runs and by the use of single stack plumbing. Storey-height prefabrication of pipework for ducts and pushfit jointing will help contain the extra costs.

Heating

Central heating resulted in additional costs per dwelling in the order of £1800 to £2300 in 1993 depending on the type and design of installation. Apart from the high installation costs there is also the disadvantage of the lack of individual control.

Electrical Installations

Electrical installations in two-storey houses are considerably cheaper than those in multistorey flats, as they are generally simpler in form with less fittings and are easier to install. Furthermore, in multistorey blocks costs are also increased by the need for heavier mains and switchgear and each flat will have to share part of the cost of the lighting installation required to serve access balconies, staircases and other common circulation areas.

Lifts

Lifts are necessary in buildings in which the entrance to any dwelling is on the fourth storey or above. This means that blocks of flats exceeding three storeys and blocks of maisonettes exceeding four storeys must be provided with lifts. In buildings of more than six storeys, two lifts are needed to maintain continuity of service during periods of breakdown and maintenance. With two lifts it may be sufficient to have stops only at alternate floors. Lifts can satisfactorily serve up to about fifty dwellings each, with a lift car capacity of eight persons. Thus if maximum utilisation of lift capacity were the sole consideration, a building with six storeys and one lift would contain fifty dwellings and a block with more than six storeys and two lifts would contain 100 dwellings. In practice the economics of lift provision is only one of many matters to be considered, but it is important. It would, for instance, be very uneconomical to design a five-storey block which contains only two or four dwellings on each floor, and incorporates a lift.

Blocks of flats have varying standards of lift provision with an average of one per thirty dwelling units. A good target would be a provision of one lift per twenty to twenty-five dwellings, which is on the low side, and amounted to a cost of about £32.00 per m^2 of floor area in 1993. Extra costs stemming from the introduction of lifts cannot readily be separated from superstructure costs in analyses, as they will include the costs of lift shaft, lift motor rooms and pits.

Additional Dwelling Amenities

Multistorey blocks of flats incorporate various facilities which are not provided in two-storey houses. Drying cabinets are often provided to compensate for loss of external drying areas and shared laundry rooms may be incorporated on the basis of about one room per twenty dwellings. Pram stores and on occasions other common facilities such as playrooms, are often provided at ground floor level. In blocks of flats exceeding four storeys in height it is customary to install a refuse chute disposal system with hoppers at each floor level usually discharging into a movable container in an enclosed chamber at ground level.

Siteworks

With two-storey housing, each dwelling is complete within its own garden, except where common unfenced grassed areas are provided at the fronts of houses. Hence the siteworks for which the local authority is responsible are limited mainly to estate roads and their paths and verges, sewers and other underground services. With multistorey design, the site preparation works usually include extensive landscaping, secondary roads to meet fire regulations, cycle and pram stores if not incorporated in the blocks of flats, and children's play areas.

Effect on Relative Costs of Contracting and Tendering Procedures

The building industry embraces a very large number of small firms and a considerable proportion of traditional house building is performed by these small firms with their limited resources and plant. The erection of multistorey blocks of flats is by comparison a complex task requiring considerable technical resources and skilled administrative ability. Only the largest firms are adequately equipped to undertake such schemes, particularly the very high blocks. Thus multistorey schemes attract a much narrower range of building firms entailing reduced competition. Tenders for building projects are often influenced by the amount of work available at any particular time and tenders for multistorey work are likely to be more sensitive to this condition.

In this country houses have proved to be much more popular than flats and so it has not been possible to build up the vast amount of design and constructional knowledge and expertise concerning multistorey flats as exists for low rise residential buildings. Furthermore, multistorey design makes possible the introduction of a whole range of new constructional processes, materials and components. Estimators may accordingly be faced with many unfamiliar problems, with little experience of past costs to draw upon and may feel obliged to price high to cover the uncertainties. Furthermore in the late seventies the popularity of high rise residential blocks dropped dramatically and few new blocks were planned, mainly for sociological and economic reasons, coupled with considerable poor construction and high maintenance costs.

Another essential difference with multistorey blocks is the much larger proportion of work allocated to nominated subcontractors. Subcontractors' work often accounts for an average of about thirty per cent of total cost rising to as much as fifty per cent or more for schemes incorporating blocks exceeding six storeys in height. In fact, lifts and central heating alone could account for much of this percentage. By way of contrast the proportion of nominated subcontractors' work in two-storey housing rarely exceeds twenty five per cent although this has tended to increase since the mid nineteen seventies.

With the earlier multistorey housing schemes there was a lack of clear budgetary direction from the inception of the schemes and this resulted in high building costs, high rents and/or excessive deficiences in the housing revenue account. Tenders were often invited on bills of quantities prepared from incomplete information leaving important details to be settled later by variation orders. Furthermore, architects complained that they were seldom allowed sufficient time by local authorities to plan schemes adequately, and the local authorities expressed concern at the inconvenience and delays stemming from the need to obtain planning and statutory approvals, rising costs and frequent changes in government policy. It is evident that contractors must price on the basis of the conditions which they expect to meet during the contract and must include an allowance for possible disorganisation arising from variations. Many building firms are improving efficiency by the introduction of management techniques and improved methods of programming and controlling building operations. Lack of precontract planning by the design team can nullify much of the benefit to be derived from these costly and time-consuming efforts of the contractor. It is evident that dwellings in high rise developments must cost considerably more than identical accommodation provided in low rise schemes. Yet, at the same time, it was equally apparent that in the nineteen fifties and sixties multistorey schemes were frequently excessively expensive and that part at least of the excessive cost stemmed from poor design work and lack of effective cost control.

IMPLICATIONS OF VARIATIONS IN THE NUMBER OF STOREYS OF BUILDINGS

Two practical examples will serve to illustrate the probable cost implications on the various elements of alternative design solutions involving variations in the number of storeys to residential and office buildings.

Example 2.1

Comparison of alternative proposals to provide a prescribed floor area of office space in a rectangular shaped three-storey block or a six-storey L shaped block.

The six-storey block will involve increased costs in respect of the majority of major elements for the reasons indicated:

Foundations. More expensive foundations will probably be needed in the six-storey block to take the increased load, assuming a soil with an average load-bearing capacity, although this will be partially offset by the reduced quantity of foundations. The irregular shape will however increase the amount of foundations relative to floor area.

Structure. It is probable that a structural frame will be required in place of load-bearing walls with consequent higher costs, and there will be an additional upper floor and flight of stairs.

Cladding. The constructional costs will increase due to the greater amount of hoisting and the larger area resulting from the more irregular shape of the block.

Roof. Constructional costs will be higher but these will be more than offset by the reduction in area of the roof.

Internal finishings. Increased area due to more irregular shape and slightly higher hoisting costs will result in increased expenditure.

Plumbing, heating and ventilating installations. Increased expenditure due to increased lengths of larger-sized pipework and ducting.

Passenger lifts. Might not be provided with a three-storey block but will be essential for the six-storey block.

Example 2.2

Comparison of alternative proposals to provide a prescribed number of flats of identical floor area and specification in two five-storey blocks or one ten-storey block.

Element	*Two five-storey blocks*	*One ten-storey block*
Foundations	Double the quantity of column bases and concrete oversite. Possibility of less costly strip foundations if load-bearing walls.	Half the quantity of column bases but they will need to be larger and deeper. Possible need for more expensive piled foundations.
Sturctural frame	Possibility of load-bearing walls. Otherwise two sets of frames but some smaller column sizes and less hoisting, so likely to be cheapest proposition.	Larger column sizes to lower six storeys as will carry heavier loads and increased hoisting will make this the more expensive arrangement.
Upper floors and staircases	One less upper floor and flight of stairs.	One more upper floor and flight of stairs. Stairs may need to be wider to satisfy means of escape in case of fire requirements and there will also be increased hoisting costs.

Element	*Two five-storey blocks*	*One ten-storey block*
Roof	Greater roof area.	Reduced roof area but savings in cost partially offset by higher constructional costs.
Cladding	Less hoisting.	May require stronger cladding to withstand increased wind pressures, and extra hoisting will be involved.
Windows	Slight advantage.	Increased hoisting and possible need for thicker glass in windows on upper floors to withstand higher wind pressure.
External doors	Double the number of entrance doors.	Might involve more doors to balconies.
Internal partitions	Slight advantage.	Some increased hoisting costs.
Internal doors and joinery fittings	Much the same.	Much the same.
Wall, floor and ceiling finishings	Little difference.	Little difference except for possibly slightly increased hoisting costs.
External painting	Some advantage.	Rather more expensive.
Sanitary appliances	Much the same.	Much the same.
Soil and waste pipes	Increased length of pipe.	May need larger-sized pipes on lower storeys.
Cold and hot water services	Double the number of cold water storage tanks and may need two boilers.	Larger cisterns, boilers, pumps, etc. and may need some larger-sized pipes and ducts.
Heating and ventilating installations	Two separate installations but some savings due to smaller-sized pipes or cables.	Cost advantage of single system but may be largely offset by larger pipes or cables and fittings.
Electrical installation	Two separate installations and intakes.	Cost advantage of single system but probably more than offset by increased size of cables.
Lifts	Two lift motor rooms but probably the same number of lift cars.	Saving from one lift motor room but may be necessary to install faster and more expensive lifts.
Sprinkler installation	Two separate sprinkler systems.	One system but some of pipework will need to be of larger size.
Drainage	More extensive and expensive system.	Some economies particularly in length of pipe runs and number of manholes.

Element	*Two five-storey blocks*	*One ten-storey block*
Siteworks	Likely to be more expensive in paths and roads but reduced ground area.	Some savings likely.
Preliminaries and contingencies	May require two tower cranes if blocks are to be erected simultaneously.	Taller tower crane needed.

COLUMN SPACINGS

Single-storey framed structures almost invariably consist of a grid of stanchions or columns supporting roof trusses and/or beams or portal frames. Sheeting rails, purlins and windbracing all help to stiffen and strengthen the structure. By increasing the lengths or spans of the roof trusses, the number of columns can be reduced and this may be of considerable advantage in the use of the floor space below with less obstruction from columns. The trusses may need to be of heavier sections to cope with the greater loadings associated with larger spans, and will need to be of different design if the spans are lengthened sufficiently. In like manner the sizes and weights of columns will need to be increased to take the heavier loads transmitted through the longer trusses, and this will partially offset the reduction in number of columns. One method of assessing the probable cost effect of varying the column spacings and spans of trusses is to calculate the total weight of steelwork per square metre of floor space for the alternative designs, when the most economical arrangement will be readily apparent.

For instance if steel stanchions 4.5 m high were provided to support steel trusses 7.5 m long at 4.5 m centres, the weight of the stanchions would be approximately 7.7 kg/m^2 of floor area. The weight of stanchions/m^2 of floor area would reduce to 5 kg for trusses of 15 m span and to 3.7 kg for trusses of 24 m span. On the other hand, with riveted steel angle trusses to ⅕ pitch and spaced at 4.5 m centres, the weight of the trusses per square metre of floor area would increase with lengthening of the roof spans as indicated now.

7.5 m long trusses – 5.3 kg/m^2
15 m long trusses – 8.2 kg/m^2
24 m long trusses – 11.9 kg/m^2

To the weight of stanchions and trusses must be added the weight of beams and purlins to arrive at the total weight of steelwork. Purlins spaced at 1.35 m centres and spanning 4.5 m between trusses would probably weigh about 8.5 kg/m^2 of floor area.

The Wilderness study group (1964) investigated the design cost relationships of a large number of hypothetical steel-framed buildings of equal total floor area and similar specification but with the accommodation arranged on one or more storeys in buildings of varying shapes with varying bay sizes, column spacings, storey heights and superimposed floor loadings. The study group confined their investigations to the functional components of roofs, floor slabs, columns, beams, ties and column foundations, collectively termed *the core*. The group produced a set of charts designed to indicate cost relationships under varying conditions of numbers of storeys (one to eight), storey heights (mainly 3 to 4.5 m), loadings (2 to 10 kN/m^2) and column spacings (3 to 12 m), but all limited to steel-framed buildings of simple design with solid *in situ* reinforced concrete floor and roof slabs.

An examination of these charts shows increasing costs with wider spacing of columns. Adopting a storey height of 3 m and floor loadings of 5 kN/m^2, a comparison is made of the effect of extreme column spacings of 4.5 m in each direction as against 12 m in each direction. The increased cost of the structure resulting from the wider spacing of columns with single-storey buildings is shown at sixty-seven per cent rising to over 100 per cent with eight-storey blocks. If the storey height is increased to 4.5 m, the increases in cost are less spectacular as they are partially offset by the extra material in the extended columns. The extra structural cost due to the more widely spaced columns in blocks with 4.5 m storey heights rises from about sixty per cent in single-storey buildings to ninety per cent in eight-storey blocks. It is interesting to note that the variations in the costs of structures due to different storey heights reduces as columns are spaced more widely apart, although the extra structural costs arising from increases in storey heights are relatively small compared with those stemming from wider spacing of columns and increased floor loadings. For example, increasing storey heights from 3 m to 4.5 m produces a six per cent addition to structural costs for single-storey buildings rising to sixteen per cent for eight-storey blocks. It must be recognised that these relationships cover only part of the costs of buildings and relate solely to steel-framed buildings with *in situ* reinforced concrete floors. Nevertheless within these limitations this study provides some extremely useful guidelines in assessing the probable cost relationships of different structural designs for a project.

A common column spacing in factory buildings is between 6 m and 9 m; spacings of less than 3 m are unusual, but those over 9 m occur quite frequently and there are some cases of spacings over 15 m. The design of factory frames is very flexible and while column spacing affects costs it is probably not the most important determinant of initial costs. The layout of the floor space depends on the size of machines, the production flow, the gangway space for internal circulation, and the space required for storing stock and work in progress. Most production line widths are quite small with a common width of between 1.5 m and 3 m. Many production lines with their gangways can be satisfactorily accommodated either singly or in multiples between columns spaced 6 m to 9 m apart. It is however often advantageous to have the column spacings wider in one direction than the other, and it appears that 6 m × 9 m spacings or multiples of these dimensions are often the most acceptable.

FLOOR SPANS

Floor spans deserve attention as suspended floor costs increase considerably with larger spans. Furthermore, the most expensive parts of a building structure are the floors and roof, namely the members which have to thrust upwards in the opposite direction to gravitational forces. As a very rough guide, horizontal structural members such as floors cost about twice as much as vertical structural members like walls.

In the upper floors of blocks of flats stiffness is an essential quality and meeting sound insulation requirements dictates a minimum floor thickness of 125 mm. In this situation the most economical spans are likely to be in the order of 4.5 to 6 m. With crosswall construction floor spans are usually within a range of 3.6 to 5.2 m. Two-way spanning of *in situ* reinforced concrete floor slabs helps in keeping the slab thickness to a minimum, and one-way spanning is only economical for small spans.

FLOOR LOADINGS

The Wilderness study (1964) has shown that variations in design of floor loadings can have an appreciable effect on structural costs. Adopting a 7.5 m grid of columns and a 3 m storey height, a comparison of structural costs for buildings with floor loadings of 2 and 10 kN/m^2 respectively, shows an increase in cost of about twenty per cent for two-storey buildings rising to about forty per cent for eight-storey buildings for the higher floor loadings. Further increases of two to four per cent occur if the storey height is increased to 4.5 m. Limited increases also arise from the wider spacing of columns when coupled with heavier floor loadings, and these increases become more pronounced in the taller blocks.

Heavy loads can be carried most economically by floors which rest on the ground, rather than by suspended upper floors. Where heavy loads have to be carried by suspended floors it is desirable to confine them, wherever practicable, to parts of the building where the columns can be positioned on a small dimensional grid. As indicated previously it is expensive to bridge large spans and it becomes quite a complex task to determine the point at which the unobstructed space stemming from larger spans equates the extra cost of providing it. Eccentric loading of vertical supports is always uneconomical and it may be worthwhile to increase a cantilever counterweight by moving the support nearer the centre of the load to reduce or eliminate the eccentricity. For this reason perimeter supports are less economical than those provided by crosswalls.

COST ESTIMATION OF STRUCTURES IN COMMERCIAL BUILDINGS

Singh (1995) has adopted a new and useful approach to the cost estimation of the structures of high rise commercial buildings, ranging from 5 to 50 storeys in 5 storey groupings, by illustrating in the form of charts the effect of different design parameters on the quantities of constituent materials, using the traditional structural systems of solid slab and beam, flat slab/waffle slab, and prestressed beam and slab. Mathematical models are established based on the statistical relationships between the quantities of constituent materials and other design variables.

Using the chart/statistical relationships, the user is able to compute approximate quantities and hence the cost of the structure, given the structural type and other design parameters. The effect of design variables such as structural type, grid formation, grade of concrete, number of storeys, continuity of structure, and shear core size are investigated, and the effect of plan shape and size of the building structure are taken into account.

DESIGN COST CRITERIA

Pell-Hiley (1974) has produced a useful schedule of design cost criteria, subdividing them into primary and secondary economic decisions as shown in table 2.3.

CONSTRUCTABILITY

The relative simplicity of constructing a building will influence the cost of a project. Constructability, sometimes described as buildability, has been defined as the extent to which the design of a building facilitates the ease of construction, subject to the overall requirements for the completed building (CIRIA, 1983). Hence the designer should have comparative ease of construction in mind at every stage of the design process, particularly in the early stages, taking a very practical approach. This necessitates a detailed knowledge of construction processes and techniques and the operational work on site, and is made much easier with the early appointment of the contractor. The principal aim is to make the construction as easy and simple as possible and to reduce waste, such as excessive cutting of components. Another aim is to make the maximum use of site plant and to increase productivity. The design details should encourage a good, logical sequence of constructional activities, and the contract documentation should show clearly that the design has been prepared to permit ease of building and the achievement of an economical project, often involving the consideration of alternative methods. Readers requiring more detailed information on this important concept and its implementation are referred to Griffith and Sidwell (1995).

On occasions a conflict may arise between ease of building and the quality of construction and aesthetic requirements, and in these cases an acceptable compromise should be sought. It will be recognised that the economics of building work is only one of the criteria to be considered and the client's best interests are served by securing the optimum balance between the various criteria described in chapter 1.

QUALITY ASSURANCE

General Background

In the late 1980s increasing concern was being expressed at the low standards of performance and quality often achieved in UK building work.

Table 2.3 Design cost criteria

Primary economic decisions

Condition	*Cost advantage*	*Cost penalty*
1 Single-storey	(a) Simple structure (b) No lifts (c) Minimal fire precautions	(a) Large foundation area (b) Large roof area (c) Paths and roadways tend to be maximum (d) Lengthy internal communications (e) Economic consequences of more land
2 Multistorey	(a) Foundation area progressively reduced (b) Roof area, ditto. (c) Paths and roadways tend to reduce (d) Peripheral drainage, ditto. (e) Economic consequences of less land	(a) Structure more complex (b) Possibility of piling as storeys increase (c) Lifts and other vertical communication including escape staircase (d) Fire precautions increase (e) Additional engineering factors, such as water boosting, exposure and air change
3 Shape	(a) Minimal wall/floor ratio to reduce overall cladding and heating losses (with capital and running cost implications)	(a) Configuration resulting in a 'dense' building shape (say over 15 m in width) will give rise to additional ventilation/air-conditioning and supplementary lighting

Secondary economic decisions

Condition	*Cost advantage*	*Cost penalty*
1 Orientation	(a) Siting to minimise the effects of solar gain	(a) Siting may conflict with other conditions such as economic drainage and services runs
2 Fenestration	(a) Decrease in window size tend to general overall savings including the effect on heating requirements (with capital and running cost implications); reduction in solar gain and the need for sun blinds	(a) Possible need for supplementary lighting (with capital and running cost implications)
3 External cladding	(a) Insulative consequences on heating requirements (with capital and running cost implications)	

This highlighted the need for structured and formal systems of construction management culminating in the establishment of the total performance concept for buildings. This concept gives clients the opportunity, from design brief through to commissioning and operating stages, to establish minimum standards of performance for all aspects of buildings, through quality assurance procedures.

The CIOB (1989) defined quality assurance as 'an objective demonstration of a builder's ability to produce building work in a cost effective way to meet the customer's requirements', while the RICS (1989) considered that quality assurance was a management process designed to give confidence to the client by consistently meeting stated objectives. BS 5750: 1987 defined quality management as the organisation structure, responsibilities, activities, resources and events appertaining to a firm that together provide organised procedures and methods of implementation to ensure the capability of the firm to meet quality requirements in accordance with Parts 1, 2 or 3 of BS 5750. This UK national standard for quality systems started the trend towards the registration of quality systems by certification and formed the basis of the European Standard for Quality Assurance, EN ISO 9000 (1994).

Quality assurance requires appropriate systems, sound procedures, clear communication and documentation that is accurate and easily understood, and prescribed standards must be set and achieved. It impinges on every aspect of the total building process and all building professionals should promote quality assurance as a common objective. Within the construction industry there are five broad sectors where quality assurance is applicable:

(1) Client: in the project brief.
(2) Designer: in the design and specification.
(3) Manufacturers: in the supply of materials, products and components.
(4) Contractors and subcontractors: in construction, supervision and management.
(5) User: in use of new building, its upkeep and repair (Griffith, 1990).

The respective duties, responsibilities and work methods encompassed by quality assurance procedures are contained within a *quality plan*, which prescribes the activities and events required to achieve the quality goal throughout the building process, and involves the co-ordination of each of the participants (CIOB, 1989).

Principal Applications

At the outset, quality assurance procedures should ensure that the client is made aware of the range of methods and services entailed in a building project and that building is a sequential process, requiring a timely decision at each stage in order to avoid delays and additional costs caused by pursuing inappropriate options.

The formulation of the quality plan at the conceptual design stage should incorporate regulatory requirements, relevant technical standards, design responsibilities, lines of communication, and verification procedures.

Contractors short listed should have proven experience in the development of quality systems on similar projects. Each tendering contractor should submit proposals covering such matters as:

(1) the relationship between time, cost and quality in successfully achieving the requirements of the client;
(2) a project quality plan, describing the methods by which the contractor's responsibilities will be monitored throughout the project, the management resources that will be employed and records that will be available;
(3) a description of how the quality plan will interface with the design team quality plans and those of individual suppliers and subcontractors (CIOB, 1989).

In the purchase of materials, services and equipment, it is essential that only satisfactory sources are selected in cost effectiveness terms and in the ability to meet specified requirements. Materials and services should be procured from organisations which:

(1) give confidence that suppliers and subcontractors have effective quality systems;
(2) ensure that every supplier and subcontractor understands that supplies or services must be to the specified requirements;
(3) ensure that all information given in the order is up-to-date and accurate.

All building sites require formal, simple and effective documentation systems to improve their efficiency, and to provide evidence that the specified quality has been achieved at first attempt. Quality assurance introduces a series of checks, usually in the form of a checklist, identifying the stages at which it can be verified that the works comply with the specified requirements, and each stage must be agreed before proceeding to the next. A master programme should be prepared in which the activities of the client, contractor and subcontractors are co-ordinated (CIOB, 1989).

The quality assurance procedures for cost control ensure compliance with the client's requirements for comparison against the cost plan. The procedures define who has authority to issue instructions under the contract, the limits of financial authority for each party and the manner in which proposals are to be costed and assessed alongside other options before a financial commitment is made. They also define the method and frequency of cost reporting to the client.

Prior to advising the client's authorised representative that he considers the work to be complete, the contractor should carry out his own detailed inspection to ensure compliance with his own quality plan. Upon being so notified, the nominated representative carries out his own inspections. If he considers that the works do not comply in all respects with the contract documents, the client's nominated representative issues a *discrepancy report*, which is commonly termed a snagging list (CIOB, 1989). The Latham Report (1994) recommended that quality assurance certification should continue to be encouraged within the construction industry as a potentially useful tool for improving corporate management systems.

3 FUNCTIONAL REQUIREMENTS AND COST IMPLICATIONS OF CONSTRUCTIONAL METHODS

This chapter is concerned with the functional requirements and cost implications of alternative constructional techniques for different types of building and of building elements. A comparison of maintenance problems and costs associated with different materials and components is made in chapter 13.

LOW AND HIGH RISE BUILDING

Residential Buildings

The relative costs of low and high rise residential buildings were compared in chapter 2 and reasons sought for the much higher cost of high rise developments over two-storey housing. Admittedly, the more intensive use of highly priced land will offset to some extent the increased costs resulting from multistorey development. In addition there may be social benefits to be gained by some occupiers through the erection of tall blocks of flats on central urban sites, such as ready access to town centre facilities and reduction in length of journey to work, although these benefits are often difficult to evaluate, as indicated in chapter 16, and there is ample evidence of sociological problems. High rise flats often cost between sixty and eighty per cent more per square metre of floor area than houses, with superstructure costs, influenced particularly by the costs of the frame and floors, accounting for just over half of the increased cost, and the costs of services including lifts taking up most of the remainder. Furthermore, the high blocks of flats show a greater range of costs than low flats or houses.

With low rise flats, the main source of higher costs in this type of flat as compared with a house is in floors, landings, staircases, balconies and finishings. These costs could be as much as two to three times greater than corresponding items in two-storey housing, while the majority of other components average from twenty to seventy per cent higher in cost. Savings in roof and substructure costs were not particularly significant. Table 3.1 shows the relative costs of low flats to houses, and is subdivided between the main component parts.

Apart from higher constructional costs, problems were also encountered in the late nineteen sixties through people not wanting to live so far from the ground, difficulties experienced by young married couples bringing up children in high flats, the government's decision to stop the payment of additional grants for multistorey flats, and general sociological and aesthetic problems. For these reasons Coventry Corporation decided in 1968 not to build any more multistorey flats, and the majority of major provincial cities subsequently ceased constructing them as the problems of tenants and management of the flats multiplied. Many post-war blocks had been demolished by the early 1990s, particularly those built of concrete slabs with corridor access. In the nineteen seventies, a move was made towards achieving high density low rise development, sometimes incorporating narrow frontage terrace three-storey houses to give densities up to 70 dwellings/ha. More emphasis has also been given to designing houses for the disabled following the year of the disabled (1980) and some authorities have built 'starter homes' of about 25 m^2 floor area for newly married couples. By the 1990s housing associations had become the main providers of social housing and the number of private houses being built reduced significantly as demand fell in the

Table 3.1 Relative costs of houses and low flats (1993)

Component	*Items*	*Three-storey flats as percentage of two-storey houses (provinces)*	*Three to five-storey flats as percentage of two-storey houses (all regions)*
Substructure	Foundations and ground floor slab including all abnormals.	125	166
Brickwork and blockwork	Structural brick work, lintels and sundry columns, including internal partitions.	101	110
Facework	Extra over cost of facings.	100	155
External fittings	Windows, external doors, porches, balcony balustrades, including painting and glazing.	126	151
Floors	Floors, staircases, balconies and landings.	254	306
Roofs	Roof structure, covering and external plumbing.	79	85
Superstructure		119	138
Floor finishings	Floor, staircase and balcony finishings, including screeds.	209	204
Plastering	Including any wall tiling.	100	116
Decorations	Internal decorations only.	100	113
Joinery	Doors, door frames, cupboards, dressers, including ironmongery.	165	153
Internal finishings		142	146
Heating	Fires, chimneys, or other forms of heating including water heating.	100	204
Plumbing	All internal plumbing services, including sanitary appliances.	147	161
Gas and electrics	Installation only.	155	234
Sundry services	Lifts, refuse disposal, etc. in three to five-storey flats.	–	–
Services		137	208
Overall relative cost		126	152

recession. Residential development is dealt with in more detail in chapter 5.

Industrial Buildings

Industrialists have a general preference for single-storey premises, but a variety of matters such as limitations on land available and increased demands for car parking space may justify a reappraisal. A wide diversity of loadings occurs in factory buildings. In a specific case, the loads to be carried may affect the choice of site, influence the building design and, if very heavy, may have a considerable bearing on the choice between single and multistorey construction.

Post-war industrial development included some multistorey factories but the majority have been singlestorey. In the new and expanding towns almost all the factory buildings were single storey, although possibly fronted with double-storey office blocks.

The nineteen eighties saw a significant demand for speculative warehouse and nursery unit industrial development, but the recession in the early 1990s severely affected manufacturing industry and resulted in a sharp decline in the numbers of new factories under construction. The main demand was for small units of the nursery or starter type which because they operated as 'seedbeds' are subject to frequent changes of ownership, and this concerns the institutions more than the lower unit rentals and higher construction costs compared with larger industrial buildings.

The design of factories is primarily concerned with two aspects – space requirements of the occupier and structural requirements. Most large industrial companies employ production and plant engineers who are able to consider alternative ways of arranging people and machinery. Site coverage is often in the order of 45 per cent after making ample provision for car parking and waiting space for large vehicles.

The roof covering is often in the form of an insulated sandwich and may incorporate rooflights where natural roof lighting is required. The surface finish of the underside of the roof will depend very much on the nature of the work to be done in the factory, and the roof may need designing to hoist and suspend machinery.

Span and bay spacing plays a significant part in design costs. The fewer the intermediate uprights the better, but careful consideration is needed to obtain the optimum acceptability of column positions, which in turn will determine structural grids. Long spans result in large quantities of structural material to ensure that roof members remain stiff and do not deflect under load conditions or under their own weight, but deepening the roof for this purpose increases significantly the volume of the building. The fewer the column positions, the greater the loading concentrated on a smaller number of points and therefore the more critical the bearing capacity of the ground. Where ground bearing capacities are low it may be preferable to use a larger number of columns rather than extensive piling or other ground consolidation. A reconciliation is also needed between the number of columns required to keep down costs and the number that will interfere with the production work in the factory.

Floors invariably take a substantial load, often in the 5–10 tonne/m^2 range. Except on small or ancillary buildings, the external walls are unlikely to be load bearing, and they are often of lightweight material such as protected metal sheeting, composition or plastics, with internal insulated linings. Sheeted buildings are generally cheaper and faster to construct but they are likely to result in higher maintenance costs. The type of external walling will also have a significant effect on the thermal and sound insulation provision. Thermal insulation is becoming increasingly important in all industrial buildings with the rising cost of energy and the increased importance attached to energy conservation. Sound insulation can be a major problem in the case of bottling plants with noisy processes, or where factories are close to residential areas.

Many other factors require detailed consideration such as the provision of closely controlled lighting levels, distribution network of services (often overhead), adequate staff facilities, and a satisfactory standard of materials, heating and ventilation.

The institutional preference is for an eaves height of 5.5–6.0 m for warehousing and distribution, reducing to 5.0 m for knowledge-based industries, which can accommodate a mezzanine level, and offices with a height of 3.7–4.3 m.

The change from manufacturing to service and knowledge-based activities with the impact of new technologies, creates the need for a different environment, encompassing innovation centres, research parks, science/technology parks, commerce/business parks, industrial parks and office parks. These new developments are designed to provide a pleasant working environment with a wide range of cultural and intellectual opportunities. The developments are envisaged as being low density, well landscaped and with above average provision for car parking, and generally occupying areas in excess of 40 ha. A non-industrial image can be obtained by reducing heights, providing more glazing and screening delivery areas. A higher standard of finishes and lighting is envisaged to reflect office standards with ample flexibility in design to meet varying needs. In 1990 there were 800 developments in the UK describing themselves as business parks (Applied Property Research, 1990; Seeley, 1992), and one of the most outstanding was at Stockley Park, West London.

Shops

The number of storeys incorporated into other types of building will often be influenced by the purpose for which the building is to be used and/or the value of the site. Shops are generally of single-storey construction for the convenience of users, whilst offices are often multi-storey to make more intensive use of highly priced central sites and to enable the occupants to be as far removed as possible from traffic noises. An ideal approach is to build offices on top of shops in the centres of large towns and cities, and flats or maisonettes above shops in neighbourhood centres.

Flexibility should be the prerequisite of every shopping complex based on the optimum structural grid. Normally 11 m clear spans permit an ample range of combinations to suit most needs. If, however, a shopping centre incorporates car parking, particularly at an upper level, then the most economical grid maintaining adequate flexibility, is to follow the accepted car parking standards of 16.5 m widths. This has the added advantage that many retailers can trade very successfully in units 8.25 m wide.

The majority of modern shopping developments are covered, often on two storeys with refurbishment programmes at about 15 year intervals. In the 1980s and 1990s there was increased emphasis on the provision of large out-of-town shopping schemes, of which one of the largest was at Meadowhall, Sheffield completed in 1990 on a site of 55 ha and containing 230 shops on two floors with a total floor area of 116 200 m^2, leisure and entertainment facilities and 12 000 free car parking spaces, at a cost of £230m.

Schools

Schools are of varying heights with a predominance of single-storey buildings amongst primary schools; secondary schools are often of two, three or four-storeys, and technical colleges and colleges of further education are often of several storeys except for single-storey workshops/laboratories. A case study undertaken for a three-form entry secondary school of 3320 m^2 floor area, showed that a two-storey building was the cheapest proposition, and that both three and four-storey buildings were significantly cheaper than a single-storey building. Factors contributing to the higher cost of single-storey buildings included the greater quantity of work below ground floor level and in roofs and drainage work, more external doors, which are relatively expensive, and more pipe and cable runs which are also costly.

The unit costs of three and four-storey blocks exceeded those of two-storey blocks with the same total floor area for several reasons:

(1) With certain elements such as roofs, and work below ground floor level, the reduction in cost per square metre obtained between the single-storey and two-storey building diminishes as the floor area increases with the provision of additional storeys.

(2) The three and four-storey buildings must be provided with two staircases whereas one is sufficient for the two-storey building.

(3) As the height of a building increases there is a marked increase in the need for wind bracing.

Offices

Tenants' requirements are an important aspect of the office development process. Developers must generally be prepared to provide good quality office buildings with high standard finishes and services. Higher rents demand or should demand higher standards. Tenants are basically looking for good, simple and reliable services, an attractive approach to the offices, a good entrance hall and flexibility in the use of the space which they are renting.

The spacing of columns across the width of the building will need determining at an early stage. For the normal 12.2–13.7 m wide office block, a generally accepted layout is two rows of columns forming a corridor down the centre of the building. This arrangement is sufficiently flexible to be used for either individual cubical offices, or if the whole floor is left open, for a single department. In landscaped open plan offices different considerations will apply, for to be effective floors must be very much wider – at least 24 m – but it will still be necessary to establish an economic structural grid for column spacing, otherwise the structural cost will be excessive in relation to the overall building cost. Technically it is possible to keep the floors completely free of columns to provide maximum flexibility, but the cost of achieving this is usually too great in relation to the potential advantages.

A ratio of between 12.5 and 15 per cent circulation space to lettable floor area is the hallmark of a well planned office building. Apart from the optimum use of space there is a persistent demand for relatively maintenance-free building materials, such as ceramics, glass, stove enamelling, anodised aluminium or plastics. There is, however, no entirely maintenance-free building, and the designer is seeking materials which have a relatively long life before renewal or resurfacing becomes necessary. An office shell may have an expected life of 60 to 100 years but the tenants could change every 10 to 15 years and internal office landscape every one to three years.

Government offices are often designed as a series of interrelated office strips 40 to 45 m in length, 12 m wide and based on a module of 1.5 m. Interiors are free of columns and capable of subdivision into single or shared offices or spaces for groups of up to 30 persons. The office strips are connected by nodes – lifts, stairs, toilets, tea points and similar fixed installations. They are intended to present a visual and environmental contrast to the office areas. The office strips will be separated by either a linear court or a central landscaped court. This arrangement provides a low-energy building using natural lighting and ventilation.

The British Council for Offices (1994) established a committee to formulate a realistic specification for office space which emphasises fitness for use and eliminates the deficiencies of many years of over design in office construction. The draft specification is aimed at high-quality urban offices but recommends that, unless the end occupant is known, the building should be completed only to shell and core standard. The main recommendations are as follows:

Critical Parameters

Plan form

Plan depth: The distance from window to window or window to atrium should be 15 to 18 m; distance from window to core should be 9 m; and no more than 25 per cent of net lettable area (NLA) should be deep space (areas more than 9 m from an internal wall or atrium).

Grid dimensions: The planning grid should be 1.5 m; column grids can be a combination of 6 m, 7.5 m or 9 m, although 6 × 6 m is not recommended.

Subdivision: Floors should be divisible into a number of dissimilar units to allow for multi-occupancy and subletting. Each unit should have direct access to lifts, toilets and escape stairs.

Occupational densities

Escape stairs: These should be designed to handle a building population density of 1 person per 10 to 12 m^2 NLA.

Lavatories: These should be provided for a building population of 1 person per 14 m^2 NLA on a 50–50 male/female ratio on a floor-by-floor basis.

Indoor climate control: This should cope with a load of 1 person per 14 m^2 NLA.

Lifts: These should handle a peak load of 1 person per 14 m^2 NLA.

Critical dimensions

The above-ceiling structural zone and services zone depend on the frame solution and the extent of services:

- 150 mm should be allowed for the ceiling and lighting
- floor-to-ceiling heights should be 2600 to 2750 mm
- the raised floor should be 150 mm unless there is underfloor ventilation where it would be increased.

Structural Considerations

Floor loading

Live load: for general areas allow 2.5 kN/m^2 over 95 per cent of the floor area, for high loading areas allow 7.5 kN/m^2.

Dead load allowances: 1 kN/m^2 for partitioning

and 0.85 kN/m^2 for raised floors, ceilings and building services equipment.

Indoor Climate Control

Air conditioning systems

- Energy consumption and life cycle cost of the system should be minimised by taking advantage of natural or low cost energy sources
- The system should be designed for independent operation on separate floors and part floors
- At the outset, consideration should be given to the cost and ease of maintenance
- The system should be capable of being upgraded to meet occupiers' future requirements
- Fresh air should be supplied to the general office at a rate of 8 to 12 l/s per person
- Extra provision of fresh air should be allowed for meeting rooms and other high occupancy areas
- Design lighting to give a heat gain of no more than 12 W/m^2
- Provide one air conditioning terminal for each 6 m of perimeter, assuming perimeter office depth between 4.5 to 6 m
- Provide one air conditioning terminal per 50 to 80 m^2
- The internal temperature should be 22°C ± 2°C.

Naturally or mechanically ventilated systems

- Design for good draught-free cross-ventilation
- Correct orientation, external shading, thermal mass and fenestration can all assist in minimising heat gain within the office space.

Electrical Systems

Lighting

- Design for a maintained illuminance level of 400 to 500 lux in open plan offices with a uniformity ratio of 0.8 over the defined task area
- Lighting should have a 15 W/m^2 maximum power consumption
- Fluorescent lights should have high frequency control gear to reduce power consumption and stroboscopic effects.

Power services

- Allow 25 W/m^2 maximum small power consumption and size risers and busbars accordingly
- Provide a sufficient number of spare ways at each riser distribution board and at the central mains panel.

Information technology provisions

- Provide dedicated vertical risers for use by occupiers complete with vertical cable trays installed
- For a building that will be multi-let, each potential tenant should have its own vertical tray.

Vertical Transportation

Lifts

- Passenger lifts should give a maximum waiting period of 30 seconds with cars assumed to be loaded at 80 per cent of capacity
- Dedicated goods lifts should be provided in buildings of more than 10 000 m^2.

Fire Protection

- Sprinkler systems must comply with Loss Prevention Council (LPC) rules and Building Control requirements
- Avoid installing hose reels wherever possible
- Install a dry riser
- Install hand-held appliances within the landlord's areas.

Security

Anti-crime design

- Provide a single public access point to the building
- Design a defensible building and surrounds, avoiding opportunities for unauthorised entry by climbing.

Defence against bombs

- Provide a portion of the building interior as a blast refuge

- Consider cladding design to minimise damage and disruption to the interior while not transferring blast load to the structure.

Parking facilities
- Provide car access control to parking and garaging
- Lay out carparking for ease of surveillance
- Avoid direct access from carparks to office floors by routing people through the main lobby.

CCTV
- Install prominent CCTV coverage of exterior, parking and common areas.

Finishes

Office space
- Emulsion paint or vinyl cloth is acceptable for walls
- Mineral fibre and metal are equally acceptable for suspended ceilings.

Lavatories
- Walls and floors should be finished with ceramic tiles, cubicles and vanities can be laminates.

Core/staircase area
- Utilitarian finishes such as painted concrete floors, painted metal handrails and painted plasterboard walls are acceptable where the stair's main function is as an escape route.

Lift car interiors
- These should be finished to the same standards as the entrance lobby.

Inadequacies in lift service are likely to have a greater impact on the tenant's opinion of the building than any other single factor. In almost any office building of more than four floors it is advisable to provide two goods lifts rather than one larger one.

Air conditioning was regarded by most developers of office buildings in the London area in the mid-nineteen seventies as a near necessity. As air conditioning systems have become more sophisticated, it is necessary to review the design of the building if air conditioning is to be provided. By the nineteen eighties, however, air conditioning became much less popular largely because of problems of control and servicing. Solarshield glass is regarded as an increasingly important part of regulating the internal temperatures of office buildings in the United Kingdom. The more that air conditioning can be controlled room by room and lighting metered and controlled area by area, the more the developer has to pay for his development. Hence it is necessary for the developer to have sufficient information from the outset as to the likely occupancy arrangements. Subdividing air conditioning and electrical controls after completion of the building is even more expensive.

Planning control exercises a number of basic constraints on the design of office buildings, particularly in relation to plot ratios, daylighting angles, car parking requirements, planning gain, and height massing and aesthetic design. The highest plot ratios in central London are 5 to 1, with 3½ to 1 being the normal standard. Outside central areas a 2 to 1 plot ratio is usually the maximum, although there is a tendency for these to be reduced. A reasonable car parking provision for most multi-purpose office buildings would be one car space per 200 to 300 m^2 of office floor area. In London in particular a developer may be expected to contribute socially desirable and less profitable and, possibly, even uneconomic uses in exchange for the gain or profits made from the planning consent for office use, known as planning gain.

In a rather different context, Leeds Crown Court set a new pattern of layout with 20 courtrooms on three operational levels and incorporating four separate circulation systems.

SUBSTRUCTURES

Adequate information on subsoil conditions is vital before a decision can be made as to the most economical type of foundation. On the majority of smaller contracts, sufficient information can be obtained by excavating and examining trial holes and an inspection will often reveal that strip foundations or simple concrete bases will be adequate. Large buildings and those on difficult sites will generally justify the cost of a borehole investigation, which may indicate the need for piled or raft foundations.

Foundations on infill city centre sites may require new foundations deeper than those of adjacent older buildings. The choice of techniques for deep excavations requires careful consideration of site conditions and comparative costs, and includes bored piles, diaphragm walls and caissons. With sloping sites extra costs arise from additional excavation and fill, stepped foundations, backdrop manholes, retaining walls and cut-off drains.

Building Research Establishment Digest 64 (1972) shows how the type of construction and subsequent use of the building are important determinants of foundation design. Strip foundations, sometimes termed footings, are usually adopted for buildings where the loads are carried mainly on walls. Pad foundations or piles are more appropriate when the structural loads are carried by columns. Where differential settlements are to be controlled within fine limits, raft foundations are often the best approach. Foundations must be designed to comply with the Building Regulations, 1991. BS 8004 (1986) includes a table of permissible allowable values under static loading for different soils, and the dead, imposed and wind loads of a building are calculated in accordance with the principles laid down in BS 6399 (1984).

BRE Digest 67 (1980) describes how a typical two-storey semi-detached house of about 85 m^2 floor area, in cavity brick/concrete blockwork external walls and with lightweight concrete block partitions, timber floors and a tiled roof, exerts a combined load approaching 1000 kN, excluding the weight of foundations. The loads at ground level in kN/m of wall would be approximately: party wall – 50, gable-end wall – 40, front and back walls – 25 and internal partitions less than 15. The use of modern materials tends to reduce total loads, although some forms of construction, such as crosswalls, may result in greater loads on certain walls. For loads of not more than 20 kN/m, strip foundations would need to be 250 mm wide in compact gravel and sand and 300 mm wide in clay and sandy clay or silty clay, increasing to 650 and 850 mm respectively with loads of up to 70 kN/m.

The four main types of foundation are now described and advice on the choice of foundation can be obtained from BRE Digest 67 (1980).

Conventional strip foundations of mass or reinforced concrete are used for many residential, educational and industrial buildings which are fairly lightly loaded on subsoils with a reasonable load-bearing capacity, such as rock or sand and gravel containing only a small proportion of clay. Where heavier loadings occur, as with multistorey structures or warehouses, these may be met by introducing larger than normal bases, or concrete pads supporting structural frameworks.

Raft foundations (rigid or flexible) are on occasions used for lightly loaded structures where some degree of settlement may be expected on the site. Settlement could arise from mining subsidence, soft subsoils or presence of fill. Rigid rafts often comprise flat concrete slabs reinforced with concrete beams acting as stiffening ribs and are very expensive. Hence it is better to use flexible rafts which usually consist of thin reinforced concrete slabs on beds of sand.

Piled and ground beam foundations are often needed for structures erected alongside rivers and estuaries and in other difficult locations. The piles are designed to support the loads either by friction from the various strata through which they pass or by bearing on a firm strata at a suitable level. *In situ* concrete piles are generally about fifteen to thirty per cent cheaper for relatively short piles, but as the length of pile increases the cost of *in situ* piles comes much closer to that of precast piles. The choice of piles is influenced by many factors including subsoil conditions, loading of the superstructure and economic considerations.

Vibrocompaction is a method of treating weak soils to improve their load-bearing characteristics. A tubular probe is lowered into the ground and contains an oscillating eccentric weight which imparts a transverse vibration to the tube and hence into the surrounding ground. The hole formed by the probe is backfilled with compacted coarse aggregate and conventional foundations are then cast on the treated ground. It often proves to be a quick and economic process and can be considerably cheaper than piled foundations or deep mass concrete blocks.

Economics of Foundation Construction

The availability of suitable plant can influence the relative costs of foundation types. Experience has shown that in areas of shrinkable clay, short piles bored with mechanical augers are competitive in cost with traditional strip foundations, even with quite small contracts, and could conceivably be about two-thirds the cost of a deep traditional strip foundation. For single house contracts, bored piles are usually slightly more costly than deep strip (trench fill) foundations (often 375 × 1000 mm) although the extra cost may be justified by the added safety factor.

British Research Establishment Digest 64 (1972) emphasises that simple costing on a labour plus materials basis may be misleading. The construction of bored piles can often continue through the winter months when the trenches for strip foundations could be waterlogged or damaged by exposure to frost. In addition to use on clay sites, bored-pile foundations with precast ground beams may be useful for low rise industrialised systems, and in crosswall house construction it is rarely necessary to use more than ten piles per house. Concrete trench fill foundations for houses are often cheaper than brick and concrete strip, with possible reductions of up to 30 per cent resulting from reduced labour, excavation and backfill, with trenches left open for a shorter period with less risk of side collapse or flooding, in depths of 900 mm or more in shrinkable clay.

Removal of subsoil water from foundation excavations is always expensive; hence foundations on sands and gravels should be kept above groundwater level wherever possible. Some general rules with regard to relative excavation costs may also be helpful.

(1) It is almost invariably more economical to utilise the existing ground configuration rather than to dig it out or make it up.

(2) Removal of surplus excavated material from a site is far more expensive than depositing and levelling the soil elsewhere on the site.

(3) Over a certain minimum (probably about 2000 m^3), the cost of bulk excavation, but not its disposal, probably reduces with quantity more steeply than any other building operation price rate.

Table 3.2 shows the relative costs for substructure work with various types of housing. It should however be borne in mind that foundation prices may be 'weighted' to secure an early return to the contractor.

Foundations as a proportion of total building costs can vary from 8 to 18 per cent and tend to decrease with increases in the number of storeys.

Basement Construction

Some general guidelines as to the relative costs of constructing basements are provided.

(1) the larger the basement the cheaper the cost/m^2;

(2) the nearer to a square shape the cheaper the cost, although the cost does not increase considerably until a ratio of three to one is exceeded;

(3) where basements are constructed in concrete, the cost/m^3 does not increase with the depth up to 4.60 m, provided the damp-proofing requirements are constant;

(4) the use of brickwork for depths up to 3 m, provided the basement walls are loaded down by the superstructure, generally shows a distinct economy over the use of concrete, but beyond this depth, this trend either diminishes or reverses;

(5) the additional cost of waterproofing concrete is relatively small and therefore worthwhile when conditions do not justify asphalt tanking;

(6) the cost of asphalt tanking constitutes a high proportion of the total basement cost and underlines the need to ascertain the proposed

Table 3.2 Social housing – relative substructure costs

Type of housing	*Weighted percentage of total costs (1993)*
Traditional brick two-storey houses	
Average value	9
Lowest value for good sites and terrace blocks	8
Highest value for poor sites and development in pairs (exceptionally difficult sites could involve extra costs of up to seven per cent)	10
Brick flats and maisonettes	
Average value	8
Lowest value for four-storey blocks on good sites	6
Highest value for blocks involving piling and for those over 30 m in length (exceptionally difficult sites or blocks with basements could involve extra costs of up to five per cent)	13
Concrete structure flats and maisonettes	
Average value	10
Lowest value for high blocks with normal foundations	4
Highest value for low blocks with piled foundations (exceptionally difficult sites or low blocks incorporating basements could involve extra costs of up to seven per cent)	16

method of waterproofing at the time of preparing preliminary estimates.

Gas Barriers

During the 1980s and 1990s the dangers of gaseous pollution of buildings, particularly from derelict and filled sites, has become more fully understood, with most attention focused on landfill gas (methane and other gases) and radon (uranium). Once detected, a building must be protected against the ingress of gas at foundation level, as required by the Building Regulations 1991.

When landfill gas is encountered, the most common method of protection is to provide a gas-proof membrane in conjunction with a reinforced concrete floor slab above a granular venting layer. This is vented either by means of a trench around the perimeter of the building or by pipes and vertical risers to release gas at roof level. As this is a costly process, the estimated cost should ideally be deducted from the purchase price of the land before the conveyance is finalised.

In 1993 radon had been traced in Cornwall, Devon, Somerset, Derbyshire, Northamptonshire and North Yorkshire. Polythene sheeting became accepted as a popular and cost-effective barrier, but it is difficult to fit it tightly around service pipes and there is always a danger of puncturing the membrane. The smallest gap or crack in the sheeting will permit gas to enter the building. One satisfactory solution is to apply an approved liquid asphaltic compound which is both a radon and damp-proof barrier.

STRUCTURAL COMPONENTS

A variety of structural forms are available from load-bearing brick walls to reinforced concrete, steel and timber frames. It is often necessary to prepare cost comparisons of different structural forms in order to determine the most economical constructional form for a particular project. The principal structural forms are now considered.

Load-bearing Brickwork

Since the 1970s it has been found relatively economical to use load-bearing brick crosswalls for buildings up to five storeys in height in place of structural frames of steel or reinforced concrete. The solid crosswalls carry all the floor and roof loads and have separating and insulating functions. These walls are able to provide structural support when the building permits the plan form to be repeated on each floor, or the load-bearing walls are so placed in relation to planning needs that they can, for the most part, continue uninterrupted from foundation to roof. This form of construction also permits flexibility in the choice of claddings to front and rear walls as they are no longer load-bearing. The optimum spacing of crosswalls is normally between 4.50 to 6.00 m centres and this places constraints on layouts. The introduction of some column and beam construction would permit the provision of large openings in crosswalls, but with comparatively small and well-defined units such as flats the crosswalls are unlikely to present any real obstacle.

A five-storey block of maisonettes and flats over ground floor shops was erected at Allestree, Derby, using 380 mm brick cavity flank walls and 215 mm brick crosswalls and staircase walls. The spacing of the crosswalls varied from 5.20 to 6.20 m. It is claimed that the brick crosswalls showed a saving of thirty-five per cent on the estimated cost of a reinforced concrete frame, and this is equivalent to an overall cost saving on the scheme of about six per cent. Although for two-storey house construction, crosswalls are not inherently more economical than more traditional forms of construction. The cheaper method will depend on many design features such as plan shape, roof and wall types and finishes, and on site conditions. Some of the factors to be considered include the supporting of timber joists on party walls and transference of roof loads to transverse walls by beams, possible buttressing of ends of walls and effect of discontinuity of operations on site.

With three-storey blocks of flats, the cost of the brick crosswall structure generally shows a similar order of costs to corresponding features of conventional cavity wall construction. Reinforced concrete crosswalls tend to show a cost higher than those of traditional brickwork. High rise flats can show as much as a sixty per cent cost range for external infilling between the crosswalls, timber cladding being the cheapest and specialist curtain wall infilling the most expensive.

To satisfy the Building Regulations 1991 regarding thermal insulation, one economical method is to use a 275 mm wall with an outer brick skin, clear cavity and thermal insulating block, 125 mm wide of Icelandic pumice, and an infill of mineral wool to give a *U*-value of 0.45.

Another approach is to use a brick diaphragm and fin wall structural system as an economic alternative to the more customary frame and cladding. The all brick structures provide an aesthetically pleasing, low maintenance building with infinite opportunities for the expression of brickwork designs in a sculptured form. They are ideally suited for large single storey buildings such as leisure centres, sports halls, swimming pools and school halls. Typically, diaphragm walls are about 440 to 777.5 mm wide overall, with diaphragms (half-brick cross ribs) spaced at about 1.2 m centres. The roof is usually designed as a horizontal plate member that props and ties the tops of the diaphragm walls, and transfers the horizontal reactions to the transverse walls. Any required level of thermal insulation can be achieved by fixing insulating batts to the inner face of the cavity leaf and the large cavity voids can be used as ducts for services or external aesthetic indentations.

Reinforced Concrete Frames

Mass concrete is a relatively cheap material which can be formed into complicated shapes using suitable formwork. However, its one significant weakness is that its strength in tension is much less than that in compression, with a ratio of about 1 : 10. Hence reinforcement, mainly in the form of steel bars, has to be provided in members which are in tension, and this will improve the tensile resistance of the concrete, resist shear forces, and increase the compressive strength of columns and beams.

Quality control is vital in this class of work and includes controlling production of the material and the handling and placing in the unfavourable conditions frequently encountered on a construction site. The most important factor affecting the strength of concrete is the proportion of air in the hardened material. Minimising the air content is achieved by maximising the workability of the concrete, ensuring effective compaction and avoiding excess water in the mix (MDA, 1992).

It is difficult to obtain a reliable independent assessment of the proportion of framed buildings using reinforced concrete although, as a rough guide, a market survey conducted by the British Cement Association indicated that 58 per cent of structural frames of two storeys and above in the UK in 1991 were constructed in *in situ* or precast concrete or in loadbearing masonry.

The main requisites to achieving speed and economy in construction are repetition, standardisation and co-ordination of the main design elements and details. Large pours require many operatives, extensive craneage and substantial supplies of formwork. Although precast concrete frames are credited with high speed erection, the design and manufacturing periods can be longer than for steel. *In situ* concrete permits alteration at a late stage in the construction process, but this characteristic is subject to abuse. However, concrete lacks adaptability if structural members need to be removed or strengthened during the life of a building.

Foundations for concrete structures may be significantly larger than those required for steel framed buildings and this can be a critical factor in the design and economics of foundations. However, foundations are normally a relatively small part of the total building costs and the economics of the total project must be the main consideration. Post-tensioned ribbed slabs or precast prestressed construction need not be significantly heavier than steel construction.

In situ concrete construction is versatile and has multidirectional continuous spanning capabilities. Concrete flat slab construction permits very flexible accommodation for services and a minimum services zone above suspended ceilings. Typical flat slab spans are 6 m to 9 m as those above 10 m are deeper and affect floor to floor heights unless prestressing is introduced. The use of shear walls surrounding enclosed spaces, lift shafts and stairs, provides moment resisting stability for flat slab structures (MDA, 1992).

The propping of concrete construction can cause problems by restricting access to recently constructed areas, but it is only required during casting and early curing and careful design can reduce propping requirements.

Reinforced concrete construction provides inherent fire resistance and enhanced detailing can provide resistances in excess of four hours. The normal prescribed cover to reinforcement will provide adequate corrosion protection in external locations.

Concrete construction is considered an important potential element of passive 'green' buildings. The heavy mass of concrete permits the development of passive heating and cooling, but will require a change in the traditional approach to floor to floor zones, with a reduction in the ceiling voids to increase the amount of exposed soffits to balance heat/cold inputs and outputs (MDA, 1992).

Economies in formwork can be obtained mainly through simplicity of shape, and repetition of units to obtain the maximum number of uses; and other ways of reducing formwork costs are as follows.

(1) Formwork should be designed to reduce labour requirements in its assembly and to permit re-use without cutting.

(2) It is often more economical to use a little more concrete than is structurally necessary to secure repetition of formwork.

(3) Column sizes should be standardised as far as possible with differing structural loads accommodated by making variations in the quantity of steel reinforcement or concrete strengths and, where column sizes must be changed, reducing one dimension at a time will save cutting.

(4) Establish uniform spacing of columns and beams, and uniform beam dimensions wherever possible.

(5) Eliminate perimeter edge beams in flat slabs wherever possible.

(6) Shapes of sections should preferably be designed to allow easy removal of forms, for

example, sides of beams could advantageously slope slightly outwards from the bottom.

Techniques of slip forming concrete with moving shuttering are used extensively for erecting tall cast in-situ industrial buildings, such as silos, cooling towers and chimney shafts, and this process has been extended to tall residential buildings, hotels, university buildings, residential institutions and the like, either as cored systems or fully-slipped form systems. The extent of the repair and maintenance work required on concrete structures will depend on the quality of materials, degree of natural exposure, and amount of mechanical or chemical attack. Rates of carbonation can be high where there are low cement contents, high water/cement ratios and poor compaction (Monk Dunstone, 1992).

Precast Concrete Structures

Precast concrete has been defined as premoulded concrete units that fulfil both the architect's visual and functional requirements and the consulting engineer's structural requirements. Precast concrete offers a wide variety of finishes which rarely require any further treatment after erection. It offers a quick form of construction, although this may be adversely affected by the speed of construction of other elements of the building. Nevertheless, precast concrete structural members can prove expensive and it is vital that they be carefully designed to ensure maximum and efficient use of handling plant and moulds. The use of precast concrete permits the production of high quality components which can be speedily erected on site.

The Ronan Point tragedy highlighted some of the problems associated with the design of large panel structures, through the possibility of a consecutive failure of floor and wall components that is 'progressive collapse'. BS 8110: 1985 lays down appropriate design rules for all large panel structures and, in addition, includes requirements for such structures where resistance against progressive collapse is an essential criterion for design. Tie forces needed for the security of large panel structures can be provided by the inclusion of reinforcement or prestressing tendons from floor to wall, peripherally, internally or as continuous vertical ties. The additional cost of these features is likely to be relatively small in most cases.

The size and shape of precast concrete units is generally governed by the limitations imposed by transportation and site erection. These limitations are normally imposed by economic and legal requirements and not necessarily by technical problems; however, further limitations on design are imposed by the need to lift and stack precast components safely and at an early stage in the precaster's yard pending delivery to site.

Designers need to consider the problems of handling in the works and special care must be taken in respect of projecting steel, edge details, finishings, lifting methods and stacking arrangements, the problems and cost of transportation, and the cost of site erection, which is usually governed by the size of crane required. The cost of transportation is generally proportional to the distance covered and weight carried, but the rate per tonne/km reduces over longer distances. For exceptionally long and wide loads, the cost can double or treble as certain statutory precautions have to be taken to ensure the safety of the load and other road-users. Finally, to ensure smooth and quick erection, tolerances, joint details and fixings must be carefully chosen. Indeed, early design decisions can have a significant effect on the speed of erection and, hence, on the viability of the project.

Structural Steelwork

Steel has a very high strength-to-weight ratio which makes it economical for a wide range of structural uses. A market survey by British Steel (1992) claimed that in 1991 steel constituted 58 per cent of the total UK market for the structural frameworks of multistorey buildings. Its main advantages comprise ready availability, rapid erection, and ease of subsequent conversion or extension.

Single-storey steel structures for workshop, storage, warehouse and farm buildings are available in a variety of standard system buildings normally consisting of either truss and stanchion frames or rigid portal frames, while longer spans will be required for buildings such as sports stadia and aircraft hangars. Bolting and welding are the two common simple and fast methods of jointing the steel members, while riveting has become almost obsolete.

The need for heavier and deeper structural members arises as the span between supporting columns is increased, accompanied by a rise in the cost of the structure per unit area covered. There will be cases where large column-free areas are required. The solution make take the form of a three-dimensional skeletal arrangement or 'space structure', consisting of individual steel components, usually of hollow section, forming single or multi-layer flat grids, barrel vaults, or domes. As the number of storeys increases, so does the need to provide wind bracing to resist the lateral wind forces (Knowles, 1991).

Since the last war some of the most successful steelbased systems in the United Kingdom have been by consortia of building clients. These include CLASP (Consortium of Local Authorities Special Programme) and SCOLA (Second Consortium of Local Authorities) – two groups of local education authorities who execute millions of pounds worth of school and other buildings each year; NENK, a system developed by the former Ministry of Public Building and Works for buildings of all types for the services; and 5M, a system for house building developed by the former Ministry of Housing and Local Government. Both Ministries now form part of the Department of the Environment which was established in 1970. In all these systems it has been stated that a steel frame was chosen as a result of user requirement studies and performance standards, from which steel was seen to emerge as the best solution. It was also pointed out that in these cases a lightweight structural steel frame offered greater flexibility in grid arrangements and allowed the easy threading through of services and other non-structural components. Structural steelwork is now seldom riveted and jointing frequently incorporates gas and arc welding.

CLASP's decision to adopt a light and flexible steel frame mainly resulted from the need to take into account ground movement due to mining subsidence. A further advantage was found to accrue from speedy erection with a steel frame for a primary school being erected within four to five days and occupying as little as one-and-a-half per cent of the total site labour involvement. SCOLA also saw the advantages associated with light steel frames of speedy erection, flexibility of internal layout and adaptability to meet changing future situations. Steel has also been used in the frames of a number of multistorey hospital buildings, and these have been selected after cost comparisons with other forms of construction, taking into account the effects on cost of piled foundations. For very tall buildings, exceeding 20 storeys high, steel framed buildings normally provide the most economical solution.

Standardisation and repetition of members and avoidance of design changes are essential to maximise time savings in the use of steel. The overall time is determined by the time needed to procure steel, prepare fabrication drawings, fabricate the steel members, deliver to site in the correct sequence, form the concrete floors and fire protect the framework. The procurement and fabrication phases of steel frames halved in the late 1980s, but site productivity did not improve to the same extent. Late changes, generally resulting from late or inadequate information, cause disruption during fabrication and major delays during site erection. It is also important that fixings for other building elements and components are satisfactorily co-ordinated during design fabrication (MDA, 1992).

Steel members can be removed, altered or added to during the life of the building to change the structural capabilities of parts of the building, and thus provide a significant degree of flexibility. Steel components and fixings also ensure great accuracy in construction, although some of this is lost with the provision of concrete floors. Both long and tall buildings necessitate special care in manufacturing tolerances, bracing systems and verticality checks.

Steel framed buildings can be as much as 30 per cent lighter than comparable concrete structures, and poor soil conditions may restrict the choice to steel framed buildings. Steel frames normally incorporate *in situ* or precast concrete floors and generally require a greater storey height than *in situ* concrete frames to accommodate services above false ceilings. A number of innovations have been introduced to give varying characteristics of beams, with regard to manufacture, shape and use, and to reduce the impact of the integration of services. Steel construction requires no propping and appropriately designed secondary beams can ensure simple and fast erection. However, the vibration damping effect of building finishes can be important with steel framed construction, and special features may be needed with very light finishes and where vibration is a critical factor (MDA, 1992).

Internal steelwork does not usually require corrosion protection but that for external steelwork can be expensive, involving the use of applied paint coatings, sprayed or dipped metal coatings, plastic coatings and sometimes concrete casings or the solid filling of the area between the flanges of steel sections, with the possibility of periodic renewal. Spray coatings of intumescent paint and other suitable coverings are generally used to counteract the very low fire resistance of bare metal.

The main economic advantages of structural steel frames for multistorey building are:

(1) overall reduction in size of structure, as the columns are smaller in section and obstruct less floor space than those required for reinforced concrete construction;
(2) precision and speed of erection;
(3) reduced foundation costs due to smaller dead weight of structure;
(4) flexibility of planning; and
(5) adaptability to requirements of possible future changes, including façade cladding and fenestration and internal layouts.

The relative economics of concrete and steel structural frameworks may also be influenced by local conditions. For instance, physical restrictions on a congested central area site may make it difficult, if not impossible, to use large steel erection cranes. Shortage of time coupled with the urgency of a particular job may not allow sufficient time for the curing of *in situ* concrete members. An abundance of sand and gravel workings in close proximity to a building site may enable concrete to be produced at an advantageous price. A good supply of skilled steel erectors in the area could result in cheaper steel erection costs. The design characteristics of a particular project may favour one particular type of construction. Hence it is difficult, if not dangerous, to generalise on the relative merits of different structural forms and it is usually necessary to make a full investigation and to cost up the alternative techniques for each particular project.

For instance by ensuring a fully agreed design brief, a minimum of subsequent changes, and full integration of structure, cladding, finishes and services, all forms of building can be constructed more quickly, and experience shows that overall project times for both steel and concrete framed buildings are often similar. The mean costs for the framed structures of medium rise buildings in 1992 ranged from £56 to 90/m^2 of GFA for *in situ* concrete, £46 to 93/m^2 for precast concrete and £66 to 102/m^2 for steel frames (MD, 1992).

Timber

The greatest use of timber framing has been in the field of industrialised building for houses and flats, and the majority of systems make use of it in the construction of walls and floors. In place of the conventional joist and floor board assembly the trend is towards prefabricated floor sections, occasionally of the stressed skin type, which are brought to the site ready for assembly. Timber frame walls possess the dual advantages of flexibility of cladding design and speed of erection. Timber walls and floors are usually required to have a fire resistance of half-an-hour in one or two-storey houses, but the walls separating dwellings have to provide protection for one hour and this has necessitated further development work.

The main impetus for the use of timber framing has come from North America and Scandinavia both of whom possess an abundance of timber, and where an estimated 90 and 80 per cent respectively of the dwellings are of timber frame construction. The main advantages of timber frame construction are that it requires no large capital investment, heavy plant or other exceptional site equipment, neither large contracts nor standardisation of house types, architectural treatment nor finishing materials. With good management and use of conventional building skills, satisfactory houses can be provided at comparable cost to traditional construction, but using one-third to one-half of the site labour and with accelerated erection times, measured in weeks instead of months.

Timber possesses high bending strength and stiffness per unit of weight, high thermal insulation and relative ease of machining, cutting, assembling, fixing and transporting. Framed panel assembly is generally a relatively simple jig operation which can be carried out in the shop with rudimentary equipment, or even on a platform built on the site. A variety of cladding materials can be fixed with ease to timber framing. There was a pronounced swing in the early nineteen eighties towards this method of construction, because of the competitive price of timber caused by the strength of sterling, coupled with the low demand for timber worldwide, and a subsequent decline in use resulting from adverse TV publicity following a BRE Report (1985) identifying many potentially dangerous faults mainly stemming from poor site construction and supervision found in timber frame houses surveyed on ten different sites between 1983 and 1985. This coincided with a significant increase in the price of timber and a general decline in its quality, which tended to reduce the relative advantages of timber frame as compared with conventional construction.

BRE emphasised that the faults are defined in the report as departures from good practice, of a kind which are commonly found in all forms of housing construction and it is clearly desirable that they are avoided; but they only rarely lead to significant failures in service. Whilst design and construction practice needed to be improved, the evidence available to BRE indicated that the performance in service of timber frame housing was no less satisfactory than that of traditional construction. An excellent publication by TRADA (1988) described and illustrated sound design and construction in timber frame houses and identified their main advantages. By the early 1990s timber frame houses had recovered some of their earlier popularity.

WALLING

Walls and partitions with associated windows and doors constitute a major item of expenditure of a building. In municipal housing these components can account for about one quarter of the total cost of brick four-storey maisonettes and about one third of the total cost of brick two-storey houses.

External Cladding

For low rise buildings, cavity walls are generally the lowest long term cost solution, provided satisfactory detailing and workmanship are secured, and adequate thermal insulation can be provided in the cavity. Stone facings are very expensive and care is needed to select a stone which is suitable for the particular environment; for example close-textured stones withstand city atmospheres better than limestones, and dense cement mortars must be avoided. Precast concrete cladding and cast stone often provide economic alternatives to natural stone.

Concrete walls, 175 mm thick, plastered internally, and using traditional formwork for multistorey flats, cost approximately fifteen per cent less than brick and concrete block hollow walls, faced externally and plastered internally, including scaffolding costs. A further reduction of about fifteen per cent may be obtained by using sliding formwork for the concrete walls. The cost of providing suitable thermal insulants must however be taken into account in the cost comparison.

The natural resistance to decay of timbers such as teak and western red cedar has led to their increased use for external claddings, although their use in housing is often limited to spandrels under windows, gable ends and other small areas of external walling. Unfortunately, the weathering of these timbers if untreated, at least in an urban environment, frequently causes them to become unsightly. An annual application of a linseed oil/paraffin wax mixture containing a fungicide or a suitable wood stain is the minimum necessary to preserve appearance, and when comparing alternative cladding costs the annual expenditure on treatment of timber cladding must be taken into account.

Similarly, when making cost comparisons of alternative metal cladddings it is essential to consider also the cost of any periodic treatment that may be required. For instance, some metals such as copper oxidise attractively upon exposure, whereas others like aluminium alloys may become unsightly. Aluminium when used externally should preferably receive anodic treatment, followed by periodic washing, to maintain a good appearance. Where the aluminium has not been anodised, more frequent washing is necessary, the actual frequency varying with the degree of atmospheric pollution.

It is often necessary to consider the use of newly introduced materials for use in external claddings and there may well be insufficient feedback of maintenance costs to make a realistic appraisal of probable total costs, when making cost comparisons with other forms of cladding.

Cladding for modern commercial and industrial buildings often falls into one of the following four categories: profiled metal sheeting, GRP cladding, GRC cladding, and curtain walling, and each of these is now examined.

Profiled metal sheeting is used extensively for cladding a wide range of modern buildings from factories to business park office buildings. Profiling provides stiffness and strength to thin materials and permits the fabrication of economic weather skins. Coated or galvanised steel, stainless steel and aluminium are all used, but the first is usually preferred because of its lower capital cost. The life expectancy of steel sheeting is directly proportional to the quality and durability of the finishes applied to it, while aluminium develops its own protective oxide film inhibiting further corrosion and the durability can be further enhanced by the application of decorative, coatings.

However, these claddings have low thermal capacity, mass and noise reduction capabilities, and hence further costs will be entailed in improving the thermal and noise insulation standards. Normally they carry no load beyond their own weight and the loads imposed by wind and snow and they offer little resistance to impact damage. Their low mass can result in economies in the provision of the structure and foundations to the building, and they can be readily erected, dismantled, re-erected, and renewed. They should not be used for complex forms of building (Josey, 1991a).

Profiled steel sheeting is obtainable in a wide variety of finishes, including hot-dip galvanised, aluminium coated, aluminium/zinc coated, and colour coated. Colour coatings include acrylics, PVCs, silicone polyesters and PVF2, usually applied over a galvanised base. BS 5427 provides details of the performance of coated products.

A BRE survey in 1993 showed a substantial number of defects in organically coated metal clad industrial buildings, often less than five years old, mainly stemming from cut edge problems, and some fixing and condensation difficulties were also experienced.

GRP cladding, or glass fibre reinforced polyester, is strong, lightweight and durable, and is well suited for use as cladding panels. It can be flat or profiled, translucent or opaque, and pigmented. It consists of a laminate of various layers of a viscous resin which sets to form a solid, and glass fibre. GRP panels are not maintenance free, and they should cleaned carefully and regularly. Furthermore, structural damage cannot be repaired effectively on site (Josey, 1991b).

GRC cladding, or glass fibre reinforced cement, is a combination of cement : sand mortar and alkali-resistant glass fibres. It combines the compressive properties of cement based mortars with tensile and flexural properties produced by the glass fibre. It permits the use of thin section claddings, often about 10 mm thick, thus reducing the load imposed upon the structure. GRC possesses many of the characteristics of concrete, but has a higher resistance to moisture penetration. Its high cement but low aggregate content results in higher moisture movement than concrete (Josey, 1991c).

Curtain walling is the concept by which the interior of a building is enclosed by a cladding which supports no loads apart from its own weight and the environmental forces which may act upon it. It encompasses a variety of different constructional methods, varying from proprietary systems providing good quality, attractive, reasonably priced sheath walls suitable for low and medium cost developments, to custom designed higher priced projects (Jocey, 1991d).

It could be helpful to the student to briefly outline the development of curtain walling since the last war. Many of the systems used in the 1950s and 1960s offered poor weather resistance, while the 1970s saw the introduction of many technical innovations with varying degrees of success. The curtain walls of the 1980s were more sophisticated and provided leakage prevention systems into which a variety of different types of panel could be fitted. In the 1980s the aim was to minimise site time in addition to maximising floor area.

Joscy (1991d) has identified two main methods of curtain wall construction, namely 'stick' and 'unit':

(1) *Stick construction* consists of a grid of mullion and transom bars into which various types of infill panel can be fitted, with most of the assembly work carried out on site. It is relatively low cost but the main disadvantage is the need for site assembly under variable conditions.

(2) In unit construction sections of wall are prefabricated and delivered to site in completed modules ready for bolting to the structure, with only the perimeter joint seals being applied on site. The main advantages are the pre-assembly under controlled factory coditions and rapid enclosure of the building with a minimum of site works, but it suffers from bulky modules.

Curtain walls are usually washed with a mild detergent and water, as recommended in CP 153: Part 1 and BS 6262 at regular and frequent intervals.

Partitions

Dry partitioning will normally show a direct cost advantage over block partitioning with two-coat plastering on both sides, despite the greater detailing required for dry partitioning, involving the preparation of drawings and cutting schedules by suppliers, the need for packing pieces, taking care to ensure that panels do not become wet and the additional cost of skim coat or 'dry' finishes. Other consequential savings of using dry partitions are as follows:

(1) Reductions in structural costs of loadbearing brickwork or blockwork, supporting frames and foundations due to a lower total dead weight of building.

(2) With the two-bedroom flats savings would accrue from the considerable simplification of construction and setting out of the floor ribs.

(3) Simplification of design procedure and consequent detailing arising from reduced load factors.

(4) Reduction in the expenditure usually necessary for the provision of temporary equipment, fuel and attendance for drying and controlling the humidity of the works.

(5) Reduction in the total cost of general conditions and preliminaries arising partly from reduced cost of building work and partly from reduced building contract period.

The design team has a large range of demountable partitions from which to choose. Selection is influenced by many factors including demountability, fire resistance, acoustic properties, appearance, cost, integration of services, weight of partitioning and feasibility of incorporating false ceilings. Where the building is to be air-conditioned, full reseach has to be conducted into air ventilation arrangements and where this is achieved by means of a false ceiling, often integrally illuminated, the layout of the partitioning needs careful thought.

Many large commercial and industrial organisations require a flexible layout when planning their office accommodation, particularly as departments can be expanded, contracted, integrated or fragmented at quite frequent intervals. In these situations a steel or aluminium framework enclosing a variety of infill materials may well offer an effective solution. With the larger offices, some firms favour open plan 'landscaped' offices which can afford advantages in working environments, flexibility and communication. A complete appraisal of open plan and enclosed office proposals for any given situation could involve a cost-benefit study of the type described in chapter 16. Another innovation is the integration of office furniture and fittings into the partitioning system. Based on a fixed module, the partitioning panels are manufactured to the same standard dimensions as the office furniture or storage units, which clip onto the framework of the partitions.

The quantity surveyor is often presented at the detailed design stage with quotations from specialist partitioning subcontractors, based on performance specifications, and on which he has to make observations and recommendations. The quantity surveyor's task would be eased considerably if the partitioning specialists would:

(1) conform to the specification or state that their system will not comply with it;
(2) produce their terms and conditions in print of sufficient size to be clearly legible; and
(3) price the offer in detail so that variations can be adjusted in a fair and reasonable manner to the satisfaction of both parties to the contract.

Table 3.3 shows the cost relationships of different types of non-loadbearing partitions.

Table 3.3 Cost relationships of non-loadbearing partitions

Type of partition	*Cost index*
100 mm aerated concrete blockwork (fairfaced)	100
Ditto. plastered both sides	165
102.5 mm cellular flettons (laid frog down and fairfaced)	130
Ditto. plastered both sides	195
100 mm sawn studwork with plasterboard both sides	120
'Paramount' partitioning (57 mm thick)	100
50 mm demountable steel partition	360
Hollow glass blockwork (80 mm thick)	1650
Terrazzo faced partition	720

Insulation

The majority of new buildings are designed to achieve reasonable standards of thermal insulation and the practical applications of this aspect will be considered in chapter 13. The *U*-value of a wall, roof or floor of a building is a measure of its thermal transmittance or ability to conduct heat out of the building; the greater the *U*-value, the greater the heat loss through the structure. The total heat loss through the building fabric is found by multiplying *U*-values and areas of the externally exposed parts of the building, and multiplying the result by the difference between internal and external temperatures. *U*-values are expressed in W/m^2 °C (heat flow in watts through one square metre of the construction for one degree Celsius difference in temperature between the inside and outside of building). Typical *U*-values are: 215 mm solid brick wall with 13 mm plaster – 2.0; 255 mm brick cavity wall (unventilated) with 13 mm plaster on inside face – 1.40; 252.5 mm cavity wall (unventilated) with brick outer skin, lightweight concrete block inner skin, cavity space and 13 mm plaster on inside face – 0.96; and, as last but 302.5 mm with 50 mm glass fibre cavity batt insulation in cavity – 0.40, as compared with the Building Regulation 1991 requirement of 0.60. The additional cost of the latter form of construction over a 252.5 mm insulated cavity wall was in the order of £4.00/m^2 in 1993. Another alternative is to use blown mineral fibre to insulate existing houses, the latter costing £450 to £700 per house in 1993.

In certain situations sound insulation can also be an important consideration. Noise is becoming a greater problem as more mechanical equipment is used in buildings, and both road and air traffic noise is increasing in volume. A study of post-war office buildings showed that on average over forty per cent of the occupants suffer annoyance from street noise, rising to about two-thirds with buildings located on main thoroughfares, and the nuisance is not mitigated by height. Surveys of noise in houses adjoining main roads in Greater

London enabled a correlation to be established between a measure of noise levels by traffic noise index (TNI) and dissatisfaction of occupants through difficulty in getting to sleep, being compelled to keep windows closed and/or being unable to entertain visitors in comfort.

With aircraft noise it has been found that by providing good double glazing and a sound attenuating ventilator unit, a reduction in loudness to about one-sixteenth of the level outside the building can be obtained. This requires the frame of the inner window to be sealed to the existing window surround with a minimum space of 180 mm between the panes of glass. This treatment when applied to three rooms per house cost about £900 per house in 1993.

ROOFING

Roof Types

The majority of multistorey flats have flat roofs. The need for lift motor rooms, tank rooms, ventilating plant on occasions, ducts and pipes, add to the complexity of roofs, and the economy of covering a number of dwellings with one roof is partially offset by the costlier construction. Building ancillary structures on roofs is both expensive and often aesthetically unsatisfactory, and their elimination would reduce costs and leave a worthwhile expanse of flat roof which might be put to other uses. By way of contrast, high blocks of flats in other European countries are normally built with pitched roofs and although this raises aesthetic problems, they have the advantages of lower first cost, longer life and provision of storage space.

One study of post-war factory buildings showed that about one-third had flat roofs, one-quarter north-light and there was also a considerable proportion of equal-pitch monitor and barrel-vault designs. The majority of modern advance factories have steel portal frame roofs. Many framed roofs are constructed of steel, and flat roofs in reinforced concrete. The material most widely used as pitched roof coverings to factory buildings is 'protected metal'. Asphalt is the most common covering to flat roofs. On average about one-fifth of the roof area is glazed, although few flat roofs have any glazing.

The loading of the roofs of single-storey industrial buildings is often less than 1.2 kN/m^2, but services suspended from the roof can often amount to an additional 0.25 kN/m^2. This twenty per cent increase in basic design load, unless suitably accommodated, can lead to overstressing of a slender roof structure.

For housing, the most common form of construction is a timber pitched roof clad with slates, clay or concrete tiles, as the flat roofs of the 1960s resulted in many failures.

Roof Finishes

The range of roof finishes available is immense; among materials available are clay, slate, concrete, metals, timber, synthetics or a combination of various types. The main factors to be considered in the selection of finishes are:

(1) the provision of an impenetrable skin which does not change its characteristics on exposure;
(2) reasonable capital cost in relation to function;
(3) low cost and ease of maintenance and repair;
(4) speed and ease of application;
(5) long life expectancy;
(6) ready availability; and
(7) suitable visual qualities, such as colour, texture, scale and applicability to the required roof form.

Concrete tiles account for ninety per cent of pitched roof coverings, presumably stemming from economy and suitability both technically and visually for low rise housing of medium to large frontage. For flat roofs there is no such clear-cut market situation, and the factors influencing choice are frequently more oriented towards the specialist skills of laying.

It is worth noting that the measured temperature range on a flat black insulated roof over a twelve-month period in this country was from 80°C to −25°C, and much higher temperatures have been recorded on south facing slopes. Surface reflection can make a worthwhile contribution to heat dispersion when high temperatures are involved. It seems evident that felts should be used on firm substrates, while single-layer plastic finishes are suitable for use in situations where building movement is likely. All flat roofs must have an effective vapour barrier and the insulation must be of sufficient thickness to prevent condensation occurring within the roof construction itself.

There is a wide range of flat roof coverings available, ranging from the more traditional type materials like bitumen felt and asphalt to less conventional coverings such as flexible heavy duty plasticised PVC (Trocal), which is available in a range of colours, and vinyl–ethylene–terpolymer (Alwitra), both of which have been granted BAA Agrément Certificates and should have an expected life in excess of 20 years. Another flat roof covering is ethylene copolymerised bitumen (Carbofol) which is especially suited for refurbishment work. The choice of new materials does make for difficulties in life cycle cost calculations, because of the problem of assessing probable future costs. The lives of conventional flat roof finishes are: lead sheet: 60 years; copper in excess of 100 years; zinc: 40 years; aluminium: 50 to 60 years; mastic asphalt: 60 years; and bitumen felt: 15 to 20 years (BCIS, 1986).

Roofing Costs

One useful method of investigating the costs of alternative roof types is now outlined.

(1) The general format and scope of the study is discussed.

(2) Details and a measured schedule for a basic design are prepared; these are discussed and general principles settled.

(3) Variations from the basic design are considered.

(4) Details and measured schedules for variations are prepared; these are discussed and co-ordinated.

(5) A schedule of rates is prepared based on an agreed priced list of materials.

(6) Schedules of measurement are priced.

(7) Cost relationships are examined.

(8) A report is prepared.

Parapet walls and perforations in roofs for dormers, stacks, and the like are excluded from the study. Measurements are taken to the outside face of external walls, although it is appreciated that wall thicknesses are a potential variable. The basic design for pitched roofs incorporates gable ends, and hipped ends are treated as variations, although it has been found that hipped ends produce an overall saving resulting from the elimination of the cost of brickwork in gables. The rates for pricing measured schedules are built up in a scientific manner in preference to extracting prices from bills which varied widely, or from price books which were not relevant to the location under consideration.

Factory building studies showed that north-light roofs of rolled-steel lattice girder construction were about ten per cent more expensive in first cost than umbrella or equal-pitch roofs irrespective of span or storey height, while monitor roofs were about twenty-five per cent more expensive than umbrella roofs.

A cost study of low regular shaped flats found that concrete flat roofs covered with asphalt were consistently higher than comparable pitched roofs, using TRADA pitched rafters and concrete interlocking tiles, the extra cost amounting on average to about 25 per cent, although the cost of softwood flat roofs covered with three layers of bitumen felt could have a small cost advantage. With pitched roofs, irregular shaped blocks resulted in considerably increased costs.

The cost relationships of various coverings to pitched roofs are given in table 3.4, based on 1993 prices.

When comparing the cost of alternative insulating materials both the cost per square metre and the U-value must be taken into account. The unit cost multiplied by the U-value will give an index for comparison purposes. Adopting this approach, expanded polystyrene generally compares very favourably with insulating plasterboard, fibreboard or insulating screeds.

Table 3.4 Comparative costs of alternative coverings to pitched roofs

Type of covering	*Cost index*
Concrete interlocking tiles	100
Concrete plain tiles	200
Clay pantiles	140
Machine-made clay plain tiles	215
Hand-made clay plain tiles	375
Medium Welsh slates (500 × 250 mm)	280
Copper	415
Aluminium	315

Flat roofs show a higher failure rate than pitched roofs, often resulting from inadequate falls, the use of unsuitable materials and poor workmanship, and expenditure on remedial work is unacceptably high. Heavy weight flat roofs cost on average about four times as much to maintain as pitched tiled roofs over the first 30 years of building life. Since flat roofs cost about 25 per cent more to build, the total costs per roof over a 60-year life are likely to be between 40 and 50 per cent higher for flat roofs than for pitched roofs. A lightweight flat roof of 1000 m^2 area could cost about four to six times as much to maintain as an equivalent heavy weight roof during the first 30 years of building life.

Average roof prices/m^2 of roof area in 1992 ranged from around £27 to £34 for factories, warehouses and housing; £45 to £48 for old people's homes, flats and sheltered housing; £55 to £58 for primary schools and health centres; £75 for offices and shops; and £106 for churches (BCIS, 1992b).

FLOORING

Sound and fire resistance requirements of suspended floors to flats must be adequately considered. In general, a concrete floor with a minimum thickness of 125 mm will meet these requirements. Plate floors are often more economical than concrete slabs and beams for multistorey flats, where spans are not excessive. There is in fact a wide variety of constructional forms available for use in suspended floors incorporating *in situ* or precast reinforced concrete, with or without beams – pot slabs, prestressed construction and steel beams. It seems evident that there is a need for a detailed investigation of the comparative costs of different forms of construction to meet varying spans and floor loadings, which has been partially met by Singh (1994).

With a regular plan arrangement precast concrete units often show distinct advantages. Erection is speedy, costly shuttering is eliminated and the units provide an immediate working area. *In situ* slabs also have their own particular advantages which can make them the best solution in certain situations; they are adaptable to variations in plan shape and section thickness and give added rigidity to the structural frame through lateral support. It is also likely that some types of suspended floor are most economic at certain spans. One investigation at Nottingham indicated that precast beam and block floors showed advantages at small spans (800 to 2400 mm), hollow beams at medium spans (4800 to 5400 mm) and prestressed tee beams at large spans (8400 to 9000 mm).

In traditional brick two-storey houses, floors, stairs and finishes account for about eight to eleven per cent of total cost. With flats and maisonettes, floor finishes on average account for about six per cent of total costs. Timber floorboards on battens and quilt are likely to be about 40 per cent more costly than vinyl tiles on screed and quilt to suspended floors. The timber boards do however, give slightly better sound insulation and allowed electrical runs to be inserted at a relatively late stage of construction.

Floor finishes also vary considerably in unit costs and the thickness of the flooring can influence structural costs as a thick finish, like wood blocks, may produce a taller building than a thinner floor covering such as vinyl tiles depending on the screed thicknesses. Cleaning and maintenance costs of floor finishes are other important considerations which should be taken into account in any cost appraisal. Cost relationships of alternative floor finishings are given in table 3.5.

The majority of stairs in multi-storey flats are of *in situ* concrete construction often with a granolithic finish. With the steady improvement in lift reliability, staircases in high blocks are needed mainly for escape in case of fire and do

Table 3.5 Cost relationships of floor finishings

Type of floor finishing	*Cost index*
2 mm thermoplastic tiles (series 2)	100
2.5 mm vinyl tiles	130
3.2 mm cork tiles	170
25 mm granolithic paving	190
25 mm softwood tongued and grooved flooring (including fillets)	370
4 mm rubber flooring and underlay	500
25 mm Iroko block flooring and polishing	710
16 mm terrazzo paving in squares with dividing strips	780

Table 3.6 Cost relationships of internal doors

Type of door (all 726 × 2040 mm) (including butts and painting)	*Cost index*
40 mm flush door, cellular core, hardboard faced	100
40 mm ditto. plywood faced	110
44 mm ditto. ½ hour fire check door, plywood faced	150
44 mm softwood purpose made 4 panel door	290
54 mm flush door 1 hour fire check, sapele veneered hardboard	435
50 mm mahogany purpose made 4 panel door	510

not require a high standard of finish, and are in many cases isolated from landings by selfclosing doors. Furthermore, if storey heights could be standardised at two, or at most three, alternatives, it should be possible to produce precast flights for use in many types of block.

DOORS AND WINDOWS

Doors

The cost relationships of the principal types of internal door are shown in table 3.6.

Windows

There is a wide range of choices available from timber to steel, aluminium and PVC-U, both single and double-glazed. Even with the same class and size of window, there can be wide variations in price according to the particular design of the window and the number of opening lights. For instance, with 1200 × 1100 mm steel windows the introduction of two opening lights over and above a single fixed light can increase the total cost of the window by twenty-five per cent, and weather-stripping can add a further ten per cent. The total cost of the window in this case includes the fixing of the window and its glazing and painting. The addition of a 75 × 75 mm painted softwood surround would increase the total cost by about twenty per cent.

Until the 1970s the use of stainless steel windows was confined almost entirely to banks, insurance offices and large department stores on account of their high initial cost. In 1967 the technique of adhesively bonding very thin stainless steel strip to aluminium alloy extrusions was introduced, and this resulted in rigid economical sections which combined the corrosion resistance of stainless steel with the flexibility of the extrusion process. At this juncture the stainless steel window became more competitive economically with those in other materials, particularly when savings on painting and other subsequent maintenance costs were taken into account. The latter aspect, which will be considered in more detail in chapter 13, prompted the use of 6000 stainless steel windows in a local authority housing project at Edmonton in North London.

In similar manner the higher initial costs of double-glazing have to be offset against savings in heating costs and the benefit of increased comfort, and this too will be further examined in chapter 13. BRE (1988) have shown that a double window, openable but weather-stripped and incorporating a 150 to 200 mm airspace, will have almost double the sound reduction qualities of an ordinary single openable window which is closed. Some useful data on double glazing for heat and sound insulation is provided in BRE Digest 379 (1993a).

Until the 1980s the greater proportion of multistorey flats had steel windows and most of the remainder used wood casements. The particular problems associated with windows at considerable heights are draught prevention, need for foolproof catches to prevent them swinging in high winds and facilities for cleaning, reglazing and general maintenance. Centre-hung windows which can be reversed for cleaning are particularly suitable in these situations. In addition there is a need to incorporate suitable extraction devices in kitchen windows to combat condensation.

The market share for PVC-U windows expanded throughout the 1980s for the replacement of decayed timber windows and in the 1990s became progressively used for new domestic and other buildings. The DOE/PSA carried out a study which showed that PVC-U windows were the most economical replacement for deteriorating windows. The windows have been granted BAA Agrément Certificates (Duell, 1991).

It is claimed by the suppliers that PVC-U windows have a long life, do not deteriorate and require a minimum of maintenance. There is evidence that PVC-U windows installed on the mainland of Europe in the early 1970s are in good condition and are retaining their properties and colour. Windows made of PVC-U combined with double glazing represent probably the most energy efficient window specification, provided frames fit properly and profiles are designed to dissipate the effects of external temperature variation across the profile.

FINISHINGS

The range of choices available for wall and ceiling finishings is probably greater than for any other component of a building, and the choice is influenced considerably by the class and use of the building. In municipal housing, finishings to walls and ceilings can account for ten to fourteen per cent of total construction costs in varyings situations. There are very large regional differences in the cost of internal decorations; for example, the application of emulsion paint to plastered surfaces can be twice as expensive in Inner London than in the Provinces.

In industrial buildings, clients are generally not prepared to spend more than 4 to 6 per cent of total construction costs on internal finishings, with the majority allocated to floor finishes, the specification being influenced mainly by function and profit. In the commercial sector, expenditure on finishings normally falls within 10 to 14 per cent of total construction costs, often with 40 to 50 per cent of this allocated to floorings, principally in the form of raised floors to accommodate mechanical and electrical (M&E) services and the needs of computers, with shallower suspended ceiling systems.

The use of plasterboard and similar sheet materials to form a dry lining to dwellings has become increasingly popular as it simplifies the work of the finishing trades and leads to earlier completion. If full advantage is to be gained, the implications of their use must be fully considered at the design stage and when planning the work on site. It is possible to fix plasterboard whereby the walls of houses can be lined at costs which generally compare favourably with that of two-coat plasterwork.

A wide range of plastic-surfaced sheet materials have been introduced for use in a wide variety of situations from wall linings, cubicles and partitions to fitted furniture and sanitary fixtures. Decorative laminates offer a number of advantages: uniform coverage of large areas; attractive appearance; good durability; good resistance to wear, biological and chemical attack, heat and moisture; and low maintenance costs. They are particularly well suited for use in the communal parts of buildings, where an attractive, hardwearing and low maintenance cost surfacing is desirable.

The acoustic properties of finishings are becoming increasingly important and acoustic materials are manufactured in three main categories: porous materials for general sound absorption but with special reference to high frequencies; resonant panels for absorption at low frequencies; and cavity resonators which can be designed to provide maximum absorption at a particular frequency. The first category is the most commonly used and suitable porous absorbents include mineral wool, foamed plastics with inter-connecting cells and proprietary tiles made of soft fibreboard.

Proprietary acoustic tiles usually have a facing which is itself acoustically neutral-selected for ease of decoration and maintenance, but perforated to allow sound to pass through to the absorbent. From a design point of view there is a wide range of choice in panels, tiles, boards and strips.

Suspended ceilings hung from the main structural soffit or fixed at perimeters create a void to accommodate electrical, mechanical and other specialist services. They can be categorised by construction type and appearance into continuous/ jointless systems; frame and tile; frame and tile with integrated services; linear strip; and louvre/ open grid. There is a very wide cost range largely influenced by the type of finish. In 1993 costs per m^2 including grids, frameworks and trims, ranged from £15 for mineral fibre tiles to £50 for perforated polyester powdercoated acoustic tiles; £85 for eggcrate tile with shadow battens; and £160 for powdercoated perforated steel plank with free issue grilles, purpose made (Moorhead, 1993). It is important to consider the ability to meet functional and performance requirements and maintenance and renewal costs when making comparisons of alternative techniques.

SERVICE INSTALLATIONS

Buildings and their environmental services have become more complex and the range of choices continues to increase. Unfortunately, the wide choice, lack of experience of new techniques and the need for assistance from more specialist skills have tended to act as constraints. In particular, environmental requirements are often considered far too late in the design process for them to make a positive contribution to the final design. This is unfortunate when viewed against the high cost of service installations which may amount to as much as twenty-five per cent of total costs on a housing scheme, 35 per cent on a commercial building and fifty per cent on a hospital project. However, after the prices of mechanical and electrical engineering services, security, fire precautions and essential information technology are added, together with the spaces needed to accommodate them and provision made for possible future changes, services could amount to as much as half the cost of a commercial building. There is a vital need for integrated design with all specialists contributing at each stage of the design process.

The way in which a multistoreyed office building is illuminated will have far-reaching implications on the structure generally, and ideally they should be considered together. The interior of a building can be lit in three basic ways: by daylight alone for most of the working day; partly by daylight and partly by artificial light, more commonly known as permanent supplementary lighting of interiors (PSALI); and permanent artificial lighting (PAL).

In terms of building form illumination by daylight will mean an office block with a depth of about 14 m; if PSALI is adopted the depth may be increased to 22.5 m, while the use of PAL will enable the depth to be increased to at least 27 m.

Each alternative has different structural implications with varying forms of fenestration. Maximum window area is needed for daylight illumination, less window area with PSALI and vision strips only with PAL. The thermal consequences of each approach are also quite different as the daylight design will involve twenty per cent more exposed external wall area, than one using PAL for the same floor area. With the latter design, mechanical ventilation at the least and possibly air conditioning will be essential. All the alternatives need costing very carefully to determine the most economical long term solution.

Plumbing and Waste Disposal

It is important to group sanitary accommodation in the same plan positions on the various floors of a building and to pay special attention to the economics of the various drainage layouts that could be adopted. When designing pipework arrangements, it should be borne in mind that it is generally more economical to use a large pipe rather than a number of smaller pipes which together produce the same cross-sectional area.

A drainage system (including both vertical and horizontal pipework) should be designed to convey waste materials away quietly, avoiding blockages, and with a limitation on the air pressure fluctuations within pipework to ensure that an adequate water seal is retained at each appliance. By the use of single stack plumbing it is possible to reduce costs by about thirty per cent over systems introducing vent piping. For example, field studies have shown that all supplementary venting could be omitted in an office building eight storeys high with ranges of five WCs and basins and a drainage stack of 100 mm diameter. In large buildings it may be an advantage to increase the diameter of the bend and drain to 150 mm.

A shower installation is likely to be about thirty per cent more expensive than a bath installation. Yet advantages of showers, such as hygienic, refreshing action, smaller space requirement, reduced water needs, and safer and easier operation for elderly people, could well outweigh the cost disadvantage. A survey conducted at a teachers' training college showed that seventy-six per cent of the respondents favoured showers.

The designer must often be concerned with the economics of different materials for rainwater goods, although admittedly he must also be very much concerned with their appearance and other characteristics. Unplasticised PVC in a range of colours and sections with supporting brackets of the same material, is commonly used for rainwater goods. There is a slight loss of resistance to impact, for example by ladders, as weathering proceeds, expansion and contraction with changes of temperature, and inability to withstand heavy snowfalls without deformation, and a probable life of about 30 years. A straight comparison of the first cost of supplying and fixing PVC-U rainwater goods is about half the price of aluminium and 40 to 45 per cent of that of cast iron.

Prices for chlorinated polyvinylchloride (CPVC), unplasticised polyvinylchloride (PVC-U), copper and stainless steel were similar in 1992, and thermoplastic polybutylene proved to be very suitable for small bore domestic pipes and fittings. Innovative hot water heating systems which were gaining in popularity in the early 1990s included domestic hot water-primary store heaters and unvented hot water systems.

A good example of the need to consider aesthetic factors as well as economic advantages is illustrated by the way in which more than 100 000 solar water heaters have been erected on the roofs of buildings in Israel to supply cheap hot water, but unfortunately they disfigure the landscape, townscape and skyline of the country.

Heating

The need for central heating in residential as well as other types of building was highlighted in the MHLG Parker Morris report (1961), which proposed that the minimum standard for an installation should be 12.8°C in circulation areas and 18.3°C in living areas when the outside temperature is −1°C, although these standards need to be kept under review. Domestic heating systems in the United Kingdom use gas, electricity, solid fuel and oil. The majority of the gas fired systems incorporated independent boilers while most of the electric installations used block storage heaters, and factory buildings use mainly oil or gas.

In cost planning heating services initial capital cost is rarely a sufficient criterion, as often the cheaper the heating system is to install the more it costs to operate. Furthermore, heating systems need to be compared on a comparable basis and the number of kJ/m^2 of gross floor area would be a useful factor.

Installation costs of different heating systems to serve a 3 to 4 bedroom house show a fairly wide spread and table 3.7 shows the average 1992 prices.

Table 3.7 Capital costs of domestic central heating systems (1992 prices)

Central heating for seven rooms comprising all appliances, distribution pipes, seven radiators and hot water service

Type of fuel	*Cost range*
Solid fuel	£2200 to £2500
Gas fired	£2300 to £2800
Oil fired	£2700 to £3100

Electricity is generally an expensive method of space heating, although the use of off-peak electricity at about 35 per cent of the normal rate has improved its position relative to other fuels. Electric central heating is versatile, produces no fumes, can be automatic in operation and is 100 per cent efficient in that no heat escapes up a flue. Off-peak electric floor heating cost about £16.50/ m^2 to install in 1992, whereas storage heaters cost about £240 to £310 each to supply and install. Heat pumps will become more attractive as energy costs continue to rise. They extract heat from a low temperature source and release it at a higher temperature and the energy used in the pumping process is added to the output. In a domestic heating system the capital cost is likely to be recovered over 5 to 7 years.

District heating permits the provision of heating services on a townwide scale and where it can be shown to meet the best socio-economic criteria, it should be seriously considered. Large-scale urban renewal, the building of New Towns and large expansions to existing towns provide ideal situations for heating services to be planned as an entity and so achieve economies of scale. The larger the scheme, the more significant will be the reduction in unit cost of the heat supplied. Not only can the heat be supplied to large numbers of dwellings and public, commercial and industrial buildings but it can also be used for ancillary purposes such as the heating of roads, shopping arcades and swimming pools. The burning of town refuse can provide a good source of heat to supplement the district heat supply as at Nottingham, while natural gas is used at Peterborough, and power station exhaust steam was used for heating residential and commercial buildings at Pimlico. In the 1980s the production of fuel from waste, usually in the form of pellets, was undertaken by a number of local authorities in Newcastle, Glasgow, Castle Bromwich, Huyton (Merseyside), Isle of Wight and Hastings, offering environmental advantages (Seeley, 1992).

It is illogical to improve heating standards without at the same time paying adequate regard to thermal insulation. If the correct standard of thermal insulation is built in at the design stage, there may well be an overall saving stemming from the cheaper heating installation; and even if the insulation is added later and at higher cost, the net outlay for a typical three-bedroom house will probably not exceed £2600.

The central heating system needed for an uninsulated house required approximately 20 m^2 of radiator surface and cost about £2300 in 1992 (gas-fired installation), based on an inside temperature of 21°C, and the running costs were estimated at £800 for a heating season of 2250 hours. If the house was fully insulated after construction, as for instance by applying 100 mm of mineral wool to the first floor ceiling, filling the wall cavities with mineral wool, draughtproofing external doors with phosphor bronze weather-strips and double-glazing all windows, then the required radiator surface could be reduced to 11.5 m^2 and the cost of the heating system reduced to about £1700, while the full cost of the insulation, as applied to an existing house, would be about £2600. In addition to the saving on the first cost of the heating system of £600 there would also be a reduction in annual heating costs approaching £300.

Some parts of the insulation work showed a better return than others and if reductions in fuel costs were the sole criteria, then roof insulation would have first priority, followed by insulating cavity walls then draught-proofing and finally double-glazing. It might be deemed advisable to confine the latter work to much used areas such as living rooms.

The comparative costs of basic mechanical engineering services in different types of buildings are given in table 3.8.

Air Conditioning

A wall radiator and an open window are no longer adequate means of heating and ventilating larger buildings. Various means of heat distribution through ceilings are available from water-heated pipes to air circulation systems and use of lighting installations. For instance, with the development of higher levels of illumination and the better thermal insulation of walls, it is possible that most, if not all, the heating needs of a build-

Table 3.8 Comparative costs of basic mechanical engineering services (prices/m² of floor area – 1993)

Type of service	*Secondary schools* £	*University arts and administration buildings* £	*Offices owner occupied* £	*Local authority flats* £	*Local authority houses* £	*Non-teaching hospitals* £
Air conditioning	–	–	140.00	–	–	–
Heating installation	46.00	43.50	–	28.50	19.80	42.00
Hot water services	8.00	9.00	4.00	6.00	4.90	14.50
Cold water services	8.50	14.00	5.20	6.80	5.40	19.10
Ventilation	8.40	–	–	2.20	–	38.00
Fire protection	1.30	2.00	3.50	–	–	2.00
Ancillary services	3.00	–	–	1.50	1.60	26.20
Total approximate cost of above services	75.20	68.50	152.70	45.00	31.70	141.80

ing can be satisfied by the lighting installations. Furthermore, an air conditioned building generally gives greater freedom of planning, particularly in the width of the building and in the positioning of toilets, and thus partially offsetting the extra cost of air conditioning.

In the late nineteen seventies developers in the United Kingdom became less favourably disposed towards air conditioned buildings, mainly because of operational problems, and they ceased to attract higher rental values, but the position changed by the mid 1980s.

The principal conditions generating the need for air conditioning are:

(1) buildings with high heat gains;
(2) buildings that are effectively sealed by the provision of double glazing and other methods;
(3) deep plan buildings with a central core distant from windows;
(4) buildings that have a high density occupancy, such as theatres, cinemas, restaurants and conference rooms;
(5) where close control of temperature or humidity is required, as in computer suites, museums and paper stores;
(6) areas of confined space and precision working, such as operating theatres and laboratories;
(7) buildings in noisy locations where windows cannot be left open;
(8) areas where the exclusion of airborne dust is essential, such as micro-chip assembly and animal houses (Silk and Frazier, 1992).

The choice and design of air treatment systems is influenced substantially by the layout and construction of the building. Consideration should be given to the loss of lettable space occupied by plant and equipment, ranging from 1.63 per cent with VRV systems to 2.95 per cent with VAV systems, in a typical office block with 8000 m² lettable floor area.

The most commonly used systems are:

Fan coil systems which may be fed with a two pipe (heating) or four pipe (heating and cooling) system. The system becomes competitive in capital cost where the units are positioned on an external wall and take in external air directly, but gives no humidity control.

Variable refrigerant volume (VRV) is becoming more popular as running costs are relatively low, but it is generally used for cooling only.

Variable air volume (VAV) adjusts the volume of air at a constant temperature. It is popular because of its adaptability to meet revised partition layouts in office buildings, but has high installation and running costs.

Constant air volume (CAV) can be used where control of individual rooms is not required and

Table 3.9 Comparative air conditioning installation and running costs for a medium sized office building (1993 prices)

Type of air conditioning system	*Installation costs/m² of GFA* £	*Typical running costs for 10 000 m² GFA per annum* £
Two pipe fan coil system with supply air and ventilation system	100–130	7 000
Four pipe fan coil system with supply air and ventilation system	110–150	9 000
Constant air volume system	130–155	8 500
Variable refrigerant volume system with inverter and supply air and ventilation system	140–165	6 500
Variable air volume system	145–180	10 000

areas or zones will run at the same temperature throughout (Silk and Frazier, 1992).

Table 3.9 shows typical installation and running costs relating to the most commonly used systems in a medium sized office of around 10 000 m^2 GFA.

A new development in the early 1990s incorporated an air conditioning installation virtually free from moving parts in its central plant. At the heart of the system is a water tank containing thousands of plastic balls filled with eutectic salts, which change state between solid and liquid at 27°C, absorbing latent heat from the water circulating between the balls or releasing latent heat into it. The system was installed in the 3330 m^2 Dirac House in phase 2 of the St John's Innovation Centre, Cambridge, and is claimed to have the following environmental and economic advantages over conventional technology. No fossil fuel chimneys and flues with emissions into the atmosphere; no cooling towers and smaller heat pumps do not create water drift, thus avoiding the Legionella problem; noise pollution greatly reduced; tank is virtually maintenance free and can be buried underground with manhole access; energy consumption transferred from day to night, allowing advantage of off-peak electricity; and smaller plant room releases lettable floor space within the building.

Lifts and Escalators

Probably the most technically complex element of construction is the lift and escalator package, and a co-ordinated design is vital. Problems are likely to occur if the engineer completes the technical design of the lift at an early stage but the architect leaves the design of the work within and around the lift installation to much later. Junctions of the building fabric and structure with lifts and escalators, including support steelwork and landing interfaces, can produce areas where the contractual and cost responsibilities are uncertain.

Lifts are generally enclosed in shafts but may be exposed to create a certain architectural effect, as in an atrium or on an external facade.

Vertical transportation should be considered essential in modern commercial developments with more than three levels above ground. Certain low rise developments, such as nursing homes, hospitals and retail centres, are dependent on lift and escalator installations for their efficient and practical operation, while wheelchair lifts are needed in many building types (Gardiner and Theobold, 1992).

With high rise residential blocks, both the height of the block and the floor layout have a major influence on the costs of lifts per dwelling. A passenger lift is usually provided in blocks of flats exceeding three storeys and in blocks of maisonettes exceeding four storeys, while in blocks over six storeys in height a second lift is also provided. Considering the economics of lift provision alone would dictate layouts incorporating at least six dwellings on each floor of a block. In fact the tender price for a lift installation varies with the height of the block, the number of floors served, the size and speed of the lift and the form of control.

The most common lift drive systems are:

(1) *Hydraulic* with a maximum economical travel distance of approximately 20 m. They are limited to a relatively low maximum number of starts per hour because of large heat gains, and

their rather low speed renders them unsuitable for heavy traffic. The pump room has to be within about 15 m of the shaft.

(2) *Traction* whereby an electric motor moves the lift car by means of cables. The motor room can be in any location adjacent to the shaft but extra cost will be incurred if it is not directly above, and if it is not at roof level ventilation or cooling will probably be required. A counterweight, operating within the lift shaft, permits the motor capacity to be reduced to a minimum. Traction lift is suitable for most applications and is generally installed where traffic is heavy and in taller buildings requiring higher lift car speeds.

There are two basic types of traction drive systems:

(1) Geared: operating at speeds of up to 2.0 m/s and usually installed in buildings of up to 11 floors.

(2) Gearless: operating at speeds of 2.5 m/s and above and normally installed in buildings of more than 11 floors.

Other types of drive systems include worm drive and scissor action, for use where heavy weights have to be moved over short distances, as with lorry lifts. Most modern lifts are computer controlled with the more elaborate systems controlling groups of lifts and recording many performance aspects and with the ability to 'learn' the actual traffic habits of a particular building. A lift installation in a typical city centre commercial development will use 1.5 to 3 per cent of the GFA (Gardiner and Theobold, 1992).

An escalator can transport many more times the number of people than a lift but will be several times more expensive in capital cost and space requirements and is not usually cost effective. For this reason they are generally restricted to low rise buildings where there is extensive open, common space.

It is customary to describe the scope and quality of an installation by its performance as shown in the following criteria:

(1) *Capacity* is the maximum carrying capacity of the lift car, quoted in number of passengers carried or the weight in kilograms. The standard capacities for passenger lifts are: 8 persons: 630 kg; 10 persons: 800 kg; 13 persons: 1000 kg; 16 persons: 1250 kg; 21 persons: 1600 kg; and 24 persons: 1800 kg.

(2) *Speed* being the maximum speed the lift car attains in m/s.

(3) *Stops* being the number of landings the lift car serves.

(4) *Travel* is the distance the lift car travels from the top of the shaft to the bottom.

(5) *Waiting time* classified as excellent <30 s; good 30–35 s; average 35–40 s; and poor >40 s.

(6) *Handling capacity* being the maximum proportion of a building's population that can be moved in a five minute period in the up-peak mode, and in an office development is usually specified as: 17 per cent single tenancy and 12 per cent multiple tenancy (Gardiner and Theobold, 1992).

Possible over-specification can be compounded by developers who sometimes assume that an extravagant passenger lift installation is essential to the success of a project. Such installations tend to be bespoke design and therefore lose the cost benefit of using model or pre-engineered lifts. For a given performance, quality and scope the cost differential between a bespoke and model installation could be more than 40 per cent and thus confirming the criticism in the Business Round Table Report (1994), when comparing UK and mainland Europe construction costs. In fact, when a detailed technical specification is supplied, tenderers are likely to submit a generally compliant bid supplemented by an alternative, and probably substantially cheaper offer.

To overcome these problems, in 1992 the National Association of Lift Manufacturers (NALM) published three documents aimed at simplifying design, procurement and installation (NALM, 1992a–c). This initiative was in response to the mounting criticism of their past poor performance coupled with the NEDO report (1990) dealing with programming and cost problems. The NALM approach should result in a generally accepted basis for specifying the installation in purely performance terms with the detailed design undertaken by the manufacturer.

The project team may be compelled to procure the lift package 12 months or more in advance of the logical design programme. Thus in some instances the design of the lifts is being processed concurrently with the piling and substructure, to ensure satisfactory space and structural provision. NALM (1992a) gives some useful guidance on this aspect.

Cost Aspects

For commercial developments in London, with a minimum floor area of 10 000 m^2 GFA, the following typical costs could apply:

Medium quality speculative: £32–£40/m^2
High quality speculative: £38–£48/m^2
Owner occupier: £43–£86/m^2.

These rates should be increased for developments of less than 10 000 m^2 GFA and decreased for provincial locations. The lift installation should be between 2 and 4 per cent of the total construction budget. Some of the design features that may be examined with a view to maximising value or reducing cost, include:

(1) *Internal lift car height*: reduce if greater than 2.2–2.3 m, producing a saving per lift of £2500 to £5000.

(2) *Landing/car door height*: reduce if greater than 2.1–2.2 m, saving £300 to £600 per car and each landing door, subject to fire certification.

(3) *Landing and car doors*: if in glass, consider stainless steel which gives better lift performance and saves approximately £500 per car and £200 to £500 per landing.

(4) *Lift car speed*: assess performance for reducing the car speed to the next lowest standard. Saving accruing by reducing from 2.5 m/s to 2.0 m/s (gearless to geared) will be approximately £20 000 per lift. By reducing from 2.0 m/s to 1.6 m/s (both geared), the saving is approximately £4000.

(5) *Landing and car door widths*: reduce if beyond manufacturer's standard range; by reducing from 1.2 m to 1.1 m will save £250 to £500 per car and each landing.

(6) *Goods lift*: consider deleting completely and allow for a top hat section and drapes to a passenger lift. The cost of the goods lift is thereby saved partially offset by the addition of approximately £2500 to the passenger lift. There should also be an increase in the net lettable floor area.

(7) *Number of lifts*: consider whether the total number of high rise lifts can be reduced by terminating the lifts serving below ground at ground level and installing a separate service to the basement.

(8) *Car and landing door finishes*: keep weight to a minimum and avoid complex designs with unusual materials; NALM (1992b) gives some useful guidance (Gardiner and Theobold, 1992).

A single 10 person lift serving six levels with directional collective controls and a speed of 1 m/s cost about £44 000 in 1993, and this increased to about £55 000 for a 21 person lift with a speed of 1.6 m/s. In blocks served by two lifts, an economy can be achieved by arranging for each lift to stop at alternate floors above the ground floor, so that each floor is served by one lift only. Faster lifts require the use of either two-speed or variable voltage motors to provide smoother acceleration and deceleration, thereby substantially increasing capital, maintenance and running costs. A 2.5 m/s lift costs over £25 000 more than a 1 m/s lift.

A 30° pitch escalator with a rise of 3.50 m and enamelled sheet steel or glass balustrades cost about £31 000 to £35 000 in 1993.

Mechanical installation accounts for fifty to sixty per cent of the total cost, the building work for ten to fifteen per cent, and the running and maintenance costs for thirty to thirty-five per cent. Goods lifts serving five levels, taking a 500 kg load at a speed of 0.25 m/s cost about £28 000 each in 1993, increasing to £33 000 for heavy duty goods lifts taking a load of 1500 kg. Annual maintenance of a 4 or 5 floor lift cost about £1600 in 1992, rising to around £4000 for a 30 floor lift while an internal escalator cost about £1300 per annum, and an external escalator double that figure.

Refuse Disposal

The following range of refuse disposal systems is currently available.

(1) Individual storage containers usually in the form of metal, rubber or plastics dustbins, with loose or captive lids or alternatively plastics or paper sacks, each with a capacity of up to $0.9\,m^3$. Since the mid 1980s the wheeled bin has been used increasingly in the UK, with a capacity of 2½ times the traditional dustbin, with smaller bins for elderly persons living on their own. These are the cheapest methods with total annual equivalent costs of provision and maintenance of containers and refuse collection and disposal at about £28.00 (1993 prices).

(2) Communal storage containers without chutes, with an individual capacity of up to $1\,m^3$, and emptied mechanically by special collecting vehicles. This offers some improvement over the traditional dustbin but is still not entirely satisfactory, as the containers need to be housed in close proximity to the dwellings they serve. Total annual equivalent costs are likely to be about £31.00 (1993 prices).

(3) Communal storage containers with chutes within 30 m of each dwelling which enable occupiers to drop refuse from upper levels into movable containers. This is a satisfactory method for use in blocks of flats and maisonettes and the comparable 1993 annual equivalent costs are likely to be about £42.00.

(4) Waterborne disposal system whereby part of the refuse is passed through a special sink unit into the sewerage system. Apart from the cost of renewal of fittings, the 1993 total annual equivalent cost would probably be in excess of £100 making this the most expensive arrangement.

(5) Other and more recently introduced systems include sack compression, container compression, onsite incineration and pneumatic conveying.

AUTOMATION

Building services became increasingly sophisticated in the 1980s and early 1990s and are very expensive on the larger projects, thereby creating the need for rigorous cost control and identification of methods of reducing costs as described for lifts. With new commercial buildings, after the costs of mechanical and electrical services, security, fire precautions and essential information technology are assessed, the spaces needed to accommodate them included and arrangements made to allow for future services, services can amount to half the cost of the building.

Modern technology enables buildings to look after themselves to a significant extent. Sensors, which can range from infra-red collectors and heat-seeking devices to humidistats and light-activated switches, can enable a single person to operate a building.

It is believed that in time most office workers will be provided with credit sized identity badges which will give them access to the building, monitor movements, put through telephone calls and turn the air conditioning, lights and heating on and off automatically when people enter or leave a room. The remote sensors are cheaper to install and operate than a traditional electrical lighting system which includes manually operated switches and socket outlets. Furthermore, it is estimated that they could produce savings of 15 per cent over a three year period.

In the event of an emergency such as a fire or bomb scare in a building containing several hundred people, they could be warned through the personal equipment. From the building's master control console the position and identity of members of staff and visitors can be logged, whereby when the fire brigade arrives they know the exact location of persons trapped in the building.

All this technological equipment has to be accommodated, which requires the provision of suspended ceilings, raised floors, ducts and control rooms which until recently did not feature in commercial building design. Wiring systems need to be incorporated in the building design, preferably with the specialist contractor involved at the design stage. A typical example of the rapid development of communications technology is the advent of the personal communication network (PCN).

The infrastructure of the new systems will consist of a large number of small 'cells' usually located on the roofs of tall buildings. Ducts will be needed to convey cables to equipment of widely varying sizes. It may be necessary to allow for point loads on roofs, and planning difficulties arising from masts and satellite dishes can be resolved by cladding with high level enclosures, which are designed to make them appear as part of the building rather than an appendage, using materials which do not interfere with the passage of radio waves (Catt, 1991b).

FIRE PROTECTION

Almost daily severe damage and often loss of life is caused by serious fires to buildings and hence it is vital to provide adequate fire protection to all buildings. BRE digest 367 (1991) describes the concept of computer modelling of fires, which provides those involved in building control and design with much enhanced facilities for predicting the consequences of fire and the planning of fire protection strategies.

When smoke and hot gases are trapped inside a building they not only threaten the safety of the occupants but also cause 'flashover', a rapid acceleration in the build-up of the fire. The way in which a fire continues to develop and spread into adjoining areas depends on the strength of the fire and the characteristics of the enclosing structure.

The theoretical computer models present a wide range of conditions to an advanced stage and they can be used with confidence to predict how the hot smoke and gases produced by a fire will be dispersed throughout a building. The theoretical models fall into two categories:

(1) *Zone modelling*: The computerised techniques are closely related to well established traditional methods for dealing with smoke movement, as described in BRE digest 260 (1987).

(2) *Field modelling*: This technique uses new methods of 'computational fluid dynamics' (CFD) to deduce the rate at which smoke will fill an enclosure.

The use of these models can lead to considerable cost savings if applied sufficiently early in the design process, particularly for large, complicated or unusual structures. In addition new or improved passive and active protection products ranging from automatic smoke blinds, intumescent seals and protective coatings, fire linings and sprayed coatings, fire barriers, automatic ventilation systems, smoke and heat detectors and fully integrated intelligent fire alarm and detection systems should be installed where appropriate.

Much of this new technology may not be applicable to old buildings, which can cost far more to rebuild than their modern equivalents. For example, BCIS considered that it cost between 43 and 47 per cent more to rebuild houses built before 1920 than those constructed between 1946 and 1979.

After the disastrous fires at Hampton Court, York Minster and Uppark House, a working party drawn from 23 organisations, including the RICS, concluded that one of the main causes of fires in historic buildings was carelessness on the part of contractors on the site. Therefore preliminaries in contracts for refurbishment or repair work on old buildings should establish safe working procedures, such as banning blowlamps, removing combustible materials from site, providing fire extinguishers and supervising the work properly with fire prevention in mind. The adoption of these measures at Windsor Castle would in all probability have prevented the fire at St George's Chapel, with the consequent serious damage to a famous national monument and the lengthy and complicated restoration work which was estimated at £60 m in 1992.

Rebuilding is also likely to need statutory consent which is likely to prolong the pre-contract period and add to the cost as well as allowing further deterioration of the valuable fabric. Furthermore, English Heritage often considers that if more than one half of a building is destroyed by fire it is preferable to demolish the remainder and start afresh (Catt, 1992a).

EXTERNAL WORKS

It is interesting to break down the cost of site-works on a housing contract to see the relative values of the various works, although it is appreciated that their distribution will vary from site to site. The average 1993 sitework costs for a typical five-person house were as follows:

Site preparation	£550
Retaining walls	150
Screen walls and fencing	850
Paved areas	450
Drainage	1100
External services	300
Landscaping	60
	£3460

External works do therefore form a significant part of total building costs and justify taking steps to reduce costs by reducing the amount of earth-work and retaining walls, restricting paved areas to a minimum, reducing pipe runs of drains and other services and seeking materials and components which perform their functions satisfactorily and at the same time show favourable costs in use figures.

On drainage work, cast iron drains are up to four times as expensive as clay drain pipes, whilst PVC-U may show a saving on clay depending on the lengths of runs and number of fittings involved. Savings in cost can often be obtained by building shallower manholes, not exceeding 1 m deep, in half brick walls in class B engineering bricks, and by omitting the spread of concrete bases beyond the manhole walls and thereby reducing the amount of concrete, excavation and fill. On occasions it may be possible to reduce drainage costs by laying both foul and surface water drains in the same trench but with 300 mm between their inverts, and by using combined manholes. Another alternative is to lay the sewers alongside houses and so dispense with long lengths of house drain. The use of drop manholes can effect considerable reductions in drain trench excavation costs on a sloping site.

There are wide variations in the cost of different forms of paving. Probably the cheapest in first cost is gravel paving (about 50 mm thick) on a bed of hardcore (usually about 100 mm in thickness). Bituminous macadam, 50 mm thick, on a 100 mm bed of hardcore would be about three times more expensive, whilst 50 mm precast concrete paving slabs on a 100 mm bed of ashes would be approximately four times more expensive than the gravel paving. A concrete carriageway is likely to be about twenty-five per cent more expensive than a bitumen macadam one. Once again it is important to consider maintenance costs in addition to first costs in order to make a really meaningful comparison, and this aspect will be further investigated in chapter 13.

Similarly with fencing there are wide cost ranges. One of the cheapest forms of fencing is post and wire, with cleft chestnut paling about one-half times more expensive than post and wire. Chain-link fencing in its turn is over twice as expensive as chestnut paling in first cost and close-boarded fencing is likely to be more expensive than chain-link depending on its construction. Apart from maintenance costs, other aspects need to be considered including the purpose which the fencing is to serve and its appearance, and these may have more influence than cost on the choice of fencing.

4 INFLUENCE OF SITE AND MARKET CONDITIONS AND ECONOMICS OF PREFABRICATION, INDUSTRIALISED AND SYSTEM BUILDING

This chapter explores the effect of site and market conditions on building costs and the way in which they account for variations in the price of similar type buildings erected in different locations. The origins and forms of prefabrication and industrialised and system building are examined and the economics of these processes critically investigated.

EFFECT OF SITE CONDITIONS ON BUILDING COSTS

Each site has its own peculiar characteristics which can have a considerable influence on the total cost of development. Some of the more important site factors are now examined.

Location of Site

The cost of building on a site in London could be as much as thirty per cent more expensive than erecting a similar building on a provincial site, due to higher wages, more costly materials and other operational costs. Some parts of the country are subject to higher rainfall than others and this can lead to greater loss of working time. Even within the same region the costs of operating on different sites can vary tremendously. For instance, a project on a remote country site may involve long lengths of temporary access road and of temporary power cable for electricity supplies and increased costs of transporting operatives and materials to the site. By way of comparison a site in a congested urban area can give rise to major problems in delivery and storage of materials and components, protection of adjoining buildings and the public, and restrictions on the use of mechanical plant. Overcoming these problems involves considerable additional costs. Furthermore, a very exposed site may make working conditions more difficult and costly and some locations may be more vulnerable to vandalism and theft and so require more costly security measures.

Demolition and Site Clearance

One site may be clear of all obstructions, whilst another may contain substantial buildings requiring demolition, heavy foundation and plant bases requiring removal, extensive paved areas which need breaking up and a number of large trees which require felling, together with the grubbing up and disposal of their roots. Some sites could contain acids, sulphates, heavy metals and other contaminants, possible emission of methane or radon entailing costly protection work, or unstable fill necessitating expensive foundations.

Contours

Few sites are entirely level and the more steeply sloping the site, the greater will be the cost of foundations and earthworks generally. The stepping of strip foundations increases their costs. Most buildings require constant floor levels and this will involve considerable excavation and fill on a sloping site. It is cheaper to form a sloping bank than to construct a retaining wall, but this may have to be balanced against space considerations. A basement boiler house ought ideally to be located in an open area to reduce excavation and tanking costs.

Ground Conditions

Where the strata is of low loadbearing capacity, it may be necessary to introduce piled or other more expensive types of foundations. Raft foundations on made-up ground or in areas liable to mining or other subsidence may be three times as expensive as normal strip foundations, whereas piled foundations used to transmit loads to a deeper load-bearing strata could be as much as five times as expensive. In these circumstances it might be advisable to consider increasing the number of storeys in the building to take fuller advantage of the higher loadbearing capacity of the piles. The cost of excavation in rock could be five to eight times as expensive as working in normal ground. Indeed bad ground conditions could conceivably increase overall building costs by as much as five per cent. The probable length of haul in the disposal of surplus soil also needs consideration.

Where groundwater level is close to the surface of the site, costly pumping operations may be needed throughout the substructural work. A wet site may also involve raising temporary sheds and offices on brick bases and more costly temporary roads. It is necessary to examine trial holes and to note the depth of topsoil, nature of subsoil, and evidence of groundwater or standing surface water. The type of strata will also influence the form and extent of support that will be needed to the sides of excavations. The information required from a subsoil investigation and the manner in which it should be presented is well described in BS 5930. With shrinkable clay subsoils it is necessary to take the foundations at least 1 m below ground level and to keep adequate distances away from trees according to their species. Other problems can arise through the compressibility of soils, such as alluvial deposits, glacial lake clays and peat, solution in soluble rocks and landslipping in glacial deposits, coal measures, Wealden clays, etc.

Planning and Building Regulations

The ratio of site value to cost of building should always be considered. In expensive central locations the aim must generally be to secure the most profitable permitted use, and coupled with this is the desirability of obtaining maximum site utilisation. The operation of planning controls through plot ratios and floor space indices is very relevant and will be further considered in chapter 14. The shape and size of the building, as well as probable cost, are affected by height restrictions, building and road improvement lines, parking requirements, light and air restrictions, landscaping conditions, access requirements and similar considerations. In like manner the probable impact of Building Regulations must always be borne in mind when working from sketch or preliminary drawings, as they can have far-reaching consequences on building costs.

Services

The position and capacity of existing services such as sewers, water mains, electricity cables, gas mains and telephone services are other important influences on site costs. Connection to a public sewer 6 m deep on the far side of a busy dual carriageway can be a costly item of work and the need to install pumping plant to drain a low-lying site has repercussions in both initial and future costs. The cost implications of combined, separate and partially separate systems of sewerage must be appreciated. On occasions it is necessary to divert existing services which cross sites, to improve and regrade watercourses which adjoin sites, and to obtain pipe easements to lay essential services across land in other ownerships.

Availability of Labour, Materials and Plant

The Chartered Institute of Building (1983) has drawn attention to the need for builders to consider the labour situation in the area of the contract and the availability of materials when preparing tenders. Investigations into the availability of materials would include local sand, gravel and ballast pits and brickworks. Contractors will have to make several important decisions at the tendering stage, each of which will influence costs.

(1) Whether all labour requirements can be met from within the organisation or whether it will be necessary to recruit for the project.

(2) Whether mechanical and non-mechanical plant already owned by the organisation is suitable and likely to be available or whether it will be necessary to purchase or hire for the project.

(3) Whether it will be desirable to sublet to specialists certain aspects of the work such as earthworks, formwork, scaffolding, structural metalwork, and mechanical and electrical services.

Other Factors Influencing Cost

A contractor can influence the cost of a project by his selection of constructional methods and by adjusting these methods to increase the effectiveness of the resources used. The free choice of method is however constrained by the design of the building, the availability of the numbers and types of resources needed for each method and by the relative cost of employing each one of these sets of resources. Another important cost aspect is the quantity of materials wasted on building sites, and the cost of wastage of materials is sometimes in excess of the allowances made by estimators and, in some cases, the total wastage of a material could have been halved by effective planning and supervision.

The existence of easements can also affect the design, construction and cost of the work and these are considered in more detail in chapter 14.

USE OF PLANT

With the continual rise in labour costs, both direct and indirect, many contractors are making greater use of plant. This requires an intimate knowledge of all types of plant, when each can be used most profitably and of the fullest possible use of plant. The latter point is significant as the use of machinery on building sites rarely exceeds seventy per cent of its working capacity. Cranes are being used increasingly on building sites resulting in higher productivity. A study in Singapore, where many high rise buildings are under construction at any one time, showed that significant numbers of tower cranes were purchased annually. The Singapore Government in the early 1980s started offering accelerated depreciation (3 years) and cheap loan financing as an incentive to greater mechanisation of building operations. The plant purchased included concrete pumps, batching plants, tower cranes, crawler cranes, material and passenger hoists, formwork systems and rough terrain forklifts.

A contractor has sometimes to choose between hiring or purchasing plant, and the decision will be influenced considerably by the likely future demand for the particular item of plant. Owning plant offers a number of advantages:

(1) plant is readily available at all times;

(2) plant may be retained on a particular site if circumstances make this desirable;

(3) plant can be transferred from one site to another without great difficulty; and

(4) in emergency situations machines can be taken off less important work as the contractor has complete control of the plant.

Nevertheless, plant hire also serves a valuable function in that it offers a wide variety of plant types to the contractor free from the liabilities attached to the purchase of plant, particularly when a contractor is short of capital. Hiring plant often ensures maximum economy with full plant utilisation and is an aid to quicker building.

The main factors to be considered when evaluating the economics of buying or hiring plant are:

(1) forecasts of commitments to assess plant requirements;

(2) availability of workshop facilities for servicing plant;

(3) length of time for which plant will be required; if less than 60 per cent utilisation is likely to be achieved, then it is better to hire than purchase;

(4) adequacy of capital available for purchase;
(5) availability of personnel for controlling and operating plant holdings; and
(6) cost of transporting plant to sites.

Costs of owned plant are made up of a wide range of items – capital costs, interest on capital, depreciation, licences and insurance, overheads, maintenance, repairs and replacements, haulage to sites, fuel and operating costs. The total annual costs have to be spread over the period of effective use to give an inclusive hourly or daily cost.

There are three principal methods of charging for plant:

(1) Percentage of the contract price, whereby total plant costs are allocated to each contract in proportion to the overall cost of the project. Its main advantage is one of simplicity of calculation but its disadvantages include lack of incentive to obtain maximum use of plant, lack of comparative information for preparing tenders and costing, and lack of information to use as a check on plant operations as a whole.

(2) Direct cost to contract, where the size and nature of the project and expected use of the plant justifies charging an item of plant completely to the contract. This method is also easy to apply but runs into difficulties if the plant is used on other contracts.

(3) Hire charge is the most usual method whereby the plant is charged to the site by the unit of time employed. The primary advantages are ease of control over economic use of plant, adequate information available for estimating purposes and for providing a check on plant operations.

In all cases an efficient costing system is necessary to ensure that realistic rates are charged for use of plant.

SITE PRODUCTIVITY

There is a clearly identifiable need to increase site productivity and so reduce the cost of building work. Various studies have also shown the need to re-appraise design and contract arrangements, particularly when compared with practice and achievements in the United States, and this has led to the more innovative contractual arrangements described in chapter 1. The Business Round Table Report (1994) also showed that lessons could be learnt from a study of procedures in mainland Europe.

Design and Contractual Procedures

With negotiated contracts, the contractor can be appointed at the outset as a member of the design/construct team. He can advise on many interrelated design aspects which have plant, component, material and labour implications, with a view to achieving the optimum solutions. Hence designs can be progressively developed, using cost planning techniques to ensure optimum value for money within the cost limit, taking into account the contractor's known comparable costs of alternative methods of construction. Other forms of contract offering similar benefits include design and build, and design and manage, while management contracting and construction management have much to offer in the way of improved productivity and cost efficiency in certain situations.

A survey of Scottish public sector housing carried out between 1974 and 1977 by the Building Research Establishment, showed labour requirements of new houses giving a wide variation between 800 and 2600 manhours, with an average of 1584 for traditional constructions and 1139 for non-traditional. Projects were initiated at Glenrothes New Town to rationalise design and reduce on-site manhours, adopting the following measures:

(1) secure the same building sequence in each house as far as practicable;
(2) the building sequence to consist of fewer and larger packages;
(3) a high degree of standardisation to be sought in the detailed construction of the houses;
(4) the construction was simplified, using traditional, readily available materials and components;
(5) designs were dimensionally co-ordinated

to reduce the cutting of materials to waste and to improve the fit of components.

In the 1980s many clients believed that the construction industry failed to offer value for money, building projects were considered to cost too much, take too long to complete and were too prone to failure, and to rapid failure at that.

A survey by the Building Research Establishment with the National Building Agency (NBA) of low rise traditional housing schemes with defects in England found that 50 per cent arose from faulty design, 42 per cent through faulty execution and 8 per cent from materials (BRE, 1982).

An examination of American performance in 1979 showed that their industry was capable of building faster and more cheaply than that of the UK (Flanagan, 1979). It was less dependent on the public sector (about 20 per cent), with its 'stop–go' policies and had, in consequence, greater stability and confidence. The client gives a more comprehensive and firmer brief and there is often a cost penalty for subsequent changes in design once work has started. The American architect is paid fees, often between 5 and 6 per cent, for a complete design service, including all engineering work. Where the design exceeds the budget figure the architect usually has to design at his own cost. In the UK, the client's frequent failure to provide precise or timely instructions often stems from his lack of awareness of the true cost of variations and delays, and sometimes consultants fail to impose sufficient commercial discipline in considering their client's requests.

American architects produce 85 to 90 per cent of the drawings at tender stage and variation orders account on average for only 1 to 1½ per cent of the total cost. There is a strong financial incentive for the contractor to build at the fastest possible rate, as he is normally paid stage payments, often on a seven day basis. Final accounts are usually settled within four to five weeks of practical completion. There is a much greater use of standard details, and specialist subcontractors generally provide a more sophisticated precontract service. Good stocks of products are held by manufacturers so there is seldom a problem to obtain delivery of materials at short notice.

In the United States the design team, with the aid of construction management, especially on major contracts, endeavour to reduce site labour costs and off site fabrication is used extensively.

In the larger American architectural practices cost information is stored on computers and is used in costing alternative designs, as well as programming and more simple unit costing.

American building contracts are let in a variety of ways. Lump sum bids with no fluctuations are obtained on the majority of small and medium-sized contracts where complete drawings and specifications are provided. The contractor is carefully chosen on his reputation and past performance and often in limited competition for a fee which is applied to the total anticipated value of the project. The contractor's fee for overheads and profit can be in the order of 2 to 4 per cent.

Methods of Accelerating Site Productivity

With the larger American commercial buildings, speed of construction has been accelerated by using standard structural steel sections with a sprayed finish for fire protection, an almost total omission of wet trades and the use of pressed aluminium cladding units fixed in two-storey units without external scaffolding. In a steel framed building, metal floor decking is invariably used as permanent shuttering. Rapid progress with services is achieved by the use of prefabricated ductwork and pipes which follow the structure two or three floors behind and ahead of all internal partitioning.

In Singapore concrete pumps were used to accelerate the placing of concrete at the Changi International Airport, achieving rates of flow up to 26 m^3/h. Use of similar equipment on a 10-storey building showed a 35 per cent saving in time and 22 per cent saving in labour for the transporting and placing of concrete, compared with the use of a tower crane.

A valuable development in the UK was the adaptation to building of the civil engineering technique of slip forming (vertical continuous climbing shuttering). Adaptations were necessary to accommodate window and door openings and other features and there were certain limitations on its use. It did however save a significant amount of construction time. Another innovation – the 'lift slab' technique – also saved time and produced other advantages such as eliminating soffit shuttering, installing services in slabs before pouring concrete, reducing the vertical handling of materials and components, and reducing interference by bad weather (Seeley, 1984b).

The output of on-site labour can be improved by better training techniques and the use of realistic incentive payments related to performance, preferably calculated on a gang basis. Improved organisational and supervisory arrangements will often result in higher productivity, coupled with greater use of specialisation, planning and programming techniques. There is a need to continually monitor and re-appraise site methods with adequate feedback of data from the site.

The following principles should be observed when formulating an incentive scheme:

(1) the scheme must be fair to all parties;
(2) the bonus earned should be closely related to the effort expended;
(3) the standard of performance required to secure bonus payment must be realistic and attainable by the average operative;
(4) bonuses should not be affected by matters outside the operative's control;
(5) bonus payments should be made as soon as possible after the work has been performed;
(6) the scheme must be relatively permanent – withdrawing a scheme because of temporary adverse economic conditions is damaging to industrial relations;
(7) operatives should be able to calculate their own bonuses;
(8) the scheme must be approved by trade unions, operatives and supervisors;
(9) there must be a suitable grievance procedure (Seeley, 1987).

Productivity cannot, however, be considered in isolation from quality control and safety aspects, both of which must be maintained. Quality control is concerned with defining the levels of acceptable imperfection, or tolerances, and in ensuring that minimum standards are achieved. The existence of a quality problem was evidenced by the National House-Building Council (NHBC) paying out £6m in 1982 on about 4000 claims. There was also an unknown number of cases where shoddy work was remedied by the builder to protect his good name. Since then there has been a growing national commitment to quality assurance with BS 5750: *Quality Systems* being updated and being increasingly applied throughout the construction industry. Quality assurance through a formal quality management system provides the opportunity to review, improve and co-ordinate the systems in operation. This improved quality of operation can help to reduce costs and can identify areas that require concentration of effort to improve quality. The work involved in correcting deficiencies in quality can be expensive in both time and cost, particularly when problems are discovered at a late stage. BS 5750 was subsequently complemented by BS 7750 dealing with environmental aspects which have become increasingly important.

The Safety Representatives and Safety Committees Regulations 1977 aim to ensure that personnel shall not be exposed to hazards of which they have no knowledge, or be instructed to work in potentially dangerous conditions over which they have no control. The Health and Safety at Work Act 1974 required employers to consult with employees' legally appointed safety representatives to ensure the health and safety of workers and to check on the effectiveness of the arrangements made. These measures were further reinforced by the introduction of the Construction, Design and Management (CDM) Regulations in 1995.

Wastage of Materials

It is universally recognised that waste of materials occurs on building sites and that some losses can be avoided. Most quantity surveyors also accept that the 'norms' used in estimating to allow for waste are nominal figures. Studies by the Building Research Establishment showed that average losses of the principal building materials were higher than the 'norms' used to allow for them in practice, and that there was considerable variability in waste levels between apparently similar sites. For example, overall waste of bricks and blocks on a site are often in the 8 to 12 per cent range, as compared with the estimator's customary allowance of 4 to 5 per cent. It seems likely therefore that many contractors could make significant reductions in waste, and in the 1980s contractors were paying more attention to waste reduction.

Waste can be classified under two broad headings: direct waste (a total loss of materials which can occur every time materials are handled, moved, stacked or stored) and indirect waste (a monetary loss related to materials and sometimes to the way they are measured, and can arise from substitution, production and operational loss and negligence). These aspects highlight the need for designers to pay more attention to ensure that dimensions fit sizes of materials available on the market, for estimators to take a fresh look at estimating practice, and for site management to exercise more effective materials control (Seeley, 1987).

The contractor's quantity surveyor can minimise loss or ensure that proper reimbursement is obtained for waste which is not the responsibility of the contractor, in the following ways:

(1) establish a regular system for waste accounting and advise on methods to reduce excessive losses;

(2) recognise that some materials descriptions in bills of quantities will result in a number of the materials failing to satisfy the requirements. For example, bricks to BS 3921 with true and regular perpends cannot be met without rejecting some over and under-sized bricks;

(3) recognise that dimensions of members indicated on drawings or in the bill may later be interpreted as minimum dimensions, and that to guarantee these, larger dimensions must be used resulting in unavoidable production waste. A typical example is an *in situ* concrete ground floor slab, where unevenness of the base results in variations in slab thickness.

MARKET CONSIDERATIONS

The level of prices submitted for a particular project may reflect the market situation at that point in time. This is governed by the volume of work in progress in the area and the relative keenness of building prices. The situation can change quite dramatically over a comparatively short period of time.

In addition many unforeseeable and external factors can influence building costs, such as national and local shortages of labour and/or materials, a credit squeeze, abnormal rainfall resulting in a sharp rise in groundwater levels, and sudden increases in the price of building materials or components or fuel. Other factors bearing upon costs include time for completion and special requirements of the building owner, such as phased completion of various sections of the work. Preliminary items may account for up to 14 per cent of the tender sum in boom periods and this may also indicate that contractors are somewhat uncertain about the requirements of the contract at the tendering stage. BCIS reported that the average allowance had dropped to 8 per cent in the depression of 1992. Tender prices can also be influenced by the way in which the contract documents are prepared and the amount and adequacy of the information which is supplied to the contractor at the tendering stage. One contractor described to the author how he put a price on some architects to meet the extra cost arising from their general lack of attention to detail and the consequent delays and disruption of the work. The method of tendering can also affect the price of a project and negotiated contracts are often more expensive than competitive tenders, but factors other than cost may influence a decision on the method to be adopted.

Many cost aspects may need considering in the feasibility studies of projects. A common problem is to assess the relative merits of converting and modernising existing buildings as against building new ones. Probable trends in prices can be an important factor. Cost limits of development projects may be set by the rents obtainable and an important aim may be to maximise the areas available for use by occupants. The periods needed for both pre-contract and construction stages may be of considerable importance to the building client, as for example where a new factory is urgently required to enable large export orders to be met on a tight time schedule.

COST IMPLICATIONS OF PREFABRICATION AND STANDARDISATION

Increased mechanisation of building work can speed up production and frequently results in reduced costs of construction. Power tools such as power floats and power saws and the shot-firing of components are suitable for use on small sites, whilst cranes and forklift trucks are better suited to larger sites. On all types of site the extensive use of prefabricated components eliminates much of the cutting on site and many of the time-consuming wet trades. Hence manufactured joinery has largely superseded site-produced joinery, plasterboard often replaces wet plaster, and there are few building projects which do not incorporate some precast concrete units as part of the structure. These units offer distinct advantages through economy in site work and independence from weather conditions, as well as the high quality and often superior finish of concrete units produced under factory conditions. Dense and lightweight, solid and hollow concrete load-bearing blocks, some with thermal insulating properties, are now being mass produced in fully automated plants often with an output of 20 000 units per eight-hour shift.

Advantages of Prefabrication

Most building processes can be better accomplished in the workshop under superior working conditions to those on the site. The ideal solution would be to produce the complete building with all its ancillary services in the workshop, independent of weather and time of year, under the best possible environmental and working conditions, and subsequently to transport the work complete to the site. Unfortunately physical and practical difficulties generally prevent this ideal from being achieved. Hence it is usually necessary to subdivide the work into construction components or elements which can be prefabricated in the workshop and then transported to the site.

Maximum efficiency and economy can be achieved by mass production methods in factories aimed at producing large outputs of selected, standardised, dimensionally co-ordinated and interchangeable components suited to a range of building types provided that there is a ready and continuing market for the products.

In 1992 the 168 bedroom, four star Hilton National Hotel at Purley Way, Croydon was completed. The £17 m development with a GFA of 10 000 m^2 was built to programme in 65 weeks, using a tunnel form structure and bathroom pods, costing no more than traditional bathrooms. A steel shuttering system, similar to that used in many mainland Europe hotel and apartment blocks, allowed the casting of *in situ* concrete floors and walls in one pour.

The advantages of using tunnel form included:

(1) savings in construction time permitting the the bathrooms to be installed at a very early stage of construction and removing the bedroom structure from the critical path, with the pods built, finished and fully equipped at the factory, and easily handled and installed on site;
(2) allowing the direct application of finishes to the walls and ceilings and carpeting to floors, thereby reducing wet trades for bedrooms;
(3) eliminating external independent access scaffolding during concreting;
(4) reducing sound transfer between rooms;
(5) permitting lower floor to ceiling heights (Sisk, 1993).

The installation of 514 en-suite bathroom pods, which were fully factory fitted including all mains services, fittings and finishings, proved to be a great advantage in the construction of student accommodation at the University College of North Wales in Bangor in 1993/94. The use of the pods added no direct cost to the project and produced considerable time savings and reduced the notorious snagging problems associated with en-suite bathrooms of traditional construction.

Standardisation

Optimum standardisation is essential for component manufacturers to obtain larger series runs to balance turnover against capital, minimise time taken in changing over machines, reduce production costs by the bulk purchase of materials, and attain better quality production of fewer varieties of component types with fast-operating cycles related to a large and sustained volume of demand. Standardisation of storey height permits the production of ranges of standardised units for staircases, refuse chutes, wall units and other components.

Standardisation is a time-consuming and costly process. It is particularly expensive in the research and development effort required to achieve solutions worthy of standardising and then to obtain agreement to the results. Where the final product is a component, it can also be expensive in capital investment for the machinery requirements. The prime factor in standardisation is repetition. Yet the normal mode of working on building sites militates against repetition where each project involves establishing a new team of men to undertake a different task to the last, and the team is subsequently disbanded when it is achieving high productivity. Common standards of communication, components, construction and procedures in design offices, in factories and on sites could lead to the same high increases in productivity as have been secured in the manufacturing industries.

The Department of the Environment found that one contractor was able to reduce the time required to build a house from over 2000 man-hours down to 1300, when he was engaged on a series of large contracts each of about 500 houses. Nevertheless, the savings resulting from standardisation are not always as extensive as might be expected. For instance a Building Research Establishment study of the effects of variety reduction in doormaking found that maximum feasible price reduction was eleven per cent on the cheapest door resulting from drastic variety reduction, increased efficiency and a massive increase in the scale of production. However, reductions in the variety of door set types (doors and frames) and painting doors on the site can yield significant price reductions. Variety reduction and the repetition which flows from it can save large sums of money in the design office, in the factory and on the site. Where it is applied intelligently, it does not restrict the designer unduly and leads to better value for money.

A study by the Building Research Establishment (1969) stressed the need for a uniform approach in the production and coding of information, and highlighted the problems encountered in retrieving and processing information in the construction industry. It showed the need for a common language and code for ease of communication and improved efficiency. The C1/SfB classification system is of great value in this connection, but this is likely to be replaced by a European coding system in the mid 1990s.

Economies of Rationalisation

As a general rule the cost per unit of a manufactured product will fall as the length of the production run is extended. After a certain level of production the rate of reduction in cost falls. For example in Finland the cost of doors fell by fourteen per cent when the production run increased from twenty to fifty but only by five per cent when from 200 to 500. Similarly, sanitary goods showed an eighteen per cent saving when the run was increased from ten to 100 but only six per cent from 100 to 400. There is a possibility that savings resulting from large extensions to production runs could be offset by increased costs resulting from larger stocks and longer delivery distances.

The logical sequence to standardisation of components is a method of production whereby all components are related to a unified set of dimensions, preferably operating on an international basis. A good approach is through modular coordination whereby all components are designed in terms of a common dimension, such as 100 mm, to permit all components to be interchangeable without the need for cutting or packing. This approach should produce savings in labour and materials, a reduction in ranges of sizes, smaller stocks, longer production runs and lower production costs. Changes of this type involve expensive retooling and problems arise from the lack of dimensional accuracy in traditional components and the tendency to be not too precise in setting out building projects.

There is a need to rationalise the dimensions of buildings and their parts and to secure precision in the manufacture of those parts, if the techniques of mass production are to be fully exploited in the building industry. The cutting away and making good, and the scribing and fiddling to make things fit, all add considerably to the cost of the building. To sum up large scale production with its resulting economies depends on the reduction in the number of sizes and shapes of products and to achieve this the process of typification and standardisation are essential.

INDUSTRIALISED/SYSTEM BUILDING METHODS

Nature of Industrialised/System Building

The aim of industrialised building was to apply the best available methods and techniques to an integrated process of demand, research, design, manufacture and construction. Basically industrialised building systems should aim to combine aesthetic value and user satisfaction with economy of materials, production methods and erection techniques. Industrialised building has been defined as the application of power and machinery and quantity production to those building processes which can effectively be undertaken in this way. This latter definition fails however to give sufficient emphasis to the important integration aspect of the process. A basic concept of industrialisation must be the organisation of the whole construction process in an integrated way, so that materials, components, plant and labour are available at the appropriate times to secure continuity both in the factory and on the site.

Historical Background

Large panel systems of construction were introduced into America and Europe at the beginning of the present century. Approximately one per cent of the dwellings erected between the wars in Great Britain were of prefabricated construction but they failed to offer any real advantage over traditional dwellings. In 1945 a Government Interdepartmental Committee approved 101 systems for use by local authorities in the construction of houses to assist in meeting the urgent post-war housing needs. They were based mainly on precast or *in situ* concrete walls, or steel or timber frames. Special grants were made available to help in meeting the cost of converting and tooling factories for the prefabrication of components. These systems proved very costly and the grants were withdrawn at the end of 1947. Typical housing costs in 1947 were: temporary bungalow – £1170; aluminium bungalow – £1600; and traditional house – £1250. During the period 1945 to 1955 twenty per cent of the houses built in England and Wales were of non-traditional construction but by 1961/62 this had dropped to six per cent.

In the early nineteen sixties industrialised methods secured increased support, particularly for use in high rise buildings, when the Government accepted the objective of substantially increasing the output of new dwellings, and industrialised building seemed to offer the only method of achieving it, faced as they were with a shortage of skilled labour. Over a five-year period hundreds of proprietary systems were brought into existence, but few had any real chance of achieving the size or continuity of orders that were essential to make them viable. The use of industrialised systems was not confined to housing; a proportion of offices, factories, hospitals, stores and schools were constructed by industrialised methods.

Classification of Systems

Industrialised building systems were often classified into two categories: closed systems and open systems.

Closed systems (contractor-based) were operated by sponsoring contractors and were often subject to the payment of royalties for production under licence covering a limited range of building types. A restricted number of structural elements, which were not interchangeable with other systems, were individually designed and manufactured by each separate sponsor of a closed system, and supplemented with standardised non-proprietary components obtained in the open market from outside manufacturers.

Another variant was the manufacturer-sponsored system in which manufacturers or distributors of building materials or components developed them to form the essential features of building systems. The items concerned commonly formed the structural elements and in some cases, such as steel frames for schools and low rise housing, may not account for more than ten per cent of the total cost of the building.

Open systems (client-based) were not subject to the payment of royalties, and were based on components designed by government departments, local authorities and other clients for buildings with similar basic requirements and performance standards, such as schools, hospitals and government office buildings. A system of standardised structural elements was dimensionally co-ordinated with a sub-system of standard non-proprietary components which were designed and selected for assembly in a variety of ways.

Examples of open systems included the *public building frame*, of precast concrete structural members designed by the Department of the Environment in conjunction with the Cement and Concrete Association; the Compendium of Hospital Building Assemblies, issued by the Ministry of Health for the industrialised hospital building programme; and the SCOLA and CLASP systems, particularly suited for schools and developed and controlled by consortia representing groups of local education authorities. Many open systems were of 'light and dry' construction often using light steel or concrete frames.

Industrialised systems can also be classified according to structural type; materials, weight and method of assembling structural elements; and building type. Structural types included the following.

(1) Load-bearing crosswall construction of concrete, bricks or concrete blocks; floors and roofs of precast or *in situ* concrete, timber or steel framing; and cladding of various types, all of which were well suited for housing.

(2) Storey height plank construction of 400 to 600 mm wide aerated concrete planks forming loadbearing walls, supporting floors and roofs of various types.

(3) Framed structures of concrete, steel or timber; prefabricated floors and roofs, with cladding produced in a factory or on the site. These systems were well suited for buildings with large rooms, such as schools, fire stations and factories.

(4) Loadbearing large panel construction for high rise housing formed of medium (2½ tonnes) or heavy weight (6 tonnes) precast concrete panels for full-size, or parts of, walls and floors, *In situ* concrete staircase walls or lift shafts gave rigidity to the structure and cladding could be formed from a wide variety of materials.

(5) Box construction of monolithic or composite prefabricated units, finished and equipped to form complete dwellings, incorporating plumbing and electrical services, doors, windows and fittings, and decorated in the factory.

(6) Composite systems with steel or concrete frames; loadbearing precast or *in situ* concrete walls and crosswalls; and cladding of precast concrete, brick, aluminium sheeting or other materials.

Advantages of Industrialised/System Building

There were certain advantages claimed for industrialised/system building which included increased productivity; lessening the adverse effects of inclement weather; a reduction in amount of non-productive work, such as scaffolding; securing better working conditions for site operatives; and the elimination of waste in both labour and materials.

There are also supplementary benefits some of which have been subject to progressive development. In a system like CLASP the components had been carefully and thoroughly designed and the designs had been subsequently modified as a result of feedback of information from sites. By using a system which had been tried and tested, considerable savings in time accrue at the working drawing stage. This permitted the designer to spend more time on clarifying his brief, carrying out user research, achieving improved planning solutions and possibly saving space and improving quality within the same cost limit and overall time scale. These developments should lead to better value for money in terms of appearance, durability and performance in the design of components. Some modifications of the CLASP system were subsequently required to overcome fire hazards inherent in the original designs, and the initial cost savings tended to reduce.

Possibly the use of the terms open and closed to describe building systems, led to confusion, because of lack of general agreement as to their main characteristics, and the fact that there was no absolute condition of complete limitation or complete freedom in many of the fields of choice. A preferable method of classifying them could have been as to whether or not they were component-based, their type of sponsorship and whether or not their development was directed towards interchangeability with other systems.

Effect on Contractual Arrangements

In the development and operation of a building system, the traditional relationships between the client, architect and contractor were modified and the role of the component manufacturer assumed greater significance. Other parties could be involved such as the corporate client, whose needs the system had to satisfy, and the system designer who could be quite separate and distinct from the designer of the individual building. The sponsor of a building system undertook a number of duties which did not occur in traditional building. These included the rationalisation of user requirements for the building type as a whole, the commissioning of design and prototype work, the co-ordination of the manufacture of components with the execution of the whole work, the advising of the designers and contractors for individual buildings on the proper use of the system, the promotion of the system and the appraisal of its performance as a basis for further development. The essence of the sponsor's duties was the provision of capital and co-ordination.

System building imposed on the design of an individual building a number of disciplines which did not occur with traditional methods of construction. These could restrict design flexibility, the choice of components and contractors, and the extent to which components could be interchanged with those outside the system. When interchangeability was provided, the design of the components was still subject to dimension and jointing restrictions. Hence designers who wished to make use of prefabricating techniques had to be prepared to submit to a working discipline based on an appreciation of operational factors.

INDUSTRIALISED AND SYSTEM BUILT HOUSING

The Labour Government in the mid-nineteen sixties provided a great impetus to the use of industrialised methods in order to secure a substantially higher output of dwellings from the available labour resources. There was considerable capital investment in plant and equipment on the basis of Government promises. Probably one of the most serious miscalculations was to believe that systems of construction producing standard house types with standard exterior finishes and allround standard architecture could compete satisfactorily with more flexible forms of construction. Some argued that rationalisation and variety reduction could produce more attractive dwellings and referred to most delightful and successful examples of standardisation in previous centuries at Bath, Cheltenham, Edinburgh and Bloomsbury (London).

High rise construction lent itself to component standardisation and repetition of elements, and there was greater opportunity for sponsors of high rise systems to persuade architects to accept system buildings in these situations than in 'close to ground' human scale development where aesthetic values were more local, personal, human and intimate. Hence a considerable swing soon developed towards the industrialisation of high rise flat construction. In the late nineteen sixties the popularity of high rise flats declined, not so much because of a failure of industrialised construction techniques to secure economy and speed in construction, as to a social reaction against the compartmentalisation of human beings. The decline was also influenced by the higher constructional costs of high rise over low rise dwellings and the abolition of the additional government grant for high rise dwellings.

For high rise development it was found generally that precast concrete systems were erected more quickly and cheaply than *in situ* concrete systems. In contrast, *in situ* concrete systems often proved the cheaper method for low rise dwellings although not the quickest. This resulted mainly from reduction in repetition and standardisation in low rise housing compared with high rise construction. Furthermore, the high initial cost of a precast concrete factory and its high overheads tend to make precast concrete units expensive. Timber and steel frames had both been used extensively for low rise housing and had the benefit of fast erection times, although steel frames were relatively costly.

It is interesting to note the criteria against which the National Building Agency judged systems for use in local authority housing contracts and on the basis of which the Agency issued certificates of suitability:

(1) system could produce dwellings at reasonable initial and maintenance costs;

(2) system was capable of producing dwellings which complied with the necessary technical and space requirements;

(3) system was sufficiently flexible to be capable of producing a satisfactory environment;

(4) condensation problems were considered no worse than with traditional construction, although this not always borne out in practice;

(5) adequate productivity;

(6) inspection of prototype revealed no serious defect.

PROBLEMS WITH INDUSTRIALISED/SYSTEM BUILDING

General Problems

There were some disadvantages inherent in factory-produced building systems. They were almost certain to give rise to high initial unit costs unless very large contracts had been secured; factory employees usually received higher wages than site operatives; these systems generally resulted in a reduction in flexibility as they were usually based on modular units, and there was lack of interchangeability between the systems. Too many projects in the housing field fell below the minimum economic threshold of fifty dwellings which was generally recognised as being necessary to achieve the full benefits of industrialisation.

Assembly operations occupied a dominant position in the industrialised building process. There was a concentration of capital-intensive machines and of highly qualified personnel. In order to obtain maximum utilisation of crane capacity it was necessary to synchronise all preceding and subsequent phases to the speed of assembly, and this required thorough and skilled programming. Some of the advantages claimed for industrialisation were based on an oversimplified analysis and many of the potential advantages could be lost by poor design details which left small but awkward pieces of work to be carried out by traditional means. The site construction could be made very much more complicated with a greater number of breaks in continuity, by the transfer of some of the work to a factory.

Industrialised building methods, coupled with long production runs, militated against modifying a design in the course of a building programme, so that user needs, as well as other aspects of the design brief, required clear definition at the outset. To achieve the essential consistency of standards, the client had to accept a more restricted range of choices than with traditional building.

Industrialised building also brought its own demands upon accuracy as it approached ever nearer to the technique of mass production. These techniques, with their attendant economies, were achieved through the standardisation of components and a greater degree of component interchangeability. If dimensional standardisation was to be achieved, dimensional accuracy was essential in both the overall manufacturing dimensions and in the correct positioning of the units during erection. The degree of accuracy was controlled through tolerances but these were not entirely effective in practice as they were difficult to specify, measure and communicate.

Problems with Housing

The Government was prepared initially to sponsor schemes of industrialised building even to the extent of a production run of 20 000 houses, but it was not prepared to enter into obligations for the continuous supply of houses over a number of years. Promoters claimed that they never had a real opportunity to substantiate the economy and viability of a system because of the uncertainty over future contracts. Furthermore, many local authorities possessed comparatively small housing sites and they only needed a few houses each of various types. There were far too many firms trying to share the market, each of whom were faced with high labour costs in design and production, the need for expensive equipment and a noticeable lack of flexibility in operation.

Industrialised building called for a systematic and disciplined approach from the building client. He would be buying a building which was largely composed of a range of mass-produced components, which could only be produced economically as part of a large, continuous manufacturing programme. In these circumstances, he might have to combine his requirements with those of others, to accept a place in a phased manufacturing and building programme and to adopt contractual methods to which he was unaccustomed.

To meet these restrictions, the larger housing authorities often found it necessary to combine in consortia, to secure a total programme adequate in size for the application of industrialised systems. However, the use of industrialised building posed the greatest problems for the smaller housing authorities. The nature of the problem became immediately apparent on consideration of housing administration arrangements in England and Wales. Even after local government reorganisation in 1974 there were about 400 housing authorities made up principally of district councils. The average local authority housing contract was for less than one hundred units and many local authorities had annual programmes of less than 500 dwellings. It was difficult for the smaller authorities to form consortia as they lacked the necessary professional staff and it needed too many constituent members to produce a viable programme. The National Building Agency was established to enable a large number of small demands to be collated into satisfactory programmes, and to give guidance on the selection of systems and contract arrangements. The Agency was however abolished in 1982.

Another major problem stemmed from the financing arrangements on an annual basis which operate in the local government sector. Local authorities are seldom prepared to formulate their building programmes for more than a year ahead, although a strong case can be made for rolling programmes extending over much longer periods. Yet another problem is associated with tendering arrangements; local authorities were often under pressure to invite tenders from a number of builders on a competitive basis, even although the Banwell Report favoured selective tendering. Changes were however taking place and some local authorities were using negotiated contracts and other modified tendering methods, permitting tenders to be obtained for the supply of components in advance of the main building contracts, or the selection of a system and a contractor at an early stage in the design process with the tender being negotiated later. Finally, the architect also had to adjust to a new discipline and to face up to the limitations on design imposed by the use of prefabricated components and a relatively limited range of claddings and finishes.

The substantial decline in the use of industrialised building systems in the nineteen seventies meant that the majority of systems had not been fully developed. The cost of industrialised dwellings for medium and most low rise housing had been higher than that of comparable traditionally built dwellings. Furthermore, savings in total construction time were not as great as anticipated. Although the shell was erected more quickly, the finishing trades and fittings often had not been sufficiently standardised and had taken as long to complete as those in traditional dwellings. Nevertheless, it would be wrong to write off system building as a complete failure. Many of the troubles experienced stemmed from the timing and scale of the system building revolution in the housing sphere.

ECONOMICS OF INDUSTRIALISED/SYSTEM BUILDING

Comparative Costs of Different Constructional Methods

Some general principles could be established relating to relative costs of constructional techniques in different situations.

(1) Loadbearing wall construction was best suited to building types with small spans, such as dwellings.

(2) Systems for buildings up to five storeys in height were more economically designed as column and beam structures.

(3) Structures over five storeys in height with increases in vertical loading were well suited to panel construction.

(4) Savings in cost stemmed from a reduction in the number of different materials used for walls, floors and façades.

High rise dwellings, with comparatively small room spans, were usually more economically constructed in concrete of boxshell construction. When using steel and reinforced concrete frames in structures subject to light loadings, it was usually preferable for the floor members to span in the opposite direction so that each bay could secure more evenly distributed loads on beams. With heavier floor loading, the units were more economically designed to span in two directions.

Large panel construction was often the most economical form of construction for high rise buildings and reduced the number of vulnerable external joints. They were, however, expensive in casting, fixing of reinforcement, provision of moulds and in handling. These disadvantages were usually more than offset by the high quality of finish, more rapid construction and improved working conditions which flowed from their use. The selection of the type of mould was conditioned by the scale of the activity and by the expected economic life of the development. Given the expectation of a reasonable economic life and supply of accurate and well-finished panels, it was possible to obtain considerable savings in the labour costs of erection, jointing and finishings which were likely to more than offset the higher costs stemming from the use of more accurate or more sophisticated moulds.

Labour Implications

Where factory-made components replaced only part of the corresponding conventional operations, this resulted in an increase in the total number of operations and greater complexity. Furthermore, in these situations it was not uncommon for the work remaining to be performed by traditional processes to be the most difficult and awkward part, and therefore very expensive. Hence, the aim at the design stage should be to replace the whole of the traditional work and to reduce the number of operations required. Furthermore, the use of industrialised methods created a demand for new skills. In this country the wage rates in the building industry were slightly below those in manufacturing industries, hence the transfer of work from site to factory was less financially attractive than in the United States where construction workers received about thirty per cent more than workers in manufacturing industries.

Material costs were frequently higher in industrialised building as compared with traditional work and it was therefore necessary to reduce labour costs. This highlighted the need for effective preplanning and organisation of all phases of system building by integrating production with deliveries to sites, keeping erection teams fully occupied and minimising non-productive time and site costs. A productivity study by the National Building Agency found in 1970 that an industrialised house took about thirty weeks to build compared with a national average for local authority housing of about fifty-three weeks.

Capital Costs

Industrialised building entailed much greater capital investment than for traditional building. In conventional building the capital investment per dwelling per day capacity was about half that in system building using large panel construction. Amortisation costs were likely to be increased since investors would probably anticipate a shorter economic life from their assets invested in industrialised building, stemming from particularity and possibility of early obsolescence. Interest charges could also be higher because of the need to attract risk capital in the face of an uncertain market. In the 1970s interest rates could have been twenty per cent where a short economic life was expected, and fifteen per cent and ten per cent for medium and long economic lives respectively.

Capital costs had to include that of development work in designing and testing the system and the production process, including prototypes where required. Fixed assets included factory buildings and plant, site equipment, vehicles, services and temporary works generally. Sufficient working capital was required to finance work and bridge the gap between receipt of income and the making of payments to labour and creditors, and could amount to as much as two to three per cent of turnover.

Another approach was to compare the investment per operative. The investment per operative in system building was approximately three times as much as that for conventional building. The cost of a factory producing precast concrete panels could vary between £¼m and £2½m, according to the scale of production and the degree of mechanisation and automation. It should also be borne in mind that up to thirty per cent of the cost of the factory could be spent on transport and trailers to carry components to sites, particularly in the case of large panels.

Building Demand

A permanent off-site factory to produce precast concrete structural elements is based on mechanised handling methods and semi-automated production line principles under controlled conditions to secure a high output ratio of accurately prefinished units. The viability of such a project is largely dependent on the maintenance of an adequate and sustained volume of demand, in order to justify the highly intensive capitalisation required for production. Indeed, many sponsors of industrialised systems have been forced out of business through failure to secure sufficient continuity of orders for economic production. Yet the demand for building, besides being dispersed on sites, is variable in volume, bespoke in character and uncertain in timing.

Admittedly off-site factories carry with them the disadvantages of the need for long production runs, high overheads and high transportation and handling costs. In some situations a better approach might be to establish an on-site factory geared to the production needs of the site, with consequent reduction of transportation and handling costs, and being capable of transfer to another site on completion of work on the original site.

Economics of Production

It is evident that industrialisation almost invariably involved increased capital investment and other indirect costs and could at the same time, result in decreased utilisation because of particularity. Hence, if industrialisation was to be economically worthwhile, these increased costs had to be offset by reduced labour costs through higher productivity. Higher productivity should be feasible through increased technical aids, improved organisation and superior working conditions. One way of reducing costs was by eliminating operations, such as by accommodating services in wall and floor units or by producing units which are self-finished.

Too often, in the early 1960s, even the sponsors of systems were not in possession of realistic production costs, and yet a full cost appraisal and feasibility study should have been undertaken before the component factory was erected. Another very real danger consisted of available orders being spread over too many firms and systems, as each manufacturer viewed the market in isolation, and this could easily result in few systems achieving optimum production. It was calculated that efficient traditional construction required about 2000 manhours to build a flat in a tall block, whilst an efficient concrete panel system of industrialised building required 500 manhours on site, 300 in the factory for concrete components and a further 300 for other factory work; a total of 1100 manhours.

Production costs of concrete components are influenced by the following factors:

(1) Installation and running costs of factory, plant and equipment, and dismantling costs in the case of temporary factories.
(2) Mould costs, which can also be affected by increased use resulting from accelerated maturing of concrete and the introduction of shift work to increase output.
(3) Material costs, which are affected by the surface treatment and final colour of the units.
(4) Labour costs, which can be affected by the integration of the casting and erection cycles.
(5) Type, size, finish and number of units.

Plain and relatively simple units can be cast economically on site. Those of intricate design are better suited to off-site production under more closely controlled conditions. Unless units of the same shape and size are used on every floor, additional costs will be incurred in varying the moulds to suit the units for use at different floor levels.

An effective computer system can assist with rapid and yet flexible control over preplanning and costing by providing programmes for:

(1) ordering materials to phased deliveries;
(2) comparing estimated costs with the current costs of all projects in progress by a detailed breakdown which forms a pre-expenditure control budget;
(3) rapidly determining payments and outstanding balances due to subcontractors on the basis of coded budget programmes; and
(4) readily obtaining monthly efficiency reports which provide detailed information about differences in costs and quantities of materials used on site, labour, plant and subcontract works.

Obviously long contracts were needed for system building because of the way in which costs rose if factories operated at a low level of output. A high proportion of factory costs were fixed, since neither the labour force nor overheads could be reduced easily when output fell below the optimum. The establishment of consortia and the grouping of orders, coupled with a restriction in the number of systems, did help to secure viability of production for a time, as did the concentration of system building factories in the vicinity of new towns and large redevelopment areas.

Comparison of Industrialised/System Building and Traditional Building

The labour requirements per dwelling with traditional building covered a wide range from 700 to 2400 manhours with an average of around 1800 manhours per dwelling. The performance with industrialised building also showed an extremely wide range with site labour requirements per dwelling varying between 700 and 1300 manhours. The difference between the labour requirements for conventional and system building was likely to increase in tall buildings because these were inherently more complicated and the work involved was better suited to industrialisation. The work directly affected by industrialisation often approached seventy per cent of the total, the remainder being siteworks, foundations and work of specialists. It would seem rather pointless using a system which did not extend beyond the structure, as the savings realised in the finishings were proportionately greater than those obtained in the structure.

Comparisons of the alternative methods of house building can conveniently be made under four main headings.

Relative labour costs. The average hourly earnings in factories in the United Kingdom producing engineering products were about 20 per cent higher than construction industry earnings.

Indirect labour costs. These were generally increased in system building because there was a greater demand for management skills and services, and for technical expertise in design and production, and they could be as much as two-thirds higher than for conventional building.

Capital charges. These were made up of the capital sum, amortisation and maintenance charges and the interest rates necessary to raise the capital. Capital investment was nearly always increased with system building, although ideally the investment should be related to the capacity of the production process. This was possible with closed systems, but with open systems it was difficult to establish because of the wide range of building types and products and the widespread use of hiring and subcontracting in the industry.

A rather more assessable index was the investment per operative as described earlier in the chapter. Capital costs could range from £2.50 to £3.50 per day per £1000 invested for conventional building, and from £2.50 to £5.20 for system building.

Utilisation. In conventional building problems of utilisation rarely arise as most firms can undertake a wide range of building work, and the general employment of subcontractors gives considerable flexibility to the industry. System building radically altered this situation in that resources were committed to the production of specific components (in open systems) or to the production of specific building types (in closed systems).

With industrialised systems contracts were usually negotiated and hence tended to be rather less competitive than traditional dwellings which were normally the subject of tenders. This could be an oversimplification, as by concentrating on design and development work the sponsor of an industrialised system could provide a very efficient and economical building. On the other hand industrialised buildings rarely seemed to function any more satisfactorily than their traditional counterparts, and often not as well, and they were generally more expensive in maintenance. Charges of monotony and poor design of industrialised buildings were rarely well founded and open systems provided kits of parts which could be assembled in a large variety of ways and so provided the designer with a reasonable range of choices. System-built structures normally made fuller use of the inherent properties of the materials than traditional buildings, although the satisfactory jointing of components often proved expensive.

THE FUTURE IN INDUSTRIALISED/ SYSTEM BUILDING AND THEIR MAIN DEFECTS

Orders placed by local authorities in England and Wales for system-built dwellings in the early seventies were limited and could be met by a relatively small number of systems. This presented a very different picture to the Government White Paper of 1965 and Ministry of Housing and Local Government circulars 21/65 and 76/65 to local authorities, which set a target of 500 000 new dwellings per year by 1970, half to be built by the private sector and half by local authorities, with forty per cent of the local authority houses to be system-built. Nearly 300 systems were introduced, although a proportion of them never developed beyond the illustrated brochure stage. The target of 500 000 dwellings per annum was never achieved, and by the early nineteen eighties had reached an all time low, offering few opportunities for system building. The exception was in some overseas countries such as Singapore, where there was high demand for public high residential blocks in the 1980s and beyond, and the Singapore Housing and Development Board (HDB) negotiated system built contracts with France in the forefront.

It is evident that there were too many systems for the available work and that much of the anticipated workload in local authority housing did not materialise. Changes in the membership of local authorities and central government policy had significant effects on housing policies. In the long term, system building could only remain healthy provided users were satisfied as to its benefits and there was sufficient flexibility to respond to variations in demand. There was also a need for the private sector to show much more interest in industrialised methods by accepting some limitation in variety in terms of basic shape and layout, in order to secure cheaper and quicker manufacture of the factory components.

Manufacturers of industrialised buildings needed relatively long periods, certainly not less than ten years, to amortise their costs; local authorities had to be able to implement rolling programmes uninterrupted by financial restrictions; and the Government needed to give directions to local authorities on the use of system building if it was to be really successful. Until local authority orders for industrialised housing were handled on a regional basis and large long term orders were given, it was bound to remain a risky and in most cases a non-viable capital venture. Another drawback to the further development of industrialised systems was the fact that they were largely restricted to the housing field. While the housing market is extensive, it still does not represent more than about one-third of the total output of the building industry and is subject to significant fluctuations. Hence the future for industrialised building is very bleak, particularly bearing in mind the extensive major defects and the general unpopularity with occupants.

Perhaps the greatest contribution that industrialised building processes have made to the construction industry was the way in which they permitted the pre-planning and analysis of actual cost. By their competition, industrialised building methods accelerated the rationalisation of traditional methods, by means of the integration of design and construction. They also tended to bring in the contractor at an earlier stage in the design/construction process.

Defects in System Building

Despite their general unpopularity with occupants, poor appearance and the many constructional and maintenance problems, and extensive demolition of tower blocks, there were still 6500 such blocks in Britain in 1994. HRH the Prince of Wales (1989) very aptly described how all over Britain local authorities were subsidised to build gaunt and unlovely towers which rose like great tombstones from pointless and windswept open spaces, like the unattractive, badly sited and over-conspicuous blocks he illustrated in Newcastle upon Tyne.

The principal defects associated with system built dwellings encompassed the breakdown of precast concrete panels, spalling of concrete, rusty and defective reinforcing bars, insecurely fixed wall ties, water penetration, excessive condensation, additional fire risks and inadequate refuse disposal systems. Dissatisfaction with high rise flats rose dramatically following the gas explosion and progressive collapse encountered in the Ronan Point flats in London in 1968, which resulted in five people being killed and 17 injured. The subsequent report on the disaster showed that the structure merely followed national design trends which had been inadequately researched, and all owners of high rise blocks with large precast concrete panels forming load-bearing walls and/or floors were requested to undertake appraisals and strengthening of the structures where necessary. In addition the high alumina cement scare of the 1970s resulted in locating, investigating and sometimes rebuilding some 50 000 structures that contained HAC in the UK costing over £70m (Seeley, 1992).

Deck access housing has been condemned as producing a combination of problems, most of which are associated with poor design and construction, and by 1990 many of these blocks had been demolished. The main problems related to construction, access and appearance. Water penetration, inadequate sound and heat insulation, poor ventilation and expensive heating were among the construction defects. Decks which gave access were often used for anti-social purposes, such as dogs' excreta and vandalism in deserted common areas. The unusual and inhuman appearance of most of the estates had contributed to the spiral of decline. On the Hulme Estate in Manchester over 3750 deck access dwellings were system built between 1967 and 1971, and structural and design faults had developed within five years of their completion. In the long term the Council were committed to eliminating deck access dwellings, approximately £1.5m was spent up to 1986 on design improvements, remedial works and disinfestation, with a further £3m required to maintain them prior to demolition, compared with the prohibitive cost of £75m required to bring them up to a standard suitable for human habitation (1986 prices). Hence many local authorities returned to the development of low rise estates, because of the many disadvantages of high rise blocks, resulting in a drastic decline in the demand for system building (Seeley, 1992).

The cost of demolishing the 1960s and 1970s damp, badly planned and crime ridden flats in concrete blocks and replacing them with low rise dwellings at Hulme (Manchester) and Waltham Forest (East London) was estimated at £147 000 per dwelling at much reduced densities in 1993.

In 1986, BRE started providing advice for owners and their consultants on methods of inspection, maintenance and renovation of the different steel-framed house types. About 140 000 metal framed houses had been built in Britain, based on some thirty different proprietary systems. The BSIF system was the most common with over 30 000 houses built in England and Wales and over 4000 in Scotland. Cases were reported of serious deterioration of the main frame or failure of claddings and fixings, and there was a history of concern about the thermal and fire performance in some of the house types (Seeley, 1987).

In 1990 urgent repairs estimated at £75m were required to colleges and former polytechnics to comply with health and safety regulations, and a further £188m in 1991 to 1992. Most repairs were needed on poorly constructed system buildings of the 1960s, which were built with a life span of only 20 years, using untried methods often leading to structural collapse and leaking roofs. The most common urgent repairs included the replacement of flat roofs, boiler systems and wiring, and rotten timber and corroded metalwork. Many buildings were also reported to have inadequate fire safety. Thus in the early 1990s these higher educational establishments were having to allocate a disproportionate amount of their resources to maintenance work.

Refurbishment of Tower Blocks

To appreciate the scale of the problem, there were in 1994 about 6500 tower blocks in Britain containing some 400 000 dwellings, and hence they still comprised a significant proportion of the nation's social housing stock. The costs associated with high rise refurbishment typically ranged between £2m and £5m per block with individual dwelling refurbishment costs showing a wide range of between £13 000 and £51 000 in the early 1990s. These costs have to be considered in the context of reductions in the funds available to local authorities, which had reduced from £12bn of capital resources in 1975 to around £3bn in 1992, both expressed in 1992 prices. Faced with falling demand for high rise residential buildings erected in the 1960s and 1970s and the major defects described earlier in the chapter, many local authorities and some housing associations produced innovative solutions in order to improve the management and performance of the buildings.

The general lack of local authority resources was compounded by the inability of other potential partners to raise the necessary finance. The private sector was impeded by the unco-operative attitude of the financial institutions and, outside London, by limited demand for the accommodation. The Council for Mortgage Lenders highlighted the reduced supply of private finance for owner occupation in high rise dwellings, because of the difficulty in mortgaging and marketing such properties. It was unlikely that housing associations could contribute in a significant way with their grants being stretched ever tighter, particularly in the areas where high rise accommodation predominated. While private finance remained a possibility, the resulting rents could be similar to the cost of servicing a mortgage for the first time buyer. For example, a West Midlands housing association calculated that a privately financed refurbishment of £15 500 per unit in 1994 would produce rents in excess of £60 per week including service charges, assuming that the block was transferred free of cost (Farr and Nevin, 1994).

The alternatives facing a local authority were demolition and redevelopment, but bearing in mind that demolition could cost £500 000 per block; transfer of ownership to another landlord; and refurbishment. However, if the local authority wished to retain the overall level of housing stock then refurbishment often represented better value for money than demolition and rebuild.

Farr and Nevin (1994) recommended that a high rise refurbishment appraisal should include the following:

- A database with accurate costings relating to the total refurbishment costs of each block to permit rational decisions to be made relating to refurbishment or demolition.
- A cost benefit analysis for each block taking into account the refurbishment cost, the identified future demand and its contribution to housing need in the area.
- A resource procurement strategy to be developed in conjunction with housing associations, the private sector and the DoE, to identify the strategy and any likely shortfall.
- The high rise strategy to be implemented within a reasonable time frame, as it is pointless attempting to market unimproved high rise buildings well into the 21st century, when much of the stock was already considered unsatisfactory in the early 1990s.

An examination of refurbishment work on tower blocks of the 1960s and 1970s, of heights varying from 12 to 32 storeys, undertaken in the early 1990s by Kensington and Chelsea LBC, Hull City Council, Waltham Forest LBC, Ipswich DC and Focus HA, Wolverhampton, revealed a wide variety of combinations of funding sources encompassing estate action, capital programme, rent increases, borrowing approval under DoE, greenhouse programme, supplementary credit, local authority HAG (Housing Association Grant), Housing Corporation HAG, private finance and housing investment programme.

The most common faults with the original blocks centred around inadequate heating (often electric underfloor heating), poor lift service, excessive condensation and mould growth, draughty steel windows, external balconies acting as cold bridges, poor thermal insulation, deteriorating fabric and/or structure, graffiti spattered surfaces and violence and vandalism.

The principal improvements encompassed the following works which were welcomed by the tenants and successfully removed the previous dissatisfaction and the difficult to let problem. Apart from attracting students from the new universities, refurbished high rise buildings appeared to be very suitable for the 50 to 74 age group, who are characteristically quiet, compatible as neighbours and often appreciate the sense of security engendered by living above ground level. Specialist provision such as laundries, clubs and playrooms in the often underused base storeys of blocks can prove attractive to residents, as can also common rooms on upper floors and roof gardens:

- New and improved lifts, often with CCT monitors.
- Security arrangements, ranging from cameras, entry phones, CCT monitoring inside and out, programmed electronic keys to tenants' outer doors to concierge service.
- Improved heating systems which can incorporate an electric cyclo-control system, district heating and associated radiators, economy 7 electric storage heaters and combined heat and power systems.
- Double glazed PVC-U windows with trickle vents, coupled with effective thermal insulation to obtain NHER ratings of 8.1 to 8.7, and heating bills reduced by up to 60 per cent.
- New external cladding of 2 mm aluminium overcladding with 75 mm mineral wool backing, fibreglass rain screen in various colours with improved insulation or brick cladding with insulation, all giving a much improved appearance.
- Kitchens modernised and extended into balcony areas.

- Structural and other improvements and increased amenities.

CLASP

Introduction

The CLASP building system derived its name from the Consortium of Local Authorities Special Programme, which comprised a group of public sector organisations who joined together to develop and maintain the building system initially for their own benefit and use. The system was first developed in the late 1950s to meet the need for a large programme of schools and other public sector buildings. Since then it has been progressively developed and improved to meet the changing technical, functional and aesthetic requirements of architects, contractors and building owners. The system has also been broadened to encompass all building types except housing, heavy industrial and high rise development (CLASP, 1990).

If has developed from its original closed system approach with limited aesthetic expression and flexibility to one which allows a wide freedom of expression, while still maintaining the technical integrity resulting from some thirty years of development. The system has been used successfully to construct over 3000 buildings in the UK and overseas, including Hungary, Portugal, France, Italy, Germany, Venezuela and Algeria. Furthermore, it is the only system of its kind to have obtained an Agrément Certificate validating a wide range of criteria including structural stability, behaviour in fire, watertightness, durability, thermal performance and maintenance. Under the provisions of the Building System Agrément Certificate 86/S13, the system is for use in the UK in the construction of buildings up to four storeys with a design life of 60 years.

In the early 1990s CLASP moved all its standard drawings on to CAD to provide an effective working drawings package and introduced other simple computer techniques of handling data. Co-ordinated project information (CPI) is used for all documentation and the National Building Specification (NBS) for the specification base.

Brief Constructional Details

CLASP is a lightweight steel-framed pin jointed building system. The frame is of RHS square tubular columns and lattice beams on a 1800 × 1800 mm grid and standard room heights of 2400 and 2700 mm with a normal floor and roof zone of 600 mm. Upper floors are of precast concrete which, together with the roof, act as diaphragms and transmit horizontal loads down to the ground slab through vertical bracing bays. The ground slab is suitably reinforced and no individual column foundations are normally required.

CLASP has a two storey pitched roof or a three storey flat roof capability in subsidence conditions, and a four storey flat roof capability on non-subsidence sites. The flat roofs consist of a metal trough decking and a built up felt finish, while the pitched roofs are of steel and timber construction with a tile covering. A number of cladding systems are available, including proprietary profiled metal sheeting, brickwork, exposed aggregate cladding panels and tiles, slates and boardings, used in conjunction with polyester coated aluminium window and door frames which are available in a variety of colours. Internal partitioning can be of PVC faced steel panels clipped into steel studs with plasterboard backing, or of plasterboard partitioning with double skinned plasterboard on each face screwed to metal studs (CLASP, 1990).

The system is designed by the CLASP Development Group which:

(a) arranges the supply of component parts;
(b) provides documentation giving guidance on planning, technical detailing and site assembly; and
(c) periodically updates the system.

The latest proactive technical development programme in the early 1990s encompassed composite construction for upper floors, rational-

Table 4.1 Typical CLASP project costs (1992)

Project description	Floor area (m^2)			Total gross floor area	Cost/m^2 (£)					Total cost/m^2 £
	GF	FF	SF		Subs'	Super'	Fittings	Services	Prelims	
Research lab; flat roof; concrete clad	677.00	403.00		1080.00	12.25	301.82	0.00	107.45	58.31	479.83
Police HQ; pitch roof; brick/metal clad	1049.00	990.00	854.00	2893.00	11.35	394.36	20.52	197.06	11.36	634.65
Primary school; pitch roof; brick clad	2186.00			2186.00	21.53	226.88	18.74	87.73	39.93	394.81
Primary school extension; pitch roof; brick clad	642.00			642.00	21.70	252.27	13.83	105.89	19.48	413.17
Small office block; pitch roof; brick clad	213.00	213.00		426.00	20.31	303.67	41.79	118.84	64.03	548.64
Small school extension; pitch roof; brick clad	242.00			242.00	24.45	278.83	19.71	126.68	23.25	472.92
Liberal study block; pitch roof; brick clad	1294.00			1294.00	17.25	238.51	15.02	82.49	16.40	369.67
Sports hall/changing; flat roof; brick/metal clad	1011.00			1011.00	12.44	296.56	12.30	104.74	19.77	445.81

The above schemes represent a sample of projects erected using the CLASP sytem. Every project has its own cost profile reflecting the function of the building and the circumstances in which it is built.

The effect of regional variations and time scale have been reduced to a minimum by correcting the data to an RV factor of 1.00 and a Tpi of 106 for the 3Q92.

The substructure cost does not include site abnormals.

Source: CLASP.

isation of the external wall, simplified foundation details, and new staircase designs.

Main Advantages

CLASP has a well established reputation in the following areas: reducing construction costs without diluting quality; the use of pre-finished components and dry construction techniques shortens construction times and improves site working conditions; maintaining technical integrity; ensuring accurate cost planning at design stage and firm cost control during the contract; enforcing rigorous quality control; majority of components are virtually maintenance free; the substantial range of component and assembly details and system documentation allows the designer more time for briefing and design; suitability for construction on ground of low bearing pressure, often enabling the cost effective use of sites hitherto considered uneconomic; and ability to withstand the effects of mining subsidence without incurring the normal additional constructional costs.

Typical CLASP Project Costs

Table 4.1 gives the cost breakdown of a selection of CLASP projects priced in the third quarter of 1992. The cheapest projects were the largest buildings, brick clad with a pitched roof, accommodating a liberal study block and a primary school (£369.67/m^2 and £394.81/m^2 respectively). Smaller school extensions of similar construction were more expensive, as was also a sports hall with a flat roof and brick/metal cladding and a research laboratory with a flat roof and concrete cladding (in the £413.17 to £479.83/m^2 range). The most expensive building in the selected sample was a police headquarters with a pitched roof and brick/metal cladding at £634.65/m^2, mainly resulting from its more costly superstructure and services.

5 ECONOMICS OF RESIDENTIAL DEVELOPMENT

This chapter is concerned with the problems associated with housing provision; the alternative forms of layout that can be employed to meet varying housing requirements and their comparative costs; the methods and economics of different forms of car-parking provision; and the considerations involved in a comparison of the relative merits of redeveloping or rehabilitating twilight areas.

BACKGROUND TO PUBLIC HOUSING PROVISION

In the period between the passing of the Housing of the Working Classes Act in 1890 and the outbreak of the first world war, local authorities collectively provided no more than 600 houses per annum on average. By way of comparison private housing provision throughout the period 1900–05 averaged 150 000 dwellings per annum. At the end of the first world war rent restrictions and high labour rates meant that the required dwellings to rent could not be provided profitably by the private sector. Hence the Addison Act of 1919 enabled local authorities to bear the cost of providing houses for the working class to the extent of the product of a penny rate (5/12p) and the Government bore the remainder. Sharply rising prices generally and inefficient administration caused average tender prices to rise by fifteen per cent within a year, and 175 000 dwellings cost the Government £200 million.

In the inter-war years a succession of Housing Acts varied the subsidy arrangements introduced by the Addison Act and the housing shortage remained sufficiently serious for rent restriction to continue. Local authorities built one million houses which was equivalent to about one-third of total provision. After the second world war the Labour Government introduced a number of additional restrictions on development, including licensing restrictions and rationing of building materials. Eighty per cent of the dwellings built between 1945 and 1951 were provided by the public sector. Orders were placed for 500 000 'prefabs' (prefabricated bungalows), but these were subsequently reduced to 170 000 due to problems of rising cost and balance of payments difficulties, followed by devaluation. The Conservative Governments of the fifties progressively reduced the public housing programme, licensing was abandoned and the private sector provided an increasing proportion of new houses. The Rent Act 1957 raised controlled rents to a level closer to that of the free market, and allowed controls to lapse from properties falling vacant. Between 1957 and 1965, over two million houses were removed from control, mainly as a result of the movement of tenants. The Rent Acts of 1965 and 1968 subsequently provided machinery for determining rents on a basis that sought to express fairly the real value of the tenancy. Local authorities borrowing activities were restricted and subsidies reduced, with the result that public sector annual completions fell from 220 000 in 1954 to an average of 100 000 in the period 1959–63.

Since the mid-fifties many private landlords have sold their residential properties as they have long since ceased to be profitable. The number of households living in privately rented property dropped from sixty-one per cent in 1947 to eight per cent in 1992. It has been estimated that over seventy per cent of the population of this country can afford to house itself with the aid of tax rebates, although these were progressively reduced in 1994/95, option mortgage schemes and other hidden subsidies. The remainder are compelled to rely on Government intervention or legal protection of one kind or aother.

The Parker Morris report (1961) established minimum space standards for local authority dwellings and this resulted in a general increase in the cost of publicly provided houses. Furthermore, heavy increases in land prices, stemming in part from the operation of betterment levy introduced by the Land Commission Act and a general scarcity of building land, resulted in speculatively provided urban dwellings costing so much to build that there was little possibility of them being rented. In the public sector the introduction of housing yardsticks in 1967 caused too much emphasis to be directed towards first costs as opposed to long-term costs. Loan payments in some local authority areas are matched or even exceeded by maintenance costs. Many local authorities have adopted a policy of selling houses in an effort to correct adverse balances on housing accounts, and this policy was extended in the late nineteen seventies and early nineteen eighties by the Conservative Government, who imposed a statutory requirement on local authorities to offer their houses for sale to tenants on advantageous terms.

LATER HOUSING DEVELOPMENTS IN THE UNITED KINGDOM

Smith (1989) usefully categorised housing in a number of different ways, such as form of tenure, type of agency carrying out the development, physical form, standard, price, age and condition, giving an indication of the complexity and variety of this type of development. In the past, housing had been broadly classified as public or private but these divisions are losing their significance as, for example, private finance is channelled into some housing associations and partnership arrangements are entered into between local authorities/housing associations and private developers.

Housing associations are charitable organisations whose main function is to provide good quality housing at affordable rents to persons in desperate need of accommodation such as the homeless, although in 1994 the government was reviewing the policy towards housing the homeless, with an emphasis on the provision of temporary accommodation for the homeless. They are largely supervised and funded by the Housing Corporation and were from 1989 onwards being encouraged to take over stocks of local authority dwellings and to become the main providers of social housing. By 1992 there had been 18 voluntary transfers of this kind. The government was also encouraging owner-occupation, including the sale of local authority houses and drastically reducing the funds available to local authorities for providing new houses or, indeed, to adequately maintain existing houses. The total maintenance backlog in housing was estimated at over £50b in 1993. The private rented sector shrank from 57 per cent in 1935, to 12 per cent in 1979 and 8 per cent of all housing in 1992, in sharp contrast to most other European countries.

The housing stock in England increased from 10.6 m in 1938, to 17.6 m in 1979 and 19.7 m in 1992. Since the second world war, about 1.5 m dwellings have been demolished under slum clearance programmes and most of the households in them rehoused in local authority accommodation. Despite a considerable improvement in the supply of basic amenities, the house condition surveys in 1981 and 1986 showed an increase in the number of houses needing major repairs and that about 1.2 m occupied houses were unfit for human habitation rising to 1.5 m in 1991. While the 1991 Scottish house condition survey showed that of the 2 035 000 homes, 94 000 (4.6 per cent) were below the tolerable standard defined by the Housing (Scotland) Act 1987 and were predominantly found in the pre-1919 stock. In addition, 423 000 (20.8 per cent) of dwellings in Scotland suffered from dampness, serious condensation or mould. Furthermore, the housing needs of single persons and the elderly were increasing rapidly in the late 1980s and the early 1990s with little action being taken to satisfy their requirements, and the number of homeless was increasing at the alarming average rate of 14 per cent annually. As central government cut back investment, local authority annual house building starts dropped from 110000 in 1975 to 13000 in 1989 and 700 in 1993/94. Housing associations provided about 55000 dwellings in 1993/94, well in excess of previous years, but even this was totally inadequate with 40000 families in expensive temporary accommodation and an estimated demand of around 100000 rented homes per year. The housing associations' ability to maintain the provision of good quality houses at affordable rents was undermined by the government announcing in 1992 that Housing Corporation grants would be progressively cut throughout the mid 1990s and this could also make the securing of private funding more difficult.

In the early 1990s the private sector was struggling to sell many of the new houses it had built because of high interest rates and a substantial downturn in the property market, resulting in a reduction of 70 per cent in private sector house starts between the second quarter of 1990 and the same quarter of 1992. The problem was compounded by the large number of mortgaged house repossessions and mortgagees with long term arrears of payments. These statistics show how housing supply is influenced to a great extent by economic and political factors and the absence of a coherent government policy on housing. The Institute of Housing (1992), now Chartered Institute, identified the main social housing requirements under the headings of investment, accountability, affordability and quality and the report deserves careful study, along with the RICS report (1986), and the reports produced by the RIBA (1985), the committee chaired by HRH the Duke of Edinburgh (1985 and 1991) and the committee commissioned by the Archbishop of Canterbury (1985). The changing situation with housing tenure categorisations in England is shown in table 5.1.

Table 5.1 Categorisation of housing tenures in England (percentages)

Stock	*1938*	*1979*	*1992*	
Housing associations	–	2	3	(597495)
Local authorities and new towns	11	29	20	(3715627)
Private renting	57	12	8	(1396768)
Owner occupied	32	57	69	(12692780)

USE OF LAND FOR HOUSING PURPOSES

Planning schemes prepared by local planning authorities restrict the land that can be developed for residential purposes. Residential areas should desirably be conveniently located in relation to workplaces, shops, schools and other essential facilities. Sites for residential development should be physically suitable – not excessively steep, although some undulation gives character, be above flood plains and be reasonably healthy. Adequate public utility services and good transport facilities are other important needs. The nature of adjoining development needs consideration as it must be compatible. Contamination can arise with landfill and on former industrial sites which are best avoided, as they can entail expensive siteworks to remove potential hazards, as provided for in part C2 of schedule 1 to the Building Regulations 1991 (Seeley, 1995). A housing association site adjoining a disused waste tip in Nottingham involved the provision of an electric surveillance system and an extraction and venting system to monitor the levels of methane and carbon dioxide discharged through a gas extraction borehole system at a cost of £250000 in 1993.

The area of land required to house a specified number of people will depend on the operative residential density. Local plans showing detailed planning proposals usually indicate the permitted density for each area of land zoned for residential use. Housing densities are expressed in a number of different ways. In pre-war planning schemes residential densities were usually expressed in houses per acre, but in more recent times it has become customary to use persons per acre or hectare (population density) or habitable rooms per acre or hectare (accommodation density), as these give the developer a greater degree of flexibility in his choice of dwelling types. A habitable room is a living room or bedroom, but not a kitchen or bathroom. The unit of bedspace has now largely superseded that of habitable rooms as a measure of housing density, as it gives a more consistent approach. Population density can be converted to accommodation density by deciding on the average number of persons per habitable room. A house with five habitable rooms and occupied by five persons would have an occupancy rate of one, but in practice dwellings are frequently under-utilised and the average family size is about three persons. Hence there is a significant variation between design and actual population densities.

The density most commonly used is that of net residential density. This is the population (or accommodation) divided by the area in hectares, including dwellings, gardens, any incidental open space (for example children's play spaces or parking spaces for visitors' cars) and half the width of boundary roads (up to a maximum of 6 m). Shops, schools and most open spaces are excluded. Gross residential density applies to a complete neighbourhood and includes all the land uses within it, and this is the more meaningful indicator of density. The fixing of net residential density standards by local planning authorities over wide areas can be deceptive and misleading; particular conditions of the site, its shape, and its relation to surroundings all influence the intensity to which it should be used. The people who are to reside on the site will also be concerned with the ancillary features that are available in the neighbourhood.

In the garden cities, in the first generation new towns and in most of the earlier post-war local authority housing schemes, net residential densities of about thirty houses to the hectare (twelve to the acre) were used. It was often argued that this was a good practical density, which could be applied to varying orientations without overcrowding and overshadowing. It permitted reasonable distances between houses, giving satisfactory street pictures with suitable depths of front garden and ample back gardens. Space was available for kitchen gardens, which are generally cultivated far more intensively than farmland by sparetime labour, which would not otherwise contribute to food production.

In 1962, the Ministry of Housing and Local Government called for higher residential densities. It was pointed out that the population of England and Wales was increasing by about 250 000 persons per annum and all available data indicated that this rate of increase would be maintained. People marry younger and live longer, and it was anticipated that greater prosperity would enable more families to have a home of their own, thus increasing the number of households. Furthermore, the occupancy rate fell from 3.30 to 2.74 persons per house between 1951 and 1980. However population forecasts in the nineteen seventies showed a relatively static future population.

The relief of overcrowding, new roads, schools and open spaces, all require more land. The plea for higher residential densities was made to reduce the total demand for land, to help preserve the countryside and protect good agricultural land. Substantial savings in land could be secured by increasing the net density to 100 persons per hectare, particularly when some development was proceeding at about sixty persons per hectare. It was further shown that net densities of 150 persons per hectare are attainable with two-storey houses and the introduction of a proportion of three or four-storey flats can produce net densities up to 225 persons per hectare. Higher densities still can be obtained by using tall blocks of flats, but these need not predominate until net densities of 350 persons or more per hectare are reached.

On the financial side it is pointed out that as the permitted density rises so the cost of land increases, although the cost of land per dwelling decreases. Building costs decrease as densities rise up to about 150 persons per hectare, because of greater use of terraced blocks, but above this density building costs rise sharply. Service costs of roads and engineering services decrease with more compact development.

With many housing schemes in the public sector the aim is to secure densities in the 100 to 150 persons per hectare range. The task of designing the layouts economically revolves around the choice of types of buildings and the proportions of each type. The director of housing or housing manager will advise on the proportions of the various sizes of dwelling that are needed. These requirements will be influenced by the range and numbers of existing dwellings; the family composition of housing applicants; the extent to which the authority's houses are under-occupied; the numbers and sizes of other dwellings being built by the authority; and the possibility of exchanges with families in privately owned accommodation.

Given the site area, the density standard and the proportions in which the various sizes of dwellings are needed, the designer can calculate the number of dwellings of each size to be provided. It is customary to build in a tolerance of about five per cent in respect of numbers and proportions. With small sites, the range of dwelling sizes should be kept small to produce economical layouts.

Because of the strong emphasis on value for money under the new financial regime in 1992/93, housing associations were being forced to seek out substantial new build projects using cheap land. As the supply of free local authority land dwindles and the depression in the private housing market lifts, housing associations may find it increasingly difficult to compete for scarce development land. It will be essential for local authorities in conjunction with housing associations to establish clearly the extent of need in their area and to translate this into a corporate strategy.

Under planning guidance note PPG3: Housing (1992), social/affordable housing is recognised as a material consideration to be specifically addressed in development plans and to be taken into account in the determination of planning applications. District councils or unitary authorities are required to adopt a corporate approach in preparing and co-ordinating development plans and area housing strategies to cover both private and social demand and to incorporate a reasonable mix and balance. Development plan policies may indicate an overall target for the provision of affordable housing throughout the plan area and may discriminate in favour of a particular form of tenure.

ASSESSMENT OF HOUSING NEED

Crude housing statistics in 1981 might seem to indicate that the overall shortage of houses has been remedied: in 1980 there was a stock of 21.0 million dwellings in the United Kingdom and a total of 20.5 million households. These figures are however deceptive in that they include an appreciable amount of housing that may be deemed unfit, and dwellings that lie vacant through changes of occupancy or to permit repairs and conversions, and they also mask wide regional variations. Indeed 4.6 million dwellings, or about one-quarter of the total housing stock in 1967, could justifiably have been classified as uninhabitable on grounds either of poor condition or of lack of basic amenities. Even in the early 1990s the number of unfit dwellings and those requiring major repairs was at an unacceptably high level (approximately 1.5 m unfit houses in England in 1991). The pressures of shortage are felt more intensely in some areas than others: Scotland and the north of England in general had a surplus of dwellings over households whereas the south-east had a deficiency in the 1980s. The agglomerated housing waiting list of the London boroughs alone stood at 234 000 in 1982. In 1981/82 there was an unusually high proportion of unoccupied private dwellings due to a considerable drop in demand, and this was repeated particularly throughout London and SE England in the recession of the early 1990s.

Increasing life expectancy coupled with the large number of births at the end of the last century produced in the nineteen fifties a population containing large numbers of elderly persons. The proportion of elderly people will continue to increase in the next ten years, but will then decline as the smaller number of persons born in the nineteen twenties and nineteen thirties reach old age. Longer life expectancy, a long-term trend towards a higher proportion of married persons in the population, and more youthful marriages, resulted in a sharp increase in the number of households and this exerted considerable pressure on the housing market in the early nineteen seventies, but this tended to decline during the following decade. In the 1980s and 1990s the demand for dwellings for single persons, particularly students, increased dramatically.

It is reasonable to assume that the proportion of married women in employment is likely to continue to rise with various implications in relation to housing. Their earnings may in the early stages of marriage assist with house purchase, and later may permit the family to move to larger or more expensive accommodation and increase the demand for such features as fully automatic heating and more labour-saving appliances, but this tendency was adversely affected by the recession of the early 1990s.

The assessment of the adequacy of housing provision is further complicated by the consistent trend towards small households over the last sixty years, which has been accompanied by a concentration on the building of five-room family houses, resulting in a mismatch between the range of dwelling sizes available in the housing stock and the range of household sizes resulting from demographic changes. The effect of salary levels and the extent of unemployment on the housing market was highlighted in the recession of the early 1990s.

Homeless households in England increased from 57 800 in December 1979 to 144 800 in March 1992, while the corresponding figures for families in bed and breakfast accommodation were 2000 and 12 200 respectively. These are good indicators of the rapidly increasing demand for rented social housing to meet the needs of unemployed or those on low incomes. The Housing Corporation (1992) assessed social housing needs in England up to 2001 at 102 500 dwellings per annum. However, the government in 1993 announced substantial reductions in social housing funding in 1994/96, with particularly severe cuts in rented housing and provision for the homeless and other disadvantaged categories. Owner occupiers were also suffering through house values falling to lower figures than mortgages, high levels of mortgage debts and numerous repossessions of mortgaged dwellings.

SOME FINANCIAL ASPECTS OF HOUSING PROVISION

The Inquiry into British Housing: Second Report (1991) contained the following very discerning and sound recommendations aimed at tackling the problems of homelessness and bad housing conditions in Britain, through a fundamental reform of the systems of housing finance.

(1) the phased withdrawal of mortgage interest tax relief (MITR), so removing the inefficiency and inequality of an indiscriminate subsidy, which favours affluent house-owners more than those on modest incomes and costs almost twice as much as housing benefit for people on low incomes;

(2) the conversion of all personal housing subsidies, especially housing benefit, into a needs-related housing allowance, which would concentrate public funds on those who need them most, and including easing the poverty trap by slowing the withdrawal of benefit as incomes rise, restoring help for younger tenants and helping older people whose modest savings bar them from benefit, poor house-owners in work who need help with mortgage costs, and owners whose properties require repairs and maintenance;

(3) the introduction of a nationwide rent-setting system based on the capital values of properties, which would give a consistent basis for rents both between landlords and within the stock of landlords, and offer adequate returns to landlords;

(4) the introduction of a whole range of new measures to reverse the continuing decline of the rented sector, embracing local authorities, private landlords and housing associations.

The overall costs of the recommendations would involve extra public expenditure of approximately £2.5b per annum, set against the saving of £9b per annum ten years later, resulting from the phased abolition of mortgage interest tax relief, providing substantial sums for investment in buildings and modernising housing.

DWELLING TYPES

There are numerous types of dwelling built by all house providers varying both in size and in physical characteristics. The majority of dwellings are still provided in houses, which can be detached,

Table 5.2 Housebuilding completions: by sector and by number of bedrooms

England and Wales	*Percentages and numbers*			
	1976	*1981*	*1986*	*1991*
Private enterprise (percentages)				
1 bedroom	4	7	12	14
2 bedrooms	23	23	28	29
3 bedrooms	58	50	40	33
4 or more bedrooms	15	21	20	24
All houses and flats (=100%) (numbers)	138477	104001	155557	132291
Housing associations (percentages)				
1 bedroom	44	58	60	42
2 bedrooms	34	28	29	38
3 bedrooms	21	12	10	17
4 or more bedrooms	1	2	1	2
All houses and flats (=100%) (numbers)	14618	17363	11055	17603
Local authorities and new towns (percentages)				
1 bedroom	32	39	46	42
2 bedrooms	26	28	30	34
3 bedrooms	38	28	21	21
4 or more bedrooms	4	5	2	3
All houses and flats (=100%) (numbers)	124512	58413	20575	8569

Source: Department of the Environment; Welsh Office.

semi-detached or built in terraced blocks, and the number of bedrooms can vary from one to four or more to suit different sizes of family. Special accommodation is needed to cater for aged persons, disabled persons, single workers, students and unmarried mothers. In the nineteen seventies in particular more emphasis was placed on the construction of flats and maisonettes, generally with a view to securing higher densities and the fuller use of highly priced land. However, most later developments reverted back to the more popular low rise developments. Some of the best planned layouts frequently contain a mixture of dwelling types.

Table 5.2 shows the percentages of different dwelling types related to numbers of bedrooms in the main categories of house providers over the period 1976 to 1991. It shows how the private sector has progressively reduced the proportion of three bedroom dwellings by increasing the other categories, while local authorities and new towns have increased the proportion of one and two bedroom dwellings at the expense of the larger houses.

House Designs

Local authorities have over the years received considerable guidance from the Ministry of Housing and Local Government, and its successor, the Department of the Environment, on the design and layout of dwellings. The guidance has been largely aimed at achieving good housing standards yet, at the same time, securing economical designs and layouts. Four-person houses generally have two bedrooms and five-person houses have three bedrooms. A variation to this general theme is the three-bedroom house which has one double and two single bedrooms, to house a family of four with two older children of opposite sexes. There are three basic designs:

(1) living room, small dining room and a working kitchen;
(2) large living room including a dining area, which may be in the form of a recess, and a working kitchen;
(3) living room and large dining kitchen.

With terraced houses, access to the rear of the house can take one of several different forms. The more orthodox arrangement consists of a common covered passageway between adjacent houses, with storage accommodation provided in outbuildings at the rear of the block. Another form of layout provides access through an enclosed store built within the walls of the house, which eliminates the draughty covered passage and provides a sound barrier between the ground floors of adjacent houses. Other designs incorporate a store at the front of the house with access through the kitchen to the rear, or alternatively access can be provided through the hall at the front and the store at the rear of the house.

Parker Morris Standards and Housing Cost Yardsticks

Since the 1960s there has been a change of approach to the design of housing accommodation stemming from the Parker Morris report (1961). In this report accommodation requirements were not based on minimum room sizes, but on functional requirements and levels of performance, with minimum overall sizes for the dwelling related to the size of family. There should be space for activities demanding privacy and quiet, for satisfactory circulation, for adequate storage and to accommodate new household equipment, in addition to a kitchen arranged for easy housework and with sufficient room in which to take at least some meals. The suggested minimum net floor area, enclosed by the walls of a dwelling, of a five-person house ranges from 97.50 m^2 for a three-storey house, 92.00 m^2 for a two-storey house and 86.40 m^2 for a flat, to 83.60 m^2 for a single-storey dwelling. For four-person houses the net floor areas range from 74.30 m^2 for two-storey houses to 66.00 m^2 for single-storey dwellings. In 1981 the Conservative Government withdrew the requirement for local authorities to adopt Parker Morris standards for new houses, which were often of a higher standard than houses built in the private sector, and rationalised Parker Morris standards became the norm with a reduction in floor space standards. Commonly adopted floor space standards in 1993 ranged from 44 m^2 for a one bedroom, 1/2 person bungalow or shared access flat to 54 m^2 for a two bedroom, 2/3 person bungalow or shared access flat, 69.5 m^2 for a two bedroom, 3/4 person house and 84 m^2 for a three bedroom, 4/5 person house. But even these floor areas were under pressure in social housing in 1994/96 with progressive reductions in government funding. The primary aim of relating space to family size and family needs remains just as relevant today as in 1961. At the same time the abolition of housing cost yardsticks and their replacement by the securement of value for money provided architects and quantity surveyors alike with scope for greater initiative to provide the type of dwellings that the occupiers wanted, even though it could result in greater rationalisation.

Table 5.3 Comparative costs of different dwelling types

Dwelling form	*Cost index*
Terraced bungalows	96
2-storey houses	100
Detached bungalows	105
2-storey flats	108
3-storey houses	111
3-storey flats	116
4-storey flats	131
5-storey flats	138
Other high rise forms	144+

Source: Colquhoun and Fauset (1991).

The operation of the latest cost procedures is described in chapter 10. Table 5.3 shows the comparative costs of different dwelling forms, while figure 5.1 shows real house price variations between 1971 and 1993.

Three-storey Houses

Three-storey houses in terraced blocks form a convenient and economical way of housing large families. Where densities are in the order of 250 to 290 bedspaces per hectare, and this type of block is used in conjunction with three-storey corner flats and four-storey blocks of maisonettes, high blocks can be avoided. Typical arrangements of three-storey houses are:

Floor	*Scheme A*	*Scheme B*
Ground floor	Hall, store and dining/ kitchen.	Hall, garage and all-purpose room.
First floor	Living room, bedroom and bathroom.	Living room and working kitchen.
Second floor	Three bedrooms.	Three bedrooms and bathroom.

Flats

Tower blocks of eleven storeys or more in height were mainly provided to accommodate smaller

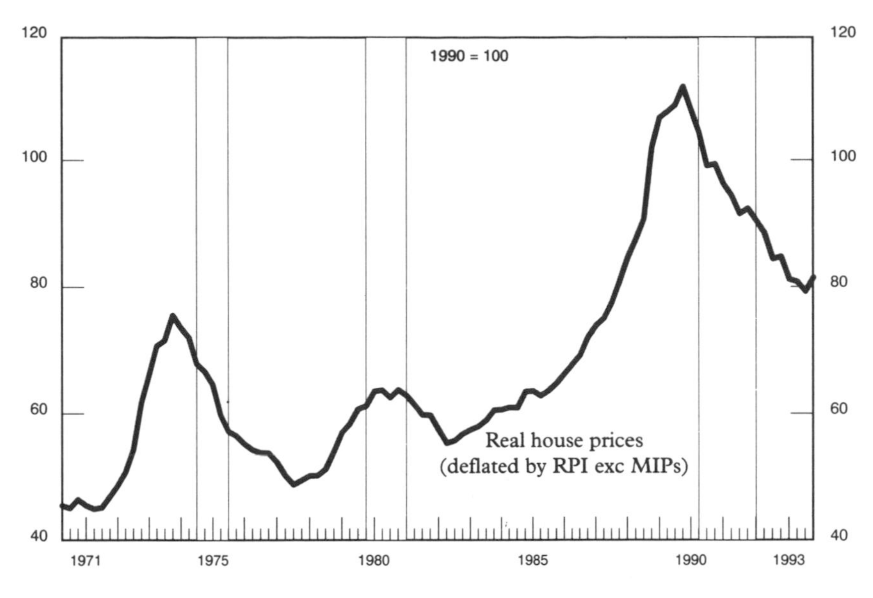

Figure 5.1 Real house prices (Source: HM Treasury, Economic Briefing No. 6, Feb. 1994)

families. A common arrangement was approximately one-half one-bedroom and one-half two-bedroom flats. Access was obtained in a variety of ways – daylighted common hall, enclosed common hall, cross-ventilated common hall or enclosed central corridor. Typical arrangements would be four flats per floor with the first three forms of access and eight flats per floor with corridor access. There would probably be two staircases and two lifts in all cases. This type of development lost its appeal in the early nineteen seventies in the United Kingdom, mainly on economic and social grounds.

In countries like Hong Kong with an acute land shortage and high population, large numbers of high tower blocks are still being built. For example, one private development completed in 1985 at Taikoo Shing comprised 10 813 flats in 51 cruciform tower blocks, each 22 to 30 storeys high, accommodating approximately 45 000 persons, sited around interconnecting pedestrianised and landscaped podiums. A typical Hong Kong 35-storey public housing block needed stabilisation piling to a ground depth of 20 to 40 m depending on the geological conditions (Hong Kong Government, 1989). Hong Kong housing developments are well described and illustrated by Seeley (1992).

High slab or linear blocks have been used where it was not practicable to house all the larger families in houses or four-storey maisonettes. The slab blocks of rectangular plan may contain primarily four or five-person living units arranged as flats or maisonettes. Access may be by balconies, corridors or staircases. The maisonettes are often arranged as three-floor units, whereby access to two maisonettes is obtained from the intermediate floor via halls and staircases, with the larger bedroom to each dwelling also located on this floor. The lower floor contains the living and dining areas, kitchen, bathroom and smaller bedroom for one dwelling, and the upper floor contains similar accommodation for the other dwelling. High slab blocks, 10 to 13 storeys in height are the predominant form of public housing in Singapore, where a significant proportion of the land surface has been reclaimed from the sea. Residential developments in Singapore are well described and illustrated by Seeley (1992).

Four-storey maisonette blocks usually consist of three-bedroom units spread over two floors, with access to upper dwellings by a balcony served by a common staircase.

Three-storey flats usually contain two and three-bedroom flats (two on each floor) served by a common staircase. All the rooms may be entered from an entrance hall or access to bedrooms may be secured through a separate inner lobby. Other variations are working-kitchens and dining-kitchens.

Elderly Persons' Dwellings

Many elderly people who are still fit prefer to live in fully self-contained dwellings. The most popular form of dwelling is the single-bedroom bungalow with a living room, small kitchen and combined bathroom and WC. There are also a large number of elderly persons who, by reason of age or infirmity, are unable to manage entirely on their own in normal type bungalows. Where these persons can be relieved of some of the burdens of normal tenancy and given friendly oversight and a limited amount of assistance by a warden, they can continue to lead independent lives and remain an integral part of the community.

In the last three decades many schemes of grouped flatlets or sheltered housing have been implemented. Basically these schemes usually provide a number of centrally heated flatlets, some communal facilities and a warden's residence. The flatlets can be designed in a number of ways: with bed-sitting rooms and kitchen fitments in cupboards, with separate kitchenettes, or with separate bedrooms. Another alternative is to locate beds in bed recesses in living rooms which can be closed off with curtains.

Baths were originally shared in the ratio of about one bath to three to five persons, and WCs in the ratio of one appliance to two or three persons. Most of the later developments incorporated individual baths and WCs. Communal facilities often include a common room, laundry, heated drying room(s) and guest room(s). The warden's quarters adjoin the flatlets and generally consist of a two-bedroom dwelling. One interesting form of development is traditional construction in a courtyard layout containing mainly two-storey cottages, but also some flats, one of which is occupied by the resident warden. The majority of older people in sheltered housing are aged 75 or over. Extra-sheltered housing has evolved to meet their needs by providing a restaurant and replacing the warden with a manager and staff to provide 24 hour cover.

Dwellings for Disabled Persons

Owing to the high cost of building work, the construction of an extension to provide suitable accommodation for a physically-disabled person is, in many circumstances, viewed as a last resort. Wherever possible, any necessary alterations are carried out within the structure of the existing dwelling, such as installing a home lift (usually of the vertical variety) or a WC under the staircase, to reduce costs to a minimum.

Most extensions contain ground floor bedrooms with or without sanitary accommodation, but some provide only a cloakroom. The layout of the dwelling and external appearance will both influence the nature of the extension. The addition of a two-storey extension incorporating a lift (£3500) and new sanitary appliances to serve the needs of a severely physically-disabled person could cost £17 000 to £19 500 in 1993, while the provision of a ground floor dialysis room could be in the order of £7000.

It is vital that disabled persons should not be placed at a disadvantage in securing suitable accommodation. Payments are available towards the cost of adaptations by means of a disabled facilities grant.

Standard House Types

In 1993 Housing for Wales introduced a range of standard or pattern book of internal house designs for use by housing associations in Wales. The proposal aimed to secure a common understanding of what space standards constituted good design for any given unit type/occupancy level, combined with a specification identifying high quality/low running cost building components which enabled planned maintenance requirements to be predicted with confidence.

In the preparation of the standard house types, evidence was sought from housing associations, volume house builders, local authorities, and numerous professional/trade bodies.

Standard house types only relate to floor plans, and a performance specification on components. External design will be determined by housing associations in liaison with their consultants/contractors and the local planning authority. Some of the broad design principles that have been established were as follows:

(1) ground floor circulation areas will be a minimum of 900 mm clear width to meet the needs of people with restricted mobility and to be accessible for wheelchair users;

(2) stairs must rise from a circulation area and not from habitable rooms, with a straight flight 864 mm wide to facilitate installation of a stair lift when required;

(3) all unit types shall conform to the furniture sizes, accommodation and circulation space requirements in *Space in the Home* (DOE Design Bulletin 6);

(4) access to rear gardens to be gained without passing through living rooms;

(5) minimum kitchen and general storage areas and kitchen unit sizes have been specified;

(6) thermal efficiency from central heating, double glazing and insulation have been stressed with its impact on running costs for occupiers;

(7) the specification of homes and external works should incorporate recommendations arising out of *Secured by Design* available from the Police Architectural Liaison Officers.

A performance specification that identifies high quality/low running cost building components and reflects the work done on whole life costing was also compiled.

From April 1994, all Welsh housing associations undertaking new developments were expected to use the pattern book of housing designs, wherever appropriate. Housing for Wales will implement detailed scheme scrutiny on projects that do not comply with standard house types.

Another set of standard designs was produced by the contractors, Lovell Partnerships, which they termed New Generation house types. The system provided 37 different templates with the use of computer aided design (CAD), and it was claimed that a 10–15 per cent change in detailing could effect a 100 per cent change in appearance.

Lovell do not accept the conventional philosophy that design and build will of necessity produce any less quality than competitive tendering with architect design. It was also considered that an improving economy combined with lower grant rates would make it increasingly difficult for housing associations to achieve the same standards.

HOUSING REQUIREMENTS OF OCCUPANTS

General Background

As a result of long-term trends in demographic structure and house building the range of house sizes in this country's current stock of dwellings is out of alignment with the range of household sizes. It is sometimes asserted that this situation can be remedied by the movement of households from larger dwellings to smaller ones, when families have reached the later stage of the family cycle; but this can only be a partial remedy as older households become less mobile. It is likely that more than one-fifth of future housing demand will be of the non-family type, catering for single, widowed and divorced persons. Of the remainder a proportion of the dwellings will cater for small households, such as newlyweds and elderly couples.

The architect concerned with mass housing often has little opportunity for direct contact with the people who will occupy his houses. Industrialised and prefabricated building methods which presupposed long production runs militated against modifying the design during a building programme, so that user needs required clear definition at the outset. Furthermore it is often argued that people do not know what they want or that the needs they express are only in terms of the types of house with which they are familiar, or of the social norms to which they subscribe. A natural extension of this line of reasoning is that housing needs would be better defined by a panel of experts, whose views would be less influenced by their immediate social environment. The more recent findings of anthropology, sociology and psychology have shown this view of 'natural man' to be untenable. If the pattern of user activities is related to the characteristics of the house plan on the one hand, and to characteristics of the user such as age and stage in the family cycle, education and social class on the other, it is possible to determine whether the user's pattern of activity is due to constraints imposed by the physical environment, or whether they reflect the needs of different categories of user. In assessing housing requirements of occupants in Hackney, an analysis was made of the features that most families wanted in a new home, such as privacy, safety, convenience and attractive setting. The mounting pressure to limit costs and secure maximum value is leading to a measure of standardisation in constructional methods.

The majority of persons in this country favour low rise dwellings and the reasons most commonly advanced for not wanting high rise dwellings are:

(1) inconvenience stemming from use of lifts and stairs;
(2) lack of garden;
(3) lack of privacy;
(4) sharing of some facilities;
(5) noise from adjoining tenants, despite sound insulation;
(6) unsuitability for young children; and
(7) height factor and possible lack of safety.

Indeed, as long ago as 1875 Octavia Hill expressed the view that even the poorest would like to have their own homes to themselves, and she found from her work in Deptford that the smallest cottages were the most popular, because in them a family was more likely to achieve its ambition of having a whole dwelling to itself, and this view was confirmed by surveys undertaken by Alice Coleman in the late 1980s.

The demand for new semi-detached houses and town houses remained relatively strong in the nineteen eighties and early nineteen nineties, although the demand in the private sector was depressed in the latter period because of the recession.

Changing occupational status, car ownership and educational attainments of a considerable proportion of the population must influence the types of accommodation required. In quite recent times there has been emphasis on the separation of the laundry from the kitchen and, in some cases, a demand for personal bathrooms, playrooms and 'dens', to provide separate space for individual activities. There has also been a trend toward open plan interiors and more living space and it has been suggested that the average family house with a present day floor area of about $84 \, m^2$ could increase to around $100 \, m^2$ by the end of the century. The possibility of increased car ownership could accentuate present access problems to houses and may point the way to more housing layouts incorporating arrangements for the separation of vehicles and pedestrians.

As well as requiring larger homes, occupants of the future will in all probability also expect better homes providing greater comfort in terms of warmth, ventilation, noise levels and privacy. Different patterns in the use of space within the home may develop in response to a more even distribution of warmth, as the proportion of centrally heated houses continues to increase.

In 1993 the National Federation of Housing Associations (NFHA) reiterated its view that affordability was the most important indicator of the success of the 1988 Housing Act, in the Federation's evidence to the House of Commons Environment Committee investigating the Housing Corporation. The Federation warned that government plans to progressively reduce average grant rates to 55 per cent in 1995/96 would worsen affordability problems, and preliminary estimates suggested that a significant proportion of tenants in working households would have to pay in excess of 40 per cent of their disposable income in rent. This would result in increased housing benefit dependency, exacerbating the poverty trap.

Tenant's Surveys

Salford University published a 'Manual for the Design and Implementation of a Tenants' Satisfaction Survey', which provided a step-by-step description of how to measure tenant satisfaction or dissatisfaction with housing, based on the University's survey for Rochdale MBC in 1989. The survey measures the degree of tenants' satisfaction with the following aspects:

housing size
heating arrangements and costs, and dampness problems
repairs service
access and security
noise nuisance and parking facilities
neighbourhood services and facilities – local transport, health centres, schools and parks
community safety issues – youth facilities, litter, vandalism, lighting and personal safety
safety on the streets, break-ins
housing area services and information
rent collection procedures
tenants' representation issues
special housing needs (Paterson, 1992).

A customer survey carried out by the Borough of Enfield in 1992 showed that most of the ten highest ranked items related to every day housing management concerns. These included the need for a clean and efficient estate, improving personal security on the estates, the time a property is vacant kept to a minimum, improved repairs service all round, and people in the right accommodation for their needs (Artingstall and Jeffreys, 1992).

Warwick DC engaged consultants on a competitive basis to undertake a comprehensive tenants' attitude survey in 1988. The Council used the information so gained to produce an action plan, which included closer consultation with tenants, increased training budget, adoption of a mission statement, and increased lighting on estates. A second survey was carried out in 1990 to establish how the action plan had been received by tenants. Tenant satisfaction levels increased significantly between 1988 and 1990, except for quality of workmanship on repair work. Warwick DC subsequently introduced their own, more selective and locally based, surveys using the lessons and skills learnt in working with the consultants (Dyas, 1992).

Many housing associations were carrying out tenant satisfaction surveys in the early 1990s to ensure closer working relationships with their tenants and to provide an improved and more efficient and responsive service. They were also setting minimum targets of performance and regularly measuring the actual achievements against the targets that had been set. Some of the targets might require subsequent adjustment in the light of experience. These were welcome moves by truly caring housing organisations.

PATTERNS OF DEVELOPMENT TO MEET VARYING DENSITY REQUIREMENTS

Higher Residential Densities

The protagonists of high density developments base their philosophy on the maximum use of highly priced land, restricting the loss of agricultural land and reducing the extent of urban sprawl. However the present average increase in agricultural productivity of 1.3 per cent per annum will more than offset the additional land likely to be needed for all urban uses. There is also a general antagonism towards high density residential development. Mothers experience real difficulties in coping with small children in high flats, and certain values associated with the privacy of the family appear to be threatened by flat life. Much has still to be learnt about the design, management and maintenance of communal spaces in and around flats. Certain types of persons such as single persons or small families whose occupation necessitates living near their work can be reasonably well suited in high rise development but they are in a minority. Economic and social objections to high rise blocks have however precipitated interesting experiments in low rise high density development which will be described later in this chapter. Nevertheless, the large proportion of childless households and the deficiency of small dwellings suggest that flats, but in relatively small numbers, have a useful part to play in the housing stock.

New towns have been attacked by architects and planners for wasting land on low density systems with huge inter-spaces – windy, draughty and miserable places where human beings are unhappy because of lack of urban feeling. Yet the urban densities of new towns are almost identical with, or higher than, those of older towns. Hence urban design character depends on the local, small conurbations of densities and not on average density across the area as a whole. Gross density offers a better measure of residential development and asserts that raising gross densities to high levels produces relatively small savings of land and in so doing involves disproportionate increase in net density. At gross densities of 150 to 160 persons per hectare, dwellings take up about half the gross residential envelope, the remainder being needed for ancillary uses; hence any increase in gross density involves double that increase in net density. It would undoubtedly help our land economy if housing and local facilities were planned together at medium gross densities of seventy-five to 100 persons per hectare.

A deeper understanding of relationships between density and other physical variables must be achieved if increased efficiency in the use of land is to be obtained. Greater theoretical experimentation with physical models is required to establish the effects of density and built-form upon standards of environment. By the use of computer graphic modes, the layouts can be adjusted to achieve a density, choice of mix and site layout based on sound financial considerations and which create a design solution which is also aesthetically and environmentally acceptable. There is scope for considerable work in this area.

The mechanism of density operates within the overall planning framework which by stipulating levels of density protects wide social interests but, at the same time, because of its generality, often inhibits development for which there may be genuine personal preferences.

As to densities of housing development, detached houses rarely exceed 20 dwellings/ha, semi-detached: 30, and terraced houses: 50. However in cities densities can be as high as 500 persons or bedspaces/ha, but even when land is scarce it is generally considered undesirable for densities to exceed 350 persons/ha. Furthermore there is often little saving with high densities because of the ancillary needs of access, car parking, waste disposal units and essential open space between and around building blocks (Seeley, 1992).

Forms of Development

The form or pattern of development will be very much influenced by the density standard prescribed for the site, although diverse variations in the form and disposition of the building blocks are also possible to achieve a specific density. The primary need is to translate the terms of the brief into a total volume or bulk of building per unit area (hectare), and then to assess the cost effects of distributing this amount of building on the site in blocks of varying height, plan area and shape, with due regard to daylighting, sunlight, access, privacy and other factors. The main objective is to minimise the proportion of the more expensive types of block and to maximise the less costly ones, and this entails a thorough knowledge of the relative costs and capacities of low, medium and tall blocks, of various plan shapes and sizes and a variety of structural forms. Furthermore, consideration must be given to running and maintenance costs in addition to initial capital costs, as dwellings in tall blocks are more expensive in both first and future costs than those in low rise developments.

In the preparation of a layout a designer is faced with a number of relatively fixed factors which have considerable influence on his design, such as nature of site, residential density to be achieved, approximate proportions of dwellings of varying types (for example, one, two, three and four-bedrooms), minimum spacing of blocks to secure satisfactory daylighting and adequate privacy, and in public housing space standards of dwellings may also be specified. Daylight indicators are used at the planning stage to ensure that the new blocks respect the light of others and permit the recommended standards of light to be attained within them.

Apart from these fixed factors there are many other matters upon which decisions have to be made by the client or designer or both. This second category of factors includes methods of accommodating large and small families (in what type of block), proportion of low rise development and whether two or three-storey, acceptability of four-storey maisonettes, height limitations on blocks, number of dwellings per floor of tall blocks, extent of provision of children's play spaces and other ancillary facilities, ratio of garages and/or car spaces to dwellings, extent of landscaping and road pattern.

It is a generally established principle that large families are best housed in low blocks and only small families accommodated in high blocks which are normally confined to a small proportion of the total housing stock. The proportion of houses that can be incorporated in a development will be largely dependent on the density adopted. In medium density developments a maximum allocation of houses is a primary aim, but with higher densities it would be unwise to include a large number of houses as this would result in the provision of many excessively tall expensive blocks to achieve the required density. Three-storey houses and flats and four-storey maisonettes have a useful role to play in medium density schemes. The introduction of one or two high blocks enables a high density to be achieved for that part of the site and so releases a larger area for cheaper and more popular low rise buildings. Where very high densities are required, it is more economical to use tall blocks with a greater number of flats per floor, possibly increasing the provision from four to eight, but developments of this kind in the future in the United Kingdom are likely to be very few.

At an overall density of about 250 habitable rooms per hectare (hrh), or possibly expressed in bedspaces or persons per hectare, it is desirable to provide a proportion of houses which does not require a density for the remainder of the site in excess of 275 to 300 hrh; this density can probably be obtained with four-storey maisonettes. A high proportion of houses entails the use of expensive high blocks. At high densities a proportion of dwellings in high blocks is generally necessary, and this can be kept to a minimum only by limiting the proportion of rooms in houses. Major advantages stemming from the use of four-storey maisonettes include increased numbers of dwellings with private gardens and with entrances at ground level.

The following general criteria warrant consideration in the design of housing schemes.

(1) Wherever the required density can be achieved without the use of high blocks this should be done on grounds of economy alone. The ideal solution is to incorporate the maximum number of two-storey houses with the balance of three or four-storey blocks.

(2) Where some high blocks are needed to secure an exceptionally high density, such as 350 hrh, the number of high blocks should be kept to a minimum and use can advantageously be made of cheaper four-storey maisonettes.

(3) Compact layouts assist in keeping the quantity of higher rise building to a minimum and in securing the greatest possible amount of low rise development to obtain the required density. It also helps to provide a more urban character to the development.

(4) More economical layouts can often be secured by the use of three-storey houses.

Popular and economical minimum frontage, three-storey blocks of flats at Harlow New Town were used on restricted infill sites and these are illustrated in figure 5.2. Each block contains six flats and four garages with two-person (one-bedroom) flats and garages on the ground floor and four-person (two-bedroom) flats on the first and second floors. The frontage of each block was kept to a minimum and was determined by the sum of the widths of four bedrooms, plus stairs and wall thicknesses. The minimum bedroom width of 2.55 m conformed to the recommendations in DB6 (*Space in the Home*) and is compatible with the width of a garage below, and the flat frontage of 6.30 m proved to be an economical span for precast concrete floor units

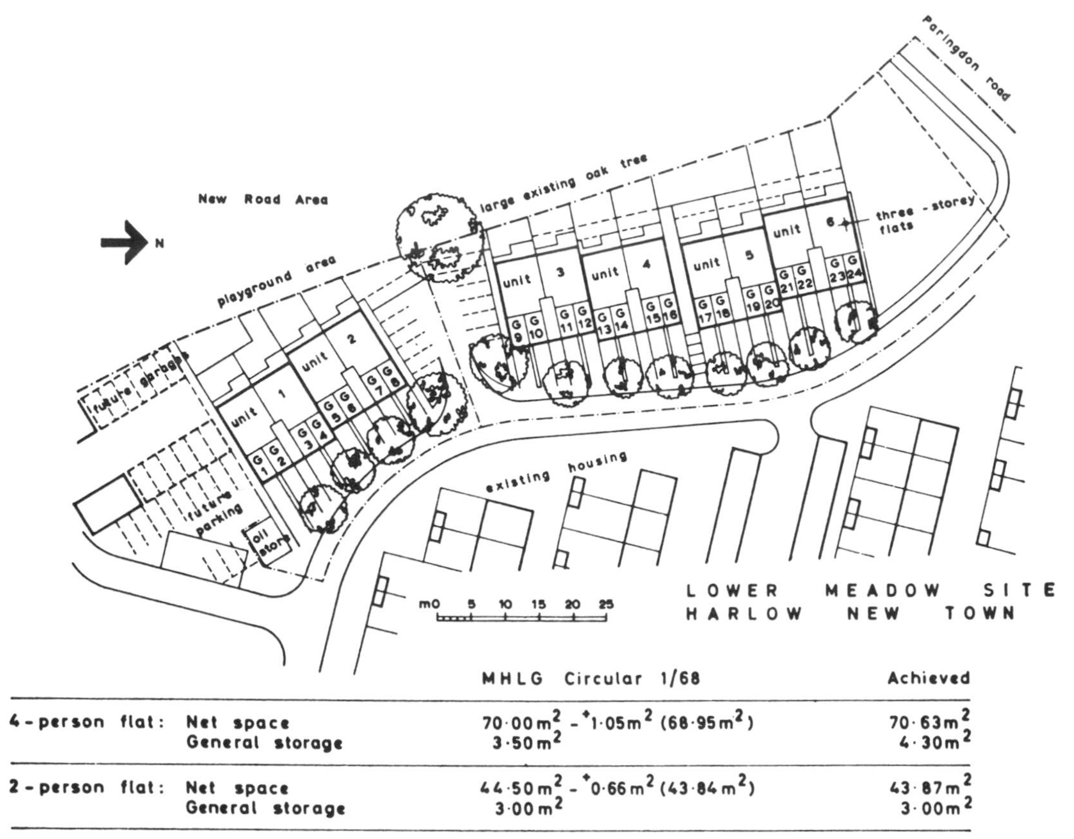

		MHLG Circular 1/68	Achieved
4-person flat:	Net space	$70{\cdot}00m^2 - {}^{+}1{\cdot}05m^2$ ($68{\cdot}95m^2$)	$70{\cdot}63m^2$
	General storage	$3{\cdot}50m^2$	$4{\cdot}30m^2$
2-person flat:	Net space	$44{\cdot}50m^2 - {}^{+}0{\cdot}66m^2$ ($43{\cdot}84m^2$)	$43{\cdot}87m^2$
	General storage	$3{\cdot}00m^2$	$3{\cdot}00m^2$

+ Maximum minus tolerance of 1½%

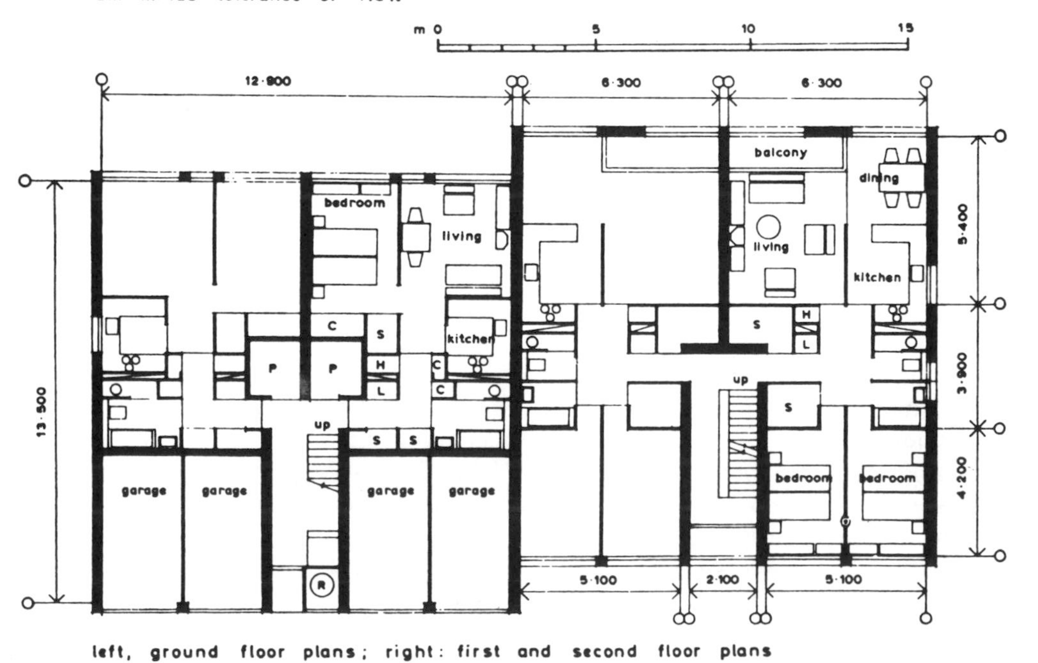

Figure 5.2 Three-storey flats infill development – Harlow New Town

supported by crosswalls. All controlling dimensions are in multiples of 300 mm.

A considerable improvement in architectural standards could, the author believes, follow from consideration and implementation of many of the main criteria advocated by HRH the Prince of Wales (1989), who has made a detailed study and investigation of the design of the built environment, and it is important for design teams to direct as much attention to appearance, as to cost, time and quality. The main themes of the Prince's study were harmony (each building in tune with its neighbour); enclosure giving a feeling of well being, neighbourliness, cohesion and continuity; materials (using local materials as far as practicable to retain and enhance local character); community (providing the right sort of surroundings to create a community spirit, with community involvement); and quality of character, fostered by attention to detail and human scale.

Housing Layouts

There are three main forms of housing estate layout: the conventional or corridor street layout; use of culs-de-sac and courts and finally various types of Radburn layout. Each will now be considered in turn.

Conventional layouts. In these, most of the houses front onto and have direct access from a development or estate road, and the back gardens to the houses are enclosed. Access to the backs of houses and enclosed gardens may be obtained through side passages, covered ways or stores to houses. The dwellings are sited on either side of a street, and separated from each other by front gardens, public footpaths and carriageway. This form of layout can become monotonous unless the houses are skilfully sited and external elevations of houses varied. There is no attempt to secure separation of vehicles and pedestrians.

Culs-de sac layouts. In the garden cities a conscious attempt was made to produce more attractive layouts by grouping houses around culs-de-sac and grassed areas. These arrangements have been incorporated into many public and private housing schemes during the last sixty years. Large numbers of culs-de sac can however produce a feeling of restlessness and long culs-de-sac without adequate turning areas can be very inconvenient. The provision of culs-de-sac in moderation can be an attractive feature and also assist in securing economical layouts by opening up awkwardly shaped interior plots of land and by improving the house to road ratio, with houses surrounding the head of the cul-de-sac.

Radburn layouts. The first form of Radburn layout was used at Radburn, New Jersey, USA between the wars, in which the houses with their private gardens backed on to culs-de-sac and fronted on to communal landscaped areas, through which footpaths gave direct access to schools, shops and amenities without crossing a road. A variety of forms of Radburn layout are now used but the primary aim of all of them is to separate pedestrians and vehicles and so secure more pleasant and safer residential environments. Indeed the choice of route for pedestrian ways should have priority over roads, with regard to directness of route, gradients and intersections with roads.

Figure 5.3 shows one form of Radburn layout at Cwmbran New Town which clearly demonstrates the dominance of the central footpath system, with the culs-de-sac so aligned that they do not interfere with the flow along the main pedestrian routes. The majority of the houses have an east/west aspect and the private back gardens face west to south for maximum sunlight.

Figure 5.4 illustrates another form of Radburn layout at the expanding town of Andover, Hampshire. It is an interesting scheme consisting of a series of closes extended laterally from short culs-de-sac off perimeter distributor roads. It possesses a compact urban character with vehicles restricted to the perimeter of the residential block. Private back gardens face west or south but quite a high proportion of houses front on to rear boundary fences of other houses.

The grouping of houses in a Radburn scheme is likely to conform to one of three basic types.

Figure 5.3 Radburn layout – Cwmbran

(1) Vehicle cul-de-sac, with a turning circle or hammerhead at the end of the carriageway and with individual or grouped garages adjoining it.

(2) Garage court, with the carriageway widened to form a single large enclosure for vehicles and grouped or individual garages.

(3) Pedestrian forecourt, whereby the head of the cul-de-sac or garage court is extended to form a paved pedestrian area from which each house is entered. Garages are grouped away from the houses. Variations on the pedestrian forecourt include the pedestrian link where a pedestrian way, lined with houses, forms a link between two vehicle access points, and the pedestrian passageway where at high densities the forecourt becomes part of a network of footways between houses.

Radburn layouts do, however, increase the complexity of schemes and involve additional external works. In many instances, however, occupiers express a preference for a more traditional layout and the extra cost of Radburn schemes is not entirely offset by improved amenity.

The relatively high cost of building land

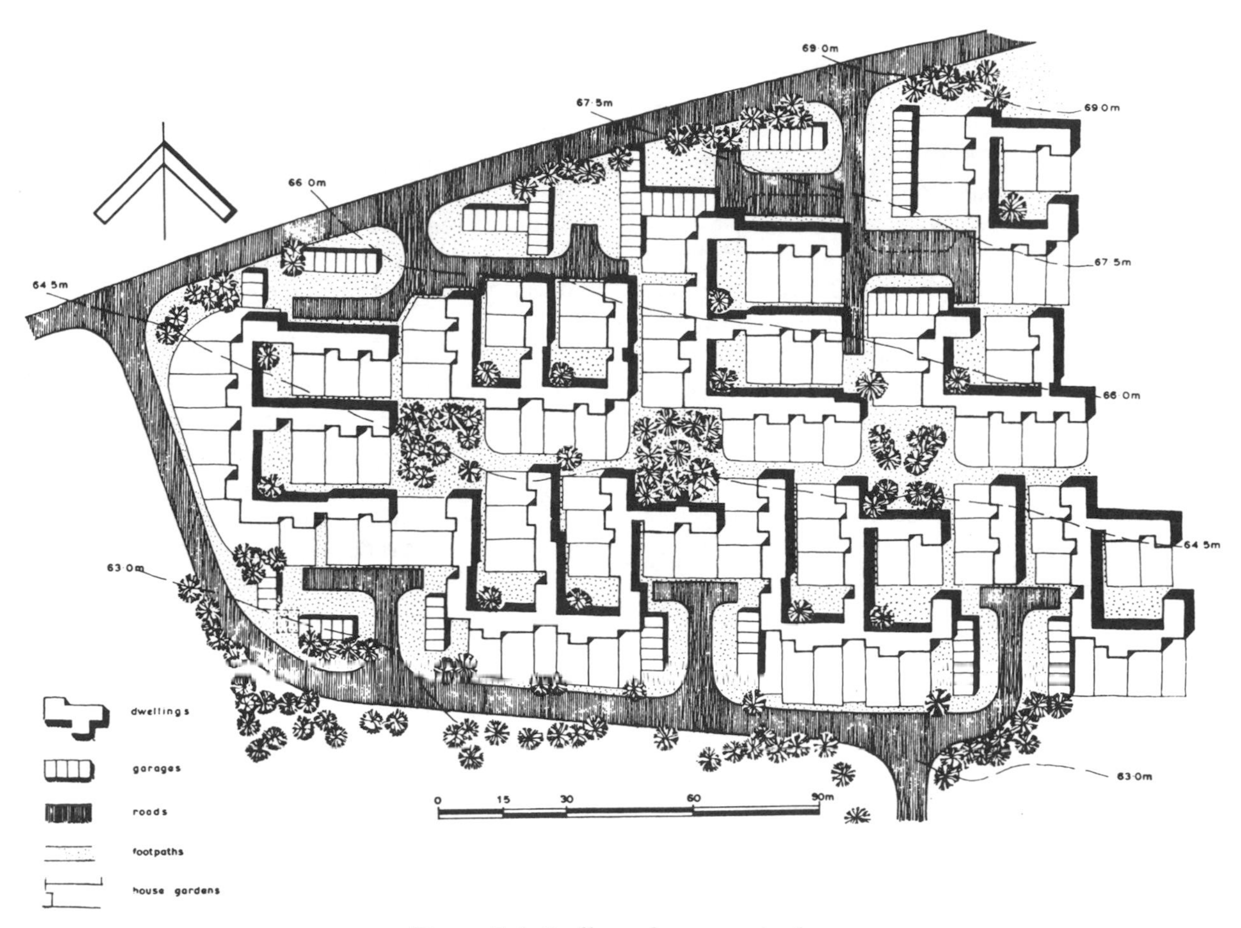

Figure 5.4 Radburn layout – Andover

coupled with the general desire to give coherence to residential neighbourhoods is likely to lead to many new developments being within the gross residential density range of 175 to 225 persons per hectare. To achieve these densities with larger homes, accommodation for two cars per household and parking space for visitors will require skilful planning and must mean that mixed rise developments will become the norm, possibly incorporating an increasing amount of patio-housing.

Other Considerations

Much sound advice on the formulation of housing layouts is contained in a Greater London Council study (1978) and in Colquhoun and Fauset (1991). In urban infill situations a primary aim is to maintain existing character while exerting individuality. In particular the retention of existing mature trees and shrubs of the right species can enhance both the functional and aesthetic qualities of a scheme and help to make the new raw development look more established.

With new development the spaces between buildings should stimulate the human senses and, in general, provide a sense of enclosure. A wide variation in the form of the spaces and the vegetation provided will enhance the quality of the environment.

It is also vital to consider carefully the hierarchical arrangement of roads ranging from local distributors to access to clusters of dwellings, and the need to integrate parking into residential areas. The planning for pedestrian movement is of prime importance, both in relation to convenience and visual attractiveness. The GLC study (1978) and DoE/DTp (1992) provide good practical advice on road and footpath layouts.

The value of landscaping of inner city estates was highlighted in a Civic Trust assessors' report on their commendation for the Meakin estate in Southwark. They stated 'Undoubtedly this landscaping scheme must have greatly enhanced the lives of the residents, particularly the children, and for the minute amount of money spent, a huge environmental gain has been achieved. The general effect has been to revitalise what must have been an extremely depressing, overbearing series of spaces, and to rescue the site from probable vandalism and eventual demolition'. This stresses the need for full consideration of environmental aspects.

Woodward and Campbell (1982) appraised an interesting inner city residential development undertaken in Covent Garden in the late nineteen seventies and illustrated in figure 5.5. The site occupies a prominent position in central London and the development is mixed – comprising housing, welfare and commercial facilities. The stipulated density of 472 pph was achieved, even although 43 per cent of the building area is non-residential. There are 102 dwellings (of which 60 are two-person flats, pairs of these normally being built over a large family flat) in eight separate blocks containing 12 or 13 units each. The mix of flats is important since two-person flats can be 20 to 25 per cent more expensive, when expressed in terms of cost/m^2, than the larger four-or-five-person flats. Nevertheless, the overall 1982 equivalent cost of £420/m^2 is high but there are a number of reasons which partially account for this.

(1) Covent Garden is an expensive area in which to build.

(2) To harmonise with existing buildings and to achieve the required density, multistorey development was inevitable.

(3) The variable mix and size of flats and the differing numbers of storeys reduced repetitive work.

(4) The high incidence of two-person flats increased the average area provided for each person.

(5) The cost was based on net habitable area rather than gross floor area.

(6) The ratio of exterior walling to net habitable floor area of 0.95 : 1 reflected the complexity of layout, but the low window to wall ratio reduced the impact of this element.

Incidentally the development is low on sunlight and daylight provision with one-third of the living rooms facing north, and the comparatively large envelope area results in increased thermal loss.

The choice of a house by a buyer often represents a compromise, reflecting what is available rather than what is desired. Hence it may be good policy to divide housing estates into areas with alternative layouts of varying densities and prices, in order to establish whether there is a demand for improved house designs. While not wishing to question the sincerity of the authors of design guides, an appraisal of purchasers' preferences would assist in the design of future estates.

Some Overseas Developments

Hong Kong

By 1989, almost half of the 5.6 m population in the Territory were accommodated in public sector housing, with nearly 0.5 m people still in temporary housing. Land formation is the key to much of the development process in Hong Kong, because 80 per cent of the Territory is very hilly terrain, and much development is dependent on the reclamation of land from the sea, river valleys and other low lying land prone to flooding in prolonged periods of torrential rain.

A typical public housing estate of 20 000 or more residents involves a lead time of about five years from the inception of the planning brief to completion. The main development stages are usually 12 to 15 months for acquisition and site clearance, 12 to 15 months for site formation, 6 months for piling and 24 to 30 months for building. A typical 35-storey public housing block needs stabilisation piling to a ground depth of 20 to

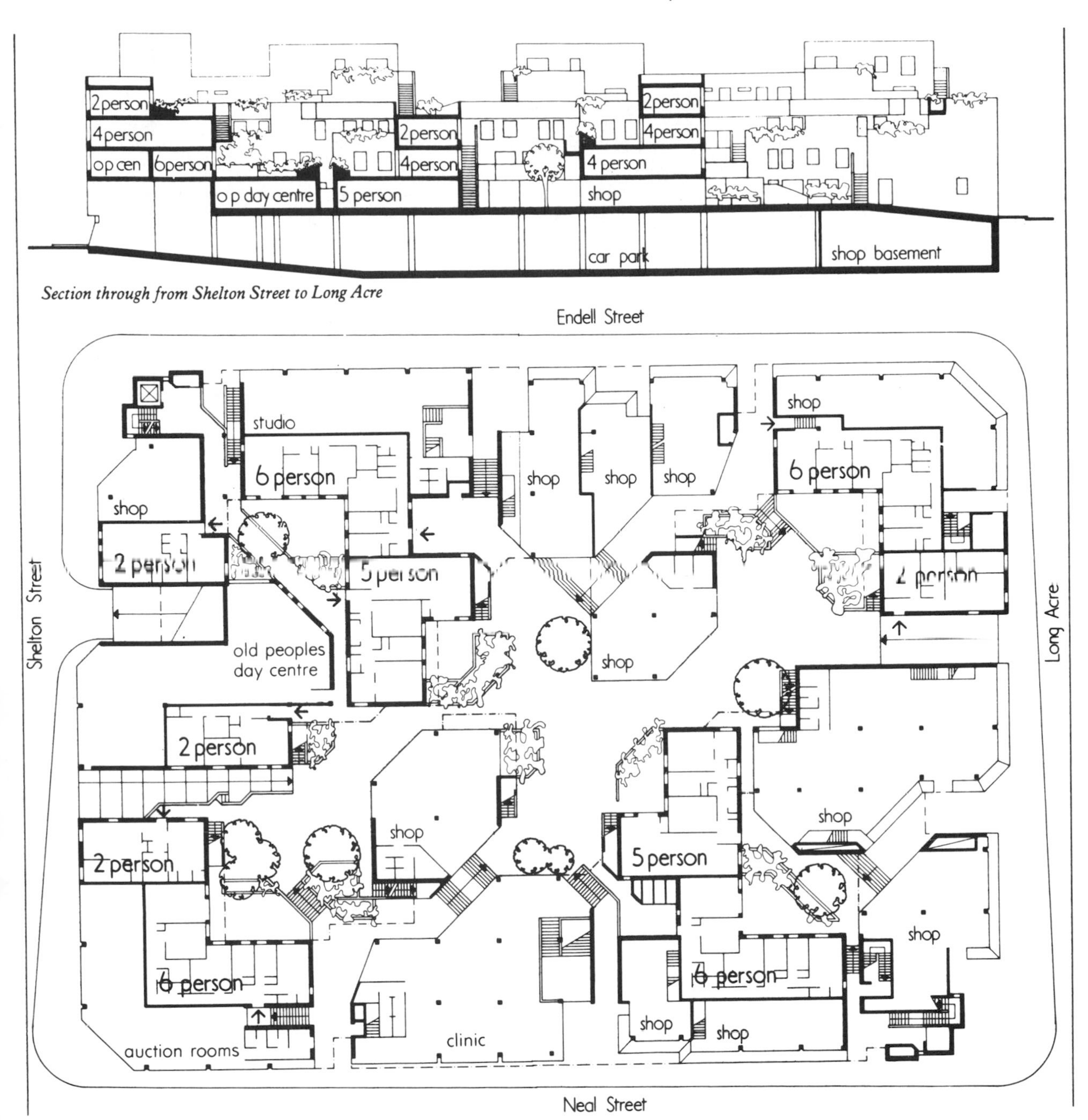

Figure 5.5 Residential development – Odham's Walk, Covent Garden (Source: Architect's Journal, 3 Feb. 1982)

Figure 5.6 Developments from Tsuen Wan to Tsing Yi Island, Hong Kong (Source: Territory Development Department, Hong Kong)

40 m, depending on geological conditions. Figure 5.6 shows very dramatically housing development stretching out from Tsuen Wan New Town to Tsing Yi Island. It illustrates the enormous tower blocks under construction with extensive ancillary works, including the dual lane carriageway and three span bridge, with the substantial water frontage and mountains in the background (Hong Kong Government, 1989).

Singapore

By 1989 the Singapore Housing and Development Board (HDB) had completed about 650 000 dwelling units, housing 2.3 m people or 87 per cent of the population. Between 1960 and 1984, the urbanised area of Singapore rose from about 25 to over 45 per cent, and there is a limit to how much land can be reclaimed from the sea on this relatively small island. Hence the gross residential density for a new town as a whole is pegged at 64 dwelling units or 280 persons/ha, while the net density of residential areas is 200 dwelling units or 880 persons/ha. At this density and given the relatively large flat sizes, the plot ratio of the built up area is around 1.6 to 2.3. The building blocks are mostly 10 to 13 storey slab blocks generally about 100 m long with 5 to 20 per cent in 4-storey buildings without lifts, and another 5 to 10 per cent in 20 to 25 storey point blocks (HDB Singapore, 1989).

It has been found in Singapore that variations in block design need not be complex and costly. By varying the small details of the block, it has often been found possible to achieve different and attractive visual impacts. The blocks can be accentuated by the different use of material, colour, column and facade detailing, treatment of roof and other components, with the blocks bent, curved and even looped around to enhance the environment, all built in an attractive landscaped setting. Distinctive roofscapes, block designs, vibrant colours and geometric shapes all help to create a character and identity for the buildings and to reflect the multi-cultural heritage (HDB Singapore, 1985).

In the new towns of the 1970s such as Ang Mo Kio, Bedock and Clementi, solar orientation was an important criterion, and there was also more flexibilty in the arrangement of blocks, resulting in a better relationship between building and street, and an attempt was made to distribute open spaces more evenly throughout the new

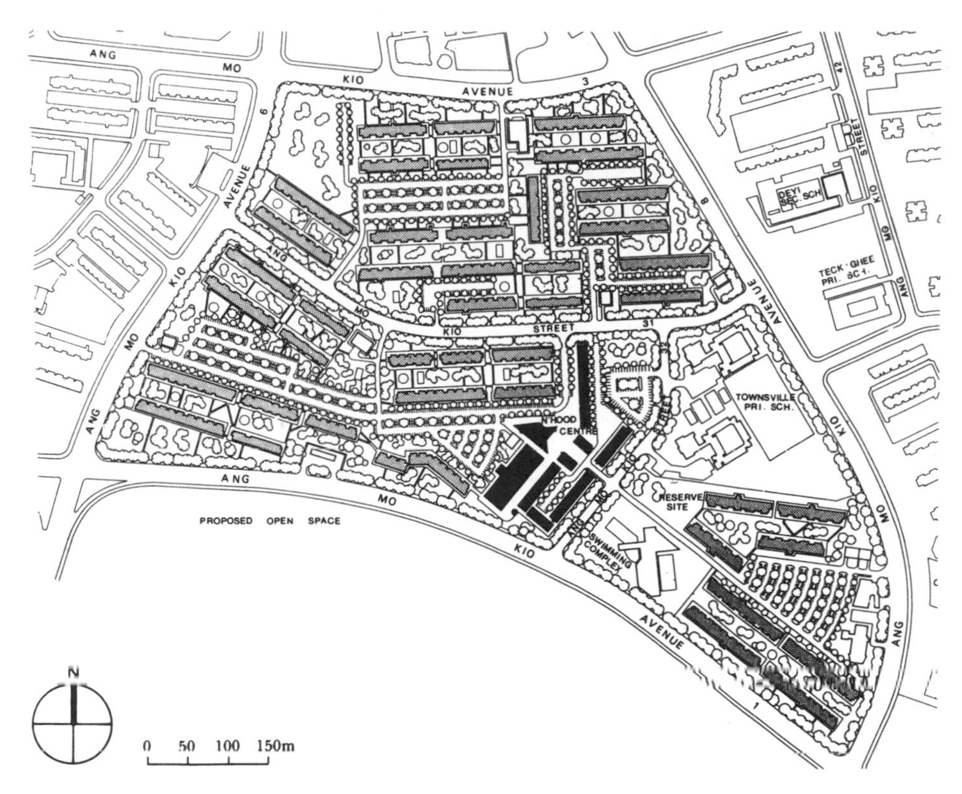

Figure 5.7 Layout plan, Ang Mo Kio New Town, neighbourhood 3, Singapore (Source: A.K. Wong & S.H.K. Yeh)

town. At the same time, some low rise blocks were built admidst the high rise blocks to provide added interest and an improved sense of human scale (Wong and Yeh, 1985). Figure 5.7 illustrates the layout plan of a neighbourhood of Ang Mo Kio New Town with its extensive landscaping and a neighbourhood centre, swimming complex and primary school, while figure 5.8 shows an aerial view of part of Ang Mo Kio New Town, with its predominantly high rise development which is so different to the UK.

In the Singaporean new towns of the 1980s, precincts were introduced as the basic planning concept. A precinct consisted of a grouping of four to eight residential blocks accommodating between 500 and 1000 households. With this smaller scale of division within neighbourhoods of 4000 to 6000 dwellings, it was possible to give the blocks a defined visual identity as a unified group of buildings, with a consistent application of selective architectural elements and themes, down to quite fine details. Thus enclosures of communal space were created connected by a network of footpaths and encompassing children's playgrounds and other communal facilities. There was also an emphasis on street architecture and continuity, as illustrated in figure 5.9, showing part of a neighbourhood in Yishun New Town, surrounding a school site (Wong and Yeh, 1985).

ECONOMICS OF HOUSING LAYOUTS

General Considerations

An IOH/RIBA paper (1983) rightly stressed the importance of adequate briefing and feedback, embracing all relevant design and maintenance personnel and, preferably, future residents especially where the development involved higher

Figure 5.8 Housing development at Ang Mo Kio New Town, Singapore (Source: HDB, Singapore)

densities and a preponderence of rented housing. The brief should, at the very least, secure agreement on site boundaries, dwelling mix, form of building, including possible use of prefabrication, industrialised housing techniques or timber frame construction, type of access, layout features, management of scheme, capital/maintenance cost budgets, heating arrangements, vehicular servicing of the development, waste disposal, children's play, and policy towards housing elderly and disabled persons.

The demand for higher densities produces the need for deeper cost investigations and tighter cost control.

The first stage of cost investigation should consider the most economical use of a particular site. In schemes of comprehensive redevelopment, consideration should be given to the best combinations of blocks in different forms and dispositions, to meet the site conditions and housing policy and accommodation requirements. In general, the rarely required very high density is achieved most economically by a combination of a few very high blocks and a majority of lower blocks. More frequently the demand will be for compact and relatively high density low rise developments. Such considerations underline the importance of cost investigation into patterns of development. After the basic development and design decisions have been made, there is still ample scope for investigating the use of various materials and forms of construction. Their relative

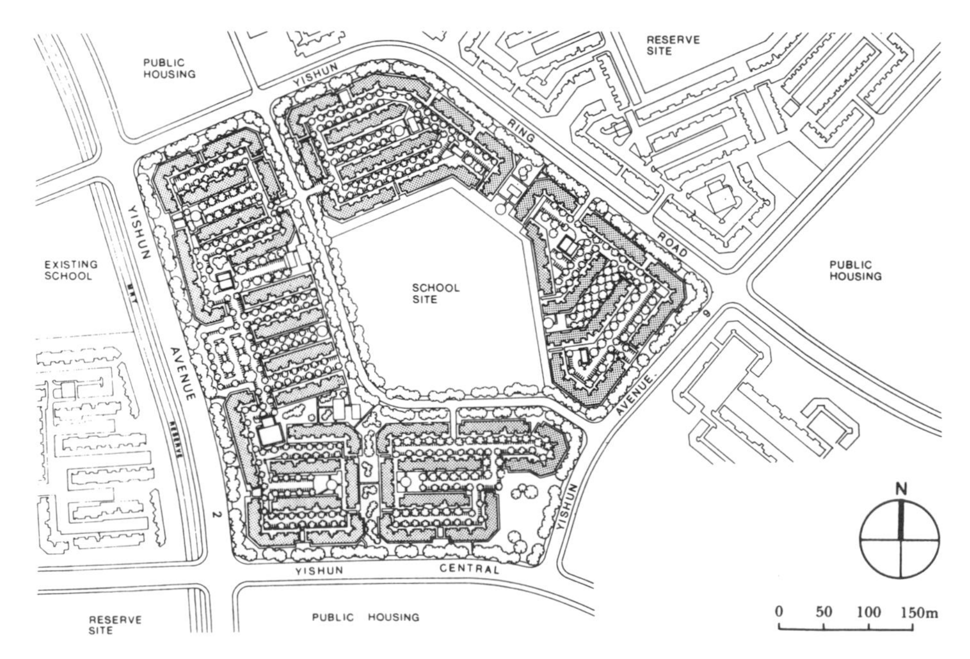

Figure 5.9 Layout plan, Yishun New Town, neighbourhood 2 (part), Singapore (Source: A.K. Wong & S.H.K. Yeh)

costs can be scheduled in a 'cost study' to enable the architect to select any desired combination in full knowledge of their cost implications. The aim is not to produce the cheapest scheme but one which maintains a proper balance of cost in relation to requirements, function, standards and appearance.

High flats generally cost about sixty to eighty per cent more per square metre of gross floor area than two-storey houses. Superstructure costs, especially those of frames and floors, account for just over half this extra cost, and service and lift costs cover most of the remainder. High flats also show a greater range of costs than low flats or houses. This was mainly attributable to variations in costs of external walls and floors. The shape of blocks and the resultant wall to floor ratios are of special significance in this connection.

A cost study at Cumbernauld New Town emphasised the need to vary housing designs to meet different ground conditions. In this particular scheme the gentler slopes were developed with three-storey, six-person, narrow-fronted terraced dwellings sited diagonally to the contours. Slopes of between one in fourteen and one in twenty were developed with stepped terraces of two-storey houses built normally, or diagonally to the contours. Whilst on the steepest parts of the site, split-level houses were introduced in terraced blocks built parallel to the contours, and stepped and staggered to provide diagonal pedestrian access with easier gradients.

Peterborough Development Corporation started planning low cost houses in 1974 at a density of about 38 dwellings/hectare. The development consisted of 90 patio style houses and bungalows, each with an integral or detached garage and additional hard standing space, arranged in 12 small courtyards around double entrance culs-de-sac.

Three house designs were incorporated in the scheme (a) 41 single storey patio bungalows with 2 bedrooms and integral garage (about 61 m^2); (b) 32 single storey patio bungalows with 3 bedrooms and detached garage (about 71 m^2); and 17 two-storey patio houses with 3 bedrooms and detached garage (about 79 m^2).

The houses were sited in a cluster layout to achieve a 60 per cent sharing of walls to reduce construction, drainage and heating costs. All bathrooms in a group of four houses back on to one another enabling sewers to be laid in each alternate court. Particular attention was also paid to privacy of tenants, child safety and reduced road costs, using single lane internal roads with passing points. The scheme proved very popular.

The DoE Development Management Working Group (1978) listed the following ways in which efficiency and value for money could be improved in local authority housebuilding.

(1) A corporate approach to housing development with an identified development manager and speedy decision procedures;
(2) setting a cost target for each scheme and working within it;
(3) monitoring the overall costs of schemes from their inception;
(4) checking for value for money against established criteria;
(5) not having preconceived ideas about how to arrive at the end product;
(6) using good simple designs frequently, rather than one-off plans;
(7) considering the builder's advice on what can be produced most efficiently.

The Working Party further recommended that local authorities should establish total cost targets for all their main types of dwellings. These targets would be set by investigating their own most economical schemes, considering the costs of nearby authorities and comparing with equivalent costs in analogous private sector developments.

BRE 350 digest (1990a) deals with the practical effects of microclimate on site layouts, showing that solar access and wind control/design are both likely to have significant effects on the size, form, massing and orientation of buildings. Shelter from northerly winds is likely to be most appropriate in general inland sites with no strong directionality. In other cases, such as where funnelling occurs, on or near coasts, or where protection from driving rain is sought, other criteria could apply. The main implications of solar access for the layout of dwellings are:

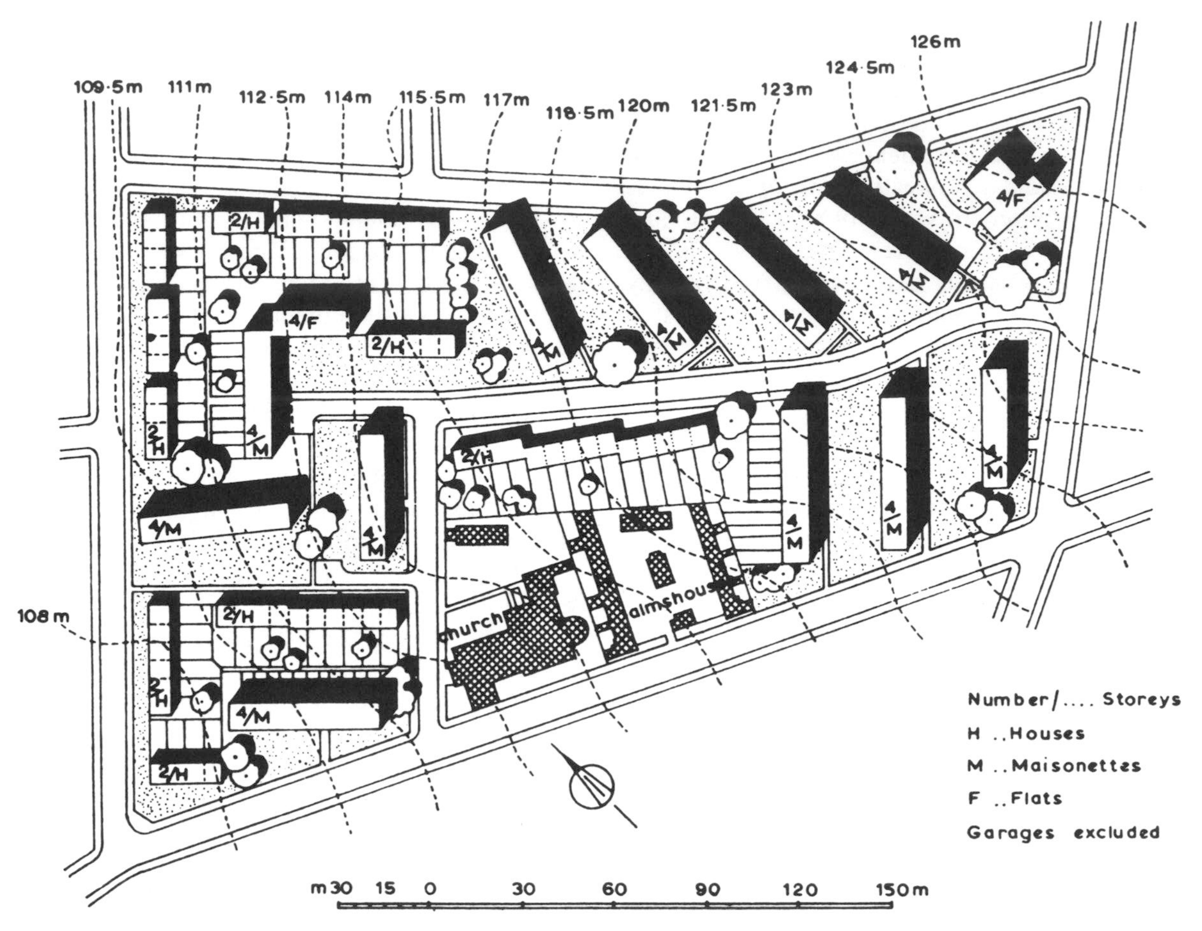

250 habitable rooms per ha: 4·95 ha:
1234 habitable rooms: 327 dwellings.

Types of dwelling and percentage required	2 - Storey houses		4 - Storey flats		4 - Storey maisonettes		Dwlgs.	Rms.
	Dwlgs.	Rms.	Dwlgs	Rms.	Dwlgs.	Rms.		
BSR. 1p. 1rm. 5%.	-	-	16	16	-	-	16	16
1 BR. 2p. 2rms. 5%.	-	-	16	32	-	-	16	32
2 BR. 4p. 3rms. 20%.	-	-	8	24	58	174	66	198
3 BR. 4p. 4rms. 25%.	-	-	-	-	82	328	82	328
3 BR. 5p. 4rms. 30%.	14	56	-	-	84	336	98	392
3 BR. 6p. 5rms. 8%.	26	130	-	-	-	-	26	130
4 BR. 7p. 6rms. 7%.	23	138	-	-	-	-	23	138
Totals	63	324	40	72	224	838	327	1234
Percentages	19·3	26·2	12·3	5·8	68·8	67·9	-	-

Figure 5.10 Residential development at 250 habitable rooms per hectare

(1) aim for maximum road length running within 15° of E–W;

(2) arrange plot shapes to allow wide, south facing frontages, to maximise solar gain through windows, although this will also increase cost;

(3) plant coniferous trees to the north of houses and deciduous trees to the south, but so sited as to avoid excessive overshadowing and possible damage by roots to buildings and services;

(4) select the dwelling type and form to limit overshadowing as illustrated in BRE digest 350 (1990a).

Cost Implications of Alternative Layouts

As one of the pioneers of housing cost investigation, the Ministry of Housing and Local Government (1958), undertook some layout and cost studies of a site in Birmingham based on three density standards. The most relevant economical layouts are illustrated in figures 5.10 and 5.11 by kind permission of the Controller of Her Majesty's Stationery Office. Figure 5.10 shows an economical layout for the site to meet a density requirement of 250 habitable rooms per hectare, with about two-thirds of the accommodation in four-storey maisonettes and about one-quarter in two-storey houses. An alternative layout which provided about forty per cent of the accommodation in eleven-storey flats or maisonettes and another forty per cent in two-storey houses was seventeen per cent more costly.

Figure 5.11 shows a suitable scheme for achieving a residential density of 350 habitable rooms per hectare on the same site of 4.95 ha. One-third of the accommodation is provided in thirteen-storey flats or maisonettes and the proportion in two-storey houses drops to about fifteen per cent, whilst four-storey maisonettes account for one-half of the accommodation. An alternative layout which reduced the proportion of four-storey maisonettes and replaced the two-storey houses with three-storey houses was marginally more expensive. Another scheme increased the proportion of two-storey houses to twenty-two per cent, and thirteen-storey flats and maisonettes then accommodated half the total number of rooms and this was about six per cent more costly than the illustrated layout. Yet another alternative incorporating five-storey flats and maisonettes was marginally more expensive with balcony access, and considerably more expensive with staircase access, and the overall effect in any case was very monotonous. A further layout incorporating two blocks of 16-storey flats gave a density of 400 hrh.

With densities up to 250 persons per hectare it is advisable to provide the maximum amount of accommodation in two-storey houses, which are approximately 16 per cent cheaper than four-storey maisonettes. With higher densities the aim should be to secure a high proportion of four-storey maisonettes, which will probably be about 12 per cent cheaper than dwellings in five to six storey blocks of flats. Increased construction costs will at the very least be partially offset by reduced land costs per dwelling. The lowest development costs are probably in the density range of 120 to 200 persons per hectare.

A scheme of mixed rise residential development at Portsdown showed that seventeen-storey blocks in calculated brickwork were only twelve per cent more expensive per square metre of floor area than four-storey blocks. A residential development in north London showed that a nine-storey block with reinforced concrete frame and *in situ* concrete foundations was sixty-three per cent more expensive in cost per square metre of floor area than two-storey houses with brick crosswalls and concrete strip foundations (exclusive of external works, preliminaries and insurances), although the foundation cost of the taller block was twenty-seven per cent cheaper than that of the houses, in terms of cost per square metre of floor area.

A final cost comparison relates to low rise development in an old person's grouped housing scheme where one-bedroom bungalows, each with a floor area of 49.5 m^2, are eleven per cent more expensive in cost per square metre of floor area than two-bedroom houses. The warden's house and communal block are 13.5 per cent more costly.

Road and servicing costs rise sharply from the

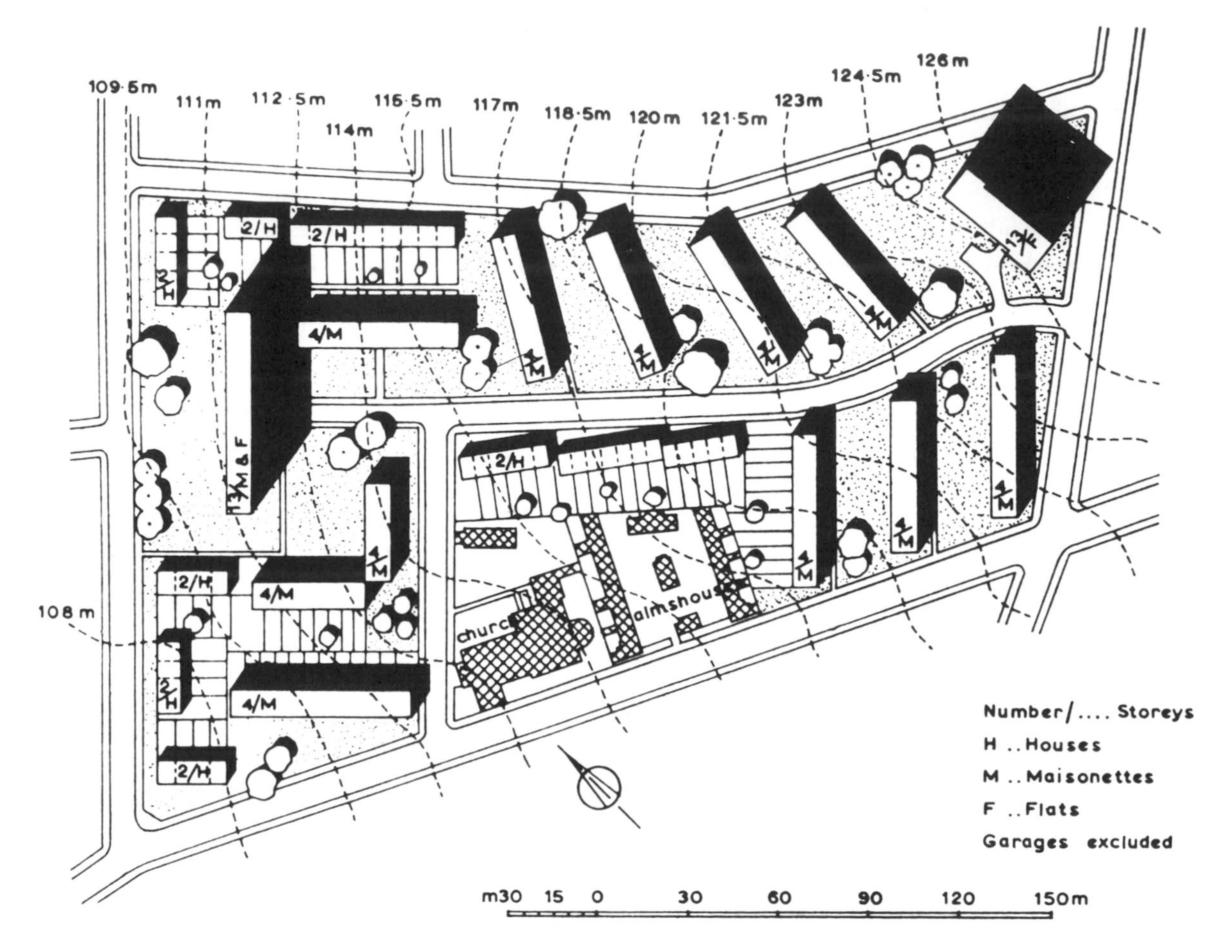

Types of dwelling and percentage required	2-Storey houses		4-Storey maisonettes		13-Storey flats		13-Storey maisonettes		Dwlgs.	Rms.
	Dwlgs.	Rms.	Dwlgs.	Rms.	Dwlgs.	Rms.	Dwlgs.	Rms.		
BSR. 1p. 1rm. 5%	-	-	-	-	24	24	-	-	24	24
1BR. 2p. 2rms. 5%	-	-	-	-	24	48	-	-	24	48
2BR. 4p. 3rms. 20%	-	-	30	90	59	177	-	-	89	267
3BR. 4p. 4rms. 25%	-	-	30	120	-	-	84	336	114	456
3BR 5p. 4rms. 30%	-	-	136	544	-	-	-	-	136	544
3BR. 6p. 5rms. 8%	15	75	22	110	-	-	-	-	37	185
4BR 7p. 6rms. 7%	32	192							32	192
Totals	47	267	218	864	107	249	84	336	456	1716
Percentages	10·3	15·5	47·9	50·3	23·5	14·5	18·4	19·5	-	-

Figure 5.11 Residential development at 350 habitable rooms per hectare

use of multistorey housing, with an effectivity ratio between 0.5 and 0.7, to detached houses beside residential roads with footpaths on both sides, where the effectivity ratio could be as low as 0.1.

A cost comparison should be made of alternative housing layouts for a site, identifying the quantity and cost of roads, paths and services per dwelling. An economical layout can be achieved by making the best use of road frontage and the judicial incorporation of culs-de-sac, closes, courtyards and similar intimate features to give a sense of neighbourliness and tranquillity, and still provide an effective traffic distribution system. The aim should also be to provide attractive layouts and interesting street pictures with a good backcloth of skilfully designed ornamental landscaping. The reduced area of roads and paths will also result in lower maintenance and operating costs. In like manner, whole life cost comparisons should be made between different road and path surfacings and between hard and soft landscaping but, at the same time, giving adequate attention to the overall appearance and the benefits to residents, in the nature of a mini cost benefit study.

CAR PARKING PROVISION

There seem to be wide variations in the standard of car parking provision in different housing schemes. For instance, at Harlow New Town the Development Corporation required 1.3 car spaces per household but the County Council insisted on raising the provision to 2.0 car spaces per household, except for old person's dwellings where the rate of provision was 0.5 spaces per dwelling unit. By way of contrast, Cumbernauld New Town decided on one car space per household in high rental areas and ⅓ space per household in lower rental areas.

In 1993 many British planning authorities recommended 1.75 car spaces per one-bedroom dwelling and 2 per two-bedroom or larger dwelling, incorporating assigned grouped garages and the remainder in unassigned hardstandings. Standards of provision should ideally be governed by experience rather than applied standards as underprovision creates frustration amongst car owners. Regard should be had to the type of dwellings, predicted rate of growth of car ownership and reasonable provision for visitors' parking.

The cheapest form of provision is that of open hardstandings which are often screened with trees for amenity purposes. They give no protection to cars from either the weather or vandals but their cost is unlikely to exceed £300 per car space (1993 costs), and their land requirement per car space is about 11.80 m^2. Some of the objections are overcome if small numbers are grouped close to dwellings.

Carports consisting of a roof and possibly one side have become quite popular in recent years. They give only limited protection from the weather and 1993 costs were in the order of £900 each (land requirement about 12.20 m^2).

Detached garages erected close to dwellings give a maximum of convenience and protection but are costly, probably in the order of £2700 (land requirement about 13.33 m^2). The cost will be reduced if the garage is built beside a house or in pairs between houses. Three-storey houses can accommodate garages on the ground floor when costs are likely to be around £3000 each (1993 prices).

On local authority housing estates it is customary to provide garages in single-storey blocks to serve groups of houses. Costs are reduced to about £2100 per garage and they permit greater flexibility in the letting of dwellings. They are less convenient for occupiers, are more vulnerable to vandalism and result in vehicular noise concentration at specific points on housing estates. Probably the most economical form of lock-up garage layout is that shown in figure 5.12 where constructional costs and apron widths are kept to a minimum. Circular blocks of lock-up garages have been provided at Cumbernauld New Town.

On a high density residential development it may be desirable to provide multistorey garages with two or three storeys, to conserve land and meet increasing car parking needs. The land requirement per car space with a two-storey car park is around 12.10 m^2. The cost is likely to be in the order of £2700 to £3300 per car space.

Building Economics

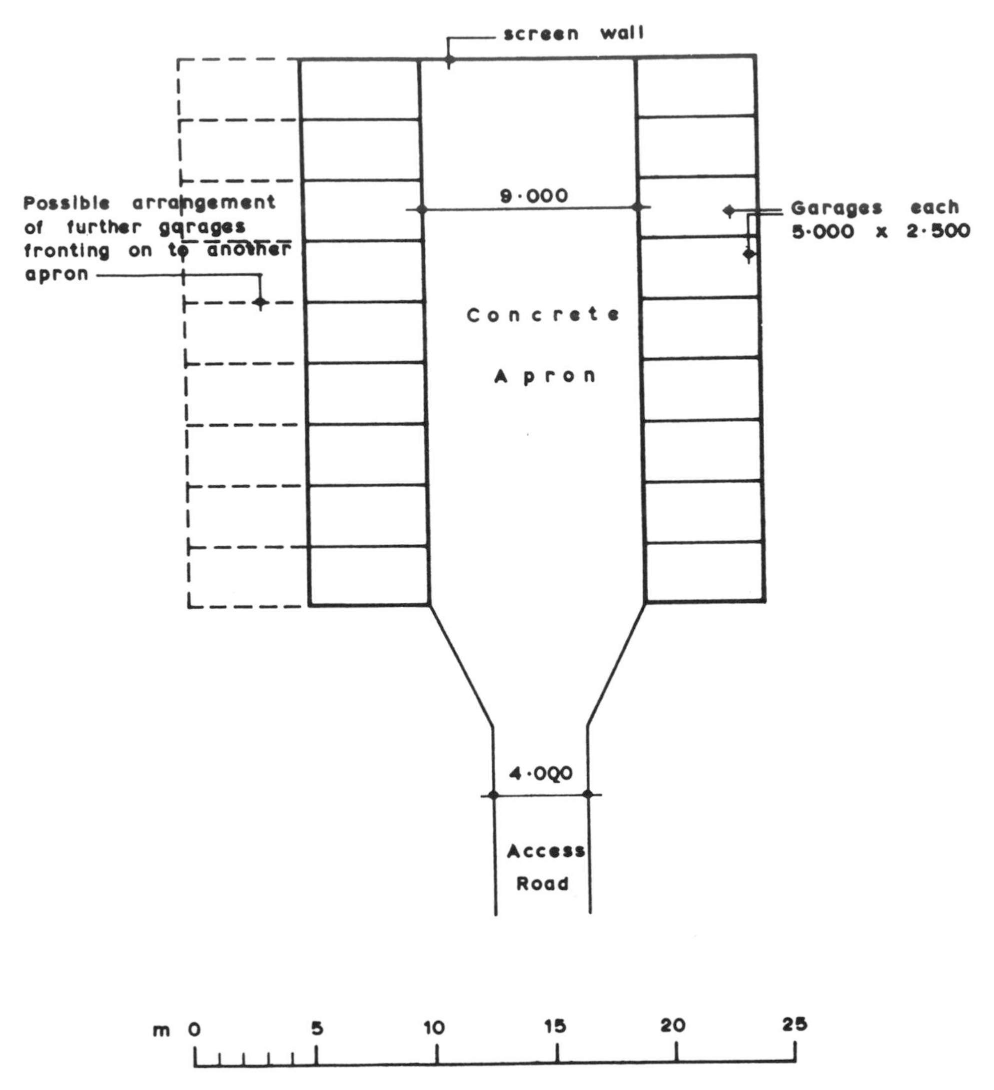

Figure 5.12 Blocks of lock-up garages

Another alternative with high rise blocks is to provide basement car parking areas under some of the blocks, particularly where they are sited on sloping ground. This is a very expensive method of garage provision and could cost as much as £4200 per car space (1993 prices).

REHABILITATION AND MODERNISATION OF OLDER DWELLINGS

Modernisation of Older Houses

In 1967 the Minister of Housing and Local Government stated that about five million houses in England and Wales were unsatisfactory to a lesser or greater degree. Approximately four million of these houses lacked one or more of the basic amenities such as internal WC, fixed bath, hot and cold water supply, and wash basin. The estimate of the number of unfit houses in England in 1982 was 3.5 million, and by 1986 had reduced to 1.2 million, although there were at least an equal number with major defects. Unfortunately the English House Condition Survey indicated 1.46 million unfit dwellings in England in 1991.

Indeed many pre-war houses lack proper bathrooms and hot water supply systems, and still in a significant number of cases the only WC in the house is entered from outside. Many of these houses have been or are being modernised by the provision of modern bathrooms, modernised kitchens, internal WCs, efficient hot water supply systems and often some form of central heating. A typical example is Nottingham Community Housing Association (NCHA), a medium sized housing association, which planned in 1993 to have all the older housing stock, including over 800 ex British Coal houses, centrally heated within five years. There are three principal methods of approach in providing minimum basic needs.

(1) An internal rearrangement on the ground floor to provide a bathroom and WC, if sufficient space is available.

(2) Conversion of the third bedroom into a bathroom; this is probably the cheapest method but reduces the amount of accommodation. Even where a house is underoccupied, the third bedroom is useful in times of sickness and to accommodate visitors.

(3) Conversion of a store adjoining the house, which, because it is usually built of half-brick walls, often requires the provision of an additional 75 mm inner skin or half-brick outer skin and suitable insulation.

The hot water supply may be obtained from a back boiler behind the living-room fire, with an immersion heater for summer use, or the installation of suitable electric or gas water heaters. Old appliances, such as shallow Belfast sinks, are replaced with new fittings and it is usually necessary to rewire the dwelling. In 1993 the cost of these minimum basic modernisation works ranged from £3000 to £6000 per dwelling, depending on the extent and nature of the work. It must be emphasised that the type of work described constitutes a bare minimum and would be below Parker Morris standards.

Apart from the modernisation of pre-war houses, many local authorities require large sums of money to rectify defects on post-war dwellings. Birmingham City Council alone estimated a repair bill of £264 m in 1980, where 17 000 flats had severe condensation and damp problems. Many post-war dwellings built by non-traditional methods have particular problems of their own such as settlement of foundations, heavy condensation, poor insulation, damp and high fire risk. An estate of about 1000 houses completed in 1970 in Southwark needed £11.5 m spent on them in 1980 to rectify major deficiencies and demolition was being considered as an alternative solution.

In the early 1990s the backlog of housing maintenance in the public sector had increased at an alarming cumulative rate, and local authorities often lacked the finance to rectify this, despite heavy rent increases. In January 1994 there were still 128 000 defective dwellings under local authority ownership as compared with 31 000 under private ownership (Minister of Housing in House of Commons, 18/1/94). This enabled central government to accuse local authorities, often unfairly, of being inefficient and to encourage the transfer of their housing stocks to tenant co-operatives, existing housing associations or newly formed companies, following a long period of permitting local authority house sales to occupants on large discounts. By 1993, the government was also considering introducing compulsory competitive tendering (CCT) for housing management services, which needs to be carefully handled to ensure the provision of an efficient and caring service.

Improvement Grants

The Housing Act 1974 and the Housing Act 1980 both aimed at conserving the national housing stock of older houses, by providing for three types of grant to persons with an interest in the property, namely intermediate grants, improvement grants and special grants.

House Renovation Grants

There are different kinds of financial assistance to suit different needs, depending on the type of property involved and the scale and type of works to be carried out. Grants may be available towards the cost of repairs, improvements, conversions of buildings, and of providing facilities and adaptations for a disabled person. In addition, minor works assistance may also be available to help owner-occupiers and tenants, in receipt of income-related benefit, who want to carry out small scale works to their homes.

The aim of the renovation grants system is to provide financial help for those who can least afford to pay for works to their properties. Therefore the amount of grant awarded will be decided by a test of the financial resources of the applicant. In the case of an owner-occupier or tenant the test will assess the household income and include some savings, while for a landlord the test will apply to the increased rental income expected from, and, in some circumstances, the likely enhanced capital value of the property.

There are four main types of renovation grants:

Renovation grant: for the improvement and/or repair of houses, including maisonettes and flats, and for the conversion of houses and other buildings into flats for letting. The main purposes are to bring a property up to the standard of fitness for human habitation (mandatory grant); to repair and/or improve a property beyond the standard of fitness, which is described later (discretionary grant); for home insulation (discretionary grant); for heating (discretionary grant); for providing satisfactory internal arrangements (discretionary grant); and for conversions (discretionary grant). Mandatory grants were under review by the government in 1994 and many builders feared that they could be reduced or even abolished.

Common parts grant: for the improvement and/or repair of the common parts of buildings containing one or more flats, ranging from purpose-built mansion blocks to small scale conversions and these are normally discretionary grants.

HMO grant: for the improvement and/or repair of houses in multiple occupation (HMOs) and for the conversion of buildings into HMOs and are usually discretionary grants.

Disabled facilities grant: for adapting, or providing facilities for, the home of a disabled person to make it more suitable for the occupant to live in, and is usually a mandatory grant.

In addition to the previous four grants *minor works* assistance may be available to those in receipt of income-related benefit for carrying out small scale works (costing up to £1000 in 1993), including insulation work.

The local authority may require that the works are carried out to a particular specification required by them. They will also require an invoice or receipt from the builder or professional adviser and, in the case of a DIY applicant receipts for the materials used. If the freehold or leasehold of the property is sold within a certain period of time after receiving a grant, the recipient may be required to notify the local authority and pay back the grant or part of it. An owner-occupier who sells the property within three years may have to repay all or part of the grant except in certain exceptional circumstances.

The local authority may decide to include the property in a group repair scheme with the occupier's agreement. Thus works which are required to the external fabric of a block of houses are done at the same time by the same contractor. The works are organised and supervised by the local authority, who will meet at least 50 per cent of the costs.

The Fitness Standard

A property, including an HMO, is fit for human habitation unless, in the opinion of the local authority, it fails to meet one or more of the following requirements and because of that failure is not reasonably suitable for occupation:

(a) it is structurally stable;
(b) it is free from serious disrepair;
(c) it is free from dampness prejudicial to the health of the occupants;
(d) it has adequate provision for lighting, heating and ventilation;
(e) it has an adequate piped supply of wholesome water;
(f) there are satisfactory facilities in the dwelling-house for the preparation and cooking of food, including a sink with a satisfactory supply of hot and cold water;
(g) it has a suitably located water closet for the exclusive use of the occupants;
(h) it has, for the exclusive use of the occupants, a suitably located fixed bath or shower and wash-hand basin, each of which is provided with a satisfactory supply of hot and cold water; and
(i) it has an effective system for the drainage of foul waste and surface water (DoE, 1990).

Repair and Improvement of Houses

Various schemes are in operation with a view to ensuring that houses are kept in a good state of repair and thus preventing them from becoming unfit and requiring substantial public expenditure. For example, Leicester City Council are endeavouring to encourage owner-occupiers to maintain their properties with local authority support, predominantly in Victorian terraced houses. Without regular maintenance improved houses will fall back into disrepair, and eventually become eligible for mandatory renovation grant aid because of unfitness.

Many owner-occupiers do not have the money to finance repairs, and those that do have no financial incentive. In fact there are disincentives emanating from the limited increase in property values after the work is carried out. Furthermore there are no compulsions or incentives to maintain the properties from lenders who fund house purchases or those making house renovation grants.

Leicester City Council set up four urban management offices catering for 12 000 properties in the older inner city areas. The residents' main concerns were lack of money, knowledge, DIY skills and tools. Hence the area management offices help with technical advice, training courses for craft skills and loan of tools. Area caretaker services offer a home maintenance service to vulnerable residents, such as elderly people, single parents and the disabled, where the only charge is for materials (Gunn, 1992).

Another valuable approach has been the setting up of care and repair services which aim to enable elderly or disabled home owners to remain in their own homes in comfort and security, by giving practical help and advice on essential repairs. These vulnerable residents can become very worried as their houses deteriorate and they benefit immensely from the help of understanding and knowledgeable people in identifying the best solution to their problems and getting the necessary work implemented. A typical example is the Newark and Sherwood Care and Repair Agency sponsored and funded jointly by Newark and Sherwood DC and Nottingham Community Housing Association (NCHA) set up in June 1991. In 1992/93 it completed 112 projects valued at £157 606 with most of them financed by a minor works grant through Newark and Sherwood DC (Newark and Sherwood, 1993).

The scale of the likely repair and improvement programme with older houses is illustrated by Nottingham Community Housing Association purchasing some 1000 tenanted properties from British Coal in 1988/89. All the houses were in varying stages of disrepair due to a lack of investment by British Coal over a long period of time, and this process is likely to be repeated many times into the mid 1990s as coalfields cease production. Initially fair rents were charged and the income thus generated was only sufficient to meet loan repayments plus a basic management and maintenance allowance, with no provision for cyclical repairs or improvements. As properties changed hands new tenants moved in with higher expectations and a 15 per cent reduction was applied to the assured rents to compensate for their poor condition.

The NCHA Corporate Plan for 1990 identified the need to invest about £8000 per property over a five year period to bring the houses up to an acceptable standard. In addition it was decided to sell properties which became vacant, with a minimum target of 20 houses per year, to help in raising funds to finance the repairs and improvements. During the period 1988–91, all houses were rewired at a cost of £634 per house and had their lofts insulated and some heating and cooking appliances upgraded in smokeless zones. The programme for 1991–95 concentrated on external repairs, central heating upgrades, and the provision of internal bathrooms and WCs.

Rehabilitation Work

Introduction

Many older houses with reasonably sound structures and other buildings, such as warehouses, chapels and schools with a wealth of character, justify rehabilitation, to permit their extended use for housing purposes. One of the main obstacles is likely to be one of cost as the rebuilding work is much more expensive than new build and some of the materials and components may need to be purpose made. A strong case can be made on aesthetic and architectural grounds to convert some of the fine buildings of years past, which are no longer required for their original use, if they lend themselves to conversion, and thus to preserve some of this rich heritage for the benefit of future generations.

Older houses can look extremely drab apart from the defects to the fabric, which can give rise to damp penetration and excessive condensation. Rehabilitated they can be made to look attractive and the faults of the original buildings cured to provide the occupants with much more comfortable and convenient dwelling units.

Financial aid for the rehabilitation of buildings is often available in the form of loans and grants from a variety of different bodies, but many of these are for non-housing uses. Possible sources for housing uses include Urban Development Grants, Housing Corporation/DOE loans and grants and City Challenge schemes, where appropriate.

Assessment of Traditional Housing for Rehabilitation

BRE Good Building Guide 6 (1990b) describes how rehabilitation work presents particular problems and a need for specialist skills throughout the design team. In practice the problems are often underestimated and the costs frequently rise unacceptably beyond the initial estimates. Findings from a 1990 BRE survey indicated that although typical rehabilitation projects achieve improvements in layout and amenity, they may be less successful in resolving more fundamental problems, some with serious implications for the long term performance of the building. Hence there is a vital need to make an accurate assessment of the property as described in the BRE Guide prior to formulating the rehabilitation work.

The assessment normally includes a standard dimension survey, condition survey and a desk assessment. Building work of rehabilitation projects is often more complex than the new build equivalent, usually involving a sequence of building operations which will maintain strength and stability during and after conversion, and the need for adequate fire and sound performance may complicate the work if the ground and first floors are being planned as separate dwellings.

Assessment of the relevant performance considerations for each building element/service should reveal most of the deficiencies which will require attention during rehabilitation. Evidence from BRE surveys indicates that insufficient attention is often directed at solid intermediate floors, balconies and canopies, bays and porches, flat roofs, parapets, chimneys, and external works, and also at the general areas of strength and stability, and durability and maintenance.

Practical Applications

After the Housing Act 1974, interest by housing associations in rehabilitation work increased significantly and by the early 1980s accounted for 50 per cent or more of the Housing Corporation's approved development programme. From 1982 to

1987, more than 10 000 houses were improved annually, with an even higher share of local authority funded schemes. However a rapid decline set in after the passing of the Housing Act 1988, with associations being unwilling to take the risks coupled with rehabilitation work as, under the new financial regime, overrun costs had to be borne by the associations, private lenders opted for new build developments and local authorities, particularly inner city councils, received less money, coupled with tenants' preference for new houses. Hence in 1992 less than 20 per cent of the houses completed using housing association grant (HAG) were rehabilitated, which is a depressing picture.

However, some associations were proceeding with rehabilitation work using a more simplified approach and there were some very attractive schemes such as the conversion of the picturesque 1895 linctus factory in Hull by Sutton Housing Trust, the Royal Free Maternity Hospital in London by Circle 33 HT, and the restoration of attractive almshouses by Warden HA in Watford and by Nottingham Community HA in Nottingham. Another approach was for housing associations to work with local authorities who had declared renewal areas, such as Black Country HA with Sandwell DC, or undertaking housing rehabilitation work as part of City Challenge schemes.

DoE in Handbook of Estate Improvement: Dwellings (1992) illustrates the vast improvements made to dilapidated structures with rehabilitation work at eleven housing estates in England and Wales, with profiles of the work carried out. The Handbook also gives sound guidance on the design and technical aspects and the precautions to be taken on site to safeguard the interests of occupants.

CIOB (1987) list the following typical items of work to be carried out in the refurbishment of dwellings:

damp-proof injection treatment
taking down internal walls
alteration of entrances
construction of single or multistorey extensions
kitchen enlargements and revised layouts
installation of new kitchen units
partial or complete replumbing, including new kitchen and bathroom units
extending electrical system and/or complete rewiring
plastering and screeding
reallocating existing areas, e.g. changing a bedroom into a bathroom or kitchen and bathroom to a kitchen/diner
improving insulation
replacing windows and/or doors
rerendering or repointing external walls
reroofing
sealing disused chimney stacks
extending drainage to accommodate alterations
wide range of repair work from plaster patches to taking up and replacing floors.

This list is not all embracing and other and more extensive works can arise, such as varying façade materials and treatments; rebuilding elevations; enclosing balconies; providing projecting porches and bay windows; replacement heating; new floor finishes; and external environmental works including hardstandings for cars.

The contractor when estimating for a rehabilitation contract should set out the order, sequence and time allowed for each stage of the construction to permit a steady progression of work, and should direct particular attention to the following aspects:

access for the contractor and others
storage space
limitation of working area
special conditions imposed by the employer
specific starting and finishing times
continuity of trades (CIOB, 1987)

Costs of Refurbishment Work

Table 5.4 shows the likely ranges of costs of some of the more commonly encountered items of major refurbishment work, categorised according to the type of dwellings in which they occur. The balcony and corridor access flats and maisonettes contain 24 dwellings per block spread over 4

Table 5.4 Costs of refurbishment work

Technical options	*Unit price £/m² GFA*	*Cost per dwelling, £ (1990 prices)*					
		Cottage estate Houses in street or group	*Walk-up linear block*			*High rise block with lift access*	
			Flat with corridor access	*Flat with balcony access*	*Maisonette with corrider access*	*Flat in tower block*	*Flat in linear block*
Strengthen existing foundations by underpinning or enlarging footings	125–450	5100–9300	3100–5300	3400–5800	4100–7000	1500–2900	1500–2900
Reinforce existing slab with reinforced concrete suspended slab on sleeper walls	80–150	3600–5200	–	–	–	–	–
Replace existing slab with reinforced concrete slab on hardcore and compressible material	90–180	3700–5700	1700–2500	1800–2600	2000–3000	600–900	600–900
Pitched roof, replace tiles, felt and battens, provide diagonal bracing, replace eaves soffit and include new ventilation grilles	35–110	1600–4400	800–2400	900–2500	900–2400	–	–
Construct new pitched roof over existing flat roof and raise party walls to underside of roof	100–200	–	2500–4000	2600–4300	2900–4700	–	–
Lay insulation over existing sound flat roof and top with shingle	30–75	–	700–1200	750–1300	800–1400	350–600	250–450
Remove existing roof covering, replace with new and top with insulation and paving slabs	70–120	–	1500–2100	1600–2250	1700–2400	650–900	600–800
Cut out and replace cracked facing bricks (10% of external wall area)	11–16	850–1200	275–400	425–600	400–575	375–525	375–550
Demolish and rebuild badly bulging brick external walls	46–86	3600–6300	1200–2200	1800–3200	1700–3000	–	–
Remove facing brickwork and clean cavity (5–15% of external wall area)	7–41	550–2150	325–1050	300–1050	300–1050	–	–
Insert new cavity wall ties	12–51	900–3800	350–1300	500–1900	450–1800	–	–
Install cavity wall insulation (mineral fibre or polystyrene beads blown into cavity) in external walls	13–25	1000–1750	350–650	550–900	525–850	–	–
Draughtstrip windows, fit trickle ventilators, and seal frames with mastic sealant	10–35	340–480	160–225	180–265	180–260	180–265	160–240
Replace existing windows with single glazed windows	110–225	1650–3500	1500–3000	1600–3200	1300–2700	1700–3200	1700–3200

Replace existing windows with double glazed windows	135–370	2200–5400	1900–4500	2000–4800	1700–4100	2100–4900	2100–4900
Replace existing front door and frame with hardwood frame and solid flush softwood door, with door viewer and security lock and chain	–	300–600	300–600	300–600	300–600	300–600	300–600
Remove existing heating installation and install new gas fired balanced or fan flued boiler, supplying hot water to domestic and heating circuits, open vented or sealed pressurised system, including radiators, pump, pipework, controls and immersion heater	25–45	2000–3000	1800–2700	1800–2700	2000–3000	1800–2700	1800–2700
Remove existing heating installation and install off-peak electric storage heaters and off-peak water heating, including hot water cylinder, pipework and wiring	19–33	1500–2300	1300–2000	1300–2000	1500–2300	1300–2000	1300–2000
Fit extract fan in kitchen and bathroom with solenoid operated shutter and humidistat control, including electrical connections and ducting to external wall where necessary	4–10	300–500	300–500	300–500	460–600	300–500	400–600
Construct false ceiling below existing ceiling, plasterboard on joists supported by wall hangers and sound insulating mineral fibre quilt above	38–55	–	1800–2500	1700–2400	650–900	2300–3200	2100–3000
Strip off existing screed and lay new screed over sound-deadening quilt	25–44	–	1300–2030	1250–2000	500–750	1650–2550	1500–2400
Lift floorboarding, insert mineral fibre quilt, lay new composite chipboard/plasterboard sheet flooring over resilient quilt, and fix extra layer of plasterboard to ceiling below	50–75	2000–3000	–	–	–	–	–
Remove existing lifts and replace with completely new system	6000–9000	–	–	–	–	3000–4500	1500–2500
Install lift in new lift shaft built externally to building	7500–20 000	–	4500–8500	3000–7000	–	4000–8000	2000–4000
Install external refuse chute and build new Paladin store	700–1350	–	350–500	175–225	125–175	350–500	200–300

Source: Davis Langdon & Everest/*DoE Handbook of Estate Improvement 3 – Dwellings* (1992).

floors, while the tower blocks contain 40 dwellings spread over 10 floors and the linear blocks have 80 dwellings spread over 10 floors.

LATEST TRENDS IN HOUSING

Environmentally Friendly Buildings

Apart from the statutory requirements for buildings, there was in 1993 an increasing awareness of the urgent need to produce more environmentally friendly buildings, although surveys of housebuilders in 1991 showed a very negative response and identified the need for more extensive and positive publicity of environmental issues and how they can best be addressed. The need to reduce chlorofluorocarbons (CFCs) – man-made chemicals mostly used in the building industry and constituting the main cause in the thinning of the ozone layer and the 'greenhouse' warming of the earth – was generally reasonably well known and assisted by BRE information paper IP 23/89, including the use of less harmful insulation materials. Another important aim was to reduce the level of greenhouse gas and particularly CO_2.

The following three straightforward techniques could assist significantly in producing environmentally friendly buildings:

(1) to produce energy efficient buildings with a minimum need to burn fossil fuels or use expensive electricity, and generating large cost savings;
(2) to ensure that materials used in buildings are environmentally friendly, such as ceasing to use tropical hardwoods;
(3) to ensure that buildings are managed so that they continue to have a low environmental impact (Levinson, 1990).

Developers of new houses designed to be more environmentally friendly can seek recognition through a BRE environmental assessment method (BREEAM), which was introduced in 1991.

Energy Efficiency in New Housing

In the early 1990s increasing importance was attached to the need to conserve energy in the home and this culminated in 15 housing associations expressing their willingness to take part in an energy cost study, which demonstrated the benefits of energy labelling, the ability to design new homes with significantly improved energy efficiency at little or no extra capital cost whilst at the same time reducing management and maintenance costs, and enabling tenants to enjoy homes that they could afford to heat. Furthermore, energy efficiency is a key option in reducing emissions of carbon dioxide, the principal global warming gas, and in preserving fossil fuel resources. From July 1993 the Housing Corporation Scheme Audit system for housing association development included an assessment of the extent to which individual associations were incorporating effective energy efficient measures into their new schemes, and thus became a factor in determining capital allocations to associations.

Most of the 15 housing associations taking part in the National Federation of Housing Associations (NFHA)/Building Research Energy Conservation Support Unit (BRECSU) new low energy housing study (1992) had already adopted insulation measures above the minimum 1991 Building Regulations standards. For example, an improved national home energy rating (NHER) of 8 (± 0.4) could be achieved for less than £200/dwelling by incorporating low cost measures as listed in tables 5.5 and 5.6, including full cavity wall insulation, well within the design cost tolerance for housing association schemes, with the average house costing about £40 000. The tendering exercise showed that energy efficient measures should be included as a standard specification, rather than as 'extras', to obtain the best prices.

An NHER value of 9 can be obtained in gas heated schemes by adopting higher insulation standards and more efficient heating services above the low cost standards, involving further additional costs of between £200 and £600. Where an association is prepared to finance additional capital repaid through rent, a payback period of 5

Table 5.5 Summary of typical recommended energy saving measures for gas heated dwellings

Measures	*1991 Building Regulations*	*NHER = 8 (low cost specification)*	*NHER = 9 (higher cost specification)*
Walls	*U*-value = 0.45	75 mm fully filled insulation *U*-value = 0.37	100 mm fully filled insulation *U*-value = 0.3
Floors	*U*-value = 0.45	25 mm insulation *U*-value = 0.45	50 mm insulation *U*-value = 0.35
Roofs	*U*-value = 0.25	150 mm insulation *U*-value = 0.25	200 mm insulation *U*-value = 0.18
Windows	Single glazing *U*-value = 5.7	Double glazing *U*-value = 2.8	Double glazing *U*-value = 2.8
Draught sealing	No requirement	Yes	Yes
Building services	Conventional gas central heating	1 or 2 bed: Homewarm 3/4/5 bed: conventional gas central heating	1 or 2 bed: conventional gas central heating plus thermostatic radiator valves 3/4/5 bed: gas condensing boilers and thermostatic radiator valves
Primary pipework insulation	No requirement	Yes	Yes
Low energy lights	No requirement	No	Yes

Source: NFHA. *Affordable new low energy housing for housing associations* (1992).

Table 5.6 Summary of typical recommended energy saving measures for electrically heated dwellings

Measures	*1991 Building Regulations*	*NHER = 6.5–7.0 (low cost specification)*	*NHER = 8 (higher cost specification)*
Walls	*U*-value = 0.45	75 mm fully filled insulation *U*-value = 0.37	100 mm fully filled insulation *U*-value = 0.3
Floors	*U*-value = 0.45	25 mm fully filled insulation *U*-value = 0.45	50 mm fully filled insulation *U*-value = 0.35
Roofs	*U*-value = 0.25	150 mm insulation *U*-value = 0.25	200 mm insulation *U*-value = 0.18
Windows	Single glazing *U*-value = 5.7	Double glazing *U*-value = 2.8	Double glazing *U*-value = 2.8
Draught sealing	No requirement	Yes	Yes
Building services	Economy 7 storage heaters	1 or 2 bed: Economy 7 storage heaters 3/4/5 bed: Economy 7 storage heaters	1 or 2 bed: Economy 7 storage heaters with auto charge control and fan assisted heaters 3/4/5 bed: Economy 7 storage heaters with auto charge control and fan assisted heaters
Primary pipework insulation	No requirement	Yes	Yes
Low energy lights	No requirement	No	Yes

Source: NFHA. *Affordable new low energy housing for housing associations* (1992).

years or less achieves a saving which can be shared equally between the tenant and the association.

Reductions of up to 20 per cent in CO_2 emissions can be made by improving energy efficiency standards in a gas heated dwelling from minimum Building Regulations standards to an NHER of 8, and up to 30 per cent by further improvements to an NHER of 9. Electrically heated dwellings produce up to 60% more CO_2 than the worst gas heated dwellings and up to 125% more CO_2 than the best gas heated dwellings.

BRE Domestic Energy Model (BREDEM) based energy rating methods and labelling schemes can provide housing associations with a technically feasible method of assessing the energy performance of their schemes against targets for improvement (NFHA, 1992).

Quality Management in Social Housing

Quality seems set to be the central issue in social housing during the 1990s and beyond. This forms a major theme of the Citizen's Charter which aims to improve the quality of services to the public and to make them more responsive to the needs of those who use them, and it extends the remit of the Audit Commission to consider quality of service as part of a regular review of standards of performance. For housing associations, the Performance Expectations of the Housing Corporation highlight the standards which associations must achieve in the delivery of their housing service. The Institute of Housing, now chartered (1992) also emphasised the need to improve internal management efficiency and to ensure that customers' views are heard.

Catterick (1992) has described how poor quality wastes resources and increases costs. It has been estimated that as many as one in four workers produces nothing at all because they spend their entire day rectifying the mistakes made by other people. Quality costs have been classified into three categories:

(1) Failure costs: these occur when the work fails to meet the prescribed objectives. Internal failure can arise from carrying out work which is not required, duplicating work or correcting work which was done incorrectly. External failure involves the consumer and represents a substantial cost in terms of time and effort and resulting dissatisfaction to the consumer.

(2) Appraisal costs are associated with checking and inspection to ensure that work is done correctly and add nothing to the service provided.

(3) Prevention costs are incurred in operating mechanisms which prevent errors from occurring in the first place and outweigh the costs of failure and appraisal.

Successful implementation of total quality management requires a defined organisational structure which harnesses the full potential of the workforce, usually involving the establishment of teams and groups. A periodic quality audit is required to ensure that a quality management system is being operated and that it is working successfully, and it will identify errors and omissions and areas needing review. BS 5750 is the national standard which promulgates the ISO 9000 series international standards for quality systems, as described in chapter 2.

European Funding

Unlike housing authorities, housing associations are represented at European level by CECODHAS (Le Comité Européen de Coordination de l'Habitat Social), the European Liaison Committee for Social Housing. That housing authorities are not represented is partly explained by the fact that the UK has the largest state housing sector in the EC, as housing in the majority of member states is provided by non-governmental bodies (Drake, 1992).

To be successful in obtaining funding from the EC, housing departments must adopt a corporate approach as in City Challenge schemes, and bids which put together a mixture of domestic and European funding are more likely to succeed. Another good approach is to combine applications from the north of Europe with those from the east and south in Objective 1 areas, where the aim is to improve the economic health of the regions whose development is lagging behind. A practical example is the linking of Islington with Barcelona, Lisbon and Budapest, involving a number of European visits and exchanges. Furthermore, Objective 2 areas in the UK, which have a declining industrial base, such as parts of Scotland and Wales, northern England and most of the Midlands, will be more likely to achieve significant levels of European funding than those outside such areas.

The European Regional Development Fund (ERDF) is targeted on those areas whose GDP falls well below that which is the average for the community (Objective 1 areas), those with a declining industrial base (Objective 2 areas), and those whose rural development is necessary because of declining investment in agriculture (Objective 5(b) areas). Objective 1 areas can claim up to 75 per cent of the total eligible costs of a project from the ERDF, while Objective 2 areas can claim up to 50 per cent. ERDF offers capital monies, while the European Social Fund offers revenue. While housing is not expressed as a priority for ERDF, leakage can occur into housing in various ways, such as through Urban Pilot Projects with a theme of urban planning or regeneration of European interest.

An ESF funded initiative, the Human Resources Initiative, can also be applied to advantage through the part of the Initiative known as Horizon, which is aimed at the homeless, refugees, immigrants, migrants and people with disabilities, where there is considerable scope for leakage into housing. The SAVE five year programme was adopted in 1991 and includes the development of methodologies for thermal insulation in building.

In 1992 local authorities that had received European funding for housing projects included Glasgow, Birmingham and Lambeth.

6 APPROXIMATE ESTIMATING

In this chapter we consider the function served by approximate estimates, the methods employed and the factors controlling their use. The various methods are compared and applied to practical examples.

PURPOSE AND FORM OF APPROXIMATE ESTIMATING TECHNIQUES

Purpose of, and Approach to, Approximate Estimating

The primary function of approximate or preliminary estimating is to produce a forecast of the probable cost of a future project, before the building has been designed in detail and contract particulars prepared. In this way the building client is made aware of his likely financial commitments before extensive design work is undertaken.

The entry of quantity surveyors with adequate techniques into the estimating field was of considerable significance in the early development of a professional role. To extend this role into that of building economist required the development of understandings and techniques of a kind that deal, not just with the items which go into the accountancy of a particular building, but with the economic and other forces, which have determined the nature and relationships of those quantities and costs, and which determine the trends they show. Indeed, economics is the study of all the forces which determine the present functioning and probable future trends of a whole industrial or financial system.

The quantity surveyor performs an extremely important role in cost assessment, giving advice as to the probable cost of a particular design proposal and variations to it, and suggesting how similar objectives could be achieved more economically. It must, however, be emphasised at the outset that no approximate estimate can be any better than the information on which it is based. Indeed, realistic approximate estimating can be achieved only when there is full co-operation and communication between architect, quantity surveyor and building client from the inception of the scheme. It is advisable to dissuade the architect from reporting forecasts of costs to the building client until some drawings, possibly no more than preliminary sketches, have been prepared and an inspection made of the site. There is a distinct possibility that the building client will endeavour to obtain an independent check on the preliminary cost figures and it is accordingly unwise to supply a high 'cover' figure. On the other hand the submission of too low a figure can lead to recriminations, as the first figure is the one that the building client will always remember. The quantity surveyor should always emphasise that an estimate based on inadequate information cannot be precise, and in such a situation he would be well advised to give a range of prices, as an indication of the lack of precision that is obtainable.

The choice of method employed will be influenced by the information and time available, the experience of the surveyor and the amount and form of the cost data available to him. It is essential to carry out a detailed site survey before a preliminary estimate is prepared as it would be quite unrealistic to assume that the site is level, and free from obstructions, fill and contamination of any kind, has a groundwater level well below the foundations, and that the soil has an average loadbearing capacity. Similarly, old drawings of existing buildings scheduled for adaptation need checking.

Classification of Approximate Estimating Procedures

The various pre-contract approximate estimating methods can be classified in the following way:

Single-purpose estimates are aimed at forecasting cost and these can be further subdivided into *preliminary estimates*, which establish the broad financial feasibility of the project, and *later stage estimates* which will produce a figure comparable with that of lowest tender.

Dual-purpose estimates are aimed at determining total costs and also the various design-cost relationships between possible variants of the projects. These estimates delve into the wider field of cost planning. The second category of estimate can be further subdivided into two groups: *primary comparative-cost estimates*, which indicate the relative costs of different design solutions which will satisfy the requirements of the building client, and *secondary comparative-cost estimates* which apply the financial yardstick to alternatives of construction, finish and service installations applicable to the selected design.

The extent to which these various estimating processes are needed for a given project is very much dependent upon the degree of importance with which the building client views financial considerations and the size of the project. For instance, buildings let to third parties, which must show a profitable return on their capital cost and owner-occupied industrial and commercial projects where capital is limited, all need to be exhaustively and skilfully costed at each stage of development of the design.

The main problems in implementing each of the main categories of approximate estimating are now considered.

Single-purpose preliminary estimates. These are often required before any drawings are prepared and are frequently computed by rather imprecise methods. These methods include unit prices such as the comprehensive price per bed of a hospital or hotel, and an all-in price per cubic metre of a building, whose location, shape, height, site works, services and other important characteristics may not have been determined at the time the estimate is required. These estimating methods are difficult to apply and should be used with the greatest care.

Single-purpose later stage estimates. These should only be prepared by single price-rate methods if the quantity surveyor is very experienced in the use of these methods, or if he has available as a starting point the tender particulars of a recent project which is similar in character and preferably designed by the same architect. In all other circumstances it is better to prepare estimates of this kind by way of priced approximate quantities which is a much more accurate method of estimating. In addition the total estimate can be broken down into convenient parts and justified if required, a process which is hardly feasible with single price-rate methods.

Primary comparative-cost estimates often use single price-rate methods as a basis for the computations. Nevertheless, it is important to make allowances for differing shapes and wall to floor ratios of alternative designs. As illustrated in chapter 2, external walling is expensive and, as a general rule, the smaller the ratio of external walling to floor area, the cheaper will be the building. The aim should be to use rectangular buildings with the sides as near equal in length as possible, provided that other aspects such as lighting and ventilation can be dealt with satisfactorily. Similarly, it is usually more expensive to construct accommodation vertically instead of horizontally at the stage when it becomes necessary to provide a structural frame, probably at the two-storey stage in the case of a heavily loaded building, and three to five-storeys with lightly loaded ones.

Secondary comparative-cost estimates are an integral part of cost planning, in comparing the costs of possible variants in construction, finishings and servicing of the selected plan. Comparisons could be made between traditional and non-traditional construction; steel frame, reinforced concrete frame and loadbearing brickwork; various types of pitched and flat roofs; different floor, wall and ceiling finishings; and different forms of heating and lighting.

In all cases the drawing numbers on which the estimate is based should be recorded on the estimate. The date of the estimate and the allowance for price fluctuations between the estimate and tender dates should also be clearly shown. Finally, all supporting data should be filed with the estimate.

UNIT METHOD

On occasions a building client requires a preliminary estimate for a building project based on little more information than the number of persons or units of accommodation that the building is to house. The unit method of approximate estimating seeks to allocate a cost to each accommodation unit of the particular building, be it persons, seats, beds, car spaces or whatever. The total estimated cost of the proposed building is then determined by multiplying the total number of units accommodated in the building by the unit rate. Thus the mathematical process is very simple but the computation of the unit is exceedingly difficult.

The unit rate is normally obtained by a careful analysis of the unit costs of a number of fairly recently completed buildings of the same type, after making allowance for differences of cost that have arisen since the buildings were constructed and any variations in site conditions, design, form of construction and materials. Variations in rates, stemming from differences in design and constructional methods, are difficult to assess and frequently there is insufficient information available to make a realistic assessment. Hence although the method has the great merit of speed of application, it suffers from the major disadvantage of lack of precision and at best can only be a rather blunt tool for establishing general guidelines, more particularly for budgetary estimating on a rolling programme covering a three to five-year period ahead. Because of the lack of precision it is advisable to express costs in ranges, with more precise costs to be determined at a later stage by more reliable estimating methods when much more detailed information is available.

The following 1994 unit rates are given as a general illustration of the application of the method to specific classes of building and are subject to all the limitations previously described.

Hospital ward accommodation	£1550 to £1750 per bed
Church hall	£460 to £560 per seat
Primary school	£1600 to £1900 per place
Secondary school	£3000 to £3150 per place
Multistorey car park	£2800 to £4500 per car space

The main weaknesses of the method lie in its lack of precision, the difficulty of making allowance for a whole range of factors, from shape and size of building to constructional methods, materials, finishings and fittings, and it is not sufficiently accurate for the majority of purposes. It does, however, serve a limited number of uses such as establishing an overall target for a cost plan or calculating a sum for investment purposes, in cases where the building client or his professional advisers have considerable experience of the construction and cost of similar buildings, such as hospitals and schools. Even under these circumstances it is necessary to use this method with the greatest care and skill and with a full appreciation of its limitations.

CUBE METHOD

The cube method of approximate estimating was used quite extensively between the wars but has since been largely superseded by the superficial or floor area method. The cubic content of the building is obtained by the use of rules prescribed by the Royal Institute of British Architects (1954) which provide for multiplying the length, width and height (external dimensions) of each part of the building, with the volume expressed in cubic metres. The method of determining the height varies according to the type of roof and whether or not the roof space is occupied. For a normal dwelling with an unoccupied pitched roof, the height dimension is taken from the top of the concrete foundation to a point midway between the apex of the roof and the intersection of wall and roof (half the height of the roof). If the roof space is to be occupied then the height measurement is taken three-quarters of the way up the roof slope, and with mansard roofs it is usual to measure the whole of the cubic contents.

With flat roofs, the height dimension is taken 600 mm above roof level except where the roof is surrounded by a parapet wall which has a height in excess of 600 mm, when the height will be measured to the top of the parapet wall. If the height of the parapet wall is less than 600 mm, the minimum height of 600 mm will still be taken. All projections such as porches, steps, bays, dormers, projecting rooflights, chimney stacks, tank compartments on flat roofs and similar features, shall be measured and added to the cubic content of the main building. On occasions it may be found that a small part of the foundations may be deeper than the remainder, and in this situation it is better to adjust the unit rate rather than to vary the cubic content of the building. Projecting eaves and cornices should be ignored when computing the volume of the building.

Where different parts of a building vary in character or function, such as a workshop with an office block frontage, then the different parts should be separately measured and priced. Basements should also be cubed separately, so that allowance can be made in the unit rate for the increased excavation and construction costs. Features such as piling, lifts, external pavings, approach roads, external services, landscaping and similar works which bear no relation to the cubic unit of measurement, should be dealt with separately by the use of lump sum figures or approximate quantities.

The assessment of the price per cubic metre of a building calls for the exercise of careful judgement coupled with an extensive knowledge of current prices and trends. Unit prices show wide variations between different classes of building and will even vary considerably between buildings of the same type, where such factors as the proportion of walling to floor area and quality of finishings and fittings vary to a significant extent. The greater the proportion of walling in relation to the cubic contents of the building, the greater will be its cost per cubic metre.

The following typical cubic metre rates in 1994 for various building types will give a rough guide:

City bank with two floors of offices over	£155 to £245
Church hall	£85 to £140
Hotel	£210 to £280
City office block with shops on ground floor (excluding shop fronts and shop finishings, but including lifts and heating)	£125 to £225
Small shop with one floor of offices over (excluding shop fronts and shop fittings)	£85 to £155

Tremendous variations can occur in the cube rates of buildings of the same type. For instance, city offices can vary from £140 to £245 per cubic metre depending on size, shape, quality of finishings, amount of partitioning, number of fittings and a whole host of other factors, and these must all be taken into account when assessing the unit rate. With single-storey industrial buildings, wide variations in storey height can occur and costs will not vary directly in proportion to height. The roof, foundations and floor remain constant and a comparatively cheap form of cladding may be used for the walls. In this situation the cube method of approximate estimating would be ill-suited and could give quite unrealistic results. A far more satisfactory approach would be to use the superficial method and to adjust the unit rate for increases in height. The cube method is, however, useful for heating and steelwork estimates.

A primary weakness of the cube method is its deceptive simplicity. It is an easy matter to calculate the volume of a building but much more difficult to assess the unit rate on account of the large number of variables which have to be considered. The cubic method fails to make allowance

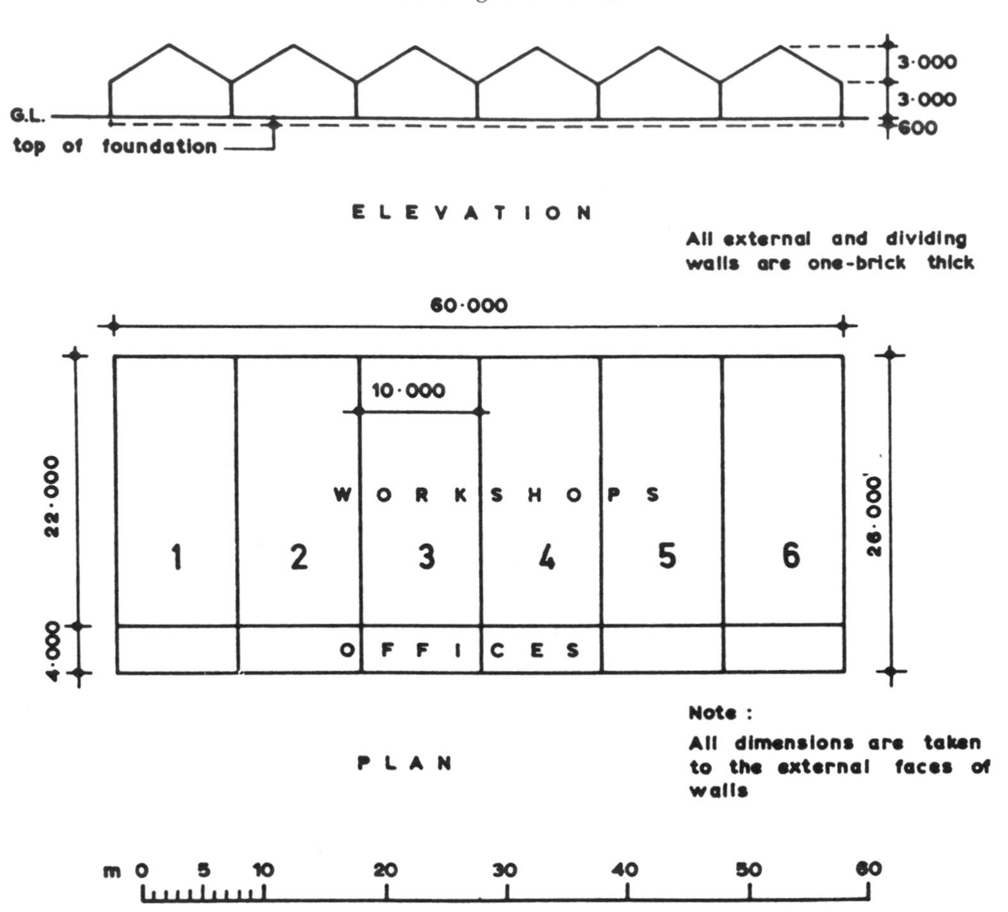

Figure 6.1 Approximate estimating – block of six unit factories

for plan shape, storey heights and number of storeys, which all have an important influence on cost, and cost variations arising from different constructional techniques such as alternative foundation types are difficult to incorporate in a single unit rate. Ideally, the building from which the basic cubic rate is obtained should be of similar shape, size and construction as the one under examination. Other weaknesses are that cubic content does not give any indication to a building client of the amount of usable floor area, and it cannot readily assist the architect in his design of a building as it is difficult to forecast quickly the effect of a change in specification on the cube unit price rate. The following example may help to illustrate the approach.

Example 6.1

Figure 6.1 shows a block of six unit factories with an office in each unit. The effective height is 600 mm (top of foundation to ground level) + 3.000 m (ground level of roof) + ½/3.000 m (roof) = 5.100 m.

Volume of six workshops = $60\,\text{m} \times 22\,\text{m} \times 5.1\,\text{m}$ = $\underline{6732\,\text{m}^3}$

Volume of offices = $60\,\text{m} \times 4\,\text{m} \times 5.1\,\text{m}$ = $\underline{1224\,\text{m}^3}$

Estimated cost of block would be

$6732\,\text{m}^3$ @ £68 = £457 776

$1224\,\text{m}^3$ @ £105 = £128 520

Total £586 296

SUPERFICIAL OR FLOOR AREA METHOD

In this method the total floor area of the building on all floors is measured between the internal faces of the enclosing external walls, and it includes internal walls, partitions, columns, stairs, chimney breasts, lift shafts, corridors and the like. Also included are lift plant, tank rooms and the like above the main roof slab. Sloping surfaces such as staircases, galleries, tiered terraces and the like are measured flat on plan. A unit rate is then calculated per square metre of floor area and the probable total cost of the building is obtained by multiplying the total floor area by the calculated unit rate. Where the building varies substantially in constructional methods or in quality of finish in different parts of the building, it will probably be advisable to separate the floor areas to enable different unit rates to be applied to the separate parts. Consideration must also be given to varying storey heights in assessing unit rates and when extracting rates from cost analyses.

This is a popular method of approximate estimating, as it is comparatively easy to calculate the floor area of a building and the costs are expressed in a way which is fairly readily understood by a building client. Furthermore, most published cost data is expressed in this form. It has advantages over the cube method as the majority of items with a cost impact are related more to floor area than to volume and it is therefore easier to adjust for varying storey heights. Nevertheless, it has a number of inherent weaknesses and, in particular, it cannot directly take account of changes in plan shape or total height of the building. Similarly, difficulties are experienced in building up unit rates from known rates for existing buildings because of the need to make allowances for a number of variables including site conditions, constructional methods, materials, quality of finishings and number and quality of fittings.

As with the cube method, special items such as piling, heating and lift installations are normally covered by lump sums which are added to the overall cost calculated on a floor area basis. The specialist lump sums may be derived from quotations obtained from specialists or be based on information arising from previous contracts. The estimated cost of external works is usually based on priced approximate quantities, an estimating process which is described later in the chapter.

A few typical 1994 unit rates per square metre of floor area follow to indicate the range of prices, but it must be emphasised that wide variations on these rates occur in practice.

Factory workshop	£480 to £770
Semi-detached house (estate development)	£390 to £510
Detached house (built singly and including central heating)	£570 to £780
Office block (medium rise, air conditioned)	£1050 to £1300
Office block (low rise, non air conditioned)	£860 to £1120
Banks (city centre)	£1200 to £1700
Shops (small)	£400 to £510
Departmental stores (including fitting out)	£1000 to £1400
Hospitals (ward blocks)	£780 to £960
Libraries	£580 to £850
Schools (secondary/middle)	£500 to £630
Hotels (city centre – 4 star)	£1170 to £1400

NOTE: Costs expressed in £ per sq ft can be converted to £ per m^2 by multiplying by ten and adding ten per cent, for example,

$$\text{£}60.00 \text{ per sq ft} = (60.00 \times 10) + 10\% = \text{£}660 \text{ per m}^2.$$

Example 6.2

To illustrate the method of approach, an estimate is now prepared for the block of six unit factories illustrated in figure 6.1 using the floor area method. In this case the measurements are taken to the inside faces of the external walls and no deduction is made for internal walls.

$$\text{Area of workshops} = 60.000 - (2/215) \times 22.00 - 215 = 59.570 \times 21.785 = \underline{1298 \text{ m}^2}$$

Area of offices $= 59.570 \times 3.785$
$= 226\,m^2$

Estimated cost of block would be
$1298\,m^2$ @ £354 = £459 492
$226\,m^2$ @ £560 = £126 560
Total = £586 052

This estimate compares reasonably favourably with the figure of £586 296 obtained by the cube method of approximate estimating.

STOREY-ENCLOSURE METHOD

Objectives

In 1954 James introduced the work of an RICS study group on a new method of single price-rate approximate estimating, termed the *storey-enclosure method*, with the aim of overcoming the drawbacks of the methods so far described in this chapter. The study group's primary objective was to devise an estimating system which, whilst leaving the type of structure and standard of finishings to be assessed in the price rate, would take the following factors into account in the measurements:

(1) shape of the building (by measuring external wall area);
(2) total floor area (by measuring area of each floor);
(3) vertical positioning of the floor areas in the building (by using greater multiplying factors for higher floors and greater measurement product for suspended floors than non-suspended);
(4) storey heights of building (proportion of floor and roof areas to external walls);
(5) extra cost of sinking usable floor area below ground level (by using increased multiplying factors for work below ground level).

Nevertheless, the following works would have to be estimated separately:

(1) siteworks, such as roads, paths, drainage, service mains and other works outside the building (these are best covered by approximate quantities);
(2) extra cost of foundations, which are more expensive than those normally provided for the particular type of building (again this is best covered by approximate quantities);
(3) sanitary plumbing, water services, heating, electrical and gas services and lifts (priced approximate quantities or price from specialist consultant);
(4) features which are not general to the structure as a whole, such as dormers, canopies and boiler flues (separate priced additions);
(5) curved work.

Rules of Measurement

The storey-enclosure method consists basically of measuring the area of the external walls, floor and ceiling which encloses each storey of the building. These measurements are adjusted in accordance with the following set of rules.

(1) To allow for the cost of normal foundations, the ground floor area (measured in square metres between external walls) is multiplied by a weighting factor of two.
(2) To provide for the extra cost of upper floors, an additional weighting factor is applied to the area of each floor above the lowest. Thus the additional weighting factor for the first suspended floor is 0.15, for the second 0.30, for the third 0.45 and so on.
(3) To cover the extra cost of work below ground level a further weighting factor of one is applied to the approximate wall and floor areas that adjoin the earth surface.

Summing up, the procedure is to take twice the area of the ground floor, if above finished ground level, and three times the area if below (measured between external walls). It will take once the area of the roof measured on plan to the external face of the walls (the same area whether a flat or pitched roof), twice the area of upper floors (to cover work above and below, including partitions), plus the appropriate positional factor; it will also take once the area of external walls above ground measured on their external faces (ground level to eaves), and twice the area below ground in basements, etc.

All the areas are multiplied by the appropriate weightings to obtain the storey-enclosure units which are then totalled. To obtain the estimated cost of the building, the total of storey-enclosure units in square metres is multiplied by a single price-rate built up from the costs of previous similar projects. The cost of external works and other special items can be added. Basically, the aim of this method is to obtain a total superficial area in square metres to which a single price-rate can be attached, and the effect of the various rules that have been outlined is to apply a weighted cost factor to each of the main parts or elements of the building.

Comparison with Other Single Price-Rate Methods

The study group considered and analysed ninety tenders for new buildings, by reference to drawings and priced bills of quantities, and compared the results achieved with those operating the cube and floor area methods of approximate estimating. The buildings were classified into a number of general types – houses, flats, schools and industrial buildings, and conversion factors were devised to make allowance for price variations arising from different contract dates, so that all were reduced to a common price time datum. Tender figures, not final account figures, were used in the investigations. The proportional cost of engineering services was assessed separately and was found to differ very widely (the proportion of the cost of services to building work varied from fourteen to thirty-eight per cent).

Table 6.1 Comparison of cube, floor area and storey-enclosure approximate estimating methods

Type of building	*Total number of cases examined*	*Number of rates within percentage grouping*		
		Cube	*Floor area*	*Storey-enclosure*
Houses	17	8	9	10
Flats	16	9	10	12
Schools	14	9	8	12
Industrial buildings	39	16	24	26

In these comparisons one of the tests used by the study group was to determine how many of the rates fell into a particular range group, above and below the tender figures. The study group adopted ten per cent plus or minus for houses, flats and schools, where ministerial planning and financial restrictions operated, and twenty per cent plus or minus for industrial buildings. Table 6.1 indicates the results of this test.

With the storey-enclosure method the prices are nearer to the tender figures and the range of price variation is accordingly reduced. The weighting factor to cope with foundations is a particularly good feature of the system, as sketch plans are often very vague at foundation level. Used with care it could result in a much more realistic method of computing and comparing the cost of building projects than any of the other single price-rate methods.

Unfortunately, it has been little used in practice, mainly because it involves more calculations than either the cube or floor area methods. Furthermore, there are no rates published for this method, hence, a quantity surveyor would have to work floor area and storey-enclosure methods

Table 6.2 Comparison of cube, floor area and storey-enclosure price rates

Type of building	*Cube method: cost/m³*	*Floor area method: cost/m²*	*Storey-enclosure method: cost/m²*
Housing association houses	£105 to £135	£335 to £425	£90 to £115
Housing association flats	£140 to £195	£440 to £600	£105 to £160
Schools (secondary/middle)	£160 to £200	£500 to £630	£175 to £210
Industrial buildings	£120 to £195	£480 to £770	£125 to £180

in parallel for a trial period and also work up storey-enclosure rates from past projects. The storey-enclosure method does little to assist either the building client or the architect, does not provide an aid to elemental or comparative cost planning and the effect of changes in specification on the price rate would probably prove difficult to assess. Nevertheless, it does not seem to be generally appreciated that although this method complicated the measurement aspect, it was accompanied by a simplification of the most difficult aspect of pricing.

Table 6.2 provides a comparison of cube, floor area and storey-enclosure price rates (1994 prices).

Example 6.3

Figure 6.2 illustrates an office block to which the storey-enclosure approach of approximate estimating will now be applied.

	Storey-enclosure units
Floors	
Basement	
Floor area = (34.000 − 760) × (13.000 − 760)	
= 33.240 × 12.240 = 406.86 m^2	
× weighting 3	
Storey-enclosure units	1 221
Ground floor	
Floor area as basement = 406.86 m^2	
× weighting 2	
Storey-enclosure units	814
First, second, third, fourth and fifth floors	
Floor area = (34.000 − 500) × (18.000 − 500)	
= 33.500 × 17.500 = 586.25 m^2	
Multiplier for first floor 2.15	
" " second floor 2.30	
" " third floor 2.45	
" " fourth floor 2.60	
" " fifth floor 2.75 × 12.25	
Storey-enclosure units	7 182
Roof	
Area = 34.000 × 18.000 = 612.00 m^2	
No multiplier: storey-enclosure units	612
Walls	
Basement	
Wall area = (34.000 × 2) + (13.000 × 2) × 3.000 = 282.00 m^2	
× weighting 2	
Storey-enclosure units	564
Ground floor — *area* — *storey-enclosure units*	
Exposed wall 180 m^2 = 180	
Retaining wall 102 m^2 × 2 = 204	
Storey-enclosure units	384
First floor to roof	
Wall area = (34.000 × 2) + (18.000 × 2) × (5 × 3.000)	
No multiplier: storey-enclosure units	1 560
Total number of storey-enclosure units	12 337

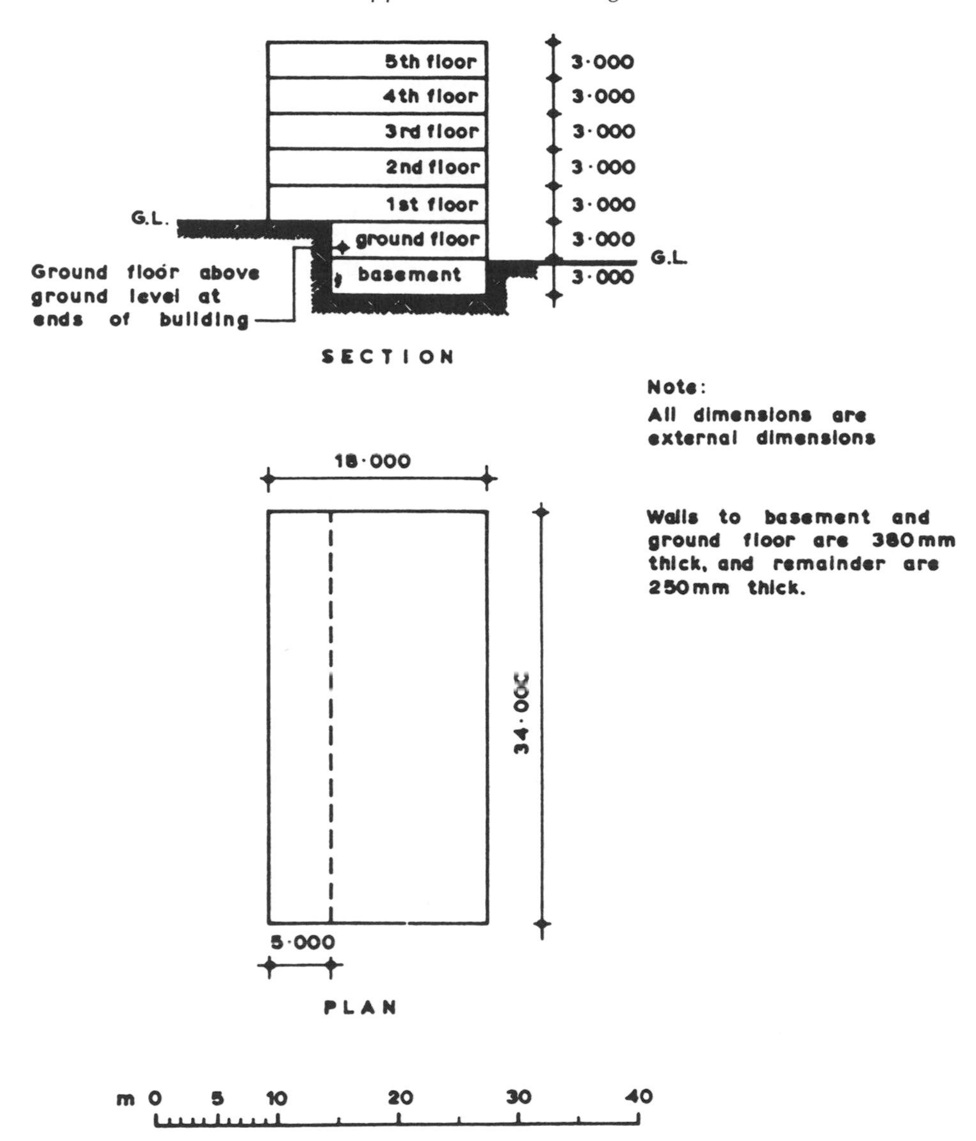

Figure 6.2 Office block – estimating by storey-enclosure method

Estimated Cost (1994 figures)

12 337 storey-enclosure units @ £275 =	£3 392 675
Estimated cost of lifts	£ 140 000
Estimated cost of external works	£ 210 000
Total estimated cost of work	£3 742 675
Rounded off to £3 750 000	

Ferry and Brandon (1991) devised a formula for calculating the optimum number of storeys for a building of given width to accommodate a specific floor area. For example where a building is to contain 9290 m^2 of floor area, with a width of 15 m and storey heights of 4.6 m. Using the formula

$$N^2 = \frac{xf}{2sw}$$

where

N = optimum number of storeys (to be determined)

$x = \dfrac{\text{roof unit cost}}{\text{wall unit cost}}$

f = total floor area (m^2)

s = storey height (m)

w = width of building (m)

the optimum number of storeys is seven.

APPROXIMATE QUANTITIES

Approximate quantities priced at rates produced at the time the quantities are computed provide the most reliable method of approximate estimating, possibly using a single price-rate method as a check. It does however involve more work than any of the methods previously described in this chapter and there are occasions when lack of information precludes its use. The method is sometimes described as *rough quantities* and the pricing document resembles an abbreviated bill of quantities. It provides an excellent basis for cost checking during the detailed design of a project.

Composite price rate items are obtained by combining or grouping bill items. It comprises a process of coalescing items in single omnibus description measurements. For example, a brickwork item measured in square metres will normally include all incidental labours and finishings to both wall faces. Doors and windows are usually enumerated as extra over the walling and associated finishings, thus avoiding the need to make adjustments to the walling. Furthermore a door item will be a comprehensive one, including the frame or lining, architraves, glazing, ironmongery and decoration.

It is good practice to build up a series of prices for composite items from a number of priced bills and to examine critically the range of variation between these prices and to establish the underlying reasons, and the relationships between the net cost of the main components, such as brick walls of various thicknesses and the gross cost of composite or all-in items, such as brick walls including finishings and all incidental labours. Sundry labours are often covered by the addition of a percentage to the composite items. The priced preliminaries in a bill of quantities must be examined before any priced rates are extracted, because of the different approaches adopted by contractors, with a view to securing comparability of rates as between different contractors.

It is desirable to develop an instinct for forecasting the total cost of small projects from an examination of the drawings prior to receipt of tenders. Afterwards these estimates can be usefully compared with the tenders and the reasons for any disparities then assessed. This process will help a surveyor to acquire a feel for building prices and to build up expert intuition.

Example 6.4

Figure 6.3 contains a plan and section of a small factory check office for which the estimated cost is to be obtained using the approximate quantities approach. The normal practice is followed of using paper with dimension columns and provision for squaring on the left hand side of the sheet, and quantity, rate and pricing columns on the right hand side. Although prices are built-up for each composite item, this would rarely be necessary in practice as a set of prices would already be established.

The priced rates for a project of this nature will vary tremendously with the particular circumstances. Considerations include whether or not it can be included as part of a larger contract, the state of the market, the time of the year when the work is to be undertaken, and particularly the availability of small builders in the locality if it is to form a separate contract. Very small projects are usually more expensive in terms of cost per unit of floor area because of the limited amount of work to be carried out in each trade and the larger proportional amount of overheads. It should also be borne in mind that it may be necessary to adjust the estimate in order to make allowance for possible increased costs occurring between the

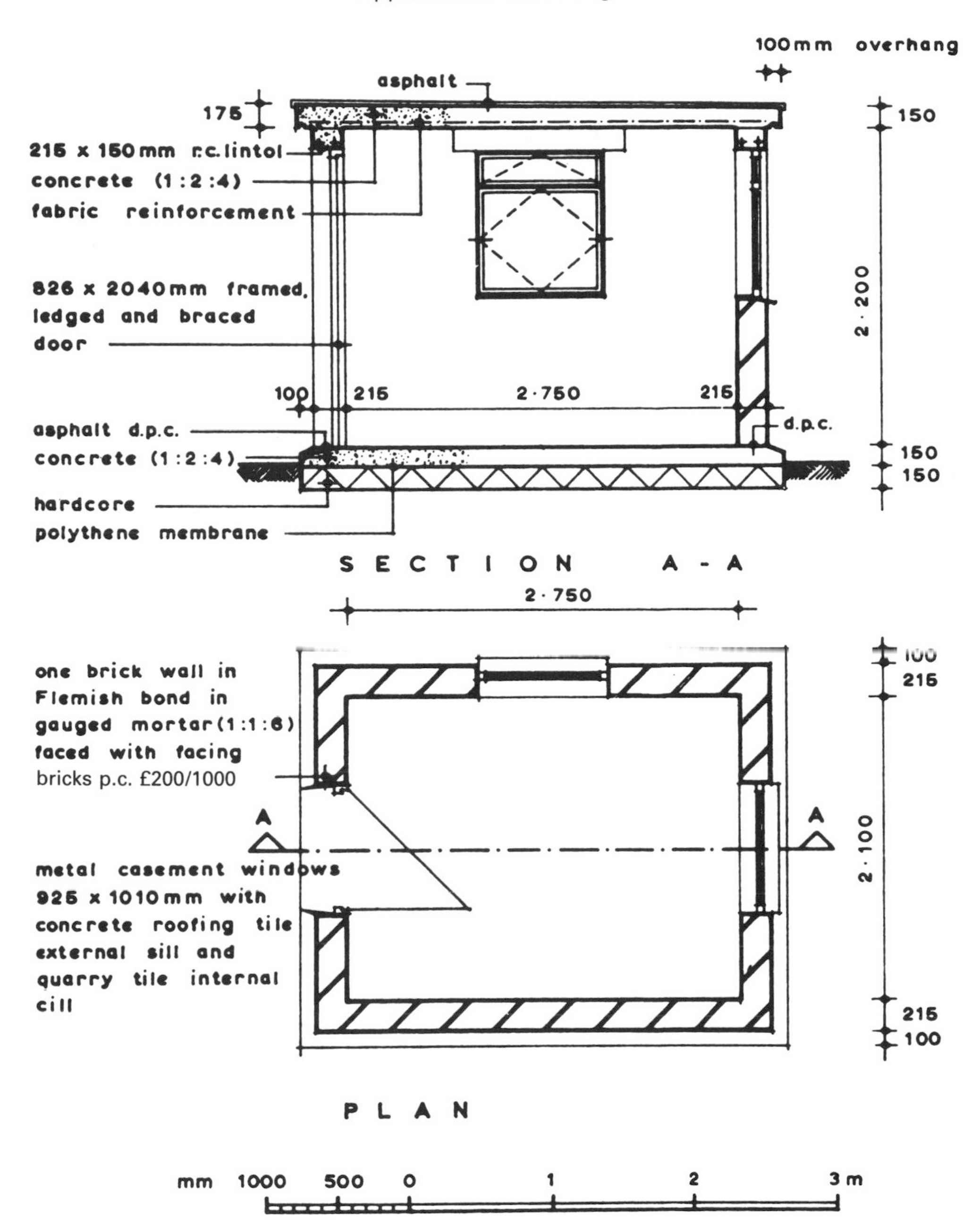

Figure 6.3 Factory check office – approximate quantities

date of preparing the estimate and the date of letting the contract.

Other composite or all-in items follow on pages 166–168, in further amplification of the approximate quantities approach to approximate estimating with 1994 guide figures (excluding preliminaries which often amount to about twelve per cent). Readers requiring further cost data on approximate quantities are referred to *Spon's Architects' and Builders' Price Book*.

Drawing No. 6.3 **Estimate for Factory Check Office** **Date: July 1994**

Ref. no.	*Dimensions*	*Extension*	*Description and price build-up*			*Quantity*	*Rate (£)*	*(£)*
			Floor Slab		*Base*			
			add	2.750	2.100			
			walls 2/215	430	430			
			proj.					
			conc. 2/100	200	200			
				3.380	2.730			
1.	3.38		Excavate oversite, remove surplus spoil, lay					
	2.73	9.23	150 mm bed of hardcore, polythene membrane					
			and 150 mm bed of concrete (1:2:4) floated to					
			smooth finish and splayed top edge to projecting					
			part of slab.			$9\ m^2$	21.00	189.00
			Price build-up					
			$1\ m^2$ excavate oversite and disposal		£ 5.00			
			$1\ m^2$ hardcore		3.10			
			$0.16\ m^3$ concrete bed		10.20			
			$1\ m^2$ polythene membrane 100 mm wide		0.70			
			$1\ m^2$ floated finish to concrete		1.00			
					£20.00			
			Sundry labours – 5%		1.00			
			Cost per m^2		£21.00			
					Walls			
			Walls		2.750			
					2.100			
					2/4.850			
					9.700			
			add corners 4/215		860			
					10.560			
2.	10.56		One brick wall in Flemish bond in gauged mortar					
	2.20	23.23	(1:1:6), faced externally with facing bricks, pc.					
			£200/1000 and flush pointed and two coats of					
			emulsion paint internally.			$23\ m^2$	65.40	1504.20
			Price build-up					
			$1\ m^2$ of one-brick wall, faced externally and flush pointed both sides		£56.00			
			$1\ m^2$ emulsion paint		2.80			
			$0.5\ m^2$ dpc.		5.00			
					63.80			
			Sundry labours – 2½%		1.60			
					£65.40			
			Metal windows					
3.	2	2	Extra over last for metal window size					
			925 × 1010 mm, including glass, painting, etc.			2	46.12	92.24
			Price build-up					
			One nr. window and fixing		£47.00			
			$1\ m^2$ glass		14.50			
			$2\ m^2$ painting		7.50			
			1 m concrete roofing tile sill		7.20			
			1 m quarry tile sill		6.20			
					c.f. £82.40			c.f. £1785.44

Ref. no.	*Dimensions*	*Extension*	*Description and price build-up*	*Quantity*	*Rate (£)*	*(£)*
			b.f. £82.40			b.f. £1785.44
			2.9 m emulsion reveal 2.40			
			1.9 m facework to reveal 2.80			
			1.2 m concrete lintel 21.20			
			£108.80			
			Sundry labours – 2½% 2.72			
			111.52			
			less 1 m^2 of brick wall with finishings 65.40			
			Extra cost of window over wall £46.12			
			Door			
4.	1	1	Extra over brick wall for framed, ledged and braced door, including frame, painting, etc.	1	52.22	52.22
			Price build-up			
			One nr. door £54.00			
			3.4 m^2 painting 15.30			
			One pair cross garnets 1.90			
			One nr. lock 10.80			
			5 m frame 45.00			
			10 m painting frame (both sides) 17.00			
			5 m emulsion reveal 4.50			
			4 m facework to reveal 5.80			
			1.1 m concrete lintel 20.00			
			174.30			
			Sundry labours, cramps, etc. – 5% 8.72			
			183.02			
			less 2 m^2 of brick wall with finishings 130.80			
			Extra cost of door over wall £52.22			
			Roof			
5.	3.38 2.37	9.23	Reinforced concrete roof, average 162 mm thick, reinforced with fabric reinforcement and covered with two coats of asphalt.	9 m^2	60.69	546.21
			Price build-up			
			0.16 m^3 RC roof slab average 162 mm thick £12.20			
			1 m^2 fabric reinforcement 1.80			
			1 m^2 wrought formwork to soffit 25.00			
			1 m^2 two coats of asphalt 16.00			
			1 m^2 emulsion paint 2.80			
			£57.80			
			Sundry items – 5% 2.89			
			Cost/m^2 £60.69			
6.			Electrical work			230.00
7.			Telephone			130.00
8.			Preliminaries			300.00
9.			Contingencies			150.00
			Estimated cost of building			£3193.87
			Rounded off to			£3200
						(£554.11/m^2)

(NOTE: NO external works are included in the estimate)

	Unit	£
Strip foundations. Excavating trench 1 m deep in heavy soil; levelling; compacting; earthwork support; backfilling; disposal of surplus material from site; concrete 11.50 N/mm^2 foundations 300 mm thick; hollow brickwork in cement mortar to 150 mm above ground level; bitumen hessian-based horizontal dpc; and facing bricks (pc £300/1000) externally. (£16.50 for each additional 300 mm in depth).	m	75.00
Hollow ground-floor construction. Excavation; disposal of surplus; hardcore 100 mm thick; concrete 11.50 N/mm^2 bed 150 mm thick; half-brick sleeper walls, honeycombed, at 2 m centres; horizontal dpc; 100 × 50 mm plates; 50 × 100 mm joists at 400 mm centres; and 25 mm tongued and grooved softwood boarded flooring.	m^2	33.00
Upper floor. 50 × 175 mm joists at 400 mm centres with ends creosoted and built into brickwork; 50 × 25 mm herringbone strutting; trimming to openings; 25 mm tongued and grooved softwood boarded flooring; plasterboard; one coat of 5 mm gypsum plaster; and two coats of emulsion paint.	m^2	32.00
Pitched roof (measured on flat plan area). 75 × 40 mm plates; TRADA trussed rafters at 2 m centres; 32 × 100 mm rafters and ceiling joists at 500 mm centres; 40 × 150 mm purlins; 50 × 125 mm binders; 25 × 150 mm ridge; 40 × 19 mm battens; felt; and concrete interlocking tiles nailed every second course.	m^2	38.20
Stairs. 900 mm wide to BS 585; treads and risers, winders, balustrade one side; plasterboard and one coat of gypsum plaster to soffit and painting; rising 2600 mm.	nr	480.00
Lavatory basin. White gvc, waste fitting, cantilever brackets and pair of taps (pc £38 complete); trap; and copper waste pipe.	nr	94.00
(Comparable costs of other sanitary appliances would be £156 for a sink; £165 for a bath; £115 for a ground floor WC and £280 for a WC on an upper floor, including a cast iron soil pipe.)		
Electrical installations. These are most conveniently priced on a cost per point basis, for example lighting points at £34.00 each and double 13-amp switched socket outlets wired in a ringmain circuit at £45.00 each.		
Drainage. 100 mm clay pipes and flexible couplings; excavating trenches average 1 m deep in heavy soil; grading bottoms; earthwork support; backfilling; removal of surplus spoil; and 150 mm granular bed and benching.	m	13.00
(£1.80 for each additional 250 mm depth of trench not exceeding 1.50 m deep and £4.70 for each 250 mm between 1.50 m and 3.00 m deep.)		
Manhole 686 × 457 × 900 mm deep internally in one-brick walls in engineering bricks on 150 mm thick concrete base; with 100 mm half-section channel and branches; concrete benchings; 610 × 457 mm cover and frame; and all necessary excavation, backfill, disposal of surplus and earthwork support.	nr	285.00
(£62.00 for each additional 300 mm of depth up to 1.20 m deep internally.)		

ELEMENTAL COST ANALYSES

Another approximate estimating method uses elemental cost analyses for previous similar projects as a basis for the estimate. The cost is computed on a superficial or floor area basis but the overall superficial unit cost is broken down into elements and sub-elements. At this lower level of division it is possible to make cost adjustments for variations in design in the new project as compared with the previous scheme. It will also be necessary to update the costs to take account of increased costs which have occurred since the tender date of the project for which the cost analysis is available. Elemental cost planning, cost analyses and building cost and tender price indices will be considered in some detail in later chapters. A cost analysis relating to a factory follows to illustrate the general method of approach, although it will be appreciated that in practice the differences between the old and new projects will generally be much greater and the problems involved preparing the estimate much more complex. This example has intentionally been kept fairly simple to avoid becoming involved in excessive detail in the costing of each element. This approach involves a close examination of the design and cost aspects of major parts or elements of the building and the information thus produced can form a preliminary cost plan. Changes in storey heights, floor loadings, column spacings or the number of storeys can have varying effects on costs as indicated in the Wilderness study (1964).

In detailed cost analyses produced by BCIS, elemental costs are supported by element unit quantities and element unit rates, such as internal walls and partitions in m^2 and sanitary appliances by number; when read in conjunction with specification and design notes, they assist in the preparation of estimates for new projects.

Example 6.5

An estimate was required in January 1994 for a single storey advance factory with a floor area of 1800 m^2 to be erected in Nottinghamshire. The cost analysis which follows relates to a factory of 1998 m^2 in Warwickshire of similar shape and specification to the proposed factory, for which tenders were received on 1 June 1990. Adjustments need to be made for the following factors.

(1) The weighted tender price indices relating to this class of building were 126 in June 1990 and 116 in January 1994 – a decrease of

$$\frac{116}{126} \times 100 = 7.94 \text{ per cent.}$$

This is then adjusted for the county factor of

$$\frac{1.02}{0.92} = \text{increase of } 10.87 \text{ per cent.}$$

The overall adjustment thus becomes: 10.87 – 7.94 = + 2.93 per cent.

(2) The new factory has an average storey height of 6.70 m (300 mm lower).

(3) The wall/floor ratio = 0.85.

(4) Facing bricks are £20/1000 cheaper.

(5) Increase in amount of partitioning by ten per cent in length.

(6) Twenty-five per cent increase in number of sanitary appliances.

(7) Fiver per cent increase in waste pipes stemming from greater number of sanitary appliances.

(8) Sprinkler installation required at estimated cost of £14 000.

(9) Ten per cent reduction in amount of siteworks (smaller site).

Cost Analysis of Existing Factory

Tender date: 1 June 1990
Contract: Lowest of six selected tenders – standard form of building contract with quantities – eight months contract period.
Floor area: 1998 m^2.
Wall/floor ratio = 0.88
Storey height: average 7.00 m

Summary of element costs

Element	*Existing factory* Cost per m^2 of gross floor area	*Existing factory* Group element total	*New factory* Cost per m^2 of gross floor area	*New factory* Group element total
1. SUBSTRUCTURE	£84.54	84.54	87.02	87.02
2. SUPERSTRUCTURE				
2A. Frame	37.10		36.55	
2B. Upper floors	–		–	
2C. Roof	44.89		46.21	
2D. Stairs	–		–	
2E. External walls	45.14		43.97	
2F. Windows and external doors	13.76		14.16	
2G. Internal walls and partitions	12.80		13.91	
2H. Internal doors	4.63	158.32	4.77	159.57
		c.f. £242.86		c.f. £246.59

		b.f. £242.86		b.f. £246.59
3. INTERNAL FINISHES				
3A. Wall finishes	1.41		1.48	
3B. Floor finishes	0.48		0.49	
3C. Ceiling finishes	1.43	3.32	1.47	3.44
4. FITTINGS	0.55	0.55	0.57	0.57
5. SERVICES				
5A. Sanitary appliances	2.54		3.26	
5B. Services equipment	–		–	
5C. Disposal installations	0.53		0.58	
5D. Water installations	2.27		2.34	
5E/G Heating and ventilating systems	–		–	
5H. Electrical installations	13.88		14.29	
5I. Gas installations	–		–	
5J. Lift and conveyor installations	–		–	
5K. Protective installations	–		7.78	
5L. Communication installations	–		–	
5M. Special installations	–		–	
5N. Builder's work in connection with services	1.20		1.24	
5O. Builder's profit and attendance on services		20.42		29.49
BUILDING SUB-TOTAL		267.15		280.09
6. EXTERNAL WORKS				
6A. Siteworks	50.98		47.22	
6B. Drainage	12.35		12.71	
6C. External services	11.85		12.20	
6D. Minor building works	7.72		7.95	
EXTERNAL WORKS		82.90		80.08
PRELIMINARIES		27.42		28.22
TOTAL (less contingencies)		£377.47		£388.39
			to nearest pound £388/m²	

NOTE: a further allowance of up to 10 per cent might be added at this stage to cover design and price risk.

The elemental costs relating to the new factory have been inserted in the summary alongside the figures for the existing factory. The techniques employed are now described.

All elemental costs are increased by 2.93 per cent to update them, and then adjusted in other ways as required.

1. *Substructure:* 2.93 per cent period/regional price increase: £84.54 + £2.48 = £87.02.

2A. *Frame:* £37.10 + 2.93% = £38.19 – 4.29% (lower storey height) = £36.55 (effect of horizontal components ignored as effect is minimal).

2E. *External walls:* £45.14 + 2.93% (period/regional costs) + 3.5% (wall/floor ratio) – 9.09% (cheaper facing bricks) = £43.97.

2F. *Windows and external doors:* £13.76 + 2.93% (period/regional costs) = £14.16 (assuming that window and door to wall ratios remain constant).

2G. *Internal walls and partitions:* £12.80 + 2.93% (period/regional costs) + 10% (increased quantity) – 4.29% (lower storey height) = £12.80 + 8.64% = £13.91.

3A. *Wall finishes:* £1.41 + 2.93% (period/regional costs) + 6.5% (increased internal partitions and reduced wall/floor ratio for external walls) − 4.29% (lower storey height of internal partitions) = £1.41 + 5.14% net = £1.48.

5A. *Sanitary appliances:* £2.54 + 2.93% (period/regional costs) = £2.61 + 25% (increased quantity) = £3.26.

5C. *Disposal installations:* £0.53 + 2.93% (period/regional costs) = £0.55 + 5% (increased quantity) = £0.58.

5K. *Protective installations (sprinkler system):*

$$\frac{£14\,000}{1800}\ £7.78$$

6A. *Siteworks:* £50.98 + 2.93% (period/regional costs) = £52.47 − 10% (smaller site) = £47.22 (alternatively computed on basis of approximate quantities).

The estimated cost of the new factory:

£388 × 1800 =	£698 400
add contingencies (about 1.6%)	11 180
Total estimated cost	£709 580

COMPARATIVE ESTIMATES

Another method of approximate estimating is to take the known cost of a similar type building as a basis and then to make cost adjustments for variations in constructional methods and materials. For this purpose it is advisable to build up costs usually related to a square metre of finished work for a whole range of alternatives, to enable speedy adjustments to be made when preparing approximate estimates. These comparative costs will also be useful for costing alternative proposals as detailed designs are developed. It is important to consider possible side effects of alternative choices as various forms of cladding may offer differing degrees of thermal insulation, and the adoption of a cheaper solution may result in greater heat losses and increased heating costs once the building is occupied. The weight of the cladding may also affect the design of the foundations. The speed of erection of any specific part of the building may have important cost implications and may also affect progress in construction of other sectors of the building. Table 6.3 contains a number of comparative prices to illustrate the form that a comparative cost schedule would take.

In practice, quantity surveyors often seek as soon as possible to produce broad/approximate quantities with 'unit costs' or 'feature costs' applied to the quantities. Hence there is a need for reliable, up-to-date information on elements, features and unit costs. Examples of features would be different types of substructures, struc-

Table 6.3 Schedule of comparative costs of different constructional methods

Typical mid 1994 London prices for medium-sized project (excluding preliminaries)

Substructure	Cost/m^2
Solid ground floor, including excavation but excluding finish	£20.00
Hollow ground floor, ditto	26.00
Upper floors	
Softwood joists and herringbone strutting (depending on sizes of joists)	12.30–18.00
Hollow tiles, 140 mm thick	30.50
Reinforced concrete, 125 mm thick, 3.65 m span, 300 kg/m^2 loading	31.00
Flat roof finishes	
Three-layer glass fibre based bitument felt roofing	9.50
20 mm mastic asphalt to BS 988 with solar paint on felt underlay	11.00
0.91 mm aluminium	32.00
0.56 mm copper	41.00
Nr 4 sheet lead	59.00

Table 6.3 Continued

Pitched roofs (measured per m² of plan area) (all of 40° pitch)	*Cost per m²*
Timber roof construction (TRADA)	17.00
Concrete interlocking tiles to 76 mm lap	17.50
Concrete plain tiles to 64 mm lap	36.00
Machine-made plain clay tiles to 64 mm lap	39.00
Hand-made plain clay tiles to 64 mm lap	59.00
Welsh slates (510 × 255 mm) to 76 mm lap	51.00
Westmorland green slates, random sizes, to 76 mm lap	125.00
External walls	
Cavity wall with outer faced half-brick skin and inner skin of 75 mm Thermalite blocks (facing bricks pc £250/1000)	52.00
One-and-a-half brick wall faced externally (facing bricks pc £250/1000)	65.00
150 mm reinforced concrete wall	54.00
Tile (machine made) hanging including battens	23.00
25 mm tongued and grooved Western Cedar boarding and battens	25.50
75 mm Portland stone facing slabs and fixing including cramps	174.00
Galvanised steel curtain walling, with proportion of opening lights, but excluding glazing and infill panels	153.00
Anodised aluminium curtain walling, with proportion of opening lights	190.00
6 mm float glass	25.50
Double glazing of two skins of 6 mm float plate in copper channel	71.00
Internal walls and partitions	
75 mm Thermalite block	10.60
100 mm Thermalite block	13.00
Half-brick wall (flettons)	18.20
Stud partition	7.20
57 mm Paramount dry partition	11.00
50 mm demountable steel partition, self-finished	59.00
Doors (726 × 2040 mm)	*(each)*
40 mm flush door, cellular core, hardboard faced, with 102 mm steel butts and painted three coats	35.00
Ditto, plywood faced	39.50
44 mm softwood purpose made four-panel door with 102 mm steel butts and painted	100.00
44 mm flush door, ½ hour fire check teak veneered hardboard faces, with steel butts and 2 coats polyurethane	55.00
54 mm flush door, one hour fire check, Sapele veneered hardboard faces with 102 mm brass butts and 2 coats polyurethane	153.00
100 × 32 mm softwood lining with grounds and painting	47.00
Ditto, in mahogany and wax polished	69.00
100 × 75 mm softwood rebated and rounded frame, including fixing and painting	63.00
Ditto, in mahogany and wax polished	93.00
Wall finishes	*Cost per m²*
Render and set in lightweight gypsum plaster	5.40
9.5 mm gypsum wallboard as dry lining	4.60
12 mm softwood wall lining and battens plugged to wall	14.50
4 mm coloured wall tiles on cement-sand backing	25.00
Floor finishes	
50 mm cement-sand screeded bed	5.30
2 mm thermoplastic tiles (grade B)	4.50
25 mm granolithic paving	8.60
2 mm vinyl tiles	5.20
15 mm black pitchmastic	7.40
3.2 mm cork tile flooring and polishing	7.70
5 mm rubber tile flooring	23.00

Table 6.3 (contd)

25 mm softwood tongued and grooved flooring including fillets	16.50
25 mm Iroko block flooring and polishing	32.00
25 mm maple strip flooring including fillets and polishing	30.00
16 mm terrazzo paving in squares with ebonite strips	35.00
Ceiling finishes	
9.7 mm gypsum lath fixed to joists (not included) and skim coat of gypsum	6.70
Suspended ceiling of 15 mm mineral fibre tiles, flame-proofed, and fixed in steel tees	23.00
Decorations	
Prepare, and two coats emulsion paint on plastered surfaces	1.25
Prepare, prime and three coats oil paint on metalwork	4.00
Knot, prime, stop and three coats oil paint on woodwork	4.30
Stain, body in and wax polish on hardwood	8.00
Pavings	
50 mm gravel paving	1.80
50 mm tarmacadam in two layers	5.00
50 mm precast concrete paving slabs to BS 368	8.80
Drains	*Cost per m*
100 mm drains including excavation average 1 m deep: clay including flexible couplings and 150 mm granular bed and haunching	18.00
PVC-U pipe, including 150 mm granular bed and haunching	18.00
Spun iron pipe to BS 437, including 150 mm concrete bed	32.00

tures, building envelopes, finishes and engineering services, both at a fairly broad, coarse level of information, and then at a more detailed, fine level of data appropriate to the progressive design development stages of projects.

INTERPOLATION METHOD

A variant of the comparative method is the *interpolation* method whereby, at the brief and investigation stages in the design of a project, an estimate of probable cost is produced by taking the cost per square metre of floor area of a number of similar type buildings from cost analyses and cost records and interpolating a unit rate for the proposed building. This method looks deceptively easy, but no two buildings are the same and it is difficult to make adjustments to the unit rate to take account of the many variables that are bound to occur between the buildings for which known costs are available and the project under consideration. In practice it often may be necessary to use a method which is a combination of both the interpolation and comparative approaches.

7 COST PLANNING THEORIES AND TECHNIQUES

There is no universal method of cost planning which can be readily applied to every type of building project. Buildings have widely varying characteristics, perform a diversity of functions, serve the needs of a variety of building clients, and their erection is subject to a number of different administrative and contractual arrangements. Hence, it is not surprising that a wide range of cost planning techniques has been devised to meet the needs of a variety of situations.

The quantity surveyor frequently acts as specialist adviser to the architect on all matters concerned with building costs and it is vital that he should be involved at the earliest possible stage in the design of the building. He can offer coniderable assistance to the architect in advising on the financial effect of design proposals and so help in ensuring that the money available is put to the best possible use and that the tender figure is close to the initial estimate.

The three main themes of one conference on integrated building design were that cost control, including life cycle costing, should be based on giving the client value for money; that design should be a team effort involving new attitudes and relationships; and that improved design is a learning process based on feedback.

PLAN OF WORK

The Royal Institute of British Architects (1973) formulated a suggested pattern of procedure for architects in the preparation and implementation of building schemes. This plan of work represents a sound and practical analysis of the operations and can be applied strictly on contracts where all information is complete and in sufficient detail to enable a contractor to prepare a tender (Turner, 1990). The plan is outlined in table 7.1, and possible activities by the quantity surveyor to provide an effective contribution in cost control are shown in figure 7.1, and all are described in the following text, pursuing an orthodox and well ordered approach.

At the inception stage of a building contract, the building client considers his building requirements and appoints an architect. The next stage in the plan of work is the feasibility stage at which an effective cost control mechanism needs to be established, and, in particular, a realistic first estimate is produced. The cost limit will be very much influenced by the floor space required, the standard of accommodation and the function of the building. The cost limit is often assessed from a comparative study of known costs of similar type buildings, sometimes referred to as the interpolation method. In other cases it may be established by means of a developer's budget calculation of the type described in chapter 15 and occasionally described as the financial method.

At scheme design stage the brief is completed and the design team develops the full design of the project. The cost plan is now formulated, which consists of a statement showing how the design team proposes to distribute the available money over the various elements or major parts of the building. Typical cost plans are illustrated later in this chapter and in chapter 10. The cost plan is used continuously throughout the detail design stage of the project as a means of checking that the detail design is kept within the agreed cost framework. The quantity surveyor translates the design team's decisions on each element into

Table 7.1 Plan of work for design team operation

Outline Plan of Work

Stage	*Purpose of work and decisions to be reached*	*Tasks to be done*	*People directly involved*	*Usual terminology*
A. *Inception*	To prepare general outline of requirements and plan future action.	Set up client organisation for briefing. Consider requirements, appoint architect.	All client interests, architect.	*Briefing*
B. *Feasibility*	To provide the client with an appraisal and recommendation in order that he may determine the form in which the project is to proceed, ensuring that it is feasible, functionally, technically and financially.	Carry out studies of user requirements, site conditions, planning design, and cost, etc., as necessary to reach decisions.	Clients' representatives, architects, engineers, and QS according to nature of project.	
C. *Outline proposals*	To determine general approach to layout, design and construction in order to obtain authoritative approval of the client on the outline proposals and accompanying report.	Develop the brief further. Carry out studies on user requirements, technical problems, planning, design and costs, as necessary to reach decisions.	All client interests, architect, engineers, QS and specialists as required.	*Sketch plans*
D. *Scheme design*	To complete the brief and decide on particular proposals, including planning arrangement, appearance, constructional method, outline specification, and cost, and to obtain all approvals.	Final development of the brief, full design of the project by architects, preliminary design by engineers, preparation of cost plan and full explanatory report. Submission of proposals for all approvals.	All client interests, architects, engineers, QS and specialists and all statutory and other approving authorities.	
Brief should not be modified after this point				
E. *Detail design*	To obtain final decision on every matter related to design, specification, construction and cost.	Full design of every part and component of the building by collaboration of all concerned. Complete cost checking of designs.	Architects, QS, engineers and specialists, contractor (if appointed).	*Working drawings*
Any further change in location, size, shape, or cost after this time will result in abortive work.				
F. *Production information*	To prepare production information and make final detailed decisions to carry out work.	Preparation of final production information i.e. drawings, schedules and specifications.	Architects, engineers and specialists, contractor (if appointed).	
G. *Bills of quantities*	To prepare and complete all information and arrangements for obtaining tender.	Preparation of Bills of Quantities and tender documents.	Architects, QS, contractor (if appointed).	
H. *Tender action*	Action as recommended in paras. 7–14 inclusive of 'Selective Tendering'.*	Action as recommended in paras. 7–14 inclusive of 'Selective Tendering'.*	Architects, QS, engineers, contractor, client.	

Table 7.1 Continued

Stage	*Purpose of work and decisions to be reached*	*Tasks to be done*	*People directly involved*	*Usual terminology*
J. *Project planning*	Action in accordance with paras. 5–10 inclusive of 'Project Management'.*	Action in accordance with paras. 5–10 inclusive of 'Project Management'.*	Contractor, subcontractors.	*Site operations*
K. *Operations on site*	Action in accordance with paras. 11–14 inclusive of 'Project Management'.*	Action in accordance with paras. 11–14 inclusive of 'Project Management'.*	Architects, engineers, contractors, subcontractors, QS, client.	
L. *Completion*	Action in accordance with paras. 15–18 inclusive of 'Project Management'.*	Action in accordance with paras. 15–18 inclusive of 'Project Management'.*	Architects, engineers, contractor, QS, client.	*Site operations*
M. *Feed-back*	To analyse the management, construction and performance of the project.	Analysis of job records. Inspections of completed building. Studies of building in use.	Architect, engineers, QS, contractor, client.	

* Publication of National Joint Consultative Committee for Building.
Source: RIBA Publications Ltd.

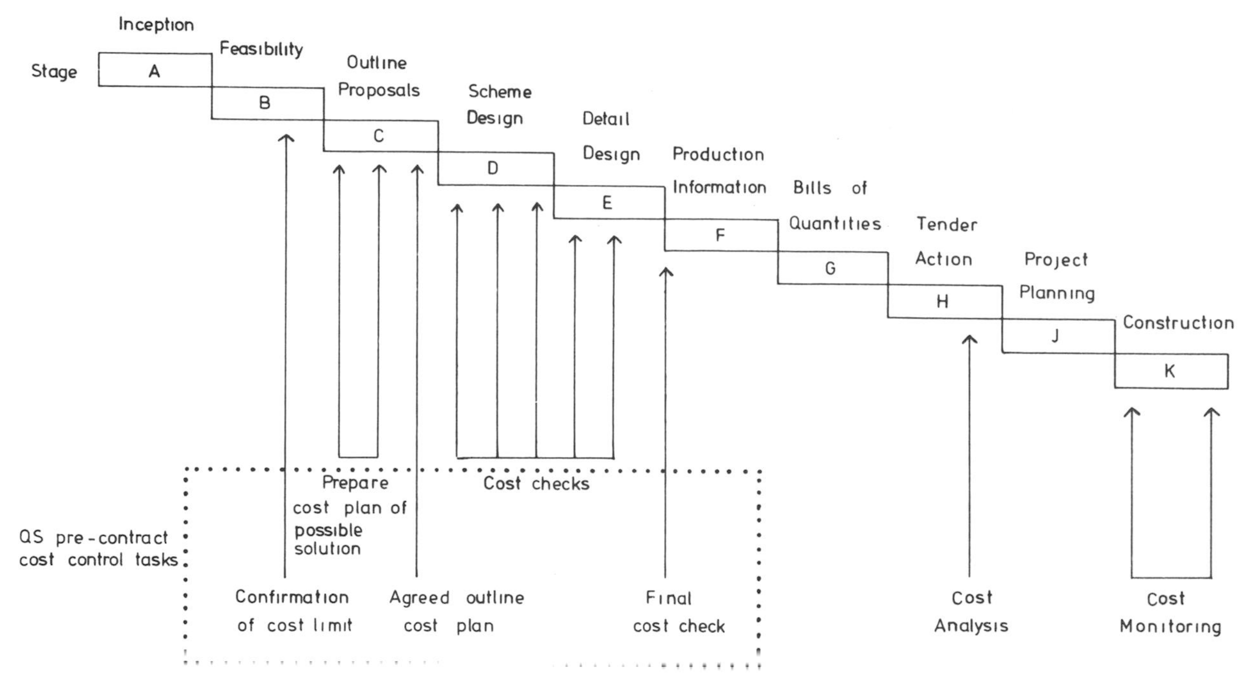

Figure 7.1 Sequence of design team's work

cost targets, to which an allowance for design and price risk, possibly up to five per cent, is added to produce cost limits.

At the detail design stage final decisions are made on all matters relating to design, specification, construction and cost. Every part of the building must be comprehensively designed, its cost checked and the design adjusted if necessary. Since detail designs and specifications are available at this stage, the most suitable and accurate estimating technique is the approximate quantities method described in chapter 6. The quantity surveyor prepares approximate grouped or composite quantities from the detailed designs for each element in turn and prices these quantities on the basis of priced bill rates for previous recent similar projects adjusted as necessary. The estimate is then compared with the cost target for the element in the cost plan and this constitutes a cost check. In this way the design team is able to exercise effective cost control. Where the estimated cost exceeds the cost target then either the element must be redesigned or other cost targets reduced to make more money available for the element in question, but leaving the overall cost limit unaltered. The quantity surveyor must keep the rest of the design team informed of any actions of this kind.

Where the estimated cost of the design element is within the cost target in the cost plan, the design should be confirmed in writing as suitable for the preparation of production drawings, often described as the production information stage. If the estimated cost is substantially below the cost target, then surplus funds may be released for other elements. Each element is investigated in turn to arrive at a logical and balanced distribution of costs throughout the major parts of the building. Finally, after checking the costs of each design element, a final cost check of all elements is made. Where a number of design elements would have to be adjusted to keep costs within the total cost limit and this would result in a building of undesirably low quality, the design team should request additional funds from the building client.

A programme or timetable of possible cost advice to be given during the development stage of a project is illustrated in figure 7.1. Cost moniroring continues throughout the construction stage and this process is described in some detail later in this chapter.

The essence of cost planning is to enable the architect to control the cost of a project (within the target) while he is *still* designing. The earlier this process is introduced, the greater the measure of control that can be exercised over ultimate cost, quality and design. Cost planning should be a continuous process, progressive checks being made from time to time in relatively more detail on perhaps smaller sections of the project as the design is finalised. Another merit of cost planning is that it introduces a positive cost checking procedure into the design stage where previously nothing systematic had existed.

Cost plays an important part throughout the design process. In the first instance, it influences the size of the project and its general form, then later it indicates the type of structure and subsequently affects the choice of services and finishings. Cost is, therefore, a continuing influence but it has two distinct phases. During the briefing stages, (A and B) the building client and architect have the joint responsibility of deciding just how much the project should cost, or alternatively of deciding what size and quality of building can be provided for a given sum. During the sketch plans and working drawings production phases (C, D and E) the architect has the responsibility of designing a building in such a way that the tender will not exceed the client's budget or cost limit. The costing process therefore has two distinct parts – firstly, the determination of total cost, and secondly the costing of the design within the total sum. The first stage (that is to decide how much to spend) is sometimes termed the approximate estimating stage, whereas the second stage (that is to decide how the budget is to be spent) may be termed the cost planning stage. Hence these two terms are not synonymous, as they have different meanings and can be used to separate aspects or phases of the costing process. They are, however, mutually supporting.

It would be useful at this stage to examine the various activities that take place during the pre-contract stages listed in table 7.1.

(A) Inception

This represents the building client's decision to build, the setting up of the administrative organisation, the appointment of an architect and other members of the design team. The architect requires a site plan, details of preliminary items, erection times and any cost limits.

(B) Feasibility

At this stage the architect and the building client are busily endeavouring to establish the client's requirements and, in particular, to distinguish between desires or whims and essential needs. The architect frequently finds it very difficult to determine just how much the building client is prepared to spend and to reconcile the two major factors of cost and quality, which are so closely related. The architect will also need to identify any site restrictions, such as preservation orders, and any probable planning conditions. Consideration should also be given by the design team to the design timetable and the tendering procedure.

Often the client is pressing for an assessment of cost before any drawings have been produced. The quantity surveyor can render valuable service at this stage by supplying cost information based on the actual cost of previous buildings of a similar type. He will make allowance for such factors as differences of location, site conditions, market conditions and quality of work, and so arrive at a provisional estimate on a comparative or interpolation basis.

(C) Outline Proposals

The building client's requirements have now been definitely established and confirmed to be viable, the site has been surveyed and the architect begins to consider the various alternative ways in which the building can be designed and constructed. With industrial projects the architect will probably decide on a single-storey building and will then determine which heights and span give the most useful and economical arrangement. With blocks of offices and flats the number of storeys will have to be decided, having regard to such matters as soil conditions, type of foundation to be used and constructional form and costs.

Some drawings will be produced at this stage and the quantity surveyor will be in a position to give general guidance on costs and, in particular, to evaluate the financial effect of different solutions to any specific design problem, and often prepares an outline cost plan.

(D) Scheme Design

During this stage the major planning problems will be resolved and the outline designs will emerge. The sketch designs will include sections and elevations, and services and finishings will be considered in addition to the form of the structural framework. For instance, if it is a framed building, it will be necessary to consider the relative merits of steel or reinforced concrete. Consultants will be brought in at this stage if not before in order that their requirements may also be investigated.

The quantity surveyor checks on his approximate estimate figure and, with the aid of extensive cost information, reappraises the initial cost plan with provisional target cost figures set down for each element or major part of the building. The quantity surveyor may adopt an elemental approach, a comparative technique or a mixture of both.

(E) Detail Design

Sketch plans are now finalised and somc working details are prepared. It is most desirable that these should be approved by the building client to avoid the possibility of future alterations. Outline schemes will be prepared by consultants and designing subcontractors and provisional estimates supplied in some cases.

The quantity surveyor will be called upon to give comparative costs of different forms of construction, materials, components and service layouts and will adjust the distribution of costs in the cost plan if required. It is to be hoped that these comparative cost studies will include probable running and maintenance costs wherever they are likely to have a significant effect on the outcome. Future costs will be examined in some depth in chapter 13. Continuous cost checks by the quantity surveyor will ensure that the development of the design remains compatible with the cost plan, and this process is sometimes described as cost reconciliation. When all the design drawings have been prepared and cost checked, a final cost review should be made by the quantity surveyor and a report submitted to the architect.

(F) Production Information

The final working drawings (production drawings) will now be prepared from which bills of quantities can be produced. Consultants, subcontractors and suppliers will be required to supply full information at this stage including realistic quotations. The quantity surveyor continues his cost checks on the data produced against the final cost plan. He will also be available to give advice to the architect on any financial or contractual matters associated with the project, including the terms and conditions of the main contract and subcontracts and on the selection of tenderers, and will be considering his work in stage G – the preparation of the bill of quantities.

A useful checklist of objectives, information requirements and procedures at each stage of the cost planning process was prepared by the Essex branch of the RICS (1982) together with supporting specimen documentation.

COST CONTROL PROCEDURE

The actual procedure involved in controlling the cost of a project depends to some extent upon the stage at which the cost limit is determined by the building client. There are three main arrangments and these are now outlined.

(1) The accommodation requirements and often the occupation date are prescribed by the

building client at the outset and the cost limit is established early in the design stage, after the sketch proposals and approximate estimate of cost are approved by the building client. Cost will be an important factor and the estimate must be realistic.

The architect and building client will begin by discussing the client's requirements, and the quantity surveyor may be asked to give general information on the known costs of similar buildings. Once the basic form of the building has been decided, the quantity surveyor should prepare an approximate estimate, in particular taking full account of shape, number of storeys and structural form.

Outline plans will usually be drawn to a scale of 1:200. The estimate may be based on cube, floor area, storey enclosure (unlikely because of lack of data) or approximate quantities methods depending on the amount of information available, with assumptions having to be made concerning finishings, fittings, services and similar matters. These estimating methods are described and illustrated in chapter 6. The sketch plans will then be prepared, usually to a scale of 1:100. The quantity surveyor should be available with comparative costs of alternative forms of construction, service arrangements and other components. By the end of the sketch plan stage, a more accurate estimate will be prepared usually based on approximate quantities. The plans and estimate are then submitted to the building client for approval.

During the detail design stage the cost of the project can be influenced appreciably by the choice of materials and constructional methods, for example, aluminium, steel or wood casements; timber, concrete or metal cladding panels; inclusion or otherwise of a parapet wall; and many other alternatives. Continual reference to a cost plan is essential as the details of each part of the project are finalised with possible adjustments to design or distribution of costs as the process of reconciliation proceeds. A final overall cost check can be made when the detail design stage is complete. It will be appreciated that the amount of cost investigation work is dependent to a large extent upon the size and nature of the project. A small and relatively straightforward scheme requires only a minimum of cost investigation.

(2) The accommodation requirements, and often the occupation date cost limits are determined by the building client. This arrangment is most likely to operate where the client has considerable experience of the past costs of similar buildings. A common example is that of educational buildings but it could also apply to commercial, industrial and residential buildings. It is important to check the accommodation requirements against the cost limit to be sure that the project is feasible, and the usual approach is to make a comparison with the known costs of recent similar buildings. The remainder of the procedure will be identical to that described in (1).

(3) Where available funds are limited and the building client wishes to carry out as much building work as possible, he may prescribe a cost limit and ask the architect for details of the size and quality of building that can be provided for this sum of money. In this case it is necessary to compute the amount of floor area that can be provided by reference to the known costs and floor areas of similar buildings. The subsequent procedure will be as described in (1).

Cost control can be defined as an umbrella term embracing all contributory stages such as cost analysis, cost planning, cost comparisons, cost checking, cost reconciliation at the tendering stage, and cost monitoring at the post-contract stage in the manner listed in the following schedule:

(1) Initial cost budget and cost plan prior to detailed design, often based on functional unit costs, elemental cost analysis, or some similar method.

(2) Preliminary cost studies, usually based on measured approximate quantities, to compare alternative materials and systems in terms of capital, operating, maintenance and depreciation costs.

(3) Cost checking detailed designs as they are produced, again usually on the basis of measured approximate estimates, to ensure that any underdesign or overdesign relative to the budget is

corrected early enough to avoid delays and abortive effort.

(4) Preparation of tender documents (for example, plans and specification; schedules of rates; bills of quantities) for pricing by tendering contractors.

(5) Checking and reporting on competitive tenders or agreeing negotiated tenders.

(6) Preparing and agreeing pre-contract variations (if required) to balance tenders with budgets.

(7) Analysing accepted tenders into functional or elemental costs (or both), comparing tenders with estimates, and modifying basic estimating data as necessary for future reference.

(8) Calculating final costs (including variations, remeasurements, PC accounts, dayworks, increased or decreased costs and authorised overtime) in two stages – rough estimates reported quickly for budget control; and accurate valuations agreed with the contractor for final payment.

(9) Checking and negotiating contractors' claims.

INFORMATION REQUIRED BY ARCHITECT AND BUILDING CLIENT

Need for Team Approach

When it is proposed to carry out private development in the central areas of large towns and cities, the cost of the land can be considerably greater than the cost of the building work. Thereore it is imperative to obtain the most suitable site available at a fair market price and to use the site for the most advantageous permitted use. In some cases the cost of the building is relatively small by comparison with the cost of the land and in the last two decades land costs have generally risen at a much faster rate than building costs. Building land in provincial towns, without services, increased from about £4000 per hectare in the early nineteen fifties to around £500 000 per hectare in the late nineteen eighties and reduced to about £350 000 in the early nineteen nineties. Typical provincial housing land could sell at about £15 000 per building plot in 1993. Land for a small shop in the centre of a provincial town in 1993 could have cost as much as £80 000 to £160 000, whereas the cost of the building might not have exceeded £60 000, a ratio of between $1\frac{1}{3}$ and $2\frac{2}{3}$ to one. It is vital that the quantity surveyor should appreciate this type of relationship and its general significance. More information on relative land values is given in chapter 14.

This set of relationships also serves to indicate the need for a team effort when formulating development proposals, particularly those connected with the implementation of costly redevelopment schemes in the central areas of large towns and cities. The team should include a valuer, architect and quantity surveyor. The valuer can advise on the best type of development and its ultimate value, and can give useful information on the most suitable type of construction and quality of finishings. The architect advises on building designs and the quantity surveyor on building costs. Together they can supply all the information necessary to prepare a complete budget of probable expenditure and revenue and are able to make recommendations on the advisability, or otherwise, of proceeding with the development before the land is purchased by the building client. The client must be satisfied that the projected development is in all respects sound and that it will produce a worthwhile return, before he makes an offer for the land.

The team of professional advisers has to consider a number of related matters, such as town planning requirements, services to be provided, operative rights of way and of light, road widening proposals and provision of access and car parking space. If the budget calculations prove to be satisfactory, a target cost will be established for the building work and this will constitute the total sum in the cost plan which subsequently emerges. The cost plan will break down this total sum over the main component parts or elements of the building.

Building Client's Main Needs

Building clients often want the best possible quality but are not prepared to pay for it. This frequently results in the architect's major problem being not one of design, but of cost. The quantity surveyor can be of tremendous assistance to the architect in the earliest days of a project, armed with his comprehensive records of historical cost information giving actual costs of essential components of buildings as constructed.

The main requirements of a client in connection with a building project can be conveniently listed as follows.

(1) The building must satisfy his needs, otherwise the architect has failed in his design function.

(2) The building should be available for occupation on the specified completion date if humanly possible.

(3) The final cost of the building should be very close indeed to the original estimate given to the building client.

(4) The building should be maintainable at reasonable cost.

The quantity surveyor is primarily concerned with the last two requirements. In particular he is anxious to ensure that the initial estimate and final account figures are closely related. The building client will always remember the initial estimate, as this will be the sum on which he has based all his calculations. The quantity surveyor needs the fullest constructional information if he is to provide a realistic estimate, and he often has to press the architect very hard indeed to obtain adequate information. In like manner the architect frequently experiences great difficulty in obtaining sufficient particulars from the building client in the early stages of a project. It is essential that the estimate prepared by the quantity surveyor should be as accurate as his skill and knowledge of the cost of previous projects will permit.

Settling the Brief

The initial brief for a building project must of necessity be a broad and flexible statement of objectives in fairly abstract terms, defining such matters as the site, building type, space requirements, general comfort standards, cost limit, timescale for design and construction, and the estimated useful life of the building. It is desirable for the brief to be drafted on the basis of a questionnaire prepared by all the design team. The next stage is to establish the external factors which define the physical limitations of the building, such as availability of public services; site survey; plot ratio, floor space index or residential densities; site coverage; building lines; maximum building height; adjoining buildings; rights of light and of way; road widening proposals; permissible points of vehicular access; extent of control of external elevations by local planning authority; and established environmental character of the area. These aspects will be considered in greater detail in chapter 14. The architect has then to consider the internal planning of the building related to both horizontal and vertical circulation networks. This will lead to a number of possible solutions which will need to be tested for general viability against the external site factors described previously.

The shape of some buildings is determined by the nature of the process carried on within them. For example, breweries are traditionally constructed on several floors so that the product will gravitate from floor to floor between successive stages of manufacture. Many factories, such as metal works, car assembly plant, paper works and steel mills, all need to be designed so that each part of the manufacturing process follows in logical sequence. Similarly offices need planning to allow maximum ease of communication for their occupants. The success of speculative ventures will be dependent largely upon the developer's ability to identify the needs of potential users.

The techniques of warehousing and distribution are changing as mechanical and electronic equipment take over work currently performed by hand and human calculation. In drawing up the brief the architect must be aware of the problems of loading and unloading vehicles by mechanical means; economical storage and handling of goods within the warehouse; and making up diverse orders related to journey planning and mechanised stock control. Costs of the handling equipment must be balanced against the cost of the building to ensure that the most economical overall scheme is being adopted commensurate with inevitable future improvements in handling techniques. Basic requirements for warehouses normally include a light, insulated weatherproof single-storey shell with as large a stanchion grid as possible, demountable partitions and high specification floor, with provision for 100 per cent expansion.

The critical relationship between architect and client/user often appears to be unsatisfactory, and this highlights the difficulties encountered in this important phase of the design process. A case study of a civic design commission undertaken by a private architect for a large local authority covering the redevelopment of an urban site found that the architect's preference for exercising his design function led to a rigid concentration on design work, which in turn resulted in a lack of awareness of the management opportunities open to the architect, and of the consequences of neglecting them. Hence the architect failed to realise how much pertinent information lay hidden in the client's briefing documents and the minutes of initial briefing meetings. Better communications between specialist sections of the client organisation and the designer could have resulted in better judged decisions in important matters such as allocation of resources, procedure and time schedule.

Planning of Services

Problems sometimes arise in reconciling service arrangements with architects' preliminary building designs. The services consultant commences with minimum statutory comfort levels as prescribed by codes of practice, Acts of Parliament and building regulations at one end of the scale, and ideal conditions at the other, to arrive at a broad environmental performance specification for discussion with the architect. This specification will include air temperatures, humidity, ventilation (natural or mechanical), lighting (natural and artificial), aspect and the thermal capacity of the structure. These could be expressed as maxima and minima to create a range of options.

The services consultant tests his performance specification against the architect's alternative diagrammatic layouts and, with assistance from the quantity surveyor, comparative cost factors will emerge. The primary objective at this stage is to find an ideal cost balance between building form and services installation, within the overall budget and the site parameters.

Subjective factors can make decision-making difficult. For example, the design team may submit that a higher level of factory lighting will increase productivity and reduce staff turnover; or that a hotel needs to be air conditioned to prevent excessive annoyance to occupants from city centre traffic noise; or that reduced glazing in an office building will produce improved internal comfort, whilst the letting agent believes that this will adversely affect its appearance and ease of letting.

The physical space needs of the services must be considered. A relatively cheap structural proposal of loadbearing crosswalls and a flat slab creates serious problems if services are to be concealed. The location of lift motor rooms (top, bottom or side), boiler plant room (roof or basement), substation with its special access requirements, escape routes from basement garages and plant rooms, and layout of basement ventiltion ducts, all have cost implications which may give rise to further options needing early evaluation. Finally, the service installation ought to be reasonably flexible to meet future changed requirements such as increased demand for power outlets or a change from space heating to air-conditioning.

Conflicting Claims

It is important that the building client shall allow adequate time for all the desirable feasibility studies to be undertaken and for many alternatives to be investigated during the design of a project, even although he is anxious to see the project designed and constructed as speedily as possible. A prime objective should be to produce a building which offers the best solution to the problem in all its aspects and this requires thorough and careful planning and adequate integration of all services and components.

At any stage one particular factor may exert a dominant influence on design. The requirements of the local authority, the desire to achieve a particular aesthetic effect, and convenience of construction are factors which may tend to exert undue influence at the expense of the best building in terms of cost to the building client, contribution to the overall environment and convenience of the user. The aim must be to balance these and other factors satisfactorily and this can only be achieved through effective team working, competence and understanding.

Design Criteria

The building client's ultimate objective is often a high investment return in net annual income and growth in capital value. Maximising the annual income is obtained by maximising the gross rental income and minimising the outgoings of all kinds. The latter include letting fees, statutory charges, energy costs, repair and maintenance costs, cleaning costs and refurbishing costs.

The building client needs to know why he is building, what market he aims to meet and what standards he seeks to achieve. A full brief should include all these aspects in detail to enable adequate feasibility studies to be undertaken.

The following detailed design criteria require special consideration:

(1) Type of tenant, where applicable, with a detailed examination of his requirements.

(2) The shape of the envelope can increase the rental value of the building significantly by the extent of its visual attraction. It can also increase or decrease the costs of servicing to each floor for lighting, power and telephones, or the rental value because of the relative ease or otherwise of fitting out, subdivision of floors and siting of offices.

(3) The appeal of the building and its commercial viability may be enhanced not only by its shape, but also by the internal appearance to user and visitor and the views from within.

(4) Probably the most important economic factor is the gross/net floor ratio. This has an important impact on potential rental value; with larger floor space generally more attractive or alternatively permitting a smaller building to provide the same floor space.

(5) Adequate attention should be paid to services such as lifts, security, parking and communications.

(6) The quality of the project should match the level of the rental.

(7) Site difficulties should be assessed at an early stage to ensure that adequate provision has been made and all likely problems solved in advance.

(8) Flexibility of design should be secured as far as practicable, for example to permit changing fuel sources, altering floor plans to cater for changed tenant demand, upgrading or refurbishment, and perhaps to permit a second investment lifetime in the one building shell.

(9) The use of mechanical/electrical controls to monitor and adapt energy usage.

(10) The adoption of finishes and equipment with extended life spans, easier cleaning/maintenance and greater durability under various operating conditions.

ROLE OF THE QUANTITY SURVEYOR DURING THE DESIGN STAGE

Building costs are now scrutinised more closely and with greater skill and accuracy as buildings have become larger, more complex and more expensive, and building clients have become more exacting in their requirements. These and other factors have compelled the architect to design with greater care and in more detail and, since the ultimate cost of a building is determined during the design stage, with particular emphasis during the early part of the period as illustrated in figure 1.1, the architect has increasingly directed his attention to cost control procedures and invited assistance from the quantity surveyor. If costs are to be effectively controlled, the quantity surveyor must be closely associated with the design process, as he is the recipient of a large volume of cost data extracted from priced bills of quantities and other sources such as the RICS Building Cost Information Service. He has available costs of complete buildings, of different structural forms, varying service layouts and a wide range of materials and components. In addition he is aware of the cost effects of variations in the shape, height and other characteristics of buildings and it is in everyone's interests that this information should be fed back to the design process.

Hence it has become increasingly apparent that the architect and quantity surveyor should be working together as a team during the design stages of a project. Where both professions are housed under the same roof as is often the case with local authorities and government departments, co-ordination of activities is that much simpler. Where practising professional firms are employed, a conscious effort is necessary to secure ample contact and co-operation throughout the design stage. Effective teamwork is necessary in order to raise individual and collective efficiency and to give better service to the building client. This entails the introduction of the quantity surveyor in a constructive role at the beginning of the design process and his active participation throughout the detailed design stage. The architect must surely benefit from the quantity surveyor's knowledge and skill in ensuring that the design is founded on a sound economic base from the outset, by the ability to make major decisions in full knowledge of their economic consequences, and by formulating the design against a cost background so that a balanced and consistent design is secured. The architect has a moral obligation to design buildings efficiently and economically. Whilst not wishing to inhibit the enterprise and creative powers of the individual designer, it is nevertheless incumbent upon an architect to take positive steps to secure overall economy in design, and the quantity surveyor can make a valuable contribution towards it.

Before the architect can settle a number of fundamental design issues he will frequently need assistance from a quantity surveyor on the various cost implications. The following examples will serve to illustrate the type of decisions that have to be made.

(1) What are the cost relationships of, say, three blocks 15 m high, two blocks 22 m high and one block 45 m high, where all three schemes will give the same floor area?
(2) At what height of building will it be necessary to introduce a reinforced concrete or structural steel frame?
(3) To what extent should precast concrete units be used as against *in situ* construction?
(4) Would it be best to use a solid reinforced concrete slab floor, hollow pots or some form of patent flooring?

The quantity surveyor needs to keep up to date on the latest constructional processes and techniques, materials and components. In conjunction with architects he will conduct cost studies into various forms of construction including internal finishings, cladding and infill panels. Probably one of the biggest difficulties facing an architect is the choice of infill panels from the wide range of available materials which range from glass and other manufactured materials of every colour and texture to concrete panels incorporating a large variety of aggregates. The architect has to consider a number of factors including appearance, acoustic properties, thermal insulation, fire resistance, durability, strength and cost. The quantity surveyor can help by advising the architect as to whether he is obtaining good value for money.

COST PLANNING TECHNIQUES

The customary method of submitting tenders based on bills of quantities which have been prepared from fairly detailed drawings means that the cost of a project is not clearly established until after the design has been finalised. This is obviously an extremely bad practice but the main deficiency can be overcome by the use of cost planning which enables the cost to be established before final design decisions are made and the cost effects of each decision can be clearly seen before the decision is implemented.

The architect and quantity surveyor should be continually examining the cost aspects throughout the design process. Typical questions are: 'Is a particular feature, material or component really giving value for money, or is there a better way of meeting the particular need?' or 'Is a certain item of expenditure really necessary?' Cost planning establishes the needs, sets out the various solutions and the cost implications of these solutions, and finally produces the probable cost of the project. At the same time a sensible relationshp must be maintained between cost on the one hand, and quality, utility and appearance on the other.

In recent years various methods of cost planning have been evolved but there is no universal system which can be satisfactorily applied to every type of project. Cost control, operating at various stages of the design process, will require different techniques according to whether the only information is about function or whether both function and the building morphology are known, or whether, finally, function, morphology and structural information is available. The method using cost plans broken down into elements has proved to be very suitable for school projects but is not necessarily the ideal system for use with industrial buildings. There are, broadly, two basic methods of cost planning currently in use, although in practice variations of these methods have been introduced.

One method has been described as the *elemental cost planning system*. This method was first introduced by the Ministry of Education (now the Department of Education and Science) and has been used with some variations by many local education authorities for the cost control of school projects.

Briefly, in the elemental system sketch plans are prepared and the total cost of the work is obtained by some approximate method, such as cost per place or per square metre of floor area. The building is then broken down into various elements of construction or functional parts such as walls, floors and roof, and each element is allocated a cost based on cost analyses of previously erected buildings of similar type. The sum of the cost targets set against each element must not exceed the total estimated cost. Cost checks are made throughout the design stage and lastly a final cost check is made of the whole scheme. Thus the system incorporates a progressive costing technique with the establishment of cost targets and the use of constant checks to ensure that the design is kept within the cost targets.

Another method is generally described as the *comparative system*. It was first introduced by the former Cost Research Panel of the Royal Institution of Chartered Surveyors and used on the Park Hill housing scheme at Sheffield. This method also stems from sketch plans but does not use a fixed budget like the elemental system. Instead a cost study is made showing the various ways in which the design may be performed and the cost of each alternative approach. The cost study will indicate whether the project can be carried out within the cost limit laid down by the building client and the cost of each of the major parts of the building. The cost study is usually based on approximate quantities and constitutes an analysed estimate.

The cost study provides a ready guide to design decisions and it enables the architect to select a combination of alternatives which will satisfy the financial, functional and aesthetic considerations. The selection thus made becomes the working plan and operates as a basis for the specification and working drawings. The quantity surveyor will need to carry out cost checks periodically throughout the design stage as with the elemental system, to ensure that the architect's proposals are being kept within the total cost limit agreed with the building client. The essential difference between these two methods of cost planning is that with the elemental system the design is evolved over a period of time within the agreed cost limit, whereas in the comparative system the design is fairly clearly established at the sketch plan stage, after the choice of various alternatives has been made, and is not generally materially altered after this stage. The elemental system has been described as 'designing to a cost' and the comparative system as 'costing to a design'.

Each of the two cost planning methods will now be examined in some depth accompanied by examples.

Elemental Cost Planning

Approximate estimate stage. The cost limit for the building will be determined either by the building client or by the joint action of the architect and quantity surveyor for the project. Approximate estimates are largely based on the known costs of similar or comparable projects.

It is often considered good policy to break down the estimate over the building elements, as in this way the architect will be aware of how much he can spend on a particular element before he settles his design and specification. The elemental costs are generally expressed as the cost per square metre of gross floor area (GFA) and constitute, in effect, a preliminary cost plan. This procedure avoids the possibility of an architect incorporating an expensive feature in his design which the target cost cannot possibly accommodate.

When the sketch plans are complete, various quantity factors are checked against those assumed in the estimate and any necessary adjustments to costs will then be made where differences occur. The quantity factors include the ratio of enclosing walls to floor area and the ratio of roof to floor area, and they have an important influence on cost.

Cost planning stage. In the cost plan the sum allocated to each element is usually expressed in pounds per square metre of floor area. The number of elements used on projects normally follows the pattern adopted by the RICS Building Cost Information Service and illustrated in table 7.2.

The elements to which costs are assigned should ideally be related to the way in which an architect builds up his detailed design. For example, an architect may consider the enclosing walls as an indivisible element which he must design as an entity, while in other cases it may be desirable to have two separate elements, such as one for windows and another for solid cladding. It is possible for different types of buildings to have different sets of elements.

In like manner it is reasonable to postulate that the element of upper floors includes the total sandwich of construction which divides one space from the space above or below it, which should form the unit for comparison by cost or performance. For instance, the sound transmission qualities are influenced by floor and ceiling finishings and so these need to be included in the upper floors element. The same criteria apply to all the main structural elements and this indicates that a good case could be made for not listing finishes as a separate element.

The cost allotted to an element may be based on a cost analysis of a similar building or on approximate quantities but in either case the architect will not be bound to the same material. The intention is to provide an allowance which is adequate to cover the functional requirements of the element related to a definite standard of quality. In this way the architect will not be tied to any preconceived ideas and will be free to make a decision at a later and more suitable stage in the design process.

The cost of elements is then converted into unit costs for the work involved, such as walling as £x per m^2, doors £y each and windows £z each, with the dimensions taken off sketch plans. Each element is examined in detail and the design and costs built up and checked against the original estimate. Some amendment to the distribution of costs between elements is bound to occur as the detailed design evolves. Close collaboration between architect and quantity surveyor in the transfer of sums between elements is essential. The introduction of target costs for elements has accelerated the use of new constructional techniques, materials and components and, in some cases, has resulted in commonly used materials being used advantageously in new ways.

The cost checking process is vital to cost planning as it provides the means by which the cost design is controlled. As each element is designed on draft drawings, these are passed to the quantity surveyor who prepares an estimate based on approximate quantities. If the estimate should exceed the cost target, adjustments are made to the design until the cost target is reached. If the estimate is less, then better quality can be provided, or the saving can be used to improve other elements. When all the elements have been satis-

Table 7.2 *Cost plan of office block*

	Element	Cost plan Date: 31.5.93 Gross floor area: 8444 m^2		Cost check Date: gfa:		Cost check Date: gfa:	
		Total cost of element	Cost per m^2 gross floor area	Total cost of element	Cost per m^2 gfa	Total cost of element	Cost per m^2 gfa
1	SUBSTRUCTURE	173 499	20.55				
2A	Frame	418 209	49.53				
2B	Upper floors	431 327	51.08				
2C	Roof	153 969	18.23				
2D	Stairs	113 498	13.44				
2E	External walls	315 923	37.41				
2F	Windows and external doors	972 231	115.14				
2G	Internal walls and partitions	408 570	48.39				
2H	Internal doors	180 013	21.32				
2	SUPERSTRUCTURE	2 993 740	354.54				
3A	Wall finishes	168 349	19.94				
3B	Floor finishes	168 429	19.95				
3C	Ceiling finishes	218 535	25.87				
3	INTERNAL FINISHES	555 313	65.76				
4	FITTINGS	22 326	2.64				
5A	Sanitary appliances	72 880	8.63				
5B	Services equipment	Included in element 5F					
5C	Disposal installations	15 940	1.89				
5D	Water installations	Included in element 5F					
5E	Heat source	Included in element 5F					
5F	Space heating and air treatment (inc. others)	767 160	90.85				
5G	Ventilating systems	Included in element 5F					
5H	Electrical installations	727 840	86.20				
5I	Gas installations	Included in element 5F					
5J	Lift and conveyor installations	133 760	15.84				
5K	Protective installations	7 078	0.84				
5L	Communications installations	19 000	2.25				
5M	Special installations	Included in element 5F					
5N	Builder's work in connection	49 217	5.83				
5O	Builder's profit and attendance	–	–				
5	SERVICES	1 792 875	212.33				
	BUILDING SUB-TOTAL	5 537 753	655.82				
6A	Site works	183 976	21.79				
6B	Drainage	32 806	3.89				
6C	External services	46 737	5.53				
6D	Minor building works	10 000	1.18				
6	EXTERNAL WORKS	273 519	32.39				
7	PRELIMINARIES	481 620	57.04				
	TOTAL (less contingencies)	6 292 892	745.25				
8	CONTINGENCIES	110 387	13.07				
9	PRICE AND DESIGN RISK	128 180	15.28				
	TOTAL	6 531 459	773.50				

Table 7.3 Detailed cost plan of substructural work to school

Quantity	*Unit rate*	*Cost (£)*
Excavate over site for removal of topsoil 225 mm thick		
3500 m^2	£1.70/m^2	5950
Excavation and disposal		
600 m^3	£6.80/m^3	4080
Excavation for foundation trench, including earthwork support, compacting, removal of surplus and backfilling		
420 m	£12.00/m	5040
Concrete (1:12) in blinding bed		
250 m^3	£53.00/m^3	13250
Concrete (1:3:6) in foundations		
350 m^3	£58.00/m^3	20300
Concrete edge beam 275 × 275 mm including formwork to both sides and two coats of bitumen paint, 150 mm high one side, trowelled smooth on top		
800 m	£19.00/m	15200
125 mm concrete slab on and including 150 mm of hardcore and ashes		
3500 m^2	£13.60/m^2	47600
Ducts. Foundation ducts size 900 × 600 mm with 150 mm concrete bottom and 100 mm sides, including 50 mm precast concrete covers		
200 m	£130.00/m	26000
Ditto, 600 × 300 mm		
52 m	£105.00/m	5460
(1994 prices)		
Total cost of element		£142880

factorily checked, the working drawings can be prepared without fear of substantial revisions, and both the architect and building client can be reasonably confident that the tender will not exceed the approximate estimate or cost limit.

Cost plans must be regarded as flexible and it must never be the aim of the quantity surveyor to instruct the architect as to where and how money can be spent. The intention of cost planning is that the quantity surveyor shall assist the architect in designing the building, by giving him full information on the cost implications of his design decisions.

Table 7.2 shows a cost plan relating to a six storey Crown office block, using 1992 prices, incorporating the breakdown of items into essential elements and their further grouping into substructure, superstructure, internal finishes, fittings, services and external works, in accordance with the BCIS format for detailed cost analyses. The first part of the table shows the initial cost plan with the total cost and cost per m^2 of gross floor area inserted against each element. Costs of groups of elements are also incorporated and totalled to give an estimated total cost per m^2 of £773.50, which includes allowances for preliminaries (8%), contingencies (2%) and price and design risk (2%). Working across the table, there is provision for cost checks to be made at suitable intervals and the allocation of costs adjusted as necessary. The gross floor area may also be changed as the detailed design is developed. Some redistribution of costs between elements is likely. There is a reasonably well balanced distribution of cost across the elements but particular attention should be paid to the five most costly elements (frame at £49.53/m^2, upper floors at £51.08/m^2, internal walls and partitions at £48.39/m^2, space heating and air treatment at £90.85/m^2 and electrical installations at £86.20/m^2), during the development of the detailed design. It will, however, be appreciated that the proportional costs between different office blocks can vary considerably, as can also their distribution between elements. For instance in the office block in question, internal walls and partitions (2G) include demountable partitions and reinforced concrete walls, while space heating and air treatment (5F) comprise fan coils for heating and cooling (air conditioning) and radiators in core areas and incorporate elements 5B, 5D, 5E, 5G, 5I and 5M, and electrical installations include the transformer, LV and HV switchgear, general lighting and power, emergency, street, area and flood lighting – they are truly omnibus items.

Table 7.3 contains a detailed cost plan for the substructure work of a school to illustrate a way in which the information can be obtained and recorded. The detailed cost plan consists of schedules of approximate quantities making up each element and incorporating the quantity of each basic item, its unit rate and the total cost involved. These costs are totalled to give the cost of the appropriate element. The unit rates will be built up from the average billed rates for similar projects suitably adjusted to take account of period increases, site differences, and other relevant factors.

Cost analyses record the costs of projects for which priced bills of quantities are available, and the costs are grouped in a similar way to those shown in the cost plan in table 7.2. Detailed cost analyses also incorporate quantity factors, such as breakdown of gross floor areas, average strengths, area of external walls, internal cube and specification notes to increase their value as a basis for cost planning new projects. The preparation of cost analyses will be examined in chapter 9, with particular reference to those produced by the BCIS.

Comparative Cost Planning

Comparative cost planning assumes that initial feasibility studies and cost advice have determined the general layout and arrangement of the building in the light of its total estimated or prescribed cost limit, and sets out to examine what could be described as a market of alternatives open to the designer in respect of each part of the building, and which are both feasible and acceptable to him. This study of alternative design solutions takes account of all the consequential effects of decisions on various parts of the building, relating to one particular part. The information concerning alternatives is set out in a manner which enables the architect to make rational decisions in the light of their individual order of cost and their cumulative effect on total cost, before he starts developing his design. Having settled his design decisions he then develops them, and only if he changes from them should the cost plan need adjustment.

The comparative method does not seek to enforce rigid cost limits for the design of particular elements, but rather to maintain flexibility of choice of a combination of possible design solutions, that will serve the purpose to be achieved. It is more concerned with the comparison of alternative possibilities within a total sum, rather than attempting to control the design piecemeal in relation to targets for limited sections of the work. Its object is not necessarily to show how cheaply a building can be produced but to show the spread of costs over various parts of the building and what economies are feasible. This enables the architect, within his cost terms of reference, to use the money to the best advantage in interpreting his design. This should lead to economy in design and will assist in the comparison of elemental cost apportionment as between one building and another.

The comparative method of cost planning differs from the elemental system in that although the building may be broken down into similar elements for the consideration of cost implications, a theoretical cost allocation to a particular element based on previous experience, is not accepted as a valid factor for controlling the design of the element. Tender pricing is less consistent for individual items even when related to similar buildings, than is the cost arrived at for the building as a whole. Instead, the cost implications of feasible alternative solutions for the elements are considered in the light of their own cost and their effect on the cost of other elements, which their adoption would involve.

Thus in the one case there is a rather arbitrary pattern of cost distribution throughout the building set up to control the design. In the other, a market of alternative solutions is established from which the architect can decide upon a combination which provides, in terms of cost, an optimum design solution for the complete building. In both cases specific cost exercises may have assisted in determining the basic design before the more detailed design is considered. Similarly, once a cost plan is set up, cost checks help in both cases to keep the development of working drawings related to it. The cost plan is only part of the whole process. Basically the elemental system is probably better suited for use with educational buildings where comparisons can be made with similar buildings which have much in common, and this would not for instance apply to industrial buildings.

The comparative system endeavours to show the architect the cost consequences of what he is doing and what he can do. It shows the effect of choice of design for one component of the building on others. It is important to use some method, such as a check list, to obtain all necessary information from the architect at the earliest stage. In deciding the order and scope of the items to be included in the cost study, the quantity surveyor should consider these important points: In what order does the architect require cost information to fit in with his development of design and production of drawings? What alternatives are both practicable and worthy of consideration?

The divisions of the building for cost study purposes will normally reflect the functional requirements of the building and generally follow the broad pattern of structure, cladding, finishings, fittings and services. Within each section, however, the subdivisions will differ according to the type of building, the constructional techniques under consideration and the alternatives which it is desired to investigate.

Table 7.4 illustrates part of a cost plan or more accurately a cost study using the comparative system. This table shows all the cost information relating to a particular element, with alternatives and the effect on cost of the choices made. It provides the cost of both initial and final solutions, with columns added to show the cost effect on other elements of the various alternatives. Costs are also expressed as cost per square metre of gross floor area for ease of comparison. Its main advantage over percentages is that the cost of an element can be adjusted without affecting the unit expression of others. Nevertheless, the percentage expression can be useful for comparison purposes and can with advantage be added when all the costs have been computed.

The rates inserted in table 7.4 include an allowance for preliminaries and insurances. This is often considered desirable to provide a common basis for assessment of rates, which might be taken from one of several bills for similar work, where preliminaries are dealt with in differing ways. Yet there may on occasions be a case for extracting and dealing separately with some aspect of preliminaries which is peculiar to the site and significant in cost. An approximate assessment of cost worked up from preliminary drawings cannot be guaranteed to include each and every item that occurs in practice. Hence it is customary to add a sum at the end of the cost plan to include such items and the usual contingency sum (possibly around ten per cent).

When the final choice is made, the columns on the right hand side of the cost study are completed and the initial cost of each selected solution has to be reassessed, where necessary, in the light of the consequential adjustments resulting from selected solutions in other sections. The final stage is the preparation of the revised cost plan, which contains the selected solutions from the preliminary cost plan, with the figures subsequently revised in the light of the later and more precise information obtained as the working drawings are developed and quotations obtained from specialists. Cost checks should be made against the elements in the cost plan as and when more up-to-date information shows changes in earlier actual or assumed data, which could have a significant effect on cost. This permits discrepancies, which might affect choice in other directions, to be considered as early as possible. If the contract is to operate on a firm price basis, then an adjustment of the revised estimated figures is needed to take account of fluctuations in the cost of labour and materials which may take effect between the date of preparation of the first cost plan and the end of the contract period. In this way an architect is able to deal with changes in the light of a reasonably sure knowledge of their effect on total cost and can act accordingly. All cost data used in the preparation of cost plans should be carefully preserved, scheduled and adjusted, as necessary, for use in preparing future estimates and cost plans. Table 7.5 lists and compares the main implications of both cost planning methods.

BUILDING INDUSTRY CODE

A sub-committee of the Technical Co-ordination Working Party at the Department of Education and Science, on which all the educational building consortia were represented, established a framework for a building industry code designed to be applicable to all forms of building and building

Table 7.4 Part of cost study using comparative method

Office block: 6680 m² gfa	*Initial solution*	*Alternatives*		*Consequential adjustment of other sections*				*Selected solution*					*Remarks*
	£	*£ per m² gfa*	£		*Add*	*Omit*	*Section Ref.*	*Adjustment from other sections*			*Net cost*		
								Ref.	*Add*	*Omit*	£	*£ per m² gfa*	
1. *Substructure* (below basement floor level). Foundations to walls and columns; hardcore under basement floor; all excavation, backfill and disposal, including that for the basement below ground level.	175 016	26.20					1				175 016	26.20 (2.98)	Figures in brackets represent percentages
2. *Basement walls and floor* (a) 250 mm reinforced concrete walls reducing to 140 mm and 102 mm brick facings at 75 mm below ground level; floor of 75 mm concrete blinding and 150 mm reinforced concrete slab with waterproofer.	157 114	23.52					2(a)				157 114	23.52 (2.67)	
(b) 328 mm brick walls (class B engineering bricks), tanking-asphalt and 102 mm brick lining; floor as (a) but with asphalt damp-course under slab and around column bases.			192 465	*Section 1* Extra excavation offset by saving in wall foundation.	1960								
(c) As (b) but 328 mm wall in London stocks.			180 681	*Section 1* Ditto	3140								
	c.f. £332 130	£49.72		*(1994 prices)*							c.f. £332 130	£49.72 (5.65)	

Table 7.5 A comparison of elemental and comparative cost planning

Elemental cost planning	*Comparative cost planning*
1. The architect may have to undertake more redrawing at the detail stage if his solutions do not meet the requirements of the cost plan.	1. More of the quantity surveyor's time is likely to be required because several alternative schemes may need to be priced.
2. The architect must have the requisite skill and ability and some knowledge of costs to keep his design details within the cost plan framework.	2. The architect will need to be forward looking and to make design decisions early in the design stage.
3. The use of the square metre of gross floor area as a suitable unit of costing assists in the comparison of different projects.	3. The pricing of alternative solutions to any design problem will indicate the cost consequences of any particular choice and is likely to identify the most economic solution to the problem. The early study of relative costs will assist the architect in determining the basis of ths design.

documentation. The main aims were to achieve greater uniformity in presentation and classification of detail, enabling wider interchange of information and greater use of computer systems. The primary facets of the code related to the type of building and the use of the spaces within it. The primary elements comprise the structure, external envelope; internal subdivisions; services and drainage; fixtures, fittings and equipment; and site. The code was used initially by educational building consortia, but has since fallen into disuse as the BCIS formats have received widespread use. The list of elements which differed from the BCIS grouping contained 21 elements ranging from general and site preparation to site furniture and surface coverings.

Within each BIC element were a number of features, each of which was a 'distinguishable unit of building, being an aggregate of parts or components which together had significance in the total building process'. Typical examples of features were doors, windows and strip foundations, and each one of them was suitably coded.

C1/SfB CLASSIFICATION SYSTEM

In addition a construction indexing system, known as C1/SfB is used extensively in the National Building Specification, where the sections are coded in accordance with the C1/SfB classification system. Furthermore, BCIS cost analyses are also given the appropriate C1/SfB classification at the head of each cost analysis according to the category of building.

The SfB system is an authorised building classification system for use in project and general related information. It has been described as a practical tool for designers, building managers, quantity surveyors and others to provide a common method of arranging information. It is claimed to make possible improved handling and co-ordination of information for design, management and cost control, including the storage and retrieval of information in practitioner's offices and information centres.

The main headings in the C1/SfB system are: 1. Civil engineering works; 2. Transport and industrial buildings; 3. Administrative and commercial buildings; 4. Health and welfare buildings; 5. Refreshment, entertainment and recreational buildings; 6. Religious buildings; 7. Educational, cultural and scientific buildings; 8. Residential buildings.

The SfB system distinguishes between two main categories of items about which decisions have to be made in any building project. They are (1) the parts of the building, and (2) the resources which are used in its construction. The system recognises parts characterised according to:

(i) the function of a part;
(ii) the type of construction used for a part.

The basic SfB system comprises three tables: (1) elements, (2) constructions and (3) resources. Since each table is comprehensive in its scope, it is possible to summarise building prices or costs for the whole project by elements, by constructions or by resources. By organising information in accordance with the three tables it is claimed the economic building is made possible because the information is available in a suitable format for the efficient management of resources.

Work was proceeding on the formulation of a universal classification for use throughout Europe, which could be operative in the mid-1990s and will involve conversion of the existing databank.

A Co-ordinating Committee for Building Project Information set up by the ACE, BEC, RIBA and RICS prepared Codes of Procedure for Production Drawings (1987a), Project Specification (1987b) and Measurement (SMM7) (1988).

COST PLANNING OF MECHANICAL AND ELECTRICAL SERVICES

Architects tend to approach the initial design of a project on the basis of the physical space requirements and the services engineer may not be called in until the working drawings are complete, possibly without serious thought being given to the space needs of the service installations, without a full analysis of the expected environmental comfort levels, without full consideration of the thermal efficiency of the structure, and possibly with an inadequate sum of money allocated to the services budget. This situation can give rise to major problems. If the desired comfort levels are to be maintained and the building budget is fixed, the building will be cheapened, the standard of services reduced or the building client asked for more money.

It has become a matter of concern amongst all who participate in the design and construction of buildings, that the engineering services element of building cost has not proved so susceptible to control as the remainder. The general lack of use of cost control techniques emanates mainly from the late appointment of consulting services engineers and the prevalence of engineering contracts based on drawings and specifications. Furthermore, with engineering services, initial capital cost is rarely in itself a sufficient criterion.

For example, mechanical services cost records should show the overall cost of heating installations in such a manner that different types of installation can be compared. Furthermore, these costs need adjusting to take account of differences in size, shape and insulation of buildings. To operate a satisfactory cost control mechanism the following techniques are required.

(1) Methods of estimating the overall costs of heating installations from minimum data.

(2) Simple methods of adjusting estimates to evaluate the cost of different building shapes and sizes.

(3) Ability to compare costs of different installations.

(4) Ways of establishing the target cost in a form capable of being cost checked as the design develops.

(5) Knowledge of detailed costs and access to a viable cost-checking procedure.

The total cost of a heating installation is made up of the quantity of heat required multiplied by the cost per unit. The quantity of heat is measured in joules and cost is conveniently related to kilojoules (kJ). It is necessary to relate this to the cost per square metre of gross floor area and to be able to add other elements to give the overall cost of the building per square metre. The conversion factor employed is the number of joules per square metre of floor. For a given type of installation, a building which requires 600 J/h per square metre of floor will cost nearly double to heat than one which requires 300 J/h. It should be borne in mind that buildings in excess of 15 m deep (front to back, containing two or more storeys) with no roof ventilation, need artificial ventilation, and the cost of the installation together with the heat loss resulting from the increased air change may be greater than the additional costs accruing from the restricted building width. The larger the building, the greater the proportion of heat loss by ventilation, and the less the difference in cost stemming from alterations in shape, if the storey height and floor area remain constant.

One useful approach is to break down the total cost of a heating installation into sub-elements:

heat source – boilers, fuel storage; distribution – pumps, pipework, calorifiers; and emission – radiators, convectors.

This form of analysis has two main advantages: it fits reasonably well into common design procedures for heating installations, enabling cost targets to be set and cost checking to take place during design; it groups together those parts of the system whose costs are likely to rise or fall together.

There is some evidence to show that as the size of a heating scheme increases so the cost of the heat source falls, but that of distribution and emission rises. In theory, distribution costs should fall as the cost of pipes does not increase proportionately to size, but this may be more than offset by the greater complexity of the control mechanism on the larger installation. The cost per emission unit is influenced by the type of unit, architectural appearance and average size of unit. The main cost of builder's work lies in the boiler room and chimney stack and it is often difficult to separate this cost from the main building work.

Ideally cost analyses of engineering services should relate the costs to the square metre of floor area and also to a specific unit for each service. The units recommended are often draw-off points for cold and hot water services, kJ (1000 BTU) for heating distribution and emission systems, kJ/h for boilers, m^3/min or sec for ventilation and air conditioning, lighting points for lighting installations and points for power installations. It is further suggested that quantity factors should also be introduced to assist in comparing the costs of engineering services in different buildings. For instance, with a heating installation a quantity factor of 0.75 indicates that either only seventy-five per cent of the building is heated or that the remaining twenty-five per cent is heated (or air conditioned) by other means and is costed separately. With cold water services, a quantity factor of 0.10 indicates a good supply of outlets with a density of one in every 10 m^2 (gross). Element unit rates equivalent to those employed in building work would be the unit rate per kJ for heating and per point for cold water services. A useful approach is to enumerate all equipment and to measure pipework, ductwork and electrical circuits in metres when preparing approximate estimates for engineering services. As indicated in chapter 9, later sources of information on the cost of heating, including the Build ing Cost Information Service, recommend the use of the kW as the appropriate unit.

There is a vital need for extensive cost feedback in the design of engineering services to secure efficient cost control. This feedback must incorporate a wide variety of cost control techniques, including summarising costs into functional unit rates; gross floor area rates; treated floor area or volume rates; rates per point, rates per kW, m^3/s, lux; unit rates for pipework, ductwork and equipment; and cost ratios of subsidiary items such as fittings, supports and insulation to principal items such as pipework and ductwork. With maintenance and operating costs, each service is costed for each 100 m^2 of gross floor area in the Building Maintenance Information analyses and these are further described in chapter 13.

It is unfortunate that whereas there is generally a sophisticated system of cost control for building elements, the monitoring of costs for services is often undertaken in a most haphazard way. Ideally the quantity surveyor should prepare preliminary estimates of the engineering services and incorporate them in the cost plan in a realistic way. Subsequently they should be measured in detail in the bill of quantities. A survey undertaken by the RICS Quantity Surveyors (Engineering Services) Committee in 1978 revealed percentage differences between tender stage and final account stage varying from +54.13% to 15.20% in the case of drawings and specification tenders for M&E work, whereas the range of bill of quantities tenders showed a much narrower range from +16.30% to −8.98%.

In hospital work, the high content of mechanical and electrical work has made it desirable to consider joint venturing. This comprises an arrangement between the main contractor and the principal M&E contractor to be jointly and severally liable for the completion of the project, and this is growing as an alternative method of executing major projects. This approach will help to eliminate possible friction between the two contractors and secure improved co-ordination.

It cannot be overemphasised that engineering services are the largest single component of construction that the average quantity surveyor is least able to deal with, and hence he may not serve the client's interests effectively. Admittedly, the complexities of engineering services make its cost management a difficult task to perform, but it must be done efficiently as these services often account for 25 to 30 per cent of large, complex projects. Many quantity surveying practices improved their engineering services skills and specialist firms of engineering services quantity surveyors emerged in the 1980s which, coupled with the introduction of specialist educational courses, combined to offer a better engineering services cost control provision.

Having defined the base line for optimum value that could possibly reduce the cost of engineering services by 15 to 20 per cent, the experienced engineering services quantity surveyor must then monitor the design development in detail to maintain tight cost control. Any changes to design resulting in an increase in the budget must be reported quickly so that the client can make a choice without delaying the project, and possibly considering alternatives which do not have a cost implication. Furthermore, the design consultant must be committed to designing within the client's budget (Kay, 1991).

It is rarely sufficient merely to give the initial capital cost, as many clients will require details of maintenance and running costs, the taxation position and any costs associated with making the scheme 'greener'. To accommodate all these aspects, the initial cost budgets prepared by the engineering services quantity surveyor are quite sophisticated. They will evaluate the design brief, where it exists, list all major design assumptions where the brief is silent, and indicate the cost adjustments for adapting the brief where these are considered necessary. Thereafter all design developments can be monitored closely and all variations identified.

For example, a bespoke specified lift installation may have a budget value of £750 000, but by rationalising the specification and procurement route this cost might be reduced to £450 000. The performance and quality will not be adversely affected and the ongoing maintenance costs could conceivably be less than with the bespoke design. Another area that can benefit from close scrutiny early in the development process is the relationship between the internal finishes, cladding and structure with the engineering services. Substantial taxation advantages can accrue from the timely review of the design options (Kay, 1991).

With a fully detailed cost estimate reflecting the rationalised design, and the services consultant developing the design within specific cost parameters, the next phase is to agree the tendering procedure. The concept of tendering on drawings and specifications that portray precisely the client's desired scheme is often accepted. However in practice the arrangements may prove to be far less satisfactory, with the lowest tenderers pricing alterations in competition, even before the order is placed. Once on site the project team could well be faced with numerous variations that are not client generated and relate to design rethinks and late inter-designer co-ordination. The designers should be requested to make clear and unambiguous statements about the status of their designs at the tender stage and, if necessary, the client should delay tendering to ensure design sufficiency (Kay, 1991).

Once the contract is placed for the final design only client variations are recognised. Through an early warning system the cost of such variations is speedily reviewed by the project team together with the programming and contractual implications. Prior to the issue of a variation order a firm cost can be agreed with the contractor in many instances. The designer is not permitted to continue developing his design post contract and cannot himself vary the works. Final accounts should be purely an arithmetical function with agreed values being stated on practical completion certificates (Kay, 1991).

There are often new design solutions and products to be investigated in order to keep abreast of current technical developments. Hence it is incumbent on engineering services quantity surveyors to be continually undertaking appropriate research to ensure that the client is provided with an up-to-date, efficient system, giving optimum value for money on a total or life cycle cost approach.

COST CONTROL DURING EXECUTION OF CONTRACT

The quantity surveyor's cost control function does not terminate at the tender stage but continues throughout the execution of the contract. At the time the contractor commences work on the site, the quantity surveyor must have carefully scrutinised the priced bills, schedules of basic rates, insurances and other relevant documents. He should at an early stage agree ground levels with the contractor and suitable arrangements for dealing with daywork vouchers and claims for increased costs. An accurate record of drawings should be maintained with revisions to drawings noted and costed and variation orders costed and filed, and at the same time the architect should be supplied with relevant cost information. The opportunity should be taken on the occasion of site visits for measurements and interim valuations to note any matters such as labour strength, plant in use, weather conditions and causes of delay, which may subsequently have a bearing on the subject matter of claims. Throughout the contract period the quantity surveyor should maintain effective cost control arrangements to keep a constant check on costs and to supply cost advice to the architect in ample time for any necessary action to be taken without adverse effects on the project.

The cost plan also has its uses during the post-contract or post-tender period. When a tender is accepted the priced bill can be analysed in a similar manner to that of the cost plan. A comparison of the priced bill and final cost plan is most valuable in that it shows up the differences between the cost plan and the tender, and so assists in preparing future cost plans. When work on site is commenced the cost analysis can be used for controlling variations. The analysed tender provides a framework of costs which can help to provide a running forecast of total cost as the work proceeds.

The main function of the quantity surveyor, once work has commenced on site is one of project financial control. It is important to ensure that any variations, claims or extras do not raise the likely final account figure above the cost limit. It is particularly necessary to monitor the financial effect of variations and these should, ideally, be costed by the quantity surveyor before they are issued.

The quantity surveyor liaises with the contractor at regular intervals for the valuation of variations, to agree remeasurement work and to discuss any claims submitted by the contractor and/or subcontractors.

Hence cost control measures operate throughout the construction period to ensure that the authorised cost of the project is not exceeded. Normally the contract sum will represent the authorised cost, but on occasions the sum can be varied during the contract period by the building client. Building clients' requirements as to cost control can vary and hence this aspect needs clarifying (Aqua Group, 1990b). Generally the contingency sum should only be used to cover the cost of extra work which could not reasonably have been foreseen at the design stage as, for example, extra work below ground level. It should not be used for design alterations, except with the prior approval of the building client.

Hence it is very important that the cost of all variations should be computed before the architect issues the variation orders, so that their financial effect can be taken into account. This requires close collaboration between the architect, quantity surveyor and other consultants, including their presence at site meetings. Similarly, early consideration should be given to expenditure against provisional and prime cost sums and the contingency fund.

It is advisable to carry out continuing cost studies in constructional areas where detailed design is incomplete, and this applies particularly to mechanical and electrical services. Additionally, the quantity surveyor will normally produce monthly forecasts of final expenditure, and to predict and monitor cash flow. In this way the quantity surveyor can check on the contractor's progress on the site and the probable consequences of any delays to completion. They form an important aspect of the financial management of capital schemes. The receipt of periodic financial reports by the building client enables him to anticipate his future financial commitments and to revise his capital budget where appropriate.

Should any of the cost information obtained by the quantity surveyor prove unsatisfactory, urgent action must be taken to rectify the situation working in close liaison with other members of the design team.

Summing up, the building client should be informed of his financial commitment and when he will be required to make payments. The design team and, in particular, the quantity surveyor must effectively control expenditure on variations, contingency expenditure, provisional and prime cost items, set against quotations, quality control, completion to time and claims. The problem of major additional requirements not provided for in the initial contract and meeting extra funding is one of paramount importance and affects all parties to the contract.

The building client can reasonably expect to be supplied with the following information:

(1) An estimate of the final account at regular intervals during the contract period, preferably monthly.

(2) A comparison of the estimate with the total allocated financial resources.

(3) If the building client's total allocated financial resources comprise components from different resource bases, an allocation of the estimate between these bases.

(4) If the comparison between the final account estimate and the resource allocation is unfavourable, he will require an explanation and an indication of remedial action.

(5) An indication of when he will be expected to pay money and approximately in what amounts.

The following detailed checks should be carried out.

Review of Valuations – check that:

(1) The computation has been prepared by the surveyor, where the quantity surveyor is responsible for reporting valuations.

(2) The gross valuation is within the approved funds and/or financial report figures and advise architect/client if there is a probable need for extra funds.

(3) The preliminaries and insurances have been calculated in accordance with the firm's policy, such as pro rata to the value of work done.

(4) The work done has not been under or over valued by reference to the architect/supervising officer's progress report.

(5) The value of materials on site does not exceed the value of materials required to be incorporated in the works for the forthcoming two or four months, depending on size of contract and stage of completion.

(6) The value of materials off site has been inspected and properly identified as the client's property.

(7) Defective work (if applicable) has been notified by the architect/supervising officer and the value deducted from the valuation.

(8) The value of M&E services has been confirmed in writing by the consulting engineers (where applicable).

(9) The value of nominated subcontractors and suppliers has been made by the surveyor (using random sample test).

(10) The value of fluctuations represents a reasonable proportion of the work done in relation to the rate of inflation.

(11) The contract completion date or extended date has not been reached or notify architect/supervising officer of contractor's potential liability for liquidated and ascertained damages.

(12) The release of retention is authorised by the issue of the appropriate certificate (that is of practical completion and of making good defects).

(13) The value of previous payments has been calculated from previous certificates as issued by the architect/supervising officer.

(14) The arithmetic has been checked.

(15) The time spent is in accordance with the manpower budget.

Review of Bill of Variations – check that:

(1) Contract sum from articles of agreement (the priced bill of quantities may have a different total especially where a contract has been signed on the understanding that variations will be made on the tender sum to reduce it to the contract sum).

(2) Total of amount certified by inspecting all certificates (not the valuations) and whether the client has made any payments direct to a nominated subcontractor and adjust amount of payments if necessary.

(3) All PC sums and PC rates have been adjusted.

(4) All provisional sums have been adjusted.

(5) All provisional quantities have been adjusted.

(6) All architect's instructions have been dealt with.

(7) All nominated subcontractors' and suppliers' documents are complete.

(8) Dayworks are properly referenced to the final account.

(9) Fluctuations are properly referenced to the final account.

(10) All figures have been arithmetically checked.

(11) Compare the final account total with the contract sum and ensure differences can be explained.

(12) Items without architect's instructions do not require an architect's authorisation and/or prepare a list and send to the architect requesting authorisation.

(13) Where the total of the contract bill is not the same as the contract sum check that the final account has been adjusted by the correct percentage.

(14) Additions and omissions have been correctly priced in accordance with the prices in the contract bill and the contract provisions.

COST CONTROL BY THE CONTRACTOR

Cash Flow

The assessment of the profitability of a particular contract consists basically of knowing precisely the value of work executed at a specific date, compared with the actual costs incurred in achieving that value of work. The difference between the two figures will be the amount available to allocate to the off-site overheads of the company, to fund its working capital and make a profit. In an adverse situation the difference may show that off-site overheads are not being covered and that no profit is being made. In the worst situation, the actual costs of construction on site may exceed the value of the work that those costs have generated.

The usual reason given for a company's difficulties is cash flow, whereas this is more often a symptom of the problem and not the cause. Many construction companies become insolvent through bad estimating and planning, ineffective contract control or inadequate site cost control. Cash flow may be defined as the actual movement of money in and out of a business. Within a construction organisation positive cash flow is derived mainly from monies received through monthly payment certificates. Negative cash flow is related to monies expended on a contract to pay wages, purchase materials and plant and meet subcontractors' accounts and overheads expended during the progress of construction. On a construction project, the nett cash flow will require funding by the contractor when there is a cash deficit; where cash is in surplus the contract is self-financing. With contracts operating under United Kingdom standard conditions with retention funds and low percentage profits on turnovers, construction firms are frequently in financial deficit for much of the contract period.

Cash flow problems can be reduced if effective procedures can be operated by the contractor in respect of the following matters:

(1) realistic monthly assessment of preliminaries from fully documented and priced preliminary schedules;

(2) increased costs under contracts with fluctuations kept up-to-date in monthly valuations;

(3) variations to the contract accurately assessed and included in valuations;

(4) daywork sheets completed and cleared for monthly payment;

(5) discounts and retention monies properly claimed against the contractor's own nominated subcontractors and suppliers;

(6) collection of all monies properly due to the contractor; and

(7) ensuring that all claims for loss and expense are fully documented, properly presented and submitted as quickly as possible.

Site Cost Control

It is cost control in the context of profit or loss that is the primary concern of the contractor's quantity surveyor. In this capacity he works closely with the site manager, who is basically the controller, monitoring performance, comparing it against pre-determined targets and taking remedial action where necessary. The factors to be controlled include the tangible physical resources of operatives, materials, machines and subcontractors. Equally important are the non-tangible items such as progress and productivity (time), cost (money), quality, safety, information, methods and the performance of subordinate management staff.

The main sources of data available to the contractor's quantity surveyor are the contract bill of quantities, estimates of cost, method statement and the master programme. During construction these four data sources will be supplemented by interim valuations; up-to-date accounts of labour, plant, materials and subcontracted work; salaries and all other site costs; and finally the programme of actual work executed compared with the assumptions upon which the tender was based. Supported by this back-up data, achievement of effective cost value comparison will involve the following activities.

(1) The calculation of the true value of work carried out on a cumulative basis to a prescribed cut-off date.

(2) The restatement of the true value of the work in terms that can be directly compared with costs. Some contractors refer to this restatement as the preparation of the earned allowances, whereby in relation to the work carried out, not more than a certain sum should have been spent on labour, plant and other components.

(3) Costs are collated to the same cut-off date as that adopted for the statement of true value, sometimes termed 'true selling value'. Costs will be adjusted to take account of liabilities for costs that have not yet been recorded but against which value of work has been taken, and for costs generated against which no value of work has yet been created.

Supplied with these three sets of data, it is then possible to ascertain whether profit or loss is being made on expenditure against labour, materials, plant, sublet work and site overheads, and whether the level of contribution is better or worse than that upon which the tender was based. The comparison may be made in terms of the whole project or, if data in terms of both value and costs can be accurately subdivided, into the construction elements making up the total project.

The computerised breakdown of interim payment applications into earned allowances is probably the most common application of computers to the financial management of contracts by contractors. A number of standard packages are available and, should the employer's quantity surveyor receive a payment application in the form of a computer print-out, he can be reasonably certain that earned allowances have been used in preparing the application.

As a management tool, the contractor's quantity surveyor is looking for consistent and inconsistent trends as between the value and cost of each of the earned allowances month by month. If a consistent trend is observed, for instance by regular over-spending on labour at a consistent level, then management will endeavour to identify some underlying reason. Examples include the all-in labour rate being higher than anticipated at the time of tender, or output being consistently lower than the constants upon which the tender was based. The inconsistent relationship between earned allowance and actual cost is more likely to have been caused by an isolated event, and again management will want to identify the reason and obtain a solution in each case. The key factor is to recognise that there is a problem at a time sufficiently contemporary with the event to be able to take effective action. On the cost side of the cost value comparison, it is essential that there is close liaison between the contractor's quantity surveyor and the site accountant on a very large project.

Financial Reporting

Good management practice dictates that reliable and regular financial reporting is necessary to control a project effectively and reports should be produced ideally on a monthly basis. A basic financial report of a contract should contain:

(1) initial tender figures and expected profit;
(2) forecast figures at completion for value and profit;
(3) current payment application by the contractor;
(4) current certified value;
(5) adjustments to the certified valuation;
(6) costs to date and the accounting period in question; and
(7) cash received to date, retention deducted and certified sums unpaid.

Cost Value Reconciliation

The cost and value of variations must be continually monitored and assessed. A contractor must ensure that he has clearly defined procedures for identifying variations and that he conforms fully with the requirements of the contract regarding notices and the supply of supporting information and particulars. After interim valuations have been made, the contractor should list all unagreed claims, variations, daywork, remeasurement, interest on overdue sums and any other disputed items. The contractor should also assess the progress achieved by monitoring the value of work carried out against that programmed. This can best be done by constructing an S curve of the forecast valuations against time and then plotting actual value against actual time.

Accurate recording of the cost of materials, plant, labour, site staff and overheads, and subcontractors' work and claims is essential in cost value reconciliation. For instance, there is frequently a delivery charge for plant and this may not have been indicated on the initial order. This can include time of the plant travelling from its depot as well as the delivery charge itself. When operators are provided the question of overtime arises and also greasing or servicing time.

A monthly reconcilation of materials delivered to the site should be made against the quantities certified in the measurement. Allowance must be made for materials rejected, used on site but not measured as in strengthening temporary roads, used off site as in minor work for adjoining landowners, and materials that are stockpiled. If the quantity unaccounted for exceeds the estimated wastage allowance, further investigation is needed. Possible causes are errors in measurement, additional work being performed without supporting variation orders, or loss through unforeseen circumstances, such as excessive penetration of granular material into a very soft subbase.

With regard to labour costs, site staff usually have the responsibility to complete timesheets indicating the number of hours worked by each operative and the amount of bonus earned in that particular period. Management will monitor the allocation of staff on the various contracts, and will record holidays, sickness or other reasons for absence and overtime payments, and check that staff and overheads charges are kept within the intended budget.

Residual credits may occur when materials or items of plant have been purchased for a certain contract but on completion still retain some foreseeable value which can be used on other future contracts. They do, however, require careful examination and evaluation (Seeley, 1993).

8 COST MODELLING

INTRODUCTION

A basic definition of a model is a procedure developed to reflect, by means of derived processes, adequately acceptable output for an established series of input data. This generalisation goes some way towards setting the parameters within which this chapter must be set. Ideally a model should be simple enough for manipulation and understanding by those who use it, representative enough in the total range of the implications it may have, and complex enough to accurately represent the system. Any model must be attempting to satisfy these seemingly incompatible criteria. The model is the black box which must be operable under a series of input data although the better models will be able to function with input data outside the original data used to formulate the model.

Models within the construction industry are often of an evolutionary nature because of the well established variability in costs and prices. It is therefore advantageous for a model to be able to react to changed circumstances and to adapt to maintain accuracy. It is the level of accuracy obtained from any model which must be the test as to its suitability. The common stages in the development of a model are shown in figure 8.1.

The chapter title of cost modelling specifically uses the term 'cost'. It is however, equally acceptable to read the term price in place of cost – always allowing for the fact that a model derived to produce an output of cost cannot be used to produce an output of price except with significant further development of the model.

The construction cost model is, by definition, a most broad area of study encompassing, as it must, all building, civil engineering and petrochemical construction works which may involve the need for estimates prior to the completion of the construction. Having stated that the range of construction types is very large then it must also be stated that building cost models will also range from complete construction works down to the smallest element of cost which might be variable.

The development of cost models for predicting single elements of cost is one which is fraught with considerable dangers because of the interdependency between so many elements within, for example, an office building. The use of a plan/shape ratio to derive a cost for external walls is, in itself, a valid start to a cost model but will not, without further complexity being introduced, adequately reflect the relationship between external and internal wall costs.

The aim of cost models is generally to represent accurately the whole range of cost variables inherent in a building design to secure improved cost forecasts and/or design optimisation. It has been argued with some justification that the traditional approach to cost modelling gives a poor representation of costs, since they are not modelled in the way in which they occur. Furthermore, quantity surveyors normally reuse data from previously completed projects as the basis for their models. Considerable adjustment to this data is needed, frequently on the basis of supposition and presumption, as no two projects are identical (Ashworth, 1986).

ACCURACY

Models within the construction industry may be one of three different forms. The hierarchy of

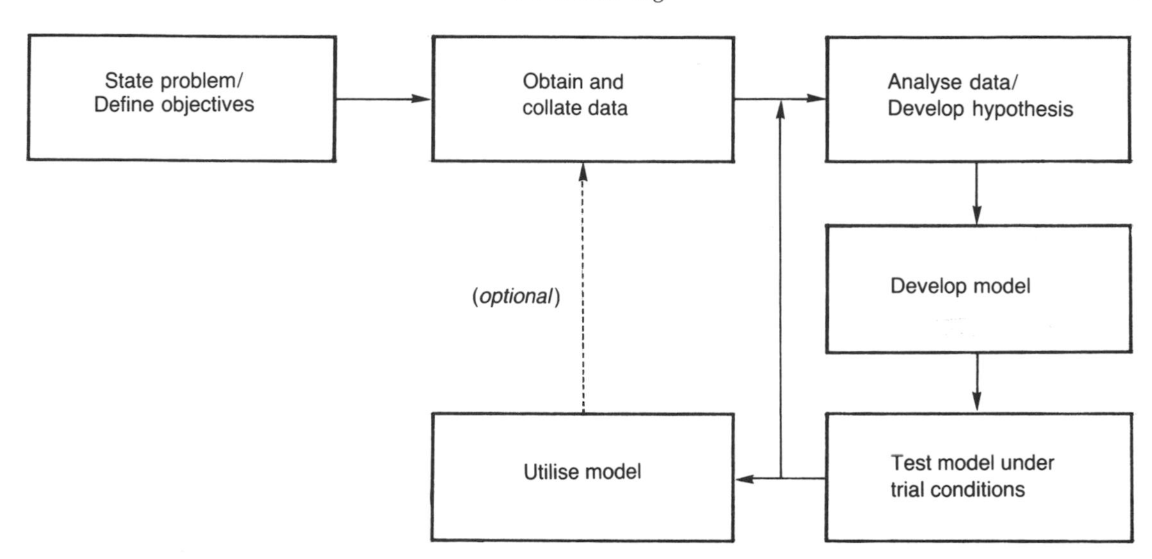

Figure 8.1 Development stages for a model

these three forms is assessed from the outputs which are produced by the models. The most basic form of model would be one where the output simply states that, of two or more options, one is the best or cheapest option. An intermediate form of model would place some degree or level of difference between options. The degree of difference will probably be expressed in percentage terms. The top of the hierarchy will produce outputs in objective terms such as total capital costs, costs per square metre of floor area or annual equivalent costs. It is this highest level of model which will be the objective of all models at their initial stage and, should this objective be unobtainable, then the lower levels will be attempted.

The definition at the start of this chapter does preclude the inclusion of bills of quantities. By its precise and exact nature, a bill of quantities cannot represent anything other than a single option and is therefore not broad enough for inclusion. The distinction becomes blurred and inexact when it is understood that a CAD system with the facility to generate information on quantities might be said to satisfy the requirements for a building cost model.

Development of cost models may be said to have commenced with the report by the Wilderness Cost of Building Study Group (1964) which produced a series of schedules detailing the costs of a steel frame for a structure. The spans, storey heights and number of storeys were varied and manually costed. This manual operation was a significant imposition on the time of the Wilderness group members and prevented the extension of the work of the group. The main findings of the group are given in chapter 2.

The introduction of micro computers gave the power to develop and use building cost models with greater speed. This increased speed of operation and development provided the flexibility necessary to secure the more effective use of models.

HISTORICAL DEVELOPMENT OF COST MODELLING

Raftery (1987) has described how the development of cost modelling has taken place since the late 1950s. The first approach was probably the work of the BCIS based on cost analyses suitably adjusted to provide cost plans as illustrated in chapter 10.

The next stage began around the mid-1970s and was characterised by the extensive use of regression analysis, aided by the growth in availability and reduction in cost of micro computers. The latest approach to modelling began in the early 1980s and centred around the following two primary characteristics:

(1) the acceptance of uncertainty and imprecision and a desire to take cognisance of this by carrying out probabilistic estimates, often based on Monte Carlo techniques;
(2) interest in artificial intelligence and knowledge-based computer systems.

BCIS On-line Approximate Estimating Package

The Approximate Estimating Package (AEP) is a suite of programs which can be used to manipulate cost analyses to form a budget estimate and cost plan. The AEP is fully integrated with the rest of the on-line service and can draw on information from a user's own data or from the BCIS data banks. Users of the AEP are relieved of routine calculations enabling more analyses to be considered at the same time. This in turn improves the reliability of answers and allows the cost implications of early design stages to be quickly determined.

The normal approach is as follows:

(1) Enter outline details of the project, such as element unit quantities, description of project and site, location, contract and tendering particulars, and size and shape of building(s).
(2) Select analyses to be manipulated from the BCIS databank or the user's own records.
(3) Define method of updating/adjusting analyses, typically using the BCIS tender price index and location study.
(4) Enter indices used for updating/adjusting analyses.
(5) Print indices used for updating/adjusting (optional).
(6) Prepare estimate either using analysis by analysis approach or element by element approach. Current practice tends to mix and match costs from all the analyses available and this approach is taken in the element-by-element option, ensuring that the data is presented at the pricing levels relating to the specific project, show relevant details from the source analyses and allow the user to insert element descriptions. Adjustments can be based on gross floor area, element unit quantity or lump sums, and allowances can be made for variations in specification.
(7) Combine estimates, decide on total and calculate cost plan.
(8) Print cost plan.
(9) Manipulate cost plan to determine the effect of a change in shape, size or specification. The revised cost plan can be printed out or used to replace the cost plan stored on disc.
(10) Forecast of fluctuations, wherein a cash flow is calculated based on the DHSS 'S' Curve formula and combined with the BCIS forecast of cost inflation to give a forecast of fluctuations payments.

Bennett and Ferry (1987) consider that the system is quick to use and enables the user either to amend the estimate for specification differences between the proposed project and each of the examples where details are known, or else to manipulate the examples with very little amendment other than the changing of element quantities. They identify the drawbacks as the lack of segregation of labour and materials and the categorisation of figures into cost planning elements rather than construction activities.

Bucknall Austin Building Cost Model

Patchell (1987) has described how Bucknall Austin identified the following four major criteria to be met by the cost planning service at the schematic or feasibility stage of the building design process:

(1) cost accuracy from very preliminary information;
(2) flexible and quick response to various options;
(3) economy of production in man and machine hours;
(4) estimating and analysis on the same basis.

To meet these criteria, the practice developed a building cost model in the form of an interactive computer program. The model uses element unit quantities as its empirical base, computed from twenty fundamental dimensions, with digitiser input directly into the cost modelling program. An appropriate set of base rates is applied to the unit quantities from the database of rates held within the model.

The selected base rates are adjusted by a global price factor which is the product of time (using BCIS tender price indices), location (using BCIS location indices), and quality (by percentage adjustment from a base of 100). The resultant projected rates are applied to the element unit quantities to form the cost plan. However, the quantity surveyor still needs to use his professional judgement to suit the project conditions.

When an entry is made, the screen displays the new total cost, area and cost/m^2 and the quantity surveyor is able to check all quantities and rates and print out any required combination, which allows great flexibility. As the model is designed for use in practice, the output also generates a full cost report and relevant working documents.

Analysis of projects is produced using the same program but in reverse order. Each entry in the database is comprehensive, storing the range of tender prices, contract type, design features, market conditions and functional units, while a detailed analysis printout calculates wall to floor ratios, internal volume, plot ratio and building footprint area, usable to gross efficiency percentages and other statistics.

The program has been used on a wide variety of projects including a complete university township in the Middle East which comprised over 1500 different building types, and the brief frequently changed the number, mix, size, shape and form of the buildings. There was therefore a need to produce a program that allowed rapid recalculation of quantities and costs.

Davis Langdon and Everest Cost Models

A series of detailed cost models have been published in *Building* covering a range of different building types. The articles give design particulars of the particular projects together with likely cost ranges, contract routes, method statement, regional factors, critical path network and other useful data. The cost model consists of a priced breakdown of all the component parts of the building with constructional details, quantities and unit costs. In addition the element cost, cost/m^2 gfa and percentage of total cost is also supplied. Another very useful inclusion is alternative design cost implications.

This helps to illustrate the varied forms that cost models can take. Here it is used to describe a very detailed cost analysis with a wealth of supporting cost and other information relating to a particular building type.

PURPOSES OF COST MODELLING

Deciding the purpose of a cost model will affect its form and the variables which will be incorporated within it. Wilson (1984) suggested the following two categories as being of significant importance:

(1) Design optimisation models which are primarily concerned with securing value for money in building and may also be used as an important component of the cost planning process and incorporate life cycle costing facets. Their main strength lies in comparing one solution with another.

(2) Tender price prediction models, where the main aim is to forecast the likely tender sum to be obtained from a contractor and also take cognisance of the contractor's estimating variability and the factors influencing market price. Because of the variable factors involved predictive models will be less reliable than design type models and some inaccuracy in forecasting is likely.

Approaches to Cost Modelling

Ashworth and Skitmore (1982) found a significant amount of improvement in estimating accuracy as the design stage approached finalisation culminating in the tender period. Wilson (1984) believed that this has given rise to the following two main approaches to cost modelling. The first method is more commonly used during the early stages of the design process and the second method after the design has been formulated.

(1) *Deductive method*. These types of model use statistical analysis of previous performance data as the basis for cost modelling, often in the form of regression analysis.

(2) *Inductive method*. This attempts to bring the more detailed approach of the contractor's estimator into the design stage and constitutes a form of simulation.

With deterministic models it is assumed that the values of all variables are either known exactly or can be predicted precisely, which is unlikely to be the case when cost forecasting for construction works, while a probabilistic model recognises that some of the variables are uncertain and attempts to counter this using probability theory, necessitating the collection of large amounts of data and a more complex approach (Ashworth, 1986).

TYPES OF MODEL

The various types of model available to practitioners throughout the construction industry were considered by Newton in 1991 who undertook a review of cost modelling in an attempt to classify the research undertaken.

The types of technique used were discovered to be (a) dynamic programming; (b) expert systems; (c) functional dependency; (d) linear programming; (e) manual; (f) Monte Carlo Simulation; (g) networks and (h) regression analysis.

Analysis of almost sixty research papers identified only six examples of manual techniques and these included the Wilderness Group (1964). In view of the fact that the manual technique has developed little since this period thirty years ago it is considered unnecessary to consider manual techniques further.

The superseding of manual methods was brought about by the advent of the availability of computers which, from mainframe through mini computer to micro computer, has brought calculating power from the specialist computer centre onto the desktop of all those involved in cost modelling.

The descriptions used to categorise cost models do include as the final type the subject of uncertainty. The models assessed are classed as deterministic or stochastic. It is the intention to consider the different types of cost model available, but with rather less emphasis on the final descriptive analysis which changes a deterministic model into a stochastic model. Stochastic or probabilistic means that the technique is concerned with controlling factors that cannot be estimated with accuracy. A stochastic simulation exercise normally culminates in a graph with the x-axis representing time and cost and the y-axis the risk level from 0 to 100 per cent. The user selects the time and/or cost for the level of risk.

A brief overview of risk analysis, as detailed by Raftery (1993) and Flanagan (1993), will suffice to outline the nature of the techniques and their application to cost models.

Dynamic programming (DP) seeks a least cost or least time path through a series of activities, usually in the form of a network, where the activities are represented by design selections within construction stages, as the basis of a time/cost optimisation technique. The technique is analagous to a multi-stage design process where at each stage decisions or selections have to be made. DP provides the means by which the network can be evaluated (Atkin, 1987).

Linear programming quantifies the relationship between two variables such as floor area and total cost and can, therefore, be plotted graphically.

Networks can form the basis of a process-modelling system. Networks very similar to PERT/CPM can be used as the basic modelling structure and costs are introduced on the network arcs in which they occur. Certain dependencies become apparent as a function of the network structure, while others can be modelled. Finally, stochastic variability can be introduced into the process and modelled in parallel to the network system (Bowen, 1987).

Risk Analysis/Monte Carlo Simulation

It is certainly not difficult to see that a significant level of uncertainty is inherent in construction cost (if less so in construction price). It may follow that a deterministic cost model used as a predictor of construction cost will only be accurate if a considerable measure of good fortune occurs. The variation in the prices of different contractors when submitting priced bills of quantities indicates that exact prediction is never likely to be achieved. It is for this reason that estimates of variance will often be given by cost models in order to limit the cost modellers' exposure to risk. This limitation of risk to the modeller has been achieved by transferring the risk to the model. Risk analyses are performed on models in a number of ways, the most common being a Monte Carlo simulation technique. In such a technique the independent variables within a model are given a degree of uncertainty by applying a level of variation to their value. This uncertainty is derived from a multitude of sources such as historical records of weather conditions. A method often used to apply the uncertainty is to define a probability profile for the variable. Examples of such profiles are rectangular, triangular and trapezoidal as demonstrated in figure 8.2.

In the simulation technique a specified number of occurrences of the variable and random number generators are applied to define the precise value for each of the occurrences and these precise values are used, in turn, in the model. From these multiple runs it is possible to produce a probability profile which can give confidence for statements concerning cost. A number of persons have developed a model to produce output for two dependent variables, usually cost and time. Given the power of current computer technology and the excellence of graphical modelling systems, it will no doubt shortly be possible to apply risk in three dimensions and, most importantly, to view the output from such systems.

As a generalisation, it is probable that the growth of the recognition of uncertainty will continue throughout construction cost modelling. The use of risk analysis and output data including confidence limits will become in time the *de-facto* standard for all construction cost or price calculations.

Statistical Models and Regression Analysis

The development of cost models using multivariate regression analysis is a complex process for which computer statistical packages such as SPSS are necessary without needing to process large amounts of complex data manually.

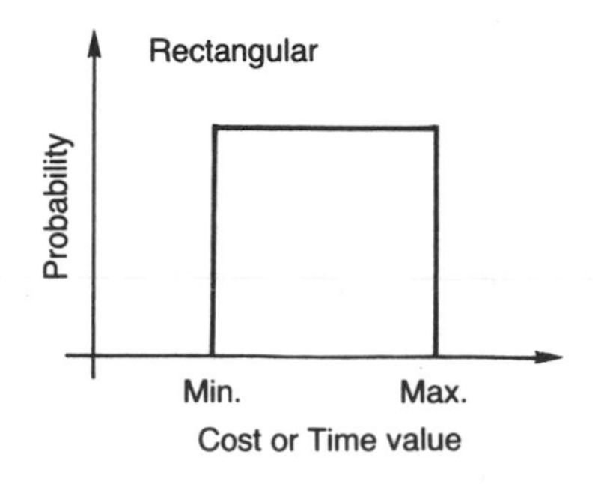

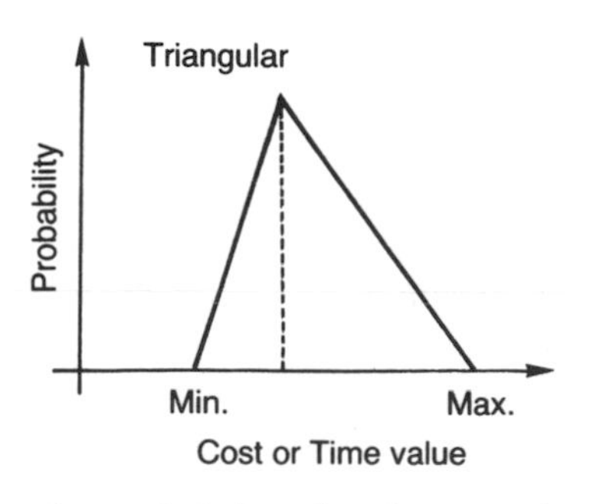

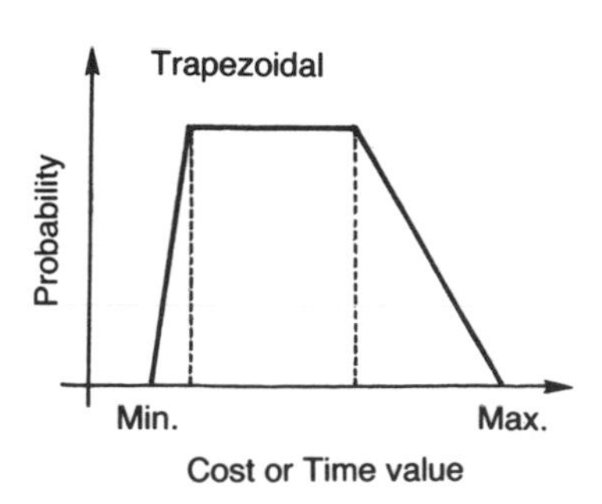

Figure 8.2 Sample probability distributions for risk analysis

Snedecor and Cochrane (1976) state that there are two principal stages in undertaking a multivariate regression analysis. The first stage is to review the validity of the independent variables and to determine whether these variables are relevant to the dependent variable or if any correlation is purely by chance. The second stage is to determine, by well established statistical methods, the hierarchy of the dependent variables and to calculate the weighting factors to produce a multivariate regression model.

Table 8.1 Example of relationship between construction cost per square metre and number of storeys

Number of storeys	Construction cost £/m²
1	600
2	550
3	500
5	550
10	600

From table 8.1 it can be seen that the construction cost per square metre of floor area is a minimum at the three storey option. The construction project may, however, be limited by other constraints such as site area and poor ground conditions and it may be thought valid to consider the costs per square metre for six and seven storey developments. Regression analysis of the data contained in table 8.1 gives an equation $Y =$ .... + ... $X +$ .... X^2 where Y is the cost per square metre and X is the number of storeys. It may therefore be derived that the cost per square metre for a six storey development is £........../m^2 while for a seven storey development is £........../m^2.

The regression equation, although multivariate, contains only one independent variable, that of the number of storeys. The second variable has been created for the single independent variable by, in this case, using the square of the variable. Equally valid would be use of second, third, fourth or more independent variables to more effectively fit the data to the dependant variable. When modelling construction costs per square metre for a particular building type independent variables such as gross floor area, plan/shape ratio and specification level may come to mind as significant enough for inclusion. The regression analysis software will be able to confirm the significance of each variable.

The three independent variables listed above are widely different in nature. The first two – gross floor area and plan/shape ratio – are objective measures where precise calculations can determine the value of these variables. The third – specification level – is one which is not readily given an objective measure. Instead the value of the variable must be derived and justified. The subjective measure of the specification level could be measured by a technique as simple as a ranking system, for instance, specification A is better than specification B which is better than specification C so the three specifications can be valued as 1, 2 and 3 respectively. This technique may be valid in exceptional circumstances but it is probably unlikely that the difference between specifications A and B and between specifications B and C are equal in scale and, more importantly, equal in cost. Some other method of measurement may therefore be required. Accordingly sampling techniques such as user satisfaction questionnaires might give a more precise measurement for use within a regression analysis.

OTHER COST MODELLING APPLICATIONS

Maver (1979) described how computer based design models have made the trade off between investment and return much more explicit. The design team generates a design hypothesis which is fed into the computer, the program models the scheme and predicts the future reality in terms of cost and performance. The design team evaluates the cost/performance profile and modifies the design hypothesis accordingly to produce the appropriate balance of investment and return.

Smith (1980) devised an approach whereby the design target parameters and cost targets were set out and compared with those contained in the budget model, as a quick way of assessing the chosen strategy. Once the strategy was agreed the design team developed the sketch plans within the framework of the strategic design plan, which itself was developed from the framework of the budget model.

Further applications of cost modelling embraced the following:

(1) to develop the more cost-sensitive design parameters, particularly those related to the functional use of spaces and the effect of depth of plan on services and frame costs;

(2) to produce project life cycle budgets requiring design teams to cost out the energy, maintenance and servicing costs of their designs and to keep these within an overall budget.

These examples illustrate the wide range of investigations for which cost modelling can be used and, as described previously, with the increased use of mini computers and micro computers, vast amounts of cost data and the design parameters of previous projects can be readily and speedily used to assist in the formulation of cost models, designs and cost plans for new buildings.

Simulated Modelling

Bennett and Ferry (1987) have reiterated how a detailed deterministic forecast of the outcome of a construction project by traditional methods is both time consuming and costly. They believe that computer simulation of the construction programme offers an alternative approach which can forecast time and costs quickly, and thus enable the consequences of alternative design or management strategies to be evaluated quickly and realistically prior to decisions being made. A further advantage claimed is that a range of probabilities can be examined rather than a single answer. At the later stages of design development the simulation of a project involves the inputting of specific design, cost and management criteria to the computer, but at earlier stages the computer must simulate much of this data for itself. The input and output of the simulation must be able to interface with traditionally produced material and a series of suitable linked packages has been developed at the University of Reading.

Network Cost Modelling System

Bowen (1987) postulated that attempts at modelling the cost of buildings have largely attempted to explain costs as relatively simple functions of different measurements of the finished building, element or component, and thereby viewing the building as a single discrete step. It is argued that such an approach ignores the fact that construction is a process consisting of separate but dependent physical activities over a period of time and subject to uncertain cost and duration. It fails to explain when and how costs are incurred (inexplicability), disregards interrelatedness (unrelatedness), and ignores uncertainty (determinism).

To overcome the above weaknesses, Bowen (1987) advocated a process-based modelling approach. It was suggested that PERT like networks can be used in a structured manner to model cost by representing expenditure as it occurs during construction activities. The proposed system can use AI (artificial intelligence) to link sub-networks representing the construction of different elements to create a complete, representative network for the entire project. For each element, different sub-networks representing different designs which are different in type, but not size or similar parameters, can be used to enable the user to compare different design/construction alternatives. Being computerised, this system can use a database containing cost information, sets of sub-networks, different default values and definitive functions.

EXPERT SYSTEMS

The development of expert systems began in the mid 1980s with the development, jointly by the Royal Institution of Chartered Surveyors and Salford University, of the ELSIE system. The system, as described by Brandon (1990), encompasses the accrued knowledge of a significant number of construction industry professionals and encompasses this knowledge within the framework of an expert system.

By using the expert system, it is possible for a user, who need not be a construction professional, to respond to questions asked by the computer software and to be led, as if by a constructional professional, through the maze of interrelated data, to arrive at a cost based upon both the information provided by the user and the assumptions made by the software where the user did not have sufficient information.

Subsequently a private company was established to market and sell the software. This company, Imaginor Systems, still based at Salford, has expanded its operations from the original office development scope of ELSIE to additional fields such as industrial buildings, as described in some detail in chapter 1.

CONCLUSIONS

Bowen (1987) postulated justifiably that conceptually the development of modelling systems for the purposes of design economics must emanate from the desire for a closer representation of reality. This implies on the one hand the acceptance and appropriate treatment of variability and uncertainty, and on the other hand a more thorough representation of construction processes, each taking place within a suitable simulation environment. Expert systems, as previously described, representing an important set of applications of artificial intelligence, and used in conjunction with operational/resource-based networks and techniques, offer the potential for a process-based algorithmic modelling system which could make a significant contribution to decision making for construction projects. Readers requiring more detailed information on building cost modelling are referred to Raftery and Newton (1995).

9 COST ANALYSES, INDICES AND DATA

This chapter is concerned with the compilation of cost analyses and other cost data for assessing costs and providing the basis for the preparation of cost plans of future building projects. It also examines the methods of compiling and applying cost indices as a means of updating past costs of buildings. The use of cost limits is also investigated.

COST ANALYSES

Nature and Purpose

Cost analysis can be defined as 'the systematic breakdown of cost data to facilitate examination and comparision'. Cost analysis should provide information for any immediate problem and can be performed in a number of ways. It can, for example, permit detailed comparisons to be made between different projects and isolate the causes of differences. These may arise from a variety of causes, such as differences in basic design or details of design; differences in regional pricing or differences in contracting conditions. Indeed, they could arise from a whole range of factors which tend to make every scheme unique in one respect or another. Cost analysis forms the basis of cost control.

Probably the forerunner of most systems of cost analysis operating today was the elemental system introduced by the former Ministry of Education and subsequently adopted by the Department of Education and Science and many other public bodies. In this system the analysis was by elements or functional parts of a building, with the aim of providing data for the establishment of cost targets for designing buildings. It is, however, one of a wide range of methods which could be used for collecting, collating and classifying the data contained in priced bills of quantities. The variety of cost problems which arise in building work makes necessary the profusion of methods of analysis, and that all problems cannot be solved by a single method, hence it is advisable to modify the analytical process to suit the needs of a particular problem.

The main difficulty in analysing a bill by elements has been the definition and demarcation of elements. In any given building the component parts do not simply or logically subdivide into the various elements or functional parts. Various parts perform more than one function; for instance, a crosswall performs two functions – load-bearing and as an internal or dividing partition. Other parts seem to fall between two elements; a concrete lintel could conceivably be included in a 'windows' element or in 'external walls'. Separation on a functional basis may on occasions give rise to difficulties; thus windows provide natural daylighting but so also do rooflights, yet one would not expect to see them in the same element. Furthermore, many elements are interdependent and a change in one may materially affect others. These problems have now been largely resolved by the Royal Institution of Chartered Surveyors' Building Cost Information Service's introduction of a Standard Form of Cost Analysis which provides standardisation of elements and will be examined in some depth later in this chapter.

With cost control, a restricted form of cost analysis, usually by elements, is required in the first place to set targets, and thereafter, a wider field of information is needed to check the cost of detailed aspects shown on working drawings against their targets. Much analysed data is obtained from priced bills and, as will be described later in the chapter, may need considerable adjustment prior to its use for cost planning. Furthermore, the pricing of a bill is the outcome of the estimating method used by the successful contractor. Pricing is not only a personal process but it also reflects the varied approaches, policies and resources of different contracting organisations. The estimating and pricing may not always truly reflect the actual costs on site. For these reasons it would be more accurate to describe 'cost analyses' derived from priced bills as 'price analyses'.

Cost studies prepared for architects mainly show major cost components and the alternatives available to the architect. Hence the method of analysing the data will often vary from project to project. For example, if the project incorporates crosswalls, then their costs should be separated from those of the infill panels; similarly, costs of load-bearing external walls should be kept separate from minor internal partitions. In some cases, finishings will have special significance as, for example, where a building is to contain an impressive entrance hall with expensive floor, wall and ceiling finishes. The cost of the finishes in this part of the building should be separated from finishes elsewhere, as it could constitute a major item of cost.

For cost planning to be effective, banks of cost or price data are needed because the average quantity surveying office cannot possess sufficient cost information to provide an adequate base for cost plans covering a wide range of building types. It was for this reason that the Building Cost Information Service (BCIS) of the Royal Institution of Chartered Surveyors, described later in this chapter, was established. One of its main functions is to publish and circulate cost analyses and other cost information to subscribing members. It is probably advisable to describe the processes as *design cost planning* and *design cost analysis* to distinguish between cost to the client and cost to the contractor.

A cost analysis shows how costs of a building are distributed over elements and groups of elements. A meaningful conclusion cannot, however, always be drawn from analyses unless full regard is paid to the quality of the work. When making cost comparisons of buildings, it is advisable to separate those parts of the work affected by site conditions such as foundations and drainage, to permit the comparisons to proceed on a similar basis. The possibility of expressing elemental costs as percentages was considered by the former Ministry of Education (now Department of Education and Science) but was rejected as it tended to conceal the actual cost per element. For example, an element costing £32/m^2 in each of two projects, one with a total cost of £370/m^2 and the other £420 would give percentages of 8.65 and 7.62 respectively. Thus the element would appear cheaper in the second project, although in fact it is costing the same in both projects. Or again an element, such as electrical services, could account for roughly the same percentage but in buildings of widely different total costs.

Another purpose of a cost analysis is to show where reductions could most beneficially be made, should the tender unfortunately prove to be too high. The greater the number of cost analyses prepared and circulated, the more extensive will be the body of available cost information and the greater the opportunity for cost comparisons leading to more effective cost control.

A bill of quantities normally provides a cost breakdown of a project on the basis of work sections in accordance with the Standard Method of Measurement of Building Works. The compilation of element costs thus becomes a process of abstracting in reverse – abstract type sheets or computer spreadsheets are given elemental headings and then the prices of items, or more usually groups of items, are transferred to the elemental sheets. The elemental costs are totalled and checked against the total cost of the project. The gross floor area is measured (from the inside face of the external walls) and the cost of elements divided by the area. It will expedite the work if cost analyses can be compiled by the same persons who prepared the original bills. An analysis for a primary school takes up to two days to prepare, depending on whether the activity has been computerised. The use of elemental bills would drastically reduce the time required to produce a cost analysis, but when used in practice the disadvantages seemed to outweigh the advantages. A compromise has been sought in the sectionalised trade bill, but this has not been used extensively.

Hence cost analyses seek to achieve various aims:

(1) to enable the design team to determine how much has been spent on each element of a building;

(2) to assess whether a balanced distribution of costs has been obtained;

(3) to permit comparison of costs of the same element in different buildings;

(4) to obtain cost data for use in planning other projects.

The costs of relatively small parts of a building can readily be seen in relation to the whole. Large sums of money are sometimes spent on the elements of structure and services which are disproportionate to the quality and efficiency which they contribute to the building. On the other hand too little may be spent on finishes to provide, for example, the appropriate acoustic quality. Adjustments in the allocation of money between different elements can often result in a better building both functionally and architecturally.

A cost analysis of a tender records the effectiveness of the cost control exercised throughout the design stage of a project. At tender stage it will usually be too late to adjust the elements which are out of balance and the reductions are likely to fall on finishes and fittings, often resulting in a less satisfactory building.

Comparisons are often made between analyses of similar projects, when elements such as floor and ceiling finishes can usually be compared directly, while others such as external walling must be adjusted to allow for the effect of plan shape, and others such as foundations, upper floors and roof must be considered in relation to the number of storeys. Elemental costs, expressed in terms of unit floor area, together with their quantity factors and element unit costs read in conjunction with specification notes, provide the means for making realistic comparisons between elements of similar projects. These comparisons can be extended to cover buildings of different types of structure and buildings of different heights by making suitable adjustments.

It is desirable to prepare costs analyses as early as possible to provide relatively up-to-date cost information. Those prepared following tender acceptance will not encompass contract variations, but they will still be sufficiently accurate for cost planning purposes.

For the preparation of a cost analysis, the following documents are required:

(1) fully priced bill of quantities, including the detailed breakdown of sums contained in prime cost or provisional sums;

(2) working drawings and specifications, for the calculation of quantity factors and the like.

The coding of dimensions and their sorting or processing by computer will shorten the time required to prepare the cost analysis.

Having obtained the cost of each element it is then expressed in terms of cost per m^2 of gross floor area for ease of comparison. Detailed cost analyses embody element unit quantities, element unit rates, specification and design notes and other relevant information relating to the project. This information amplified by reduced photocopies of plans and elevations provide well detailed documentation of the project and its costs. Offices having a programme of similar projects often prepare a shortened form of cost analysis for many of their projects, as they already have the detailed information in the source documents should it be needed.

Cost Analysis of Educational Buildings

Vast experience in the preparation of cost analyses has been obtained in connection with educational buildings. This stems from the lead given by the former Ministry of Education in 1950 and has been assisted by the large and continuous volume of school building work and the similarity of many of the buildings. The detailed analyses contain a wealth of information covering not only the cost of elements related to a square metre of gross floor area, but also quantity factors such as wall/floor ratios and element unit quantities (for example area of each type of internal partition in square metres and number of lighting points), and specification and design notes describing the main constructional methods and materials used. This form of analysis is invaluable in making cost comparisons between buildings with varying quantitative and qualitative factors. Case studies involving the practical application of cost comparison techniques follow in chapter 10.

STANDARD FORM OF COST ANALYSIS

The Standard form of cost analysis was first published by the Building Cost Information Service (BCIS) of the RICS in late 1969, with the full support of the main users of cost data, for controlling building costs during the design stage. Previously, cost analyses had been prepared on a number of different forms and on the basis of a variety of instructions and element lists which detracted from their value for cost comparison purposes. Quantity surveyors are now able to compile their records of building costs knowing that whatever their source, the analyses have been prepared on the same principles and that the historical data derived from them will be comparable.

The information obtained from these analyses will assist and improve cost planning advice and will enable the design team to control more effectively the cost of a building during the early design stages, to ensure that the building client receives value for money in terms of aesthetics, space use and constructional form against initial capital and subsequent running costs. The primary purpose of cost analysis must be to provide data which allows comparison to be made between the cost of achieving various building functions in one project with that of achieving equivalent functions in other projects. To this end the costs are analysed by elements which group together items fulfilling a specific function. The standard form of cost analysis is in various stages of detail related to the design process; broad costs are needed during the initial period and progressively more detail is required as the design is developed. Information in this form is essential to quantity surveyors' cost control techniques. The Building Cost Information Service (BCIS) collects and disseminates cost information within the quantity surveying profession and the standard cost analysis encourages this by facilitating the establishment of a substantial library of cost or price data, expressed in metric terms and to two decimal places of a pound per square metre of gross floor area, although there is not an adequate supply of cost analyses for the more uncommon and prestigous buildings, as described later, which unfortunately places a restriction on the value of the otherwise excellent service despite all the effort expended by those operating it.

In the standard form the element list has been divided into six groups, five of which cover the building and the sixth external works. The substructure is a single element group, while the superstructure comprises frame, upper floors, roof, stairs, external walls, windows and external doors, internal walls and partitions, and internal doors; internal finishes embrace wall, floor and ceiling finishes. Fittings appear as a collective single element. The largest single group covers services, which is subdivided into fifteen elements or components, some of which have limited application.

The principles and definitions incorporated in the standard form were formulated by a working party of the RICS. They should help to remove some of the problems and ambiguities which can arise, particularly when analysing a complex building.

Two types of standard cost analysis are provided – a concise cost analysis and a detailed cost analysis. The concise cost analysis occupies only half a page and provides background information about the project and costs of the six element groupings, including the cost/m^2 of gross floor area, with separate figures for preliminaries included in the elemental groupings or given as a single item, and comparable UK 1985 (index

Table 9.1 Concise cost analysis: thirty sheltered flats

BCIS Code: C – 2(3) – 2440
New build

CI/SfB
843.

Project Tender Price Index 107 Base: 1985 BCIS Index Base
Indices used to adjust to base date 100 103

Sheltered Housing – 162

Total Project Details | *BCIS On-line Analysis No. 13064*

Job title: 30 Sheltered Flats, Holt Road
Location: Cromer, Norfolk
Client: Orbit Housing Association
Date for receipt of tender: 28-Feb-92 Date of Tender: 17-Feb-92
Contract period (months) – stipulated: 18 offered: 14 agreed: 14
Type of contract: JCT private contract 1980 edition
Fluctuations: Firm
Selection of contractor: Selected competition
Number of tenders – issued: 8 received: 8
Tender amended – Addendum Bill

Measured Work:	675 546
P.C. Sums:	387 000
Provisional Sums:	115 600
Preliminaries:	104 400
Sub-total:	1 282 646
Contingencies:	40 000
Contract sum:	1 322 646

Analysis of Single Building

No. of storeys: 2(3)
Gross floor area: 2440 m²
Functional unit:

Type of construction: 24 No. 1 person flats, 6 wheelchair flats, communal facilities and warden's flat. *In situ* concrete strip foundations, precast concrete floors. Some structural steel framing. Brick/block external walls. Pitched timber roof with concrete slates. Softwood windows. Block and plasterboard demountable partitions. Flush doors. Plaster and tiles to walls; carpet and vinyl to floors; plasterboard and Armstrong suspended ceilings. Fittings. Electric storage heating. Sanitary, ventilation and electrics. Laundry and kitchen equipment. Lift. External works.

Element	*Total cost of element*	*Cost per m²*	*Total cost of element inc. prelims.*	*Cost per m² inc. prelims.*	*Cost per m² inc. prelims at 1985, UK mean location*
Substructure	57 046	23.38	62 101	25.45	24.70
Superstructure	492 944	202.03	536 622	219.93	213.44
Internal finishes	110 394	45.24	120 175	49.25	47.80
Fittings	32 900	13.48	35 815	14.68	14.25
Services	287 541	117.84	313 019	128.29	124.51
Building sub-total	980 825	401.98	1 067 732	437.60	424.69
External works	197 421	80.91	214 914	88.08	85.48
Preliminaries	104 400	42.79	–	–	–
TOTAL (less contingencies)	1 282 646	525.67	1 282 646	525.67	510.16

BCIS – 1993/94.

base mean UK location) costs per m² including preliminaries. The 1990/93 recession resulted in a very small variation between the 1985 and 1993 price levels, although they appeared likely to rise by about 5 per cent per annum in 1995 and 1996. Examples of two concise analyses are given in table 9.1 (30 sheltered flats) and table 9.2 (warehouse).

A comparison of these two cost analyses shows some interesting variations, as the base date in the last column (1985) is the same in each case. The wide variation in the unit costs of buildings of different types is immediately apparent – £510.16/m² for small flats in two–three storey blocks as compared with £173.00/m² for a single-storey warehouse. These prices will also be affected by regional variations as described later in the chapter. Despite its overall much lower cost, the warehouse substructure costs are over 40 per cent more than those of the flats because of the entirely different form of construction needed to take the heavier loads (pad foundations on piles as compared with strip foundations). On the other hand there are very few fittings in the warehouse, and

Table 9.2 Concise cost analysis: warehouse

BCIS Code: A – 1 – 6884	CI/SfB 284.2
Project Tender Price Index 87 Base: 1985 BCIS Index Base Indices used to adjust to base date 100 101	Purpose-built warehouses/stores – 37

Total Project Details — *BCIS On-line Analysis No. 12194*

Job title: Warehouse, Hunslet Business Park
Location: Hunslet, Leeds, West Yorkshire
Client: Douthwaites Florist Sundries Ltd
Date for receipt of tender: 7-Aug-91 Date of Tender: 28-Jul-91
Contract period (months) – stipulated: 6 offered: 6 agreed: 6
Type of contract: JCT Intermediate form of contract
Fluctuations: Firm
Selection of contractor: Selected competition
Number of tenders – issued: 7 received: 7
Tender amended – Provisional Sum reduction

Measured Work:	1 055 897
P.C. Sums:	–
Provisional Sums:	64 000
Preliminaries:	88 432
Sub-total:	1 208 329
Contingencies:	19 500
Contract sum:	1 227 829

Analysis of Single Building

No. of storeys: 1
Gross floor area: 6884 m²
Functional unit:

Type of construction: Single storey warehouse. Reinforced ground bearing slab laid over existing concrete, ground beam/strip footings, pad foundations all supported on contractor designed piles. Steel frame. Facing brick/block insulated cavity external walls. Coated steel cladding to walls and roof; patent glazing curtain walling, rooflights and roller shutter doors. Amenity area: load-bearing block walls, flat plywood roof, sanitary installation, plaster to walls, vinyl sheet to floor, part suspended, part plasterboard ceilings. Extensive external works, landscaping and drainage.

Element	*Total cost of element*	*Cost per m²*	*Total cost of element inc. prelims.*	*Cost per m² inc. prelims.*	*Cost per m² inc. prelims at 1985, UK mean location*
Substructure	227 487	33.05	245 450	35.66	35.15
Superstructure	600 619	87.25	648 047	94.14	92.79
Internal finishes	4 922	0.71	5 311	0.77	0.76
Fittings	2 372	0.34	2 559	0.37	0.36
Services	24 001	3.49	25 896	3.76	3.71
Building sub-total	859 401	124.84	927 263	134.70	132.76
External works	260 496	37.84	218 066	40.83	40.24
Preliminaries	88 432	12.85	–	–	–
TOTAL (less contingencies)	1 208 329	175.53	1 208 329	175.53	173.00

BCIS – 1992/93.

services costs are minimal, whereas in the case of the flats they account for about one-quarter of total costs.

The detailed cost analyses normally occupy about 3 pages as the example of the advance factories contained in table 9.3. These analyses contain considerable information about the form of the project, site and market conditions, contract particulars, accommodation and design features, floor areas suitably categorised, internal cube, external wall area, wall/floor ratio and storey heights. There is a summary of contract cost information and element costs giving the costs of all elements and element groupings, expressed as both total element cost and cost/m^2 of gross floor area, shown with preliminaries taken separately and alternatively with preliminaries apportioned amongst the elements. In addition the element unit quantity and element unit rate are incorporated against each element, where appropriate, and this is a useful feature when comparing cost analyses for similar projects. The costs of preliminaries and contingencies are also shown. Specification and design notes

Table 9.3 Detailed cost analysis: advance factories

BCIS On-line Analysis No. 10315

CI/SfB
282.1

Advance factories – 22 – a

BCIS Code: A – 1 – 1998

Job Title:	Shed B Redevelopment, Little Heath Industrial Estate	Indices used to adjust costs to 1985 UK mean location base
Location:	Little Heath, Coventry, West Midlands	TPI at tender 131; at 1985 mean 100 Location factor 0.95
Client:	City of Coventry	
Date for receipt:	21 June 1990 Date of tender: 1 June 1990	

INFORMATION ON TOTAL PROJECT

Project details:
2 single storey blocks, rectangular on plan in steel framed construction each comprising 3 factory units. External works include demolition of warehouse, cladding to gable of adjacent factory, roads, paths, parking, brick and timber enclosures, landscaping, services, drainage and skip compound.

Site conditions:
Level demolition site with bad ground conditions. Excavation above water table. Founding level average 8 m below floor level.

Market conditions:
Competitive. JCT 80 Contract with contractor's design supplement. Tenders received for PC sums incorporated into analysis with balance figure included in external works.

Project tender price index: 118 (base: 1985 BCIS Index Base)

Tender documentation: Bill of Quantities

Selection of contractor: Selected competition

Number of tenders – issued: 6
received: 6

Type of contract: JCT local authority contract 1980 edition

Cost fluctuations: Firm

Contract period – stipulated by client: –
– offered by builder: 8 months
– agreed: 8 months

Competitive Tender List
782 225*
796 000
798 360
803 115
824 241
830 612

ANALYSIS OF SINGLE BUILDING

Accommodation and design features:
2 rectangular single storey blocks each with 3 factory units. *In situ* concrete piles. RC beams and beds. Steel frame. Facing brick/block walls with steel cladding/metal stud, plasterboard. Proprietary composite panel cladding to roofs including double skin translucent sheeting. Steel windows and overhead doors. Flush doors. Block internal walls with metal stud/plasterboard over. Fittings. Prodorglaze to walls; vinyl tiles to WC floors; plasterboard ceilings to toilets. Sanitary, disposal, hot and cold water and electrical installations. External works.

Areas:

Basement floors	–
Ground floor	1 998 m^2
Upper floors	–
Gross floor area	1 998 m^2
Usable area	–
Circulation area	–
Ancillary area	–
Internal divisions	–
Gross floor area	1 998 m^2

Functional units:

Percentage of gross floor area
1 Storey construction 100.00%

Floor space not enclosed	–	*Storey Heights:*			
Internal cube	13 986 m³	Average	below ground floor	–
External wall area	1 753 m²		at ground floor	7.00 m
Wall to floor ratio	0.88		above ground floor	–

BRIEF COST INFORMATION

TOTAL CONTRACT	
Measured work	563 208
Provisional sums	9 050
Prime cost sums	127 140
Preliminaries	54 787 – being 7.83% of remainder of contract sum (less contingencies)
Contingencies	28 040
Contract sum	782 225*

* See market conditions text

BCIS – 1991/92.

CI/SfB
282.1

Advance factories – 22 – b

ELEMENT COSTS

Gross internal floor area: 1998 m²

Date of tender: 1 June 1990

Element		*Preliminaries shown separately*				*Preliminaries apportioned*		
		Total cost of element	*Cost per m² gross floor area*	*Element unit quantity*	*Element unit rate*	*Total cost of element*	*Cost per m² gross floor area*	*Cost per m² at 1985, UK mean location*
1	SUBSTRUCTURE	168 907	84.54	1 998 m²	84.54	182 138	91.16	73.25
2A	Frame	74 132	37.10	1 998 m²	37.10	79 939	40.01	
2B	Upper floors	–	–			–	–	
2C	Roof	89 698	44.89	2 137 m²	41.97	96 725	48.41	
2D	Stairs	–	–			–	–	
2E	External walls	90 184	45.14	1 588 m²	56.79	97 249	48.67	
2F	Windows & external doors	27 492	13.76	165 m²	166.62	29 646	14.84	
2G	Internal walls and partitions	25 575	12.80	891 m²	28.70	27 578	13.80	
2H	Internal doors	9 245	4.63	24 No.	385.21	9 969	4.99	
2	SUPERSTRUCTURE	316 326	158.32			341 106	170.72	137.18
3A	Wall finishes	2 811	1.41			3 031	1.52	
3B	Floor finishes	960	0.48	68 m²	14.12	1 035	0.52	
3C	Ceiling finishes	2 863	1.43	75 m²	38.17	3 087	1.55	
3	INTERNAL FINISHES	6 634	3.32			7 153	3.58	2.87
4	FITTINGS	1 102	0.55			1 188	0.59	0.47
5A	Sanitary appliances	5 068	2.54	30 No.	168.93	5 465	2.74	
5B	Services equipment	–	–			–	–	
5C	Disposal installations	1 056	0.53	30 No.	35.20	1 139	0.57	
5D	Water installations	4 534	2.27	18 No.	251.89	4 889	2.45	
5E	Heat source	–	–			–	–	
5F	Space heating & air treatment	–	–			–	–	
5G	Ventilating systems	–	–			–	–	
5H	Electrical installations	27 741	13.88			29 914	14.97	
5I	Gas installations	–	–			–	–	

ELEMENT COSTS (contd)

Element		*Preliminaries shown separately*				*Preliminaries apportioned*		
		Total cost of element	*Cost per m² gross floor area*	*Element unit quantity*	*Element unit rate*	*Total cost of element*	*Cost per m² gross floor area*	*Cost per m² at 1985, UK mean location*
5J	Lift & conveyer installations	–	–			–	–	
5K	Protective installations	–	–			–	–	
5L	Communications installations	–	–			–	–	
5M	Special installations	–	–			–	–	
5N	Builder's work in connection	2 390	1.20			2 577	1.29	
5O	Builder's profit & attendance	–	–			–	–	
5	SERVICES	40 789	20.41			43 984	22.01	17.68
	BUILDING SUB-TOTAL	533 758	267.15			575 569	288.07	231.47
6A	Site works	101 855	50.98			109 834	54.97	
6B	Drainage	24 667	12.35			26 599	13.31	
6C	External services	23 703	11.86			25 560	12.79	
6D	Minor building works	15 415	7.72			16 623	8.32	
6	EXTERNAL WORKS	165 640	82.90			178 616	89.40	71.83
7	PRELIMINARIES	54 787	27.42			–	–	
	TOTAL (less contingencies)	754 185	377.47			754 185	377.47	303.31

CI/SfB
282.1

SPECIFICATION AND DESIGN NOTES

Advance factories – 22 – c

1	SUBSTRUCTURE	225 mm RC floor slab on Visqueen 1200 DPM on 50 mm concrete blinding on 150 mm hardcore bed. RC ground beams between pile caps. Contractor designed piling: 168 No. 450 mm bored RC piles, average 8.00 m (average £314.78 each); cavity wall construction from top of pile caps/ground beam to 150 mm above ground level.
2A	Frame	Steelwork (90.74 t) comprising 32 No. portal frames spanning 23.30 m fixed at 4.75 m centres, approximately 6.20 m to eaves and 7.80 m to ridge, wind bracing, purlins and cladding rails.
2C	Roof	Briggs Amasco Perfrisa composite metal panel roof cladding with 6.5% in translucent panels. 150 mm galvanised steel downpipes and 750 mm girth box gutters.
2E	External walls	Part cavity wall of half brick facing, 70 mm cavity with 30 mm Wallmate insulation and 140 mm block inner skin (636 m² @ £71.62/m²). Part wall cladding of Briggs Amasco Colorclad C32, plasterboard lining on HEP galvanised steel support sections and 80 mm glass fibre insulation (952 m² @ £46.73/m²).
2F	Windows & external doors	W20 section galvanised steel polyester powder coated windows glazed with 6.4 mm laminated glass and with steel security grilles. 12 No. 907 × 628 mm, 12 No. 1237 × 1218 mm (26 m² @ £206.91/m²); (grills 26 m² @ £76.64/m²). Crawford 342 insulated sectional overhead doors (6 No. @ £2505.64 ea.); 12 No. solid ply faced doors (£413.59 ea.).
2G	Internal walls and partitions	100 and 190 mm block walls (412 m² @ £28.22/m²); stud partitions (479 m² @ £29.12/m²).
2H	Internal doors	Solid ply faced flush doors in softwood frames (24 No. @ £357.29). Decoration £670.

3A	Wall finishes	27 m quarry tiles to sills (£11.87/m); tiled splashbacks (14 m @ £12.61/m); Prodoglaze to blockwork (£2314).
3B	Floor finishes	68 m^2 vinyl tiles to WC floors (£14.12/m^2); remaining floors self finished.
3C	Ceiling finishes	Joisted ceiling with plasterboard and chipboard decking (75 m^2 @ £36.89/m^2).
4	FITTINGS	Hat and coat rails, mirrors and sink base units.
5A	Sanitary appliances	12 No. WCs, 12 No. wash hand basins, 6 No. stainless steel sink tops.
5C	Disposal installations	Osma Clearbore 40 system.
5D	Water installations	Copper hot and cold water services: 30 No. cold draw off points (£44.70 each); 18 No. hot water draw off points (£53.09 each). Electric water heaters in toilets (6 No. @ £372.91 each).
5H	Electrical installations	Electric lighting and power.
5N	Builder's work in connection	Normal builder's work in connection with services.
6A	Site works	Site preparation. Paved areas, planted areas, roads and car parking, boundary walls, fencing and gates; external signs.
6B	Drainage	Concrete circular manholes. Hepsleve stoneware pipes.
6C	External services	Incoming services mains.
6D	Minor building works	Work to existing factory gable.
7	PRELIMINARIES	7.83% of remainder of Contract Sum (excluding Contingencies).
8	CONTINGENCIES	4.01% of remainder of Contract Sum (excluding Preliminaries).

CREDITS

SUBMITTED BY:	City of Coventry
CLIENT	Coventry City Council
ARCHITECT/QS/ENGINEERS	Coventry City Council

Note: Outline drawing omitted because of lack of clarity.

are included for each element to help with the adjustment of element prices when preparing cost plans for new buildings. Photostated reduced copies of drawings illustrate the general form of the building, as shown in the examples in chapter 10.

The detailed cost analysis of the advance factories shows the superstructure accounting for about 42 per cent of total costs and the services approaching 5.5 per cent. In more complicated buildings, services costs could amount for as much as 30 to 40 per cent of total costs. Substructure costs will be affected considerably by site and ground conditions.

Useful cost analyses are also published from time to time in the *Architects' Journal*, the *Building Economist*, formerly published by the Builder Group, and other technical journals. They refer to completed buildings and hence the prices are several years out of date. On the other hand they often contain excellent drawings and photographs of the finished building, specification notes, unit quantities and rates, and a cost comment drawing attention to cost aspects of significance.

The more important contents of the standard form of cost analysis and the method of approach to be adopted in each case are now outlined.

General Principles

(1) The elemental costs are related to the gross floor area and also to a parameter more closely identifiable with the element's function (element's unit quantity).

(2) Supporting information on contract, design/shape and market factors is defined so that the costs analysed can be fully understood.

(3) Professional fees are not included in the analysis and contingency sums are shown separately.

(4) In detailed analyses, design notes are inserted against each element.

Coding

The BCIS reference code classifies buildings by form of construction, number of storeys and gross floor area in square metres. There are four constructional types: A – steel-framed; B – reinforced concrete framed; C – brick; D – light framed steel or reinforced concrete. Hence a two storey steel-framed building with a gross floor area of 6000 m^2 would be coded as A – 2 – 6000. The CI/SfB code is also given as shown in the accompanying examples.

Supporting Information

Detailed cost analyses contain the following supporting information.

(1) A brief description of the building with particular reference to any special or unusual features affecting overall cost.

(2) Site conditions including access, proximity of other buildings, construction difficulties associated with topographical, geological or climatic conditions, and existing site conditions, such as the existence of woodland or existing buildings.

(3) Market conditions, including level of tendering, availability of labour and materials, keenness and competition.

(4) Contract particulars such as type of contract, basis for pricing, project tender price index, method of tendering, firm price/fluctuations, number of tenders issued and received, contract periods and list of tenders.

(5) Design/shape information including general description of accommodation with gross floor area of each type, internal cube, external wall area, wall to floor ratio, approximate percentages of building having a different number of storeys, and storey heights.

Detailed Analysis

The principal expressions used in the detailed cost analysis and the method of computing the various factors and costs are now examined in some detail.

Specification and design notes give the requirements of the element and the specification notes describe the form of construction and quality of material sufficiently to explain the costs in the analysis.

Element unit quantity and rate. The cost of an element is also expressed in suitable units which relate solely to the quantity of the element itself. For example, with floor finishes the element unit quantity is the total area of the floor finishes in m^2 while with a heat source it is kW.

All areas must be the net area of the element, for instance external walls will exclude window and door openings. Cubes for air conditioning and similar systems are measured as the net floor area of the part of the building concerned multiplied by the height from floor finish to underside of ceiling finish.

The element unit rate is the total cost of the element divided by the element unit quantity. Hence the element unit rate for floor finishes is the total cost of the floor finishes divided by their net area in m^2. Where various forms of construction or finish exist within one element, they are listed separately in the specification and design notes section with the total cost and area, as illustrated in table 9.3.

Floor areas. The gross floor area is made up of all enclosed spaces fulfilling the functional requirements of the building measured to the internal structural face of the enclosing walls. It includes the area occupied by partitions, columns, chimney breasts, internal structural or party walls, stairwells, lift wells, and the like, as well as lifts, plant, and tank rooms above roof level.

Gross floor area is subdivided between *usable*, *circulation*, *ancillary* and *internal divisions*. Usable area is that fulfilling the main functional requirements of the building, such as office or shop space. Circulation relates to entrance halls, corridors, staircases, lift wells and the like. Ancillary covers lavatories, cloakrooms, kitchens, cleaners' rooms, lift, plant, tank and similar rooms. Internal divisions relate to partitions, columns, chimney breasts and internal structural or party walls. The sum of the areas in the last four categories will be equal to the gross floor area.

In the case of residential buildings reference is sometimes made to net habitable floor area which covers the floor area within the enclosing walls, including partitions, chimney breasts and the like, but excluding balconies, public access areas, communal laundries, drying rooms, and lift, plant and tank rooms.

Roof and wall areas. Roofs are measured by the plan area (flat) across the eaves overhang or to the inner face of parapet walls, including rooflights.

The external wall area is measured on the outer face of external walls and overall windows and doors.

Wall to floor ratios. The wall to floor ratio is obtained by dividing the external wall area by the gross floor area to two decimal places. They can vary from 0.25 or less for large warehouses to 1.00 to 2.00 or more for housing.

Related matters. The storey height is measured from floor finish to floor finish, except in the case of single storey buildings and the top floor of multistorey buildings, where it is taken from floor finish to underside of ceiling finish. The internal cube is measured as the gross internal floor area of each floor multiplied by its storey height.

A functional unit is expressed as net usable floor area (offices, factories, public houses, etc.) or as the number of units of accommodation (seats in churches, school places, persons per dwelling, etc.).

Elemental divisions/design criteria. It might be helpful to mention that decorations are included in the finishes element and that windows and external doors include lintels, sills/thresholds, cavity damp-proof courses and work to reveals of openings, in addition to the window/door, frame, ironmongery and glazing.

COST LIMITS

General Background

Most new buildings have a cost limit which is the sum of money which the client considers is the maximum that he is able and/or willing to pay for the building. For instance, with an industrial organisation the limit may be determined by reference to the amount that the building is likely to contribute to the profits of the firm. The cost limit set by a private house builder will be influenced by market forces and possibly by the amount which the client has available or can borrow in order to finance the project. When charitable or similar organisations are involved the limit is likely to be decided by the organisation and may be based on the amount that it is estimated can be raised by grants, loans, donations, gifts and from other sources.

In the public sector the government has from time to time imposed limits on the amount which various authorities can spend upon different types of constructional work, to secure a reasonable balance between different projects, minimum constructional standards and value for money. Some limits have in the past tended to be rather inflexible and hence in more recent times government departments have tended to replace them with block grants, which allow the client authority greater discretion.

Cost limits and allowances established for buildings financed out of public funds generally take the form of cost targets and are often based on user accommodation, such as the square metre of usable floor area for schools. The documents detailing cost targets are extremely complex and encompass a wide range of different circumstances, and they therefore require careful study prior to their implementation.

Housing

Up to 1981, government funds to local authorities for housing purposes were determined by housing yardsticks, but these were abolished as they had several major deficiencies. For example, they concentrated on initial costs which tended to result in high maintenance costs and they were not updated sufficiently frequently. DoE replaced the housing yardsticks with cost criteria which are intended as indicators and not cost limits. The public housing programme was up to the late 1980s the largest user of the UK's building resources. Furthermore, prices were increasing with the raising of housing standards, the erection of more houses in highly priced locations and in redevelopment schemes, increased provision for cars and the greater use of more costly pedestrian segregated layouts. In the absence of effective cost control, the total housing expenditure could reach excessively high proportions resulting in a substantial reduction in the housing programme and/or large rent increases. Unfortunately both of these conditions prevailed in the early 1990s largely because of drastic cuts in public sector funding by central government, resulting in severe financial restraints.

Similarly with housing associations, who received much of their funds in the form of Housing Corporation grants, the grants were being progressively reduced in the 1990s, resulting in reductions in quality and size of dwellings and escalating rents which poorer families could no longer afford. Hence the tenants of housing associations were becoming increasingly dependent on housing benefit and the poverty gap was widening. The Housing Corporation was insisting on value for money and prescribed a range of performance expectations, thereby making the task of housing associations, by now the main providers of social housing, increasingly difficult, as they endeavoured to fulfil their stated aims of providing good quality houses at affordable rents. A set of tables provided the basis for determining the cost of housing association schemes funded by the Housing Corporation. In practice, numerous housing associations will be bidding in competition annually for a diverse range of housing schemes, and in 1993/94 were on average receiving about one third of the Housing Corporation grants for which they applied.

Hospitals

The Department of Health and Social Security (DHSS) provides design guides on basic areas for hospitals including circulation space. Cost allowances are prescribed for different forms of accommodation and oncosts for differing site conditions are also included. Additional allowances are available for providing high blocks and lifts, constructing abnormal foundations and other abnormal features. The cost allowance details give different cost allowances for various sizes and types of wards, in addition to administrative areas, operating theatres, laboratories, out-patients and teaching facilities.

The major problems encountered with hospitals in recent years stemmed largely from the long time needed to design and construct these large complicated buildings. Technical advances in medicine during the long construction period have often resulted in many changes to the building work, making the estimation of the finished cost of construction very difficult and overruns on time likely.

Schools

Since the early post-war years there has been in school building a steady process of refinement of layout, economy of specification, dual use of space and general elimination of waste, through the ingenuity, innovation and co-operation displayed by the design teams and educationists, with some stimulus coming from the application of school building cost limits. These cost limits imposed by the Department of Education and Science (DES) varied with the type of school, the number of pupil places, abnormal site conditions, the way in which the buildings were erected, whether in one phase or more, the provision of kitchen and dining facilities, caretaker's residence and other related matters.

However in the 1970s unrealistic cost limits, with lower capital costs, showed a disregard for future running and maintenance costs and produced increasingly stereotyped architectural solutions and strict limitation of variations in plan shape, size, height and elevational treatment, highlighting the need to keep a proper balance between costs and other design factors. Hence these formal cost limits have since been withdrawn and DES determines cash allocations nationally and controls the amount of work commenced in schools each year, depending on government policy. Lump sums are allocated for projects and the cost of the building work must be contained within them, although some extra allowances are available. Area-based cost units operate which vary with the type and size of school.

Universities

The Higher Education Funding Council (HEFC) has to be consulted before any university project is commenced, and the Council has to agree the type and size of project and a provisional expenditure limit together with the area of the building which will be prescribed by the Council's officers.

After a university building has been designed the expenditure limit is adjusted to make allowance for inflation and any substantial abnormal costs. The final expenditure should not exceed the adjusted limit. No official unit costs are issued by HEFC and the Council's officers consider each project separately.

BUILDING COST/PRICE INDICES

Purpose of Cost Indices

The cost of any building design is determined primarily by the cost of labour and materials involved in its erection. Variation in the cost of either of these basic factors will influence the cost of an item of work, both absolutely and relative to the cost of the entire structure. The detailed labour and material content of every building differs and these variations must be taken into account by the adjustment of cost data used for cost planning purposes. The compilation of indices of building costs is the most satisfactory method of approach. The cost indices first introduced for building work were based on labour and material price variations, suitably weighted to take account of the approximate relevant quantities of each used in a particular building type. There are however weaknesses in this approach which will be discussed later in this chapter.

The main function of building cost indices is to assess the differences in levels of tenders at varying dates. They may be used to compare building costs where tenders were obtained at different dates or to adjust analyses of the past costs of buildings to current prices. Future trends in price levels may be assessed, albeit rather imprecisely, by a study of cost indices and, having regard to possible future changes in labour and material prices, output of work and other related factors.

Problems in Selecting Cost Indices

It is not always easy to decide which building cost or price index to use in a particular situation. In addition the composition and sample size of the index have a considerable influence on the reliability and quality of the assessments and advice given.

The terms *building costs* and *tender prices* often lead to confusion as building cost is commonly used to refer to the cost to the client. Building costs are the costs incurred by the builder in performing building work and embrace such factors as wages of operatives, prices of materials, plant costs, rates, rents, overheads and taxes. In contrast, tender prices represent the price a client must pay for a building, and include building costs but also take account of market considerations and allow for profits and the builder's forecast of cost changes throughout the contract period. Growing demand for building work may cause tender prices to rise at a faster rate than building costs, while decreased demand may have the opposite effect.

The variety of cost and price indices used in the construction industry reflect:

(1) the diversity of construction work and the need for different sectors to have dedicated indices;

(2) the multiple requirements of practitioners involved in updating tenders, measuring changes in contractors' costs and revaluing output at constant prices;

(3) the many difficulties confronting cost evaluation in a field as diverse as construction, where there is no fully comparable basis of cost measurement over time;

(4) the need for different series of indices covering different time periods (Tysoe, 1991).

The basic cost or price measurement problem arises from the great diversity of work carried out by the construction industry, ranging from major building and civil engineering works to repairs and maintenance. New construction projects vary according to type, size, design specification, complexity and method of construction. Even similar projects vary because of different site conditions which influence both costs and price. Each project tends to be unique and, faced with the task of measuring changes in costs and prices over time, there is no single standard of comparison. In addition, construction projects often take a considerable period of time from start to finish and costs may be computed at different stages of the construction process.

When constructing an index Tysoe (1991) recommends that the following four factors should be considered:

(1) the purpose of the index;

(2) the selection of constituent items, weighted as necessary;

(3) the choice of weighting reflecting the relative importance of the items;

(4) the choice of the base year, preferably a recent year when there were no unusual occurrences.

Source and Nature of Cost Indices

There are several published cost indices which are intended to provide an empirical guide to changes in building costs/prices as illustrated in later examples.

Forecasting of future building costs is likely to be somewhat hazardous even with the benefit of building cost indices. Price movements can be very erratic on occasions. For example an examination of table 9.4 shows that in 1992, general building costs based on increases in the cost of labour, materials and plant rose by 3.5 per cent, while actual building prices based on accepted tenders fell by 4.5 per cent, although the largest reduction in tender prices was in 1990 when they fell nearly 9 per cent, despite an increase of building costs of 6.3 per cent. Over the whole period of the recession from 1990 to 1993, all-in tender prices fell by 23.9 per cent and building costs rose by 18 per cent, an almost unbelievable situation as contractors struggled to secure contracts in a very depressed market, often submitting unrealistically low tenders, by cutting oncosts to the bone, rationalising their organisations and using surplus profits made in the boom period of the late 1980s.

Tender Price Indices

These are the most useful cost indices for quantity surveyors and the most popular is probably that produced by the BCIS and illustrated in tables 9.4 and 9.5. There are, however, alternative tender based cost indices produced by Davis, Belfield and Everest (table 9.6) and those prepared by the Department of the Environment (table 9.7).

As well as measuring changes in basic costs, tender price indices indicate the feelings of the industry about its current and future workload. When demand for the industry's services is high, not only do contractors' margins increase but so also do the margins charged by materials' suppliers and producers and the money paid to attract labour. When demand is low, as during 1990–92, all these factors fall. It is the difference in the rate of change in the indices that is significant, whereas the absolute difference between the two indices is not meaningful (BCIS, 1993).

Table 9.4 Tender price and building cost indices

1985 mean = 100

Quarter		*Tender price indices*						*General building cost index*
		All-in TPI		*Private sector TPI*		*Public sector TPI*		
		Index	*No in sample*	*Index*	*No in sample*	*Index*	*No in sample*	
1984	i	94	37	91	20	95	13	92
	ii	96	29	97	25	92	10	94
	iii	95	41	104	30	93	13	96
	iv	96	68	94	32	96	30	96
1985	i	97	82	98	46	99	31	98
	ii	102	61	101	36	105	17	99
	iii	99	84	101	33	99	37	101
	iv	103	54	100	40	100	17	102
1986	i	101	77	101	39	99	36	102
	ii	102	78	103	44	101	30	103
	iii	103	70	105	33	99	27	105
	iv	105	79	104	52	104	25	106
1987	i	108	88	110	40	105	31	107
	ii	106	87	108	71	104	29	108
	iii	109	94	112	81	109	34	110
	iv	117	81	120	69	118	34	111
1988	i	119	95	121	90	118	28	112
	ii	121	70	123	65	121	25	114
	iii	128	110	131	83	127	41	117
	iv	128	86	131	81	122	31	118
1989	i	134	92	136	78	132	37	120
	ii	134	80	137	76	128	26	122
	iii	138	85	137	64	139	31	126
	iv	134	95	135	55	131	39	127
1990	i	135	95	137	74	129	32	128
	ii	131	72	130	54	134	24	131
	iii	126	81	129	52	125	29	135
	iv	123	72	124	37	121	34	136
1991	i	117	77	116	31	115	39	138
	ii	114	82	115	39	114	30	139
	iii	114	109	113	45	113	39	141
	iv	111	86	110	29	111	29	142
1992	i	112	75	110	33	111	31	142
	ii	108	55	108	32	108	16	143
	iii	106	63	107	23	107	31	146
	iv	107	58	110	23	104	24	147
1993	i	108	59	107	29	105	17	148
	ii	108	35	112	27	106	7	149
	iii	109	17	116	7	100	5	151 (P)
Forecast								
	iv	109						151
1994	i	109						153
	ii	110						154
	iii	111						156
	iv	112						157
1995	i	113						159
	ii	114						161
	iii	117						163
	iv	118						163

(P) Provisional

Source: *BCIS Quarterly Review of Building Prices*, December 1993.

Table 9.5 Building cost indices for different types of buildings

Base 1985 mean = 100

Quarter		*Cost indices*				
		General building cost (exc. M&E)	*General building cost*	*Steel-framed construction cost*	*Concrete-framed construction cost*	*Brick construction cost*
1984	i	92	92	92	92	92
	ii	93	94	93	94	94
	iii	96	96	96	96	96
	iv	97	96	96	96	97
1985	i	98	98	98	98	98
	ii	99	99	99	99	99
	iii	102	101	101	101	101
	iv	102	102	102	102	102
1986	i	102	102	102	102	102
	ii	103	103	103	103	103
	iii	105	105	105	105	105
	iv	106	106	106	105	106
1987	i	106	107	107	106	107
	ii	108	108	108	107	108
	iii	110	110	111	110	111
	iv	111	111	111	110	111
1988	i	111	112	112	111	112
	ii	113	114	114	113	114
	iii	117	117	117	116	117
	iv	118	118	118	118	118
1989	i	119	120	120	120	120
	ii	122	122	123	122	122
	iii	126	126	126	126	126
	iv	127	127	127	127	127
1990	i	128	128	129	128	128
	ii	130	131	131	130	131
	iii	135	135	135	135	135
	iv	136	136	136	136	136
1991	i	136	138	138	137	138
	ii	137	139	140	138	139
	iii	140	141	142	140	142
	iv	140	142	142	140	142
1992	i	140	142	143	141	143
	ii	141	143	144	141	144
	iii	144	146	147	144	147
	iv	145	147	148	145	147
1993	i	145	148	149	146	148
	ii	147	149	151	148	149
	iii	149*	151*	152*	150*	151*
Forecast						
	iv	149 est.	151 est.			
1994	i	151 est.	153 est.			
	ii	152 est.	154 est.			
	iii	154 est.	156 est.			
	iv	155 est.	157 est.			
1995	i	157 est.	159 est.			
	ii	158 est.	161 est.			
	iii	161 est.	163 est.			
	iv	163 est.	165 est.			

* Provisional.

Source: BCIS.

Table 9.6 Davis Belfield and Everest Indices

Base 1976 = 100

Price index		*Tender cost index*		*Building cost index*	*Mechanical cost index*	*Electrical cost index*
1993	1	227		370	333	373
	2	242		371	334	377
	3	233		373	336	378
	4*	238		374	337	378
Forecast		*Min.*	*Max.*			
1994	1	240	244	377	339	379
	2	242	247	378	340	379
	3	246	252	389	343	382
	4	250	257	392	349	389
1995	1	254	262	394	352	391
	2	259	268	397	354	392
	3	265	275	409	358	397
	4	269	279	412	365	406

* Provisional

Source: *Building Economist*, January 1994.

The BCIS index is based on at least 80 tenders nationally, to achieve stability as far as practicable, with the number of projects included in each quarter shown against the index figures. While the PUBSEC index relates to rates in bills of quantities for accepted tenders of £50 000 or more in value for public sector building works excluding housing, civil engineering, mechanical, electrical and alteration works. In the PUBSEC smoothed index (published solely from Q3/1992), the lowest and highest 10 per cent of projects by number have been excluded to remove unrepresentative outlying tenders. Tender prices represent the price a client pays for a building and they reflect the prevailing market conditions. In times of boom tender prices may increase at a faster rate than building costs, while in times of depression, as in 1990–92, the reverse will apply, as shown in table 9.4. The varying workload influences the contractor's short term expectations.

A tender price index is produced by the examination and analysis of priced bills of quantities for accepted tenders and comparing the bill rates against a base schedule of rates. In reality, bills of quantities are repriced by BCIS using a base schedule of rates and the base tender figure compared with the actual tender figure to produce a project index. A large number of bills of quantities are indexed in this way and the resulting project index figures are first adjusted to eliminate average differences due to location, contract size and method of procurement at the base date and then averaged to produce the published index.

Full bill repricing involves a substantial amount of work and time and to reduce this to reasonable proportions, a sampling process is sometimes used. The index is compiled by repricing at base rates the major items incorporating the largest price extensions in each work section of the bill of quantities for each contract, and continuing until the total value of repriced items amounts to 25 per cent of the total for the work section. Studies were carried out by the Department of the Environment, who pioneered the system, using various levels of sampling of items in bills of quantities. It was found that to achieve the same level of reliability, using as little as a 25 per cent sample of the total bill value, it was only necessary to slightly increase the number of bills analysed, mainly because a relatively small number of billed items account for a high proportion of the total cost. Because of the wide

Table 9.7 PUBSEC index (tender price index of public sector building non-housing)

Base 1985 = 100

Year & quarter		Firm price (Factor = 1.00)			VOP (Factor = 1.00)			All-in (Factor = 1.00)	
	Raw index	*Smooth index*	*SS*	*Raw index*	*Smooth index*	*SS*	*Raw index*	*Smooth index*	*SS*
1990 Q1	129	130	81	133	130	9	130	130	90
Q2	129	127	62	119	122	5	128	127	67
Q3	121	123	49	116	118	3	120R	122	52
Q4	121	121	55	119	120	9	121	120	64
1991 Q1	119	119	69	127	122	9	119	119	78
Q2	115	115	61	113	115	7	115	115	68
Q3	111	112	86	105	107	4	111	112	90
Q4	109	110	79	104	105	8	109	110	87
1992 Q1	110	109	57	107	104	4	109	108	57
Q2	106	106	58	98	101	1	105	105	59
Q3		104	64		99	5		104	69
Q4		104	55		100	3		104	58
1993 Q1		106	64		101	2		105	66
Q2		108P	40		103P	4		107P	44
Q3		110P	22		103P	1		108P	23
Q4		111*			104*			109*	
1994 Q1		113*			106*			111*	
Q2		114*			107*			112*	
Q3		116*			109*			114*	
Q4		118*			110*			116*	
1995 Q1		120*			112*			118*	
Q2		121*			114*			119*	
Q3		123*			115*			121*	

To convert these indices to other base multiply by: 1975 = 100, ×2.184; 1990 = 100, ×0.794.
Smoothed indices only were published from Q3/1992.
P = Provisional; R = Revised; SS = Sample size; * = Indicative index.

Source: DoE, *Public Sector Building Works: Quarterly Building Price and Cost Indices*. HMSO.

scatter of tender prices BCIS aims to include 80 projects in each quarter to achieve stability, but this is not always possible. The BCIS tender price indices are restricted to new buildings with contract sums exceeding £50 000 (to be adjusted periodically) and priced in competition or by negotiation.

The BCIS All-in Tender Price Index (TPI) is based on a random sample of accepted tenders for new building work with contract sums over £50 000 which have been priced in competition or by negotiation. The index covers both public and private sectors.

The Private Sector Tender Price Index is based on an analysis of a random sample of private sector projects excluding housing. The index has been produced by BCIS for DoE. BCIS also produce the Private Commercial TPI and the Private Industrial TPI, which are both detailed in the BCIS Quarterly Review of Building Prices.

The Public Sector Tender Price Index is based on all projects in the public sector excluding public housing. The index provides a direct comparison with the All-in Private Sector TPI.

A tender price index has a number of advantages over a building cost index:

(1) it represents the cost to the building client as opposed to the contractor;

(2) it does not rely on the summation of a

number of other indices so that any inherent inaccuracies are not compounded;

(3) it provides an indication of the tendering climate;

(4) it helps to measure the effectiveness of cost planning;

(5) the individual tender price index can be used to evaluate specific price determinants such as location, building type and method of construction within certain constraints; and

(6) it can be used to set realistic cost limits.

There are, however, some problems inherent in the use of tender price indices, as the tender price often differs from the final cost of a project. For example, the contract may contain fluctuations clauses for labour and materials and the entries against prime cost and provisional sums are approximations. Furthermore, the unit rates contained in a bill of quantities may not be realistic as they are affected by so many factors. These variables include different wage rates, labour outputs, plant usage, overheads, profit and the contractor's experience and skill. The way in which these factors operate in a particular contracting organisation determine the competitiveness of its pricing. Contractors also adopt different techniques in building up their unit rates, including the method adopted for the pricing of preliminaries, and this multiplicity of pricing strategies introduces considerable variability into unit rates. One study showed the standard deviation of individual rates in the order of 15 per cent of the mean and that for the index of a complete bill of quantities as approximately 12.5 per cent of the mean.

The basis of construction of the index must be known if it is to be used intelligently. The choice of index will be influenced by the user's needs. It should be borne in mind that short term changes of an index may not be very reliable.

Movement in Tender Prices

The contractor's capacity to maintain a downward trend in tender prices in the early 1990s stemmed from a continued assault on subcontractors' prices, reduced oncosts and profit margins, and the increasing use of casual and East European labour engaged at rates considerably below national wage agreements. However, there was evidence in 1993 that fear of the adverse consequences of securing work at rock bottom prices was beginning to have an effect. Tender results increasingly showed only one contractor prepared to drop significantly below the rest, in contrast to the general scramble which had dominated the tender process from 1990 to 1992, as firms competed fiercely to maintain turnover. The large contractors were still surviving mainly on money made on contracts won prior to the recessionary years, but a further fall in 1992 of 6 per cent in the DoE's output figures suggested that pressure will be increased to secure work at economic prices.

The Davis Langdon & Everest year-on-year forecast for *Building* to the first quarter of 1994 was for a maximum rise of 2 per cent, which would still leave tender prices below the level recorded in the third quarter of 1986. In 1995 the rise was forecast to be 2.5 to 6 per cent as continued recovery in housing and greater investment in the infrastructure started to lift the construction industry.

In 1994 there was evidence that suicidal tendering was coming to an end with the emergence of a two-tier contracting market. Cyril Sweett established that directors of contracting organisations were willing to place a premium on risk when making adjustments in tender settlements with estimators, instead of adopting a blanket approach to discounting tenders before submission. Projects perceived as risky or difficult attracted less discount than mainstream work. Previously standard practice had been to deduct between 5 and 10 per cent before submission, but in 1994 this applied only to bread and butter work. While cut-throat competition remained for traditional work, discounts were regularly halved on risky contracts.

Hence tenders were increasingly becoming closer to breakeven point on projects where the risk element was greater, such as design and build, heavily phased contracts, and tender enquiries without bills of quantities. Where a 5 per cent discount might have been applied, less attractive projects were reduced to 2.5 per cent. It was anticipated that tender inflation would be around 5 per cent in 1994 followed by an overall increase of around 6 per cent in 1995, although subsequently this seemed unlikely and highlights forecasting problems.

Building Cost Indices

Some cost indices are related solely to fluctuations in the cost of labour and materials, although, as described earlier, there may not be a very close correlation between price fluctuations and tender prices over a period of time. It is also necessary to use an index which is related to the type of construction and class of building under consideration. Even then the proportions of materials in the same building type may vary appreciably between different projects.

The BCIS General Building Cost Index is based on a cost model of an average building. The model was derived from the analysis of 80 bills of quantities. These bills were broken down into the work categories defined in the NEDO formula method of calculating fluctuations under a building contract. The inputs to the index are the work category indices prepared by the Property Services Agency (PSA) for use with the NEDO formula. The indices allow for changes in the costs of nationally agreed labour rates, material prices and plant costs (BCIS, 1993).

A variety of different building cost indices are shown in table 9.5 (including indices for steel framed, concrete framed and brick construction) and table 9.8 (housing projects). Some quantity surveyors prefer to use these as they assert that it enables them to start from a firmer factual base and then to make suitable adjustments to suit local conditions.

Table 9.6 shows mechanical and electrical cost indices and the mechanical indices show lower rates of increase than those operating for general building work, while the electrical services are higher in 1993 but the forecast shows the mechanical index dropping below the building cost index in mid-1994 and then retaining the same relative positions up to the end of 1995, but the differential is only 1.5 per cent. RICS Building Maintenance Information (BMI) produces a series of maintenance cost indices and these cover general maintenance, redecoration, fabric maintenance, services maintenance, cleaning costs and energy cost indices for different fuels.

NEDO Price Adjustment Formula

Another set of cost indices are those issued monthly by the Property Services Agency cover-

Table 9.8 Housing cost index

	1985	*1986*	*1987*	*1988*	*1989*	*1990*	*1991*	*1992*	*1993*
January	367.15	386.09	406.53	428.93	459.94	498.35	534.28	551.94	557.59
February	367.95	387.15	408.87	430.00	461.96	499.62	534.72	551.97	558.10
March	369.11	388.34	410.60	433.42	464.40	502.43	537.38	548.90	558.05
April	372.57	390.69	413.14	435.97	468.23	505.21	537.49	550.16	561.32
May	374.60	391.10	413.79	436.15	468.39	505.85	537.57	549.81	561.47
June	374.90	391.23	413.96	437.40	468.40	505.90	538.33	549.53	562.13
July	384.28	399.48	423.35	450.74	487.61	528.12	550.68	556.19	561.91
August	385.94	400.66	425.19	452.84	490.22	528.60	551.65	556.87	562.77
September	385.91	401.37	425.24	453.49	491.21	528.87	551.73	556.95	563.33
October	385.59	401.95	425.94	455.42	494.23	530.21	554.23	557.00	563.56
November	385.80	402.64	426.66	456.11	494.72	530.22	554.70	557.25	563.67
December	386.02	403.02	427.71	455.69	494.74	530.28	550.66	557.17	563.56

Base: December 1973 = 100.00.

Source: *Building*.

ing price fluctuations in 34 work activities and also skilled labour, unskilled labour and plant. The application of the NEDO Price Adjustment Formula offers considerable benefits over the traditional method of calculating the cost of fluctuations, particularly in the reduction of the amount of work involved often resulting in quicker reimbursement to the contractor. NEDO has been replaced by the Joint Forecasting Committee for the Construction Industries.

Validity of Building Cost Indices

For some years there has been a growing concern with regard to the degree of reliability of available building price indices. This concern stems partly from the lack of consistency between various published indices and partly by their failure to indicate the movement in building prices that persons in the building industry felt, from their own experience, to have occurred. An additional and significant reason was the wide coverage of the majority of indices, without regard to specific locality or building type.

The BCIS issues data on regional cost differences and these are examined later in the chapter.

There are serious problems associated with the construction of index numbers based on labour and material costs, notably concerning productivity and plant, overhead costs and profit. Regarding indices based on full repricing of bills, quantity differences and variability of pricing between bills could lead to high coefficients of variation, although price variability might be reduced by stratification into geographical areas. The cost of constructing indices by full repricing of bills would be dependent upon the size of sample chosen and would probably be very high. Although the use of computers for repricing and the use of full bill repricing for cost control purposes might reduce the cost considerably. Furthermore, it seems likely that indices based on short lists of items selected from the bills reflect reasonably well the trends in prices derived from indices based on a full bill as described earlier in the chapter in connection with tender price indices.

Application of Building Cost Indices

Building cost indices are often used to bring a tender figure for a previous similar project up to current prices, so that it can be used for comparison purposes as part of cost planning. An example may serve to illustrate this use.

The calculation to adjust a tender or a cost analysis rate to allow for time variations, can be carried out using the following formula:

$$A = \frac{(B - C)100}{C}$$

where A = percentage change, B = index figure at date to which tender is being adjusted, C = index figure applicable to the current proposal.

An example will help to illustrate its use in practice.

Date of tender of existing project to be used for comparative purposes: July 1992. Anticipated date of tender of new project: August 1995. From BCIS tender price index:

$$1992\ (3) = 106 = C$$

$$1995\ (3) = 117 = B$$

$$A = \frac{(117 - 106)100}{106} = 10.38 \text{ per cent}$$

A quicker way of obtaining the percentage variation is to divide 117 by 106 = 1.10377, and then to move the decimal point two places to the right and deduct 100. In practice the adjustment is further shortened by using the multiplier of 1.10377.

Assuming the tender price in July 1992 was £2 300 000 then updating to the August 1995 price would be £2 300 000 × 1.10377 = £2 538 671.

It will also be necessary to make some allowance for increasing costs during the period between the date of preparation of the estimate and the probable tender date. This sum is often referred to as *design risks* and will be calculated on the basis of the likely time period and current price trends, including market conditions.

On occasions it will be necessary to assess the effect of price increases in materials on the estimated cost of a project, particularly where the rate of increase is substantial. In order to do this it is necessary to know the approximate breakdown of the cost of the project into labour and materials, and a further breakdown of the cost of materials so that the proportion by value of each major material or component to be used on the project can be assessed. This cannot be done with any great precision at the approximate estimate stage and would involve a large amount of work at the bill stage. It is therefore customary to use analyses already prepared for previous projects of a similar type. The reader may find the analyses in tables 9.9 and 9.10 useful as a general guide.

A further example may serve to illustrate the use of these schedules in assessing the probable cost effects of increases in the price of materials. A house is estimated to cost £95 000 and the following price increases have been notified. It is required to determine the probable effect of these increases on the estimated cost of the house.

Bricks + 6.8 per cent; cement + 7.4 per cent; timber + 3.6 per cent; paint + 4.5 per cent; electrical goods + 7.2 per cent; metalwork + 5.8 per cent.

The additional material costs will consist of

Bricks	6.8% of 6.3 = 0.43	
Cement	7.4% of 2.3 = 0.17	
Timber	3.6% of 5.6 = 0.20	
Paint	4.5% of 0.4 = 0.02	
Electrical goods	7.2% of 1.1 = 0.08	
Metalwork	5.8% of 1.4 = 0.08	
	Total	0.98 per cent

This represents a percentage addition of 0.98 on the contract. The probable total additional materials cost is £95 000 × 0.98 = £931.

Location Factors

The cost of a building is affected by its location, in addition to a whole range of design factors. The BCIS has identified a number of localised variables

Table 9.9 Analysis of groups of elements in typical school

Group of elements	*School (partly steel framed and partly brick construction) percentage*
Substructure	6.4
Superstructure	40.3
Internal finishes	8.2
Fittings	8.8
Services	25.2
External works	11.1
Total	100.0

Table 9.10 Weighted analysis of house and siteworks (operative date: 1 January 1994)

Craft operative	213.84
Labourer	193.06
Plumber and mate	59.54
Electrician	16.72
Sand and aggregate	19.72
Cement	22.98
Reinforcement	2.42
Membranes	1.84
Precast concrete	5.12
Bricks	63.05
Blocks	17.22
Roof tiles	14.23
Bitumen felt	8.03
Timber	55.76
Thermal insulation	4.35
Doors	7.11
Windows	14.66
Kitchen fittings	9.41
Ironmongery	2.55
Metal sundries	11.78
Rainwater and waste	2.91
Sanitary fittings	11.28
Copper tubes and fittings	7.59
Hot water cylinder	2.09
Cold water cistern	1.22
Boiler	8.82
Radiators	8.96
Plaster and plasterboard	11.38
Glazing	7.95
Paint	3.62
Drainage	14.27
Electrical goods	11.33
Site overheads	74.17
Office overheads	90.91
TOTAL	1000.00

Source: BCIS.

Table 9.11 Regional factors

Region	*Factor*	*Range*	*Standard deviation*	*Sample size*
Northern	0.97	0.71–1.56	0.11	220
Yorkshire and Humberside	0.96	0.72–1.40	0.10	252
East Midlands	0.92	0.70–1.35	0.10	239
East Anglia	0.97	0.75–1.36	0.10	192
South East (excluding Greater London)	1.03	0.78–1.49	0.11	821
Greater London	1.11	0.79–1.71	0.14	427
South West	0.94	0.64–1.21	0.09	322
West Midlands	0.94	0.65–1.21	0.09	290
North West	0.99	0.73–1.32	0.09	378
Wales	0.94	0.72–1.26	0.10	149
Scotland	1.09	0.74–2.17	0.15	346
Northern Ireland	0.75	0.58–1.00	0.08	50
Channel Islands	1.41	1.07–1.68	0.18	23

Source: *BCIS Quarterly Review of Prices* (December 1993).

which include market factors such as demand and supply of labour and materials, workload, taxation and grants, and the physical characteristics of a site – its size, accessibility and topography. Not even identical buildings built at the same time but in different localities obtain identical tenders.

While all these factors are particular to a time and place, certain areas of the country tend to have different tender levels to others. The location factors given in table 9.11, which is a condensed version of a BCIS table (1993) attempt to identify some of the general differences using information derived from the BCIS tender price index. The regions chosen are administrative areas and are not significant cost boundaries as far as the building industry is concerned. It is stressed that even within counties or large conurbations, great variations in tender levels are evident and that in many cases these will outweigh the effect of general regional factors. Although the BCIS publishes separate county factors, these have been omitted from table 9.11 to conserve space. However, the regional factors do show the range of locational price variations.

The statistical reliability of the regional and county factors is dependent on the number of projects included in each sample and the variability of the factors. A mean average based on a small number of widely varying figures is less reliable than one based on a large number of closely related figures. The factors are averages and therefore any individual project is unlikely to coincide exactly with the average rate, but the factors provide useful general guidelines. The mean, the range, standard deviation and sample size are given in each case.

An explanation of some of the more commonly used terms may be helpful to the reader.

Mean is the sum of the figures divided by the number of figures. It is the average price paid for the buildings in the sample.

Mode is the figure which occurs most often. It represents the most common price of buildings in the sample. The mean and mode figures provide indications of the 'average' costs of the buildings surveyed and therefore show the most likely cost for each type of building.

Range is the upper and lower figures of the sample.

Standard deviation is the square root of the mean of the squares of the deviations from the mean of the sample. This gives an indication of the dispersal of the figures within the range and around the mean. In approximate terms, assuming a normal distribution, 70 per cent of the figures will be ± 1 standard deviation of the mean, 95 per cent with ± 2 deviations and 99 per cent within ± 3. Therefore, a small value of standard deviation in relation to the value of the mean indicates a more reliable mean figure.

Sample size is the number of buildings of each type included in the survey. The higher the number in the sample, the more reliable are the results likely to be.

The following example will serve to illustrate the use of a county location factor in the preparation of a rough approximate estimate for a new project.

Rough approximate estimate of an advance factory of 2000 m^2 to be located in Derbyshire, forecast to go to tender during 2nd quarter 1995

Basis of estimate is C1/SfB 282.1 advance factory (mean 3rd quarter 1992) (BCIS cost analysis)	£254/m^2
Location factor (Derbyshire)	0.93
Cost/m^2 at 3rd quarter 1992	£236/m^2
Gross internal floor area (2000 m^2)	2000
Mean building price of advance factory, Derbyshire	£472 000
Allowances for any necessary adjustments and additions	–
	£472 000

Allowance for inflation to 2nd quarter 1995
BCIS tender price indices:
3rd quarter 1992 = 106
2nd quarter 1995 = 114 (forecast)

Adjustment – $£472\,000 \times \frac{114}{106} = £507\,623$

Rough approximate estimate at 2nd quarter 1995	£508 000

(answer is rounded off as it is very much an approximation with the forecast index for 1995 and the imprecise nature of the locational factor)

APPLICATION AND USE OF COST ANALYSES

The purpose of an elemental cost analysis is to show the distribution of the cost of a building among its elements or main functional parts in terms which are meaningful to both designers and building clients and so to permit the costs of two or more buildings to be compared. The cost analyses can be used to fulfil four main purposes.

(1) To enable designers and building clients to appreciate how the cost is distributed among the functional components of a building.

(2) To help designers and building clients to develop ways in which costs can be allocated to produce a better balanced design.

(3) To allow prompt remedial action on receipt of high tenders, by indicating the components where reductions in cost are possible.

(4) To assist in the cost planning of future projects by comparison of cost analyses.

In comparing cost analyses for different buildings, variations in the cost of specific elements between the different projects often have to be analysed and explained. If, for instance, internal doors cost £17.56/m^2 of gross floor area in one office project and £26.44/m^2 on another; then reasons would have to be sought for this considerable variation in cost. Possible reasons are that there are more internal doors in relation to floor area on the second project (quantity); that the doors are of higher quality in the second project (quality); and that the second project relates to a more recent tender in a period of rising prices and so all the prices in this analysis are higher than those for the first project (price level).

When adjusting element rates in a cost analysis of an existing building as a basis for a preliminary cost plan for a new project, it is necessary to consider each of these three factors. Adjustments for period price variations between the tender date of the previous project and the present day will apply to the rates for all the elements, and will probably be made on the basis of the appropriate BCIS tender price indices. Quantity and quality factors will vary from one element to another and each item will have to be considered individually.

For instance, supposing that the element cost of external walls in the analysis of the previous building was £52.45/m^2 and the wall/floor ratio was 1.120. The new building has a wall/floor ratio of 0.680 and it can be assumed that the wall construction will be similar in both cases. The external wall rate for the new building will be much lower as there is much less wall in relation to the enclosed floor area, and the wall/floor ratios can be used to determine the external wall rate for the new building.

New element rate for external walls =

$$£52.45 \times \frac{0.680}{1.120} = £31.84$$

A further example relates to internal doors

	Previous building	*Proposed building*
Floor area	12 000 m^2	30 000 m^2
Number of internal doors	25	80
Cost of internal doors/m^2 of gross floor area	£21.80	$£21.80 \times \frac{12\,000}{25} \times \frac{80}{30\,000} = £27.90$

Adjustments frequently have to be made to the elemental cost rates to take account of differences in quality. All-in unit rates of detailed cost analyses, priced approximate quantities or price books may be used individually or collectively to determine differences in cost stemming from variations in the quality of elements. For instance, assuming that the previous building contained 40 mm softwood doors and the proposed building is to have 40 mm veneered flush doors, then the doors in the new building would be about one-third more costly than those in the existing building and a further adjustment will be needed to the new internal door element rate

$$£27.90 \times \frac{4}{3} = £37.20$$

It is generally considered preferable to update element rates before adjusting them for quantity or quality. This sequence of events has the main advantage of producing, at the intermediate stage, a current element unit rate which could prove useful in preparing cost plans for other buildings. However, the reliable use of 'floor area/cost per m^2 rates' is often limited, particularly without the context specification level, procurement aims and procurement methods, as price differentials arise from these factors.

COST DATA

Type of Information Required at Different Stages of Design

The cost information required by designers differs in its form and general characteristics according to the stage of cost planning at which it is required. The cost information required at each stage is now considered.

Preliminary estimate. This is required at the inception of a scheme, or more probably at the feasibility stage. Only the broadest of information is available such as the location and type of building, approximate amount of accommodation required and possibly a general quality standard and an outline drawing. The estimate is likely to be prepared on a floor area basis or unit method by comparisons of cost data available for previous similar projects possibly subdivided into groups of elements. The data used will be obtained from within the quantity surveyor's own organisation and from published sources such as BCIS and various technical journals.

Comparative costings. These are necessary when alternative building designs or building techniques are under consideration. A variety of estimating methods and sometimes a combination of them are used – in particular, a comparison of costs of past buildings, floor area and approximate quantities are popular methods. The quantity surveyor's own cost records, priced bills and published cost data will form the principal sources of cost information at this stage.

Initial cost plan. The earlier figures must be confirmed as soon as possible since they are influenced considerably by plan, shape, storey height, use of building and other related factors.

Elemental approximate estimates operate on the basis of the synthesis of cost by functional elements (synthesis being the composition, putting together or building up of separate elements and is the opposite of analysis). The elemental rates may be computed from rates for the same functional elements extracted from cost analyses of previous projects and it will be advisable to use all-in unit rates for some elements, such as internal partitions, windows, internal doors, finishes and sanitary appliances, where the type and amount of the component parts can vary so widely. Adjustments must also be made for variations in design criteria, including soil conditions, loads and spans. Once again the quantity surveyor's own cost records, published data from a variety of sources, but particularly the BCIS and approximate quantities, will all play their part. On occasions it is good policy to use one method to check costs obtained by another.

Cost checks. As the working drawings are developed cost checks are made to ensure that the design is keeping within the cost budget. As the design is finalised, comparisons of smaller and smaller parts of the design in one or more alternative constructions may be required so that the architect can control the design within the overall target and give value for money. Alternative designs will require costing in more detail to compile a more precise cost plan. Whether these costs are submitted as a cost market of alternatives from which selections can be made, or whether they are used to make comparisons with elemental cost targets, they require careful and realistic assessment. In general, the measurement and costing of all-in rates for general contractor's work and specialist quotations for specialists' work will form the basis of the computation. The same sources of cost information will be relevant as in the previous stage. Finally, the priced bill of quantities from the accepted tender is analysed in the same form as the cost plan, both as a check of it and as the recorded analysis of tender cost. Final accounts are rarely used as a basis for cost analyses as the information is too dated.

It will be appreciated that to obtain reliable cost information from as many sources as possible entails standardisation of elements and cost analysis format, and this is being achieved through the use of the BCIS standard form of cost analysis. It is generally recognised that the proper purpose of cost analysis is to isolate and give a value to each element which performs the same function, irrespective of the building in which it appears. Only then can comparisons be made between one building and another in a meaningful way, and elemental targets have any real value.

One critique on *Building Cost Information* postulated that within the context of pre-contract estimating three stages can be defined, each of which required a different kind of cost data.

Order of cost estimate. The initial estimate of cost which may be prepared from schematic diagrams and area schedules. At this point the client wishes to know the range of cost which a particular project may incur, and this information may normally be wanted in hours rather than days.

Outline cost plan. An estimate prepared at initial sketch plan stage which indicates both an accurate total target cost which the client can reasonably insist upon his professional advisers to maintain, and a broad elemental division of this target sum for the guidance of architects and engineers. This will normally be prepared in a matter of days.

Full cost plan. Prepared in parallel with the preliminary working drawings, and giving an itemised breakdown and outline specification of the cost allocation to design elements. This will normally be prepared in a matter of weeks.

Sources of Cost Information

It might be helpful at this stage to consider some of the principal sources of cost information available to quantity surveyors.

Department of Education and Science. The Ministry of Education in the early nineteen fifties, faced with an urgent school building programme and a limited allocation of funds, formed a development group which, amongst other things, evolved a system of cost planning and control, details of which were published in Building Bulletin 4. This document laid down standards and provided cost and other information of considerable assistance to education authorities and their architects in the rational design and construction of many efficient and architecturally pleasing schools, and was revised and reissued in 1972 by the Department of Education and Science.

The Wilderness Group. The Wilderness Group, composed of ex-members of the RICS Junior Organisation Quantity Surveyors' Committee formed a Cost of Building Study Working Party which published a paper on the storey-enclosure method of approximate estimating in 1954, and this was followed by an investigation into building cost relationships of various design variables, which has proved to be of assistance in costing alternative designs of steel-framed buildings. The latter approach has been further developed and refined by Singh (1995) to assist in the cost estimation of the structures of commercial buildings.

Technical journals. The *Architects' Journal* commenced the publication of cost analyses for building projects in 1954, based on the Ministry of Education elemental approach. These articles provoked some criticism on the grounds that the information provided might be dangerously misleading in inexperienced hands. There can be no doubt that these analyses provided some useful information for architects and quantity surveyors and stimulated considerable interest in the costs of buildings. Other journals such as Building subsequently published information on building costs.

Building Cost Information Service. The RICS Quantity Surveyors' Cost Research Panel was formed in 1955 by the Council of the Royal Institution of Chartered Surveyors on a recommendation of the Quantity Surveyors' Committee. The Panel undertook a vast amount of research work into many aspects of building costs and the results of many of the investigations were the subject of papers in *The Chartered Surveyor*.

The Quantity Surveyors' Cost Research Panel was superseded by the Building Cost Information Service in 1965. This service is available to subscribing members who receive cost analysis forms which they complete and return as they are able for contracts selected by them as suitable. These returns when edited are published as part of the Building Cost Information Service (BCIS), to provide a library of cost data for comparison purposes, and are made available to subscribers. This service also includes valuable information on cost indices, average building prices, cost trends, location factors, rebuilding costs and other related matters. This constitutes useful background information and the cost analyses are of particular value where they cover special type buildings with which the quantity surveyor concerned has had no previous direct experience. The original BCIS brief and amplified cost analyses were subsequently replaced by concise and detailed cost analyses.

Sustaining central databanks is critical for all BCIS publications. Maintaining an up-to-date computerised databank for subscribers on-line requires more resources but provides a much better service from a central computer via a telephone line. The system can be used to produce elemental cost plans and cost analyses can be selected using various search criteria, such as building type, area and number of storeys.

Building Maintenance Information. This service which operated initially from the University of Bath was introduced in 1971 to assist its subscribers in obtaining value for money from efficient property occupancy and economic building maintenance. The service makes a regular distribution of up-to-date data on cost indices, labour and materials costs, maintenance techniques, legislation, statistics, publications, case studies, occupancy cost analyses, design/performance data, desk design appraisals and research and development papers. Further reference is made in chapter 13 to the work of this service, which now operates under the auspices of the BICS.

Housing and Construction Statistics. These are published quarterly by the Department of the Environment and provide a wealth of information on public sector building tender prices, output price indices, housing costs and prices, building workload, stock of dwellings, housebuilding performance, building society mortgage advances and many other related matters.

Other sources. Building price books provide useful information on current measured rates in London, together with other helpful data such as all-in or comprehensive rates and constants and costs of labour and materials. This data forms a useful guide but needs adjusting to local conditions.

Quantity surveyors can assemble cost data to good advantage. The office copy of the priced bill can have cube, floor area, storey-enclosure and similar rates inserted in it. The preliminaries bill can contain a full description of the construction of the building with its overall dimensions, cubic content, floor area, site restrictions, order of works and other peculiarities. Final accounts should be filed with a set of small scale drawings and the accepted estimates and final accounts of all the specialist subcontractors. The author also sees considerable merit in the preparation of cost analyses, amplified where possible, for each project passing through the quantity surveyor's office, as part of his own library of cost/price information.

Reliability of Cost Information

The main source of cost information, contractors' prices, is itself subject to considerable variation – from contractor to contractor, district to district, project to project and over periods of time. This does not invalidate its use but does mean that the user must interpret the cost information in its appropriate context and adjust it as necessary before reuse. Computing average costs from a number of projects has little relevance and it is better to base an assessment on a limited number of suitable schemes where extensive background information is available. It is necessary to know the conditions surrounding the prices which are to be used for cost planning, and how these conditions differ from those applicable to the new project.

Many factors should be considered when collecting and analysing cost data. These factors include the size of project; climate of building industry and relative keenness of tendering; stability of prices and availability of labour and materials; whether fixed price tender; special requirements as to speed or order of completion; method of pricing preliminaries; location of project; type of contract; and time of year when executed.

Much of the cost information used by the quantity surveyor is obtained from priced bills of quantities and it is advisable to use average prices for a particular project rather than those associated with the lowest tender. All quantity surveyors are fully aware of the wide variations in billed rates for even the most commonly encountered items of building work, sometimes referred to as the vagaries of tendering. One has only to examine the price per cubic metre for excavating a foundation trench not exceeding 0.30 m wide and 1.00 m deep or the price per metre for providing, laying and jointing 100 mm British Standard clay pipes, in half-a-dozen priced bills for the same project, to appreciate how great these differences can be. Strangely enough, although wide differences may occur in the rates submitted for individual billed items, the differences between the various tender figures are often quite small.

Skilled judgement is therefore necessary when extracting and using cost information from priced bills of quantities, obtained as a result of competitive tendering. Price differences in billed rates are not necessarily due to mistakes in estimating, but are more likely to stem from the use of different techniques in pricing and in performing the work on site. The treatment of preliminaries varies considerably between contractors; some cover them in the preliminaries bill, some spread them over the billed rates, while others adopt a combination of both approaches. If the preliminaries are spread then billed rates for like items are comparable. Preliminaries may however embrace much more than the standing overheads of the contractor's establishment and varying overheads such as foreman, use of plant and site huts. They may reflect major cost factors such as access conditions, restricted site working, phasing of the work, labour shortage and similar matters, and these are just the sort of items for which allowance should be made by the quantity surveyor. One of the major processes in cost planning is estimating the anticipated cost of a project, and separation of the cost of preliminaries is probably beneficial in this context.

Furthermore, Southwell (1971) described how the builder's price is the building owner's cost, and when we are using data provided by bills of quantities for cost analysis, it is cost to client in respect of the elements of design, rather than the cost to contractor, with which we are concerned. Southwell emphasised that the distinction between cost and price is not merely an academic terminological issue. The fact that there is a lack of coincidence between the unit of production cost and unit of design cost accounts for some part of the wide discrepancy between unit rates in bills of quantities. Other causes include the difficulty of predicting operational times, alternative operational methods, and all other situational aspects, such as site peculiarities, regional differences, tendering conditions and market levels. Southwell made a study of the reliability of the unit rate as an indicator of the true costs of the specification. He asserted that the unit cost is made up of three main components: firstly, standard costs within determinable parameters; secondly, additional costs due to specific design requirements; and thirdly, abnormal requirements due to factors which are indeterminate at forecasting stage, but which can be expressed as probabilities.

Cost data should never be transferred indiscriminately from a bill to an estimate. Some process of modification and adjustment is needed before it can be applied to other projects, and this involves both the study of the factors affecting the original pricing and of the characteristics of the new project. Allowances must be made for differences in tender dates, contract conditions, method of procurement, design team, regional differences, type of contractor, site conditions, availability of local labour and materials and any abnormalities in pricing. Indeed it is not always possible to determine whether billed rates have been subject to abnormal pricing, especially in the early 1990s when many tenders included merely minimal oncosts and profit margins.

Southwell (1971) also referred to the need to make adjustments for differences in tendering keenness and price levels generally. The only indication we have of the general keenness in tendering is the closeness of the tender figures. In some cases a very low tender is so exceptionally low that it throws doubt upon its use as a basis for cost analysis. For these reasons, probably the most reliable information for cost planning and cost investigation is that based on approximate quantities prepared for each design and priced by an experienced quantity surveyor, who will make allowance for relevant abnormal and local conditions. The pricing of approximate quantities should represent an intelligent forecast of the level of prices contractors may offer for that work. Furthermore, when estimating future costs in should be borne in mind that building costs sometimes increase at a faster rate than rises in price of labour and materials. This is difficult to explain but may be due to various factors such as higher oncosts and profit margins, higher additional payments to operatives, or even lower rates of output. While at other times when the market is extremely depressed, tender prices fall well below the levels anticipated as a result of rising labour and material costs. Some contractors in these circumstances appear willing to submit tenders with no profit margin in order to keep their organisations in business – a truly desperate situation.

Some quantity surveyors are opposed to the centralised processing of cost information as through the Building Cost Information Service (BCIS), using the argument that even detailed cost analyses are insufficient and that all the necessary background information can only be secured through close familiarity with a project and access to the contract documents. If this view is accepted then the only place for a library of cost information is within the quantity surveyor's own organisation.

There is little doubt that the quantity surveyor must be prepared to build up extensive cost records of projects passing through his office and to prepare cost analyses tabulated in a suitable form. Without this information he cannot prepare reliable estimates of cost of future projects or give the wide range of cost information which building clients require. Different classes of work call for different methods of cost planning and analysis, and quantity surveyors will often need to exercise considerable skill and ingenuity in developing systems which are particularly suited to the type of work in which they tend to specialise. Thus quantity surveyors engaged mainly on educational buildings are likely to use quite different costing techniques to those concerned with power station contracts. More data is needed to establish the relationship between a procurement method and the initial estimate compared with the final cost of a building, in order to identify the probability of price risk attached to each method of procurement. More research is also needed on the relationship between price certainty and the quality of production information made available to the construction team. In all cases, however, the cost planning service can be regarded as a natural extension of the quantity surveying function.

In order to provide the data needed in cost planning, a store of cost information must be built up from analyses of completed work, and there is benefit to be derived from presenting bills in a form that can be more readily analysed than the conventional arrangement by works sections, or from preparing bills by some process that permits a conventional presentation to be easily re-sorted for the purpose of analysis, which has been assisted by computerisation.

The increase in the use of non-traditional procurement routes has restricted the amount of tender price information that is available from the analysis of bills of quantities, based on firm design of the whole of a contract.

Cost Prediction

When a new building is being planned there are usually several different designs and/or construction types that could be adopted. The choice may be difficult if cost is an important factor in the decision, because the relative costs of the various options may change considerably during the period between planning and construction. Banks and Kroll (1980) devised a system of calculating the construction costs of a range of designs under a range of future economic conditions.

The method used a mathematical model which covered the costs associated with purchasing materials and erecting a house shell from them on a prepared site. Material costs were expressed as total requirements of labour, energy and transport by means of national input–output tables. Site costs and any prefabrication cost were expressed in terms of labour, plant, and profit costs and the cost of capital to finance the operation.

The model gave the total cost of each design in terms of a very small number of economic variables, principally labour and energy prices, profit levels and interest rates. A forecast of the future movements of this limited group of variables was sufficient to give an estimate of the future construction cost. The building materials incorporated were restricted to timber, bricks, cement, aggregate, concrete products, gypsum, glass, plastics, steel and aluminium.

As an illustration of the working of the model the calculation of the cost of materials is described. The material i has a quantity of $m(i)$ units and the quantities of energy, labour and transport making up one unit of material i are $c(i,1)$, $c(i,2)$ and $c(i,3)$ respectively, all at say 1990 prices. The energy, labour and transport price indices relative to 1990 are $x(1)$, $x(2)$ and $x(3)$ respectively and the profit level in the production of materials is p. Under these conditions the cost of sufficient material to build the house is:

$$c(i) = m(i).\{c(i,1).x(1) + c(i,2).x(2) + c(i,3).x(3)\}(1 + p).$$

This represents a genuine attempt to predict possible changes in building costs but will unfortunately still be subject to the vagaries of the market.

It has, however, to be accepted that building cost forecasting can only be based on predictions and that these are subject to error. Every effort will be made by the quantity surveyor to restrict the sources of error to a minimum. Past trends are not always a sound basis for future predictions and quantity surveyors will frequently need to apply considerable intuition and skill in forecasting future costs. The use of sophisticated formulae and mathematical models will not necessarily provide the correct solutions and it is often advisable to adopt several different approaches and to then compare the alternative outcomes.

Prediction errors can arise from incorrect assumptions and care should be taken to check the soundness of the assumptions used. Another possible source of error relates to sampling, where the size of the sample is too small or weighted too heavily in one direction. There is always a danger that errors may be cumulative, for example if two alternatives are being compared, the errors might increase the total estimate of one alternative while, at the same time, reducing that of the second alternative, thus compounding the cost differential. Fortunately, in many instances errors tend to be compensating so that the dangers are not so great as might appear at first sight.

The use of sensitivity analysis, as described in chapter 8, can help in this situation. Cost estimates are calculated on the basis of minimum and maximum values for each significant factor, such as interest rates, to provide a range of results. This approach will often show that the order of preference changes at some point in the range of each factor and concentrates attention on this very significant aspect. Preliminaries and profit on average on building projects dropped from 14 per cent in 1989 to 8 per cent in 1993.

The BCIS suggested that in 1993/94 contractors would have to raise their margins in order to produce profit from reduced workload, rather than buy turnover at a loss. Fixed price tenders which did not allow for the certain rise in materials prices that will accompany an upturn could prove to be economic suicide for contractors.

The first signs of a hardening in tender prices was considered likely to come from subcontractors, and this could stave off the spiral of decreasing margins leading to bankruptcies that otherwise could occur. The alternative situation would be to share work among fewer contractors with survivors obtaining the work that eliminated their competitors and forcing better margins out of clients, resulting in a chaotic marketplace.

Average Building Prices

The BCIS in its quarterly review of building prices (1993) analyses the tender prices of approximately 5000 buildings, encompassing some 150 different building types. The aim of the study is to show the variations in prices between buildings of different function and the range of prices which occurs for buildings of each functional type. The figures used are contract sums excluding external works and contingencies, with preliminaries apportioned by value. They are expressed in £/m^2 of the gross internal floor area. The figures are adjusted to constant current prices using the latest BCIS indices. It must, however, be borne in mind that prices can vary from one region to another and between individual sites, as can also the form of design and type of construction. The figures therefore, can only be a general guide to the level of building prices and skilful interpretation is always necessary in their application, as evidenced by the wide price ranges for buildings shown in table 9.12, with the highest often three to five times the lowest and in extreme cases there is a ratio exceeding 11 to 1.

A selected sample of average building prices is given in table 9.12, based on the third quarter of 1993, with supporting statistics showing the means, modes, ranges, standard deviations and sample sizes for each type of building listed. The prices for the third quarter of 1993 were, however, generally about 6 per cent below the comparable figures for the first quarter of 1992, emphasising once again the depressed state of the market in the early nineteen nineties.

A comparison of the costs of different buildings shows some surprising results such as the unit prices of public conveniences being up to double those of air conditioned offices, because of the high wall/floor ratio and large number of expensive appliances and plumbing work. The cheapest buildings listed are multistorey car parks with their open type construction. Other expensive buildings from the selected list in table 9.12 include law courts, restaurants, theatres, banks, building society branches and computer buildings, followed by office buildings, fire stations, police stations, hospitals, nursery schools, churches and chapels, and universities.

Table 9.12 Average building prices

CI/SfB	Building type	3rd Quarter 1993			Standard deviation £/m²	Sample size
		Mean £/m²	Mode £/m²	Range £/m²		
125	Multi-storey car parks	153	121	88–229	43	20
282.1	Advance factories, generally	227	207	113–644	69	278
282.2	Factories, purpose built, generally	330	214	123–1155	170	138
284	Warehouses/stores, generally	266	229	82–827	126	196
315	Local administrative buildings	521	508	236–778	121	34
317	Law courts	737	696	342–1099	176	33
320	Offices (steel framed)	584	451	120–1473	224	202
320	Offices (concrete framed)	673	537	326–1464	219	236
320	Offices (brick construction)	487	388	258–1192	150	141
320	Offices (air conditioned)	700	539	120–1343	243	179
320	Offices (not air conditioned)	503	480	258–1055	144	208
338	Banks/building society branches	746	686	280–1559	219	40
342	Shopping centres	368	323	154–738	156	35
344	Hypermarkets & supermarkets	431	339 & 643	85–824	181	70
345	Shops, generally	352	260	127–892	161	69
345.1	Shops with domestic, office accommodation	389	336 & 537	196–1037	151	39
372	Fire stations	637	611	302–1337	182	78
374	Police stations	675	635	373–1181	207	42
412	Hospitals, generally	689	603	274–1675	205	378
421	Health centres, generally	487	460	226–992	123	279
447	Old people's homes	473	431	270–977	126	134
512	Restaurants	746	749	375–1674	307	15
517	Public houses, licensed premises	570	445	332–1101	168	55
524	Theatres	756	–	513–1126	220	8
532	Community centres, generally	478	464	186–850	138	128
532.1	General purpose halls	501	510	193–978	160	47
562.12	Gymnasia/ sports halls	448	466	266–781	117	88
630	Churches, chapels	594	549	268–1084	166	74
711	Nursery schools/creches	611	590	303–1073	177	103
712	Primary schools	471	433	226–958	109	415
712.1	Middle schools	372	354	284–487	48	19
713	Secondary schools (high schools)	454	368	246–1562	178	111
714	Sixth form/tertiary colleges	530	480	239–905	178	16
721	Universities	623	442	297–1985	445	13
762	Public libraries	556	533	346–1140	141	78
766	Computer buildings, generally	884	648	534–1698	337	25
810	Housing, mixed developments	344	341	87–760	86	269
810.1	Estate housing, generally, 2 storey	313	294	163–690	81	555
810.1	Ditto., 3 storey	345	241	184–954	165	33
816	Flats, apartments, generally, 3–5 storey	408	379	182–925	119	241
820.1	'One off' housing, detached	529	372	186–2101	311	59
843	Sheltered housing, generally	396	359	208–837	96	585
852	Hotels	555	515	292–1115	164	34
916	Conference centres	515	–	319–877	203	6
941.1	Public conveniences	1043	1142	512–1789	307	38

Source: *BCIS Quarterly Review of Building Prices* (December 1993).

Cost Information Service for Engineering Systems

As long ago as 1965 the Heating and Ventilating Contractors Association authorised an investigation into the setting up of a cost information service for engineering systems for buildings. It was considered that 'the growing application of cost control techniqus to buildings, the greater sophistication of engineering systems and their ever-increasing proportion of building cost made it essential that the engineering services industry should have up to date and detailed information about costs upon which it can budget and cost plan'. The investigations have taken place and questionnaires covering the cost of engineering services were circulated to engineering service offices in 1970.

The main problems were described by Cox (1971) as the collection of insufficient data and, in some cases, lack of design information and incomplete price breakdown. It was envisaged that designers and designer-contractors, for whom the service was intended, should subscribe towards the operating costs. Cox (1971) believed that the cost planning of engineering services was the responsibility of the engineer and not the quantity surveyor. This attitude could prevent beneficial co-operation and co-ordination between the engineering and quantity surveying professions which seems so desirable in the sphere of cost planning. However, in the late nineteen seventies and nineteen eighties quantity surveyors became increasingly involved in the cost planning of mechanical and electrical services, and a number of specialist quantity surveying building services practices and educational courses were established. The RICS identified the need for the establishment of reliable feature comparisons and tender price indices for environmental engineering services by BCIS, which were scheduled for completion by 1995.

COST RESEARCH

One of the primary weaknesses in the building industry is undoubtedly that of communication. Cost planning improves direct communication between the building client, architect and quantity surveyor, and on occasions valuation and building surveyors also join the team. There is often, however, a wide gap between the designers and the contractor, as the contractor is frequently not appointed until the design is complete and the design team lacks the contractor's expertise on site problems and the practical application of constructional techniques. Hence information on important aspects such as the use of heavy plant and likely costs of new or unusual forms of construction may not be available. It has also been found that building clients often possess a considerable fund of knowledge of costs in use but that this is not always made available to the design team.

Co-ordinated programmes of research can assist in bridging this gap and in making the cost control mechanism more effective. Mention has already been made of the considerable volume of research undertaken by the Royal Institution of Chartered Surveyors mainly through BCIS and various Government departments, including the Department of the Environment and the Department of Education and Science.

Research shows that the contractors' tenders which subsequently form the quantity surveyors' cost data are generally accurate to about ±6 per cent. The quantity surveyor using such information is unlikely to achieve the same degree of accuracy. If prime cost and provisional sums are discounted, quantity surveyors' estimates are generally accurate to about ±12 per cent. A further factor which can distort cost predictions using historical data in a time of recession and/or high inflation, are decisions to be made concerning increased costs. The contractor's estimating procedure is often very different to the quantity surveyor's techniques for preparing a cost plan and this will account for some of the discrepancies which occur between cost plans and tenders. More research is needed in this important area.

Flanagan and Norman (1982) examined the range of bids on a sample of 129 projects. The bidding range was expressed as a percentage by calculating $\frac{\text{high bid} - \text{low bid}}{\text{low bid}} \times 100$. The results showed a very wide bidding range, generally in excess of 20 per cent. The coefficient of variation, which expresses the standard deviation as a percentage of the arithmetical mean, showed how variable are the data which are summarised in the calculation of the mean.

Two distinct groupings emerged in the sample exhibiting very different tendering characteristics – projects up to £2m and those over £2m. In the first grouping the majority of bids tended to be much closer to the lower bid, the first mode was in the range 0.2 to 0.3, and the median was approximately 0.3. For projects above £2m value, the first mode lay in the range 0.3 to 0.4 and the median was approximately 0.45.

It is recognised that tendering behaviour is likely to be affected by seven principal factors:

(1) size and value of project and construction or managerial complexity required to complete it;
(2) regional market conditions;
(3) current and projected workload of the tenderer;
(4) type of client;
(5) type of project;
(6) method of procurement; and
(7) composition of the design team.

The quantity surveyor is advised to treat a low bid with some caution when using it as the dominant source of price data for estimating, as it will often be a poor indicator of true price. The quantity surveyor should take into account bidding range, the relationship between the lowest and second and third lowest bids, and all the relevant information on the historical project being used as the data base.

Computers are being used increasingly to assist in the assembly of cost data, and this is well evidenced by a study undertaken at the University of Hong Kong in 1980. This particular study considered aspects of the construction industry in Hong Kong with particular reference to construction costs, resources and the capacity of the industry. A technique was developed for the rapid analysis of a construction project into its component resources such as materials, labour and plant, and the corresponding costs. A unit rate synthesis matrix (168 × 99) was first built into a computer program, incorporating 168 types of work items common to building projects and 99 different material, plant and labour resources. Items of work in a bill of quantities for any building were then matched with the selected work items in the matrix and the corresponding quantities obtained. Once data on quantities and resource prices were fed into the computer, the program was able to produce a complete analysis of resources and cost. Twelve multistorey Hong Kong building projects were analysed in this way. The program can be extended to produce cost indices, elemental cost analyses, cost predictions for new designs, cost plans, resource programs and cost control of building work on site.

A RICS QS working party (1990) found that:

(1) The collection and analysis of data was an expensive activity. Some of the larger organisations that employ quantity surveyors were devoting considerable resources to cost research, data collection and comparison. It was almost certain that each organisation used a slightly different format for recording and comparing its data, thereby making inter-organisation comparisons more difficult. It was probable that smaller organisations did not have either the resources or access to sufficient information to be able to carry out their own data collection and/or research.

(2) The number of analyses prepared and submitted to BCIS was relatively small, particularly for certain building types, such as prime location and/or high quality office buildings and high-tech industrial facilities. There was an almost total lack of prestige projects in all sectors, principally because subscribers to BCIS and/or their clients appeared reluctant to provide such confidential information. By comparison, publications such as *Building* and the *Architects' Journal* had greater success in featuring high profile projects. A basic shortage of tender price information tended to restrict the usefulness of BCIS for cost planning and cost/price prediction, other than in general terms. Early cost estimating remained the area most commonly needed by surveyors for cost advice on a project by project basis and is an area where further work could usefully be undertaken.

One of the major shortcomings of a cost analysis prepared by a third party was the validity of the allowances made for the work covered by prime cost and provisional sums. This highlighted the need to provide information on specialist features such as curtain walling, cladding, lifts and engineering services.

(3) Insufficient numbers of quantity surveyors had consistently analysed data which was suitably classified, and could not therefore demonstrate the validity of their data, particularly in the early stages of projects.

The working party (1990) concluded that:

(1) The purpose of collecting and analysing cost/price data is principally to improve price prediction, as part of a supporting function in the management of quality, time and cost of projects. The availability of sufficient reliable cost/price data should improve general communication on projects, establish cost parameters and a framework for cost control and measuring/monitoring the scale of risk of projects. However, cost is only one factor in the effectiveness of an 'information manager' who would aim to manage quality, time and resources as these, when combined, incur 'cost' and result in 'price' to a client. Data collection is expensive and its relative usefulness should continually be reviewed. It cannot be overemphasised that quantity surveying clients continue to require improved cost/price advice.

(2) Cost/price information that is generated outside an organisation that seeks to use it is often viewed as less satisfactory than that which is generated internally. This is because familiarity with information usually provides a background to its neutrality or 'blandness' and helps in making the judgements that are usually required for the skilful use of the information. Continual review and assessment of definitions of data, and of data structures, is needed to provide a framework for analysis and to give confidence in the use of 'commonly available' data.

(3) The reliable use of 'floor area/cost per m^2 rates' is often very limited, particularly where the context, specification level, and procurement aims and methods are not known by the user of the information, as there are price differentials that arise from these factors.

The working party (1990) identified the following items as needing further exploration:

(1) The establishment of an integrated data base that will permit data at a 'coarse' grain to be compared with levels of increasingly 'fine' grain data.

(2) The degree of price predictability that it is possible to achieve, given the amount of information available at concept/feasibility stages. The BCIS subscribers questionnaire showed that the most urgent development required was 'developing tender forecast techniques'.

(3) The uses made of cost/price prediction within the management of time and cost by project managers and the design and construction team.

(4) The degree to which cost control procedures, as opposed to cost monitoring, are currently used within private and public procurement, taking into account different procurement methods.

(5) The establishment of any relationship between a procurement method and the initial estimate compared with the final cost of a building, in order to indicate the probability of price risk attached to each method of procurement.

(6) The relationship between the quality of the brief, initial cost at feasibility stage and at project completion, taking into account construction capital cost and all aspects of project funding, benefits/penalties of the procurement method adopted.

(7) The establishment of any relationship between price certainty and the timing and form of the client's consultant team's appointments.

(8) The relationship between price certainty and the quality of production information made available to the construction team.

(9) The opportunities that computer aided design (CAD) may offer in improving the analysis and prediction of cost.

(10) The relationship of cost/price data to integrated data bases in general.

The working party (1990) believed that the following developments were very desirable and should be capable of implementation by 1995/96:

(1) The establishment of a central source of 'feature' and 'unit cost' cost/price information, available to the profession generally, that allows comparison of major construction and environmental engineering elements, irrespective of building type. Cost/price information would require analysis in accordance with standard criteria of functional/specification/performance indicators and would reflect the procurement method used,

all provided within the framework of integrated data provision. The surveying profession needs up-to-date information, often not available from the analysis of tenders, that reflects locational and market conditions and provides cost information on 'features'.

(2) The improvement of concise cost/m^2 analyses, amplified by more contextual information, as for example on procurement methods.

(3) The establishment of reliable 'feature' comparisons and tender price indices for environmental engineering services, as described earlier in the chapter.

(4) The establishment of cost/price comparisons in the European context, both for building types and for major construction and environmental engineering services elements.

(5) The establishment of life cycle costing data for operating costs, annual and intermittent maintenance, replacement and alteration costs, and the expansion of cost data on rehabilitation schemes.

(6) The provision of additional on-line software to make the programs quicker and more user friendly and the supply of BCIS on-disk as an alternative to down-loading on-line.

10 PRACTICAL APPLICATION OF COST CONTROL TECHNIQUES

In this chapter cost planning techniques are applied to a variety of practical situations involving the computation of building costs for a range of building projects.

EXAMPLE 10.1

A preliminary estimate of cost is required for a two-storey factory building to be erected in a small East Anglian town, as an extension to an existing light engineering factory. It is envisaged that work on site will commence in September 1994 and that the contract period will be twelve months. A cost analysis on the BCIS concise analysis format of a West Midlands factory (table 10.1) is to be used as a basis for the estimate, as the form and construction of the proposed factory is likely to be very similar.

The following factors are to be taken into account in the preparation of the estimate.

(1) The location of the new factory is rather remote and it is anticipated that the contractor's employees will have to travel 30 km to the site. There is a shortage of contractors in the region due to extensive development in progress.

(2) The site is flat and of filled ground of many years standing, but with a low bearing value.

(3) The external dimensions of the new building are to be 80 m × 20 m and the storey heights 5 m on the ground floor and 3.75 m on the first floor (West Midlands factory had storey heights of 4.8 m and 3.6 m).

(4) External works are estimated at £115 000 and fittings at £7500.

(5) No office accommodation is required as the present accommodation is adequate.

(6) Wall/floor ratio of West Midlands factory is 0.522.

Solution

The first step is to assess the probable cost effects of the different features of the two projects.

Time and locational factors. The BCIS building cost index for steel-framed buildings for the third quarter of 1992 was 147, and the price index at September 1994 (forecast) is 155, the small variation of 5 per cent being due to the intervening recession. Probably a more realistic approach in the 1990–93 recession would be to use tender price indices instead of building cost indices, as tender prices did not follow building costs, as shown in chapter 9. The locational factors are 0.95 for the West Midlands and 0.99 for East Anglia giving an increase of 4 per cent. Further adjustment is also needed to take account of the lack of competition in the area and the greater distance that building operatives will have to travel to the site, resulting in increased travel costs, which could be in the order of 9 per cent. An overall weighting of 18 per cent would seem appropriate to meet these additional costs.

Substructure. It can be reasonably anticipated that the ground floor will need to be a fully suspended reinforced concrete slab and that stanchion bases will require support from piles about 6 m long. It would therefore be prudent to increase the substructure cost to £40/m^2 of gross floor area.

Table 10.1 Cost planning example 10.1 – concise cost analysis of factory

BCIS Code: A – 2 – 3491

Project Tender Price Index 88 Base: 1985 BCIS Index Base
Indices used to adjust to base date 100 116

CI/SfB
282.12
Advance Factories/Offices – mixed facilities (class B1) – 28

Total Project Details

BCIS On-line Analysis No. 10729

Job title: Industrial Development
Location: Stoke on Trent, Staffordshire
Client: Private Client
Date for receipt of tender: 31-Aug-90 Date of Tender: 21-Aug-92
Contract period (months) – stipulated: 6 offered: – agreed: 6
Type of contract: JCT private contract 1980 edition
Fluctuations: Firm
Selection of contractor: Selected competition
Number of tenders – issued: 8 received: 8

Measured Work:	467 274
P.C. Sums:	479 275
Provisional Sums:	35 000
Preliminaries:	19 886
Sub-total:	1 001 435
Contingencies:	25 200
Contract sum:	1 026 635

Analysis of Single Building

No. of storeys: 2
Gross floor area: 3491 m^2
Functional unit: 2859 m^2 usable floor area

Type of construction: 2 storey factory with offices, component assembly and warehousing. RC pad and strip foundations, ground slab and stairs; PCC upper floor. Steel frame. Brick and profiled cladding to external walls and roof. PC sum for windows. Block internal walls, WC partitions. Flush internal doors. Fair faced walls, plaster to offices; PVC or carpet to floors; suspended office ceilings. Fittings. PC sums for gas heating, electrics, ventilation, plumbing and sanitary services. Fire alarms. External works include macadam paving, timber fencing, services and drainage.

Element	*Total cost of element*	*Cost per m^2*	*Total cost of element inc. prelims.*	*Cost per m^2 inc. prelims.*	*Cost per m^2 inc. prelims at 1985, UK mean location*
Substructure	104 471	29.93	106 588	30.53	26.34
Superstructure	434 705	124.52	443 512	127.04	109.59
Internal finishes	95 870	27.46	97 812	28.02	24.17
Fittings	9 234	2.65	9 421	2.70	2.33
Services	205 955	59.00	210 127	60.19	51.92
Building sub-total	850 235	243.55	867 460	248.48	214.35
External works	131 314	37.62	133 975	38.38	33.11
Preliminaries	19 886	5.70	–	–	–
TOTAL (less contingencies)	1 001 435	286.86	1 001 435	286.86	247.46

Superstructure

$$\text{gross floor area} = 2 \times (80.000 - 510) \times (20.000 - 510)$$

where 510 is twice the external wall thickness.

$$= 2 \times 79.490 \times 19.490$$
$$= 3099\ \text{m}^2$$

$$\text{external wall area} = 2/(80.000 + 20.000) \times (5.000 + 3.750)$$

$$\text{(measured on external face)} = 2/100.000 \times 8.750 = 200.000 \times 8.750$$
$$= 1750\ \text{m}^2$$

$$\text{wall to floor ratio} = \frac{1750}{3099} = 0.565$$

Hence the wall to floor ratios of both buildings vary considerably resulting from the less efficient plan shape of the new building coupled with the increased height of the external walls. The external walls cost will increase because of the higher wall/floor ratio, and the cost of the frame and internal walls will increase due to the rise in storey heights. The external walls will increase by 0.565/0.522 = 8.2 per cent. The percentage increase in storey heights is 8.750/8.400 = 4.2 per cent and the frame, internal walls and wall finishes should be increased proportionately. The frame, internal walls and partitions and stairs will account for about 60 per cent and the external walls about

25 per cent of superstructure costs, and wall finishes about 20 per cent of internal finishes (based on an examination of available comparable detailed analyses). Hence increases in costs/m^2 will be:

$$\text{Superstructure} - £127.04 \times \frac{60}{100} \times \frac{4.2}{100} = £3.20$$

$$\text{and} \quad £127.04 \times \frac{25}{100} \times \frac{8.2}{100} = \frac{£2.60}{£5.80}$$

$$\text{Internal finishes} - £28.02 \times \frac{20}{100} \times \frac{4.2}{100} = £0.24$$

External works are estimated at £115 000, fittings at £7500 and services at £180 000, and the costs/m^2 of gross floor area will be:

$$\text{external works} = \frac{£115\,000}{3099} = £37.11$$

$$\text{fittings} = \frac{£7500}{3099} = £2.42$$

$$\text{and services} = \frac{£180\,000}{3099} = £58.08$$

It is now possible to build up a total building cost/m^2 of gross floor for the new factory and to adjust this total by an addition of 1 per cent to cover time and locational factors.

Groups of elements	*Cost/m^2 of gross floor area* (including preliminaries)
Substructure	40.00
Superstructure (£127.04 + £5.80)	132.84
Internal finishes (£28.02 + £0.24)	28.26
Fittings	2.42
Services	58.08
External works	37.11
Total	£298.71

Total building cost/m^2 based on cost analysis of previous factory	£298.71
add 18 per cent (time and locational factors)	53.77
Adjusted current rate	£352.48
Estimated cost of new factory = 3099 × £352.48	£1 092 336
add contingencies (2.50%)	27 308
Total estimated cost	£1 119 644
Rounded off to	£1 120 000

The use of a BCIS concise cost analysis as a basis for the preparation of an approximate cost estimate for a new building of similar type and function, illustrates the problems encountered, which result mainly from the very limited information supplied on this form of analysis. A considerable proportion of all BCIS cost analyses are of this type, subdividing the costs into the six main elemental groupings and showing these costs related to the m^2 of gross floor area (BCIS 1992b).

In order to make the comparative cost study in Example 1 more meaningful, data has been included on the wall/floor ratio and the storey heights of the existing building which are not included on the concise cost analysis. Some of the calculations have also been rendered more difficult than would have been the case if a detailed cost analysis had been used as the basis for the example. It was, however, decided to use a concise cost analysis in this instance to highlight the difficulties which would arise in practice. In general concise cost analyses provide broad cost comparisons of buildings to show the cost ranges and varying distribution of elemental group costs, and they do provide a useful service within these rather narrow parameters. Example 2 illustrates the use of a BCIS detailed cost analysis to perform a similar function on a rather more complicated cost study.

EXAMPLE 10.2

The second cost planning example covers the preparation of a first cost plan for a four-storey block of flats, to be built in London, from sketch designs which have reached quite an advanced stage of development (figure 10.1), and with the help of a detailed cost analysis (table 10.2) and drawings (figure 10.2) relating to a three storey block of flats in Cornwall. Basic information and brief specification notes covering the new project follow.

Basic Information

The block of flats is to be built for a housing trust on a steeply sloping site on the outskirts of London. It is anticipated that construction will commence in early June 1994 and a contract period of twelve months will be specified with competitive firm price tenders. The proposed accommodation consists of fourteen dwellings made up of one-bedroom flats and bedsitters, with access from a central common landing, and with six garages on the ground floor.

Brief Specification

Substructure: Concrete strip foundations with cavity brickwork; 150 hardcore, 50 concrete blinding, damp-proof membrane and 100 reinforced concrete bed.

Superstructure
Upper floors: 225 reinforced *in situ* concrete.

Roof: Built-up felt roofing and pre-screeded wood-wool slabs on 50 × 175 and 75 × 225 softwood joists. PC sum of £1100 for rooflights.

Stairs: reinforced *in situ* concrete; felt-backed PVC treads, risers and strings; mild steel rail with hardwood handrail and kneerail.

External walls: 255 brick and block cavity wall faced externally.

Windows and external doors: Softwood standard windows with clear glass and 50 hardwood glazed door and entrance screen (2 nr); up-and-over metal garage doors with timber frames.

Internal walls and partitions: One-brick walls and 100 and 75 lightweight concrete block partitions.

Internal doors: 35 hollow core flush doors in softwood linings (52 nr).

Internal finishes

Wall finishes: 12 gypsum plaster to brick and block walls and emulsion paint; ceramic tiles (50 m^2).

Floor finishes: Felt-backed PVC paving.

Ceiling finishes: 12 gypsum plaster and emulsion paint.

Fittings and furnishings

Fittings: Fitted kitchens (14 nr), wardrobe fronts (28 nr) and sundry shelves.

Services

Sanitary appliances: PC sum of £10 500 for appliances.

Disposal installations: PVC rainwater goods, wastes, overflows and soil stacks.

Water installations: Copper tubing, 98 nr draw-off points, immersion heaters and tanks of 3600 litres capacity.

Heat source: PC sum of £14 500.

Space heating: Ducted heating system

Ventilation systems: PC sum of £2200.

Electrical installations: PC sum of 32 000.

Builder's work in connection with services: Normal builder's work in connection with service installations.

(*continued on page 256*)

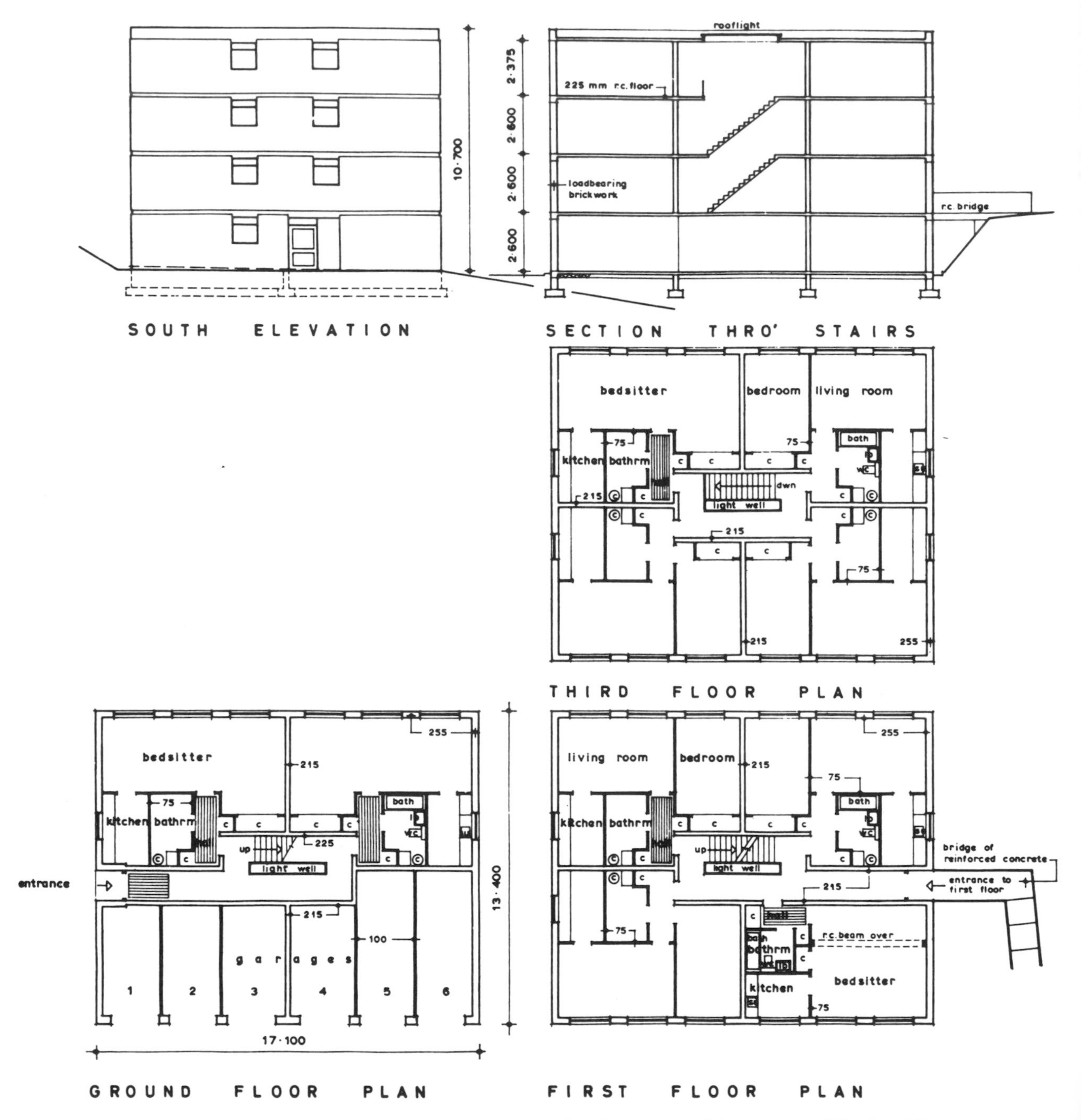

Figure 10.1 Cost planning example 10.2: sketch design of four-storey block of flats

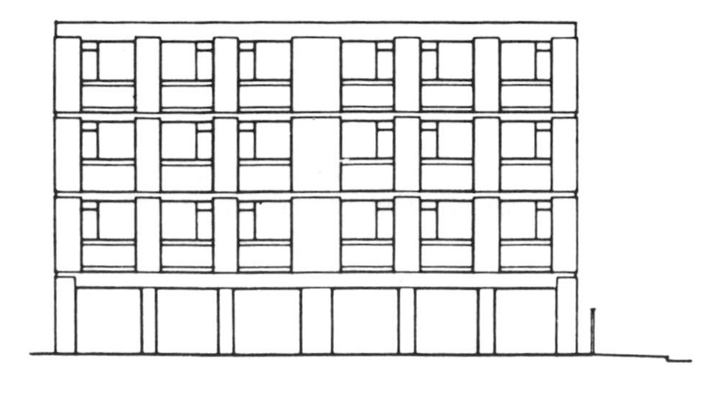

EAST ELEVATION

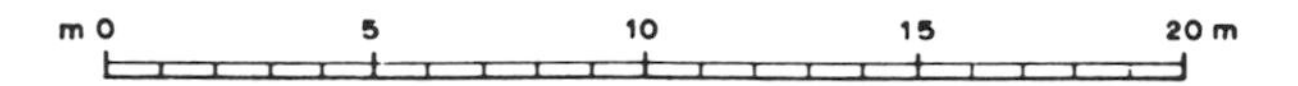

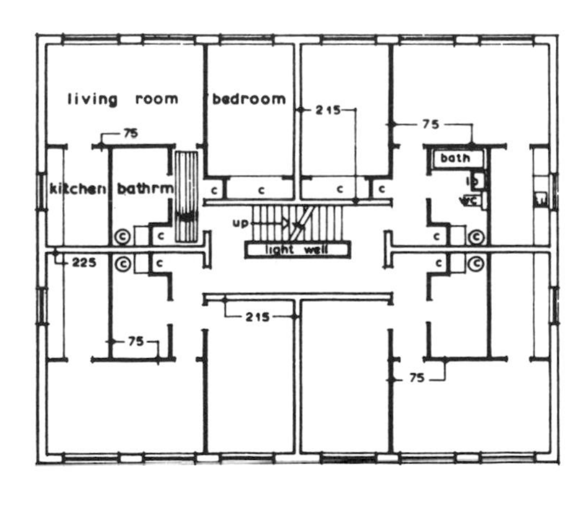

SECOND FLOOR PLAN

Figure 10.1 (continued)

Table 10.2 Cost planning example 10.2 – detailed cost analysis of block of flats (3 storey)

Job title: Flats
Location: Cornwall

Client: Local authority
Tender date: 14 November 1990

INFORMATION ON TOTAL PROJECT

Project and contract information

Project details and site conditions:
9 nr self-contained one bedroom flats having a gross floor area of 499 m^3 in a three-storey block, built of load-bearing brick walls, reinforced concrete first and second floors and tiled pitched roof, on a level site 800 m from the city centre.

Market conditions:
Conditions were competitive and there was an adequate supply of labour and materials.

Type of contract: JCT Standard Form (L.A. Edition): firm price
Selection of contractor: Selected competition

Tender documentation: Bill of quantities

Contract period		
	stipulated by client	12 months
	offered by builder	12 months
	agreed	12 months
Number of tenders issued	6	
Number of tenders received	6	

Competitive tender list
£
262 668
271 456
276 237
285 821
292 440
298 365

ANALYSIS OF SINGLE BUILDING

Design shape information

Accommodation and design feature: 9 nr self-contained flats in three-storey block. All flats are one bedroom flats. Block size – 21.440 × 8.130 × 9.750 to ridge with one projection on each long face.

Areas		
Basement floors	–	m^2
Ground floor	171	m^2
Upper floors	328	m^2
Gross floor area	499	m^2
Usable area	410	m^2
Circulation area	62	m^2
Ancillary area	–	m^2
Internal division	27	m^2
Gross floor area	499	m^2
Floor spaces not enclosed	–	m^2
Roof area	206	m^2

Functional 18 nr persons

$$\frac{\text{External wall area}}{\text{Gross floor area}} = \frac{410}{499} = 0.82$$

Internal cube = 1260 m^3

Storey heights
Average below ground floor – m
at ground floor 2.600 m
above ground floor 2.600 m and 2.375 m

Design/Shape

Percentage of gross floor area:	
Below ground floor	– %
Single-storey construction	– %
Two-storey construction	– %
3-storey construction	100 %

Brief Cost Information

Measured work	£155 847		
Provisional sums	£6 598		
Prime Cost sums	£69 885		
Preliminaries	£18 338	being 7.89%	of remainder of contract sum (less contingencies)
Contingencies	£12 000		
Contract sum	£262 668		

Functional unit cost excluding external works at tender date: £11 770 per nr of persons

(continued)

Table 10.2 (continued)

ELEMENT COSTS

Gross internal floor area: 499m² — Tender date 14 Nov. 1990

	Element	*Preliminaries shown separately*			*Preliminaries apportioned*		
		Total cost of element £	*Cost per m² gross floor area* £	*Element unit quantity*	*Element unit rate* £	*Total cost of element* £	*Cost per m² gross floor area* £
1	SUBSTRUCTURE	£14 581	£29.22	171 m²	85.27	£15 728	£31.52
2A	Frame	–	–	–	–	–	–
2B	Upper floors	13 648	27.35	344 m²	39.67	14 725	29.51
2C	Roof	16 077	32.22	237 m²	67.84	17 346	34.76
2D	Stairs	7 031	14.09	–	–	7 585	15.20
2E	External walls	22 914	45.92	310 m²	73.92	24 720	49.54
2F	Windows and external doors	13 079	26.21	99 m²	132.11	14 112	28.28
2G	Internal walls and partitions	16 073	32.21	555 m²	28.96	17 340	34.75
2H	Internal doors	11 123	22.29	122 m²	91.17	12 001	24.05
2	SUPERSTRUCTURE	£99 945	£200.29			£107 829	£216.09
3A	Wall finishes	19 167	38.41	1207 m²	15.88	20 679	41.44
3B	Floor finishes	11 582	23.21	456 m²	25.40	12 495	25.04
3C	Ceiling finishes	6 297	12.62	436 m²	14.44	6 796	13.62
		£37 046	£74.24			£39 970	£80.10
4	FITTINGS	£9 141	£18.32	–	–	£9 865	£19.77
5A	Sanitary appliances	5 609	11.24	–	–	6 053	12.13
5B	Services equipment	–	–	–	–	–	–
5C	Disposal installations	2 734	5.48	–	–	2 949	5.91
5D	Water installations	9 780	19.60	–	–	10 554	21.15
5E	Heat source	8 044	16.12	–	–	8 678	17.39
5F	Space heating and air treatment	5 349	10.72	–	–	5 773	11.57
5G	Ventilating system	–	–	–	–	–	–
5H	Electrical installations	14 891	29.84	–	–	16 068	32.20
5I	Gas installations	1 437	2.88	–	–	1 552	3.11
5J	Lift and conveyor installations	–	–	–	–	–	–
5K	Protective installations	–	–	–	–	–	–
5L	Communication installations	–	–	–	–	–	–
5M	Special installations	–	–	–	–	–	–
5N	Builder's work in connection	3 293	6.60	–	–	3 553	7.12
5O	Builder's profit and attendance	–	–	–	–	–	–
5	SERVICES	£51 137	£102.48			£55 180	£110.58
	BUILDING SUB-TOTAL	£211 850	£424.55			£228 572	£458.06
6A	Siteworks	12 575	25.20	–	–	13 568	27.19
6B	Drainage	3 633	7.28	–	–	3 917	7.85
6C	External services	1 677	3.36	–	–	1 811	3.63
6D	Minor building works	2 595	5.20	–	–	2 800	5.61
6	EXTERNAL WORKS	£20 480	£41.04			£22 096	£44.28
7	PRELIMINARIES	£18 338	£36.75			–	–
	TOTALS (less Contingencies)	£250 668	£506.34			£250 668	£502.34

Table 10.2 (continued)

		SPECIFICATION AND DESIGN NOTES
1	SUBSTRUCTURE	Concrete strip foundations; concrete block walls; 150 hardcore bed; 150 concrete bed.
2B	Upper floors	150 reinfored concrete first and second upper floors.
2C	Roof	Pitched tiled roof of concrete interlocking tiles on battens and felt (226 m^2); timber roof construction with roof trusses. 150 reinforced canopy to front entrance covered with 25 thick, two coat mastic asphalt (9 m^2).
2D	Stairs	Precast concrete steps faced with terrazzo on exposed surface and with mild steel balustrades.
2E	External walls	Hollow walls of Lungfield multi rustic facing bricks and Lignacite solid thermal blocks. Vertical tile hanging (45 m^2).
2F	Windows and external doors	Standard metal windows to wood surrounds with clear sheet glass (78 m^2), and fixed lights with float glass (10 m^2), including painting and ironmongery. Softwood glazed screen incorporating 1 nr 50 thick panelled softwood glazed door (11 m^2), including painting and ironmongery.
2G	Internal walls and partitions	Brick hollow wall (37 m^2), Lignacite block hollow walls (65 m^2) one brick wall faced in Dapple Light facings one side (58 m^2), half brick walls (133 m^2), 100 Lignacite block walls (210 m^2), 75 Lignacite block walls (52 m^2).
2H	Internal doors	Sapele faced flush doors (72 nr), hardboard ½ hour fire check flush doors (9 nr), including painting and ironmongery.
3A	Wall finishes	Cement and sand render (41 m^2) gypsum plaster (1166 m^2) emulsion paint (1150 m^2).
3B	Floor finishes	Granolithic (13 m^2), vitrified tiles (42 m^2), vinyl asbestos tiles (401 m^2).
3C	Ceiling finishes	5 thick gypsum plaster on plasterboard (146 m^2), 15 thick gypsum plaster to concrete base (290 m^2), emulsion paint (436 m^2).
4	FITTINGS	Fitted kitchens (9 nr), sundry shelves and fittings.
5A	Sanitary appliances	PC sum of £10 500 for sanitary appliances.
5C	Disposal installations	PVC rainwater goods, wastes, overflows and soil stacks.
5D	Water installations	Copper pipework to BS 659, 54 draw-off points, instantaneous water heaters and combination hot water storage units.
5E	Heat source	Gas fired warm air units (9 nr).
5F	Space heating	Sheet metal ducting and asbestos cement flue pipes.
5H	Electrical installations	PC sum of £32 000 for 18 nr consumer distribution points, 99 nr socket outlets, 61 nr lighting points, 9 nr TV outlets.
5I	Gas installations	27 nr draw-off points.
5O	Builder's work in connection	Normal builder's work in connection with service installations.
6A	Siteworks	Retaining and screen walls, fencing, pavings, and parking bay.
6B	Drainage	Surface water: 100 dia. PVC-U pipes to 4 nr soakaways. Foul drains: 100 dia. PVC-U pipes and 4 nr brick manholes.
6C	External services	Water and electric mains.
6D	Minor building works	2 nr detached stores and lines enclosure.
	PRELIMINARIES	7.89% of remainder of contract sum (excluding contingencies).

(continued from page 251)

External works

Sitework: PC sum of £35 000 to cover access road, parking areas, paths, landscaping and fencing.

Drainage: PC sum of £7500.

External services: PC sum of £4500 for water and electric mains.

SOLUTION

Time and Locational Factors

In making comparisons with element rates in the cost analysis of the three-storey block of flats, allowance must be made for the 3½ years difference in contract dates (November 1990 to June 1994), any regional price differences and any likely variation in the market situation.

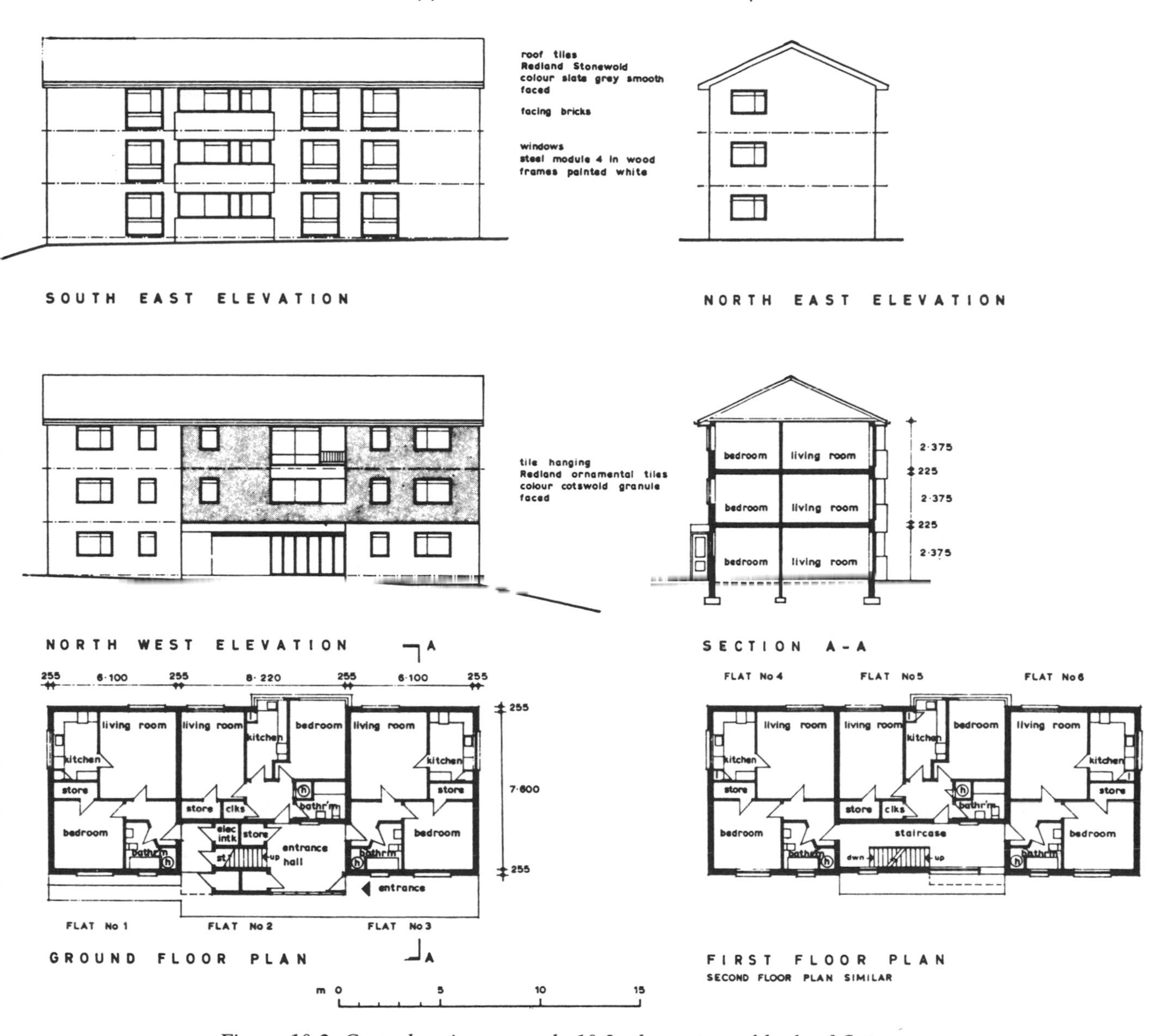

Figure 10.2 Cost planning example 10.2: three-storey block of flats

The tender price index for brick buildings is 121 for November 1990 and is estimated at 110 for June 1994. An examination of the BCIS regional and county factors indicates that building prices in greater London are likely to be about 22 per cent higher than in Cornwall (regional factors were 1.11 and 0.91 respectively in December 1993). The market situation in the south-west is quite keen as there is a shortage of work in the region, and there is also ample evidence of very keen competitive tendering in outer London with very small differences between tenders. It does not therefore seem that any adjustment of prices would be needed to cover varying market conditions. The total percentage addition to the tender rates for the existing block of flats would be 22 − 9(110/121) = 13 per cent.

Quantity Factors

Gross floor area

	Length	*Width*
	17.100	13.400
less twice external wall thickness 2/255	510	510
	16.590	12.890

gross floor area = 4 × 16.590 × 12.890 = 855 m^2 made up of 213.75 m^2 on each floor.

Roof area. The roof area is measured to the inside face of the parapet walls and will also be 213.75 m^2.

External walls. The external wall area is measured on the outer face of the external walls without any adjustments for windows or doors. The perimeter girth is 2 × (17.100 + 13.400) × 10.700 = 652.7 m^2. The wall to floor ratio =

$$\frac{\text{external wall area}}{\text{gross floor area}} = \frac{652.7}{855} = 0.763$$

Windows. The total window area calculated from the plans and elevations is 174.3 m^2, and the window to floor ratio is 174.3/855 = 0.204.

Internal load-bearing walls

Lengths:	
ground floor	= 43.500
first floor	= 45.700
second floor	= 37.800
third floor	= 37.800
total length	= 164.800

Area of internal load-bearing walls = 164.800 × 2.375 = 391.4 m^2

$$\frac{\text{Area of internal load-bearing walls}}{\text{Gross floor area}} = \frac{391.4}{855} = 0.458$$

Partitions

Lengths:	*100 mm*	*75 mm*
ground floor	21.500	33.000
first floor		64.600
second floor		68.800
third floor		68.800
Totals	21.500	235.200

Areas of partitioning: 100 mm = 21.500 × 2.375 = 51 m^2

75 mm = 235.200 × 2.375 = 559 m^2

$$\frac{\text{area of partitioning}}{\text{gross floor area}} = \frac{610}{855} = 0.713.$$

Note: The windows, internal walls and partitions to floor ratios are not often shown on detailed cost analyses, but they can usually be calculated from the drawings accompanying the cost analyses if required.

Costs of Elements

The next step is to calculate the estimated costs of each of the elements expressed in terms of cost per square metre of gross floor area. The costs can in some cases be computed from the costs of the same element shown in the cost analysis of the existing block of flats in table 10.2, with adjustments as necessary for time and locational, quantitative and qualitative factors. In other cases it may be simpler and more realistic to compute the costs from approximate quantities (element unit quantities) priced at current rates. On occasions it can be a valuable exercise to calculate the cost by both methods, with one acting as a check on the other.

(1) *Substructure*. The element unit rate in table 10.2 should be increased by 13 per cent to cover time and locational factors and a further fifteen per cent to cover the more expensive construction stemming from thicker strip foundations and reinforced slab: £85.27 + 13 per cent + 15 per cent = £110.81

Cost of element = 213.75 × 110.81 = £23 686
(ground floor area)

$$\frac{\text{Cost of element/m}^2}{\text{of gross floor area}} = \frac{23\,686}{855} = £27.70$$

(2A) *Frame*. Not applicable.

(2B) *Upper floors*. The element unit rate in table 10.2 can be adjusted by the addition of 13 per cent (time and locational factors) and 30 per cent (increased thickness of part of floor slabs from 150 to 225 mm and increased hoisting costs): 39.67 + 13 per cent + 30 per cent = £58.28

Cost of element = 641.25 × 58.28 = £37 372
(floor area of three upper floors)

$$\frac{\text{Cost of element/m}^2}{\text{of gross floor area}} = \frac{37\,372}{855} = £43.71$$

(2C) *Roof*. The roof construction for the existing block of flats is not comparable and it is therefore necessary to calculate an all-in rate for the built-up felt roofing on woodwool slabs and timber joists. A suitable all-in rate would be 56.00/m^2 of roof area plus £550 for rainwater disposal.

Element cost then becomes 213.75 × 56.00
= 11 970 + 550
= £12 720 + cost of rooflights (£1100)
= £13 820

$$\frac{\text{Cost of element/m}^2}{\text{of gross floor area}} = \frac{13\,820}{855} = £16.16$$

(2D) *Stairs*. The form of construction covered by the cost analysis in table 10.2 is rather different from that specified for the new project. Hence it is more satisfactory to compute a new element unit rate from first principles, possibly using other cost analyses and price books as a guide. The cost is estimated at 3 flights at £2000 = £6000

$$\frac{\text{Cost of element/m}^2}{\text{of gross floor area}} = \frac{6000}{855} = £7.02$$

(2E) *External walls*. The external walls in both projects are of similar construction but it is necessary to make adjustment for the time and locational factors, the changed wall to floor ratio and the extra cost of building brickwork to the fourth storey. Thus the all-in unit rate of 73.92/m^2 of net wall area can be taken as a basis with suitable adjustments.

(i) Add 13 per cent to cover time and locational factors

73.92 + 13 per cent = £83.53

(ii) There is a considerable improvement in the wall to floor ratio and this will be reflected when the element cost/m^2 of gross floor area is calculated, by using the net external wall area. Alternatively the adjustment could be made using the respective wall/floor ratios.

(iii) Increase the unit rate by a further five per cent to take account of the additional cost of building the brickwork to the fourth storey

£83.53 + 5 per cent = £87.71

The net area of external brickwork is calculated as follows

Gross area of external wall		652.7 m^2
less window area:	174.3 m^2	
door area:	32.9 m^2	207.2 m^2
Net external wall area		445.5 m^2

Cost of element = 445.5 × 87.71 = £39 075

$$\frac{\text{Cost of element/m}^2}{\text{of gross floor area}} = \frac{39\,075}{855} = £45.70$$

(2F) *Windows*. For convenience of computation windows and external doors are separated. The cost analysis in table 10.2 covers metal windows while the specification of the new project relates to wood windows. It is therefore advisable to build up a new unit rate. The cost of the sub-element can then be estimated as

174.3 m^2 × £75.00 = £13 073

$$\frac{\text{Cost of sub-element/m}^2}{\text{of gross floor area}} = \frac{13\,073}{855} = £15.29$$

(2F) *External doors*. Variations in specification again make it advisable to compute the new sub-element rate from all-in rates, and to break down the sub-element into its two main constituent parts, namely (1) two glazed doors with screens (2) metal up-and-over garage doors.

(1) 50 mm hardwood doors and screens (2 nr) = 5.6 m^2 × £276 =	£1546
(2) Metal up-and-over garage doors 27 m^2 × £106 =	£2862
Total cost of sub-element	£4408

$$\text{Cost of sub-element/m}^2 \text{ of gross floor area} = \frac{4408}{855} = £5.16$$

The total cost of the windows and external doors element becomes £17 481 and the cost/m^2 of gross floor area = £20.45

(2G) *Internal walls and partitions*. There is a wide variety of walls and partitions used in the existing block of flats and so once again it is advisable to calculate the new rate independently. Suitable all-in rates for the internal walls and partitions have been used.

One brick load-bearing wall 391.4 m^2 × £39.00 =	£15 265
75 mm hollow block partition 559 m^2 × £12.50 =	£6 988
100 mm hollow block partition 51 m^2 × £15.70 =	£801
Total cost of element	£23 054

$$\text{Cost of element/m}^2 \text{ of gross floor area} = \frac{23\,054}{855} = £26.96$$

(2H) *Internal doors*. It is possible to adjust the element unit rate for the previous project, by making allowance for the time and locational factors and the varying qualities and thickness of door, as the number of doors is not proportional to the floor area.

Previous rate: £91.17 + 13 per cent (time and locational factors) = £103.02

Deduct ten per cent to allow for variations in quality of doors

£103.02 − 10 per cent = £92.72
Cost of element = £92.72 × 88 = £8159

$$\text{Cost of element/m}^2 \text{ of gross floor area} = \frac{8159}{855} = £9.54$$

(3A) *Wall finishes*. The specification for both projects is very similar for this element and the simplest approach would be merely to adjust the previous elemental rate for time and locational differences. The new rate then becomes £38.41 + 13 per cent = £43.40.

The total element cost is £43.40 × 855 = £37 107

Alternatively the cost could be computed from all-in rates.

(3B) *Floor finishes*. On account of the rather wide variations in the types of floor finishes, it is advisable to calculate the new rate independently. A suitable all-in rate for felt-backed PVC paving would be 15.50/m^2.

To arrive at the net area of flooring, it will be necessary to deduct the space occupied by internal load-bearing walls and partitions and the floor area of the garages.

Gross floor area		855 m^2
less floor area of garages	87 m^2	
Space occupied by internal walls and partitions, say 10% of 855 − 87 =	77 m^2	164 m^2
Net floor area		691 m^2

Cost of element = 691 × 15.50 = £10 711

$$\text{Cost of element/m}^2 \text{ of gross floor area} = \frac{10\,711}{855} = £12.53$$

(3C) *Ceiling finishes*. It will be in order merely to adjust the previous elemental rate for time and locational differences. New rate then becomes

£12.62 + 13 per cent = £14.26
Cost of element = 855 × £14.26 = £12 192

(4) *Fittings*. Owing to varying specification particulars it is advisable to build up a rate for this element from the three component parts:

(i)	70 nr. kitchen fittings at £120 each (5 per dwelling)	£8 400
(ii)	Shelving, say, $20\,m^2 \times$ £42	£840
(iii)	Wardrobe fronts = 28 nr at £165 each	£4 620
	Total cost of element	£13 860

$$\text{Cost of element/m}^2 \text{ of gross floor area} = \frac{13\,860}{855} = £16.21$$

(5A) *Sanitary appliances*. A prime cost sum of £10 500 is to be provided.

$$\text{Cost of element/m}^2 \text{ of gross floor area} = \frac{10\,500}{855} = £12.28$$

In the absence of a prime cost sum the cost could be calculated by pricing up the appliances to be provided.

14 nr. lavatory basins at £100	£1 400
14 nr. sinks and drainers at £200	£2 800
14 nr. baths at £320	£4 480
14 nr. WC suites at £125	£1 750
Total cost	£10 430

Another approach would be to adjust the elemental cost for the existing building for the increased number of similar appliances = £5609 × 14/9 = £8725 + 13 per cent (time and locational factors) = £9859

(5C) *Disposal installations*. The rate in table 10.2 will need adjusting on two counts; firstly time and locational factors and secondly, the extra height of the new building increases the amount of branch pipework more than proportionately, as appliances on ground floor are connected direct to drains. Allow an extra 10 per cent to cover the second aspect.

New rate then becomes
£5.48 + 13 per cent + 10 per cent = £6.81

Total cost of element is 855 × £6.81 = £5823

(5D) *Water installations*. The proposals provide for 98 nr draw-off points with a similar specification. It would seem reasonable to average the cost of the cold and hot water draw-off points in the previous project to give an average cost per draw off point, and then to increase this rate to take account of time and locational differences. Average cost per draw-off point on the previous project is

$$\frac{£9780}{54} = £181.11$$

Adjusted rate = £181.11 + 13 per cent = £204.65

Total cost of element then becomes
98 × £204.65 = £20 056

$$\text{Cost of element/m}^2 \text{ of gross floor area} = \frac{20\,056}{855} = £23.46$$

(5E) *Heat source*. A prime cost sum of £14 500 is to be provided.

$$\text{Cost of element/m}^2 \text{ of gross floor area} = \frac{14\,500}{855} = £16.96$$

(5F) *Space heating*. The specification particulars are similar for the two schemes and the elemental rate in the previous scheme can be adjusted for time and locational factors.

New element rate becomes
£10.72 + 13 per cent = £12.11

Total cost of element is £12.11 × 855 = £10 354

(5G) *Ventilation system*. A prime cost sum of £2300 is to be provided.

$$\text{Cost of element/m}^2 \text{ of gross floor area} = \frac{2300}{855} = £2.69$$

(5H) *Electrical installations.* A prime cost sum of £32 000 is to be provided.

$$\frac{\text{Cost of element/m}^2}{\text{of gross floor area}} = \frac{32\,000}{855} = £37.43$$

(5N) *Builder's work in connection with services.* It is reasonable to adjust the rate for the previous scheme.

New element rate becomes
£6.60 + 13 per cent = £7.46

Total cost of element = £7.46 × 855 = £6378

(6A) *Siteworks.* A prime cost sum of £35 000 is to be provided.

$$\frac{\text{Cost of element/m}^2}{\text{of gross floor area}} = \frac{35\,000}{855} = £40.94$$

(6B) *Drainage.* A prime cost sum of £7500 is to be provided.

$$\frac{\text{Cost of element/m}^2}{\text{of gross floor area}} = \frac{7500}{855} = £8.77$$

(6C) *External services.* A prime cost sum of £4500 is to be provided.

$$\frac{\text{Cost of element/m}^2}{\text{of gross floor area}} = \frac{4500}{855} = £5.26$$

Preliminaries and insurances. To be calculated at 8 per cent of the remainder of the contract sum.

Contingencies. To be calculated at 3 per cent of the remainder of the contract sum.

The estimated elemental costs and costs of elements per square metre of gross floor area will now be summarised and totalled and in doing this the first or initial cost plan will also be produced (see table 10.3 on p. 263).

NOTE: No sum has been included to cover price and design risks as the element rates have been increased to cover probable cost increases up to the date of tender. An alternative approach would have been to work on price levels current at the date of the estimate and to have included a percentage (possibly about three per cent) to cover possible increases in building prices during the design period.

Tender prices became very competitive in 1990/93, and BCIS prices in December 1993 for flats (3 storeys) showed a mean of £345/m^2 and a mode of £241/m^2. The range of 3-storey flat prices was very wide indeed (£184 to £954/m^2), but this could include flats with large variations in the quality of design, construction and facilities. The prices also range countrywide and greater London prices will be above the average.

EXAMPLE 10.3

The third example in this chapter is concerned with one of the later cost planning processes in the design stage of a building project. The project is a social club building and is illustrated in figure 10.3. It is a single-storey building with a gross floor area of 470 m^2. The initial cost plan for the scheme is illustrated in table 10.4, together with the entries relating to the first set of cost checks undertaken as the design developed. The building client has requested that the scheme should desirably be kept within a budget figure of £275 000, which is equivalent to £585.11 per square metre of gross floor area, but does not favour any reduction in the size of the building. The example will endeavour to show the way in which costs of elements are investigated and steps taken to reduce the estimated cost of the project, together with the appropriate communications to the architect.

As the design of the social club has developed and working details and detailed specification requirements have been prepared, cost checks have been carried out to confirm that it is feasible to provide the elements within the cost targets listed in the initial cost plan and, in view of the building client's request, to look for economies at the same time. Slight adjustments have been made to the costs of a number of elements and their consequences, in relation to the costs per

Table 10.3 Cost planning example 10.2 – cost plan: four-storey block of flats, greater London Gross floor area: 855 m² (anticipated tender date: June 1994)

	Element	Total cost of element £	Total cost of groups of elements £	Cost of element per m² of gross floor area £	Cost of groups of elements per m² of gross floor area £
1	SUBSTRUCTURE	23 686	23 686	27.70	27.70
2A	Frame	–		–	
2B	Upper floors	37 372		43.71	
2C	Roof	13 820		16.16	
2D	Stairs	6 000		7.02	
2E	External walls	39 075		45.70	
2F	Windows and external doors	17 481		20.45	
2G	Internal walls and partitions	23 054		26.96	
2H	Internal doors	8 159		9.54	
2	SUPERSTRUCTURE		144 961		169.54
3A	Wall finishes	37 107		43.40	
3B	Floor finishes	10 711		12.53	
3C	Ceiling finishes	12 192		14.26	
3	INTERNAL FINISHES		60 010		70.19
4	FITTINGS	13 860	13 860	16.21	16.21
5A	Sanitary appliances	10 500		12.28	
5C	Disposal installations	5 823		6.81	
5D	Water installations	20 056		23.46	
5E	Heat source	14 500		16.96	
5F	Space heating	10 354		12.11	
5G	Ventilating system	2 300		2.69	
5H	Electrical installations	32 000		37.43	
5N	Builder's work in connection with services	6 378	101 911	7.46	119.20
5	SERVICES BUILDING SUB-TOTAL		£344 428		£402.84
6A	Siteworks	35 000		40.94	
6B	Drainage	7 500		8.77	
6C	External services	4 500		5.26	
6	EXTERNAL WORKS		47 000		54.97
	TOTAL (less preliminaries)		391 428		457.81
7	PRELIMINARIES (8%)		31 314		36.62
	TOTAL (less contingencies)		422 742		494.43
8	CONTINGENCIES (3%)		12 682		14.84
	TOTAL		£435 424		£509.27

square metre of gross floor area, are duly recorded in table 10.4 under the heading of cost check 1. Significant changes have been made to the costs of rooflights, windows, internal doors, ceiling finishes, fittings, fire-fighting equipment, special installations and external works, and details of the changes are now shown in the form of a report to the job architect (pp. 268–9).

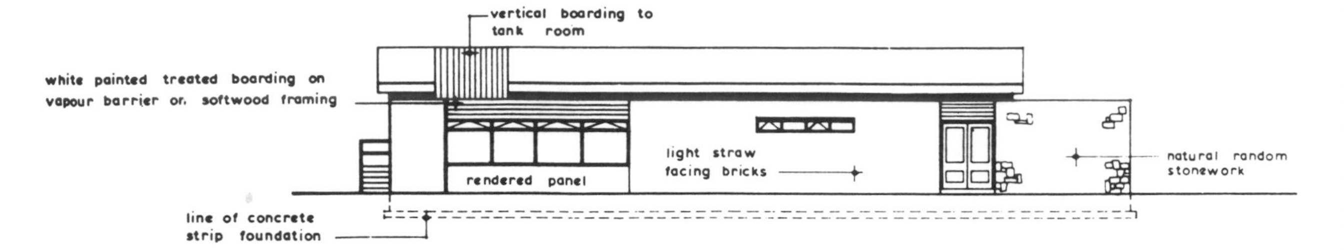

EAST ELEVATION

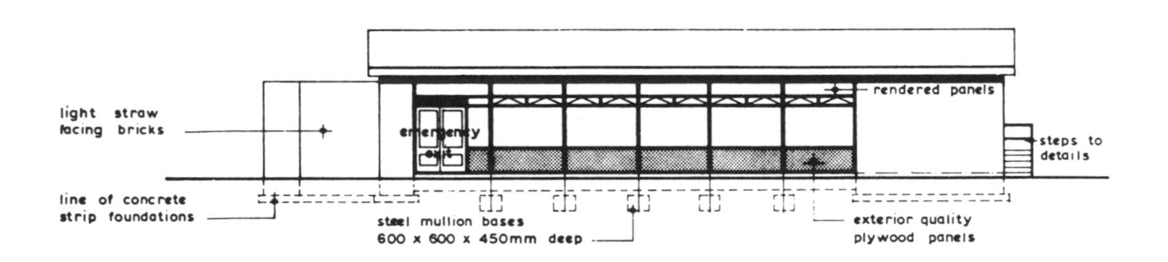

WEST ELEVATION

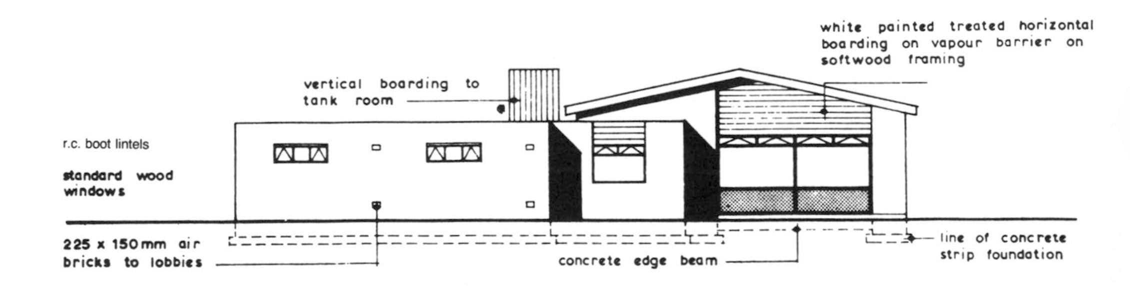

NORTH ELEVATION

Figure 10.3 Cost planning example 10.3 – social club

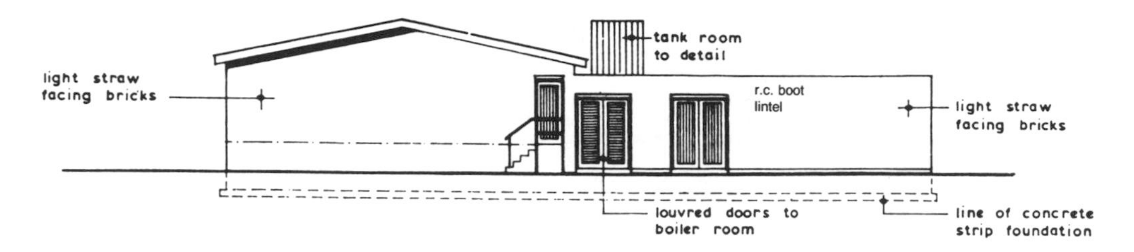

SOUTH ELEVATION

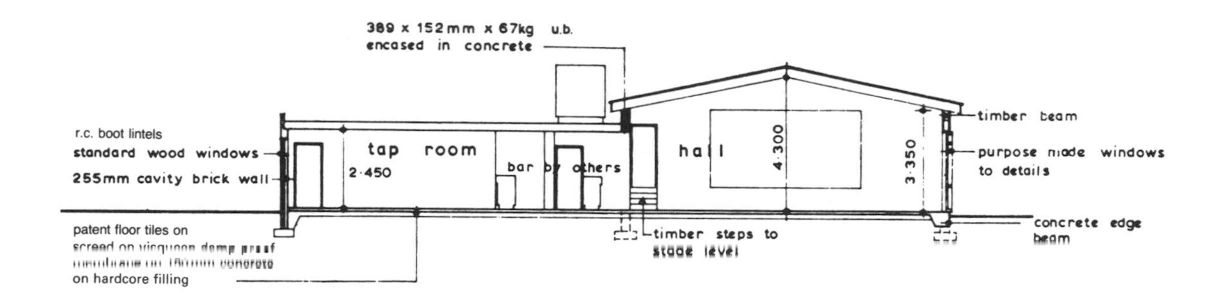

SECTION A - A

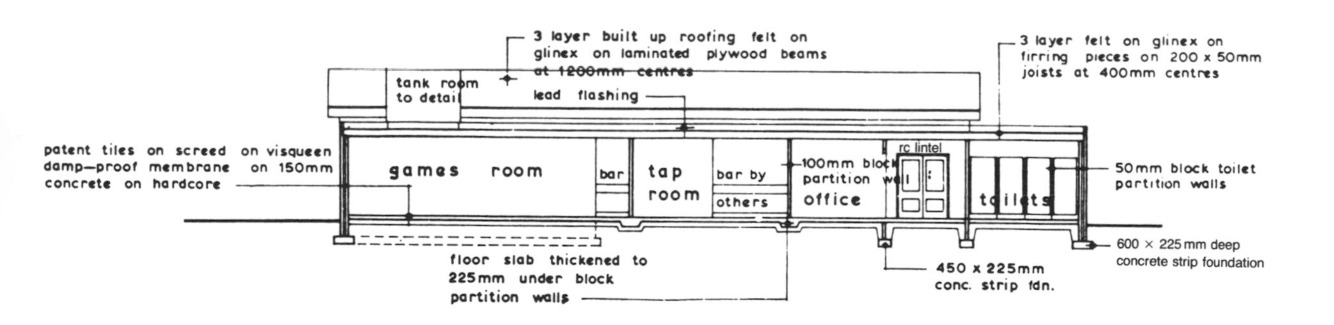

SECTION B - B

Figure 10.3 (continued)

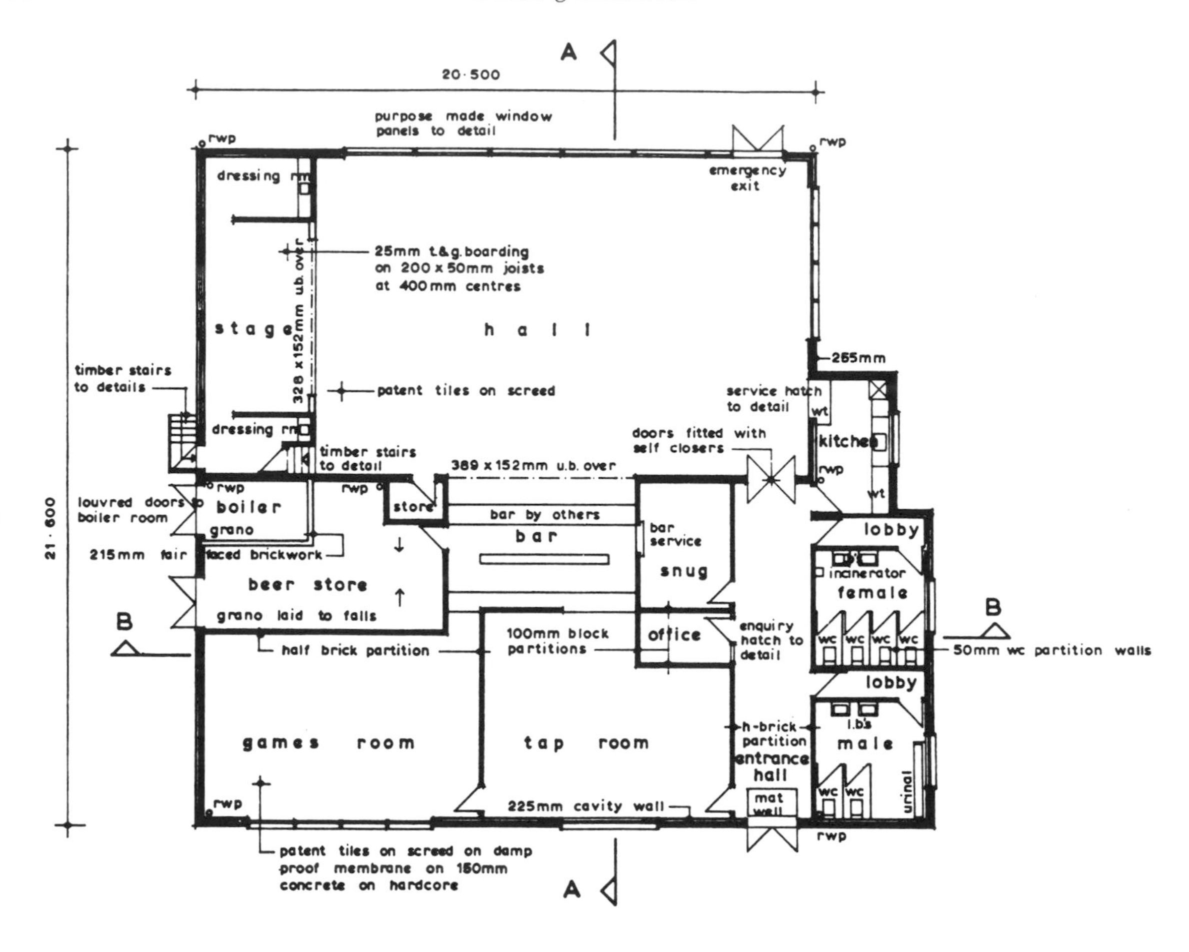

Figure 10.3 (continued)

Table 10.4 Cost planning example 10.3 – initial cost plan and record of cost checks of social club

Element		*Gross floor area: 470 m²* *Initial cost plan*		*Cost check 1*		*Cost check 2*	
		Total cost of element £	*Cost of element/m² of gross floor area* £	*Date*	*Cost of element/m² of gross floor area* £	*Date*	*Cost of element/m² of gross floor area* £
1	Substructure	24 703	52.56	4.7.94	52.32		
2A	Frame	2 312	4.92	4.7.94	5.04		
2B	Upper floors	1 429	3.04	4.7.94	2.92		
2C	Roof	42 563	90.56	7.7.94	84.80	4.8.94	54.24
2D	Stairs	780	1.66	7.7.94	1.66		
2E	External walls	28 219	60.04	20.7.94	59.92	4.8.94	57.41
2F	Windows and external doors	9 616	20.46	20.7.94	23.00		
2G	Internal walls and partitions	10 819	23.02	13.7.94	22.76	4.8.94	8.59
2H	Internal doors	7 266	15.46	13.7.94	14.24		
3A	Wall finishes	10 660	22.68	18.7.94	22.44		
3B	Floor finishes	17 202	36.60	18.7.94	36.48	4.8.94	33.51
3C	Ceiling finishes	8 187	17.42	18.7.94	19.32		
4	Fittings	6 495	13.82	20.7.94	3.70		
5A	Sanitary appliances	4 418	9.40	25.7.94	9.40		
5C	Disposal installations	517	1.10	25.7.94	1.10		
5D	Water installations	[illegible]	1.40	[illegible]	[illegible]		
5E	Heat source						
5F	Space heating	35 429	75.38	25.7.94	75.36		
5G	Ventilating system						
5H	Electrical installations	25 841	54.98	25.7.94	54.72		
5I	Gas installations	263	0.56	26.7.94	0.56		
5K	Protective installations	2 143	4.56	26.7.94	0.96		
5M	Special installations	3 252	6.92	27.7.94	3.38		
5N	Builder's work in connection with services	2 322	4.94	27.7.94	4.94		
5O	Builder's profit in connection with services	4 183	8.90	27.7.94	8.80		
6A	Siteworks	21 319	45.36	28.7.94	34.80		
6B	Drainage	11 524	24.52	28.7.94	24.40		
6C	External works	5 198	11.06	28.7.94	11.00		
	Preliminaries	9 804	20.86	28.7.94	20.80		
	Contingencies	3 196	6.80	29.7.94	6.80		
Price and design risk		5 997	12.76	29.7.94	12.76		
	Totals	£308 195	£655.74		£623.90		

To: Architect
From: Quantity surveyor 1.8.94

Social Club

Cost checks have been made of all elements of the above project and a summary sheet (table 10.4) shows the current position. It has been possible to reduce the total cost expressed in terms of the square metre of gross floor area from £655.74 to £623.90, but this is still outside the client's budget limit. Details of the cost changes arising from the amendments to design, as agreed with you, are now listed.

	Savings per m² of gross floor area £	*Excess per m² of gross floor area* £
(2C) *Rooflights*. These have been omitted with a consequent saving of £2700. There will be no corresponding increase in the cost of the roof element, as the cost of the additional area of roof is offset entirely by the saving in cost resulting from the omission of forming and waterproofing the roof around the rooflights.	5.75	
(2F) *Windows*. The purpose-made windows are proving to be more expensive than was originally anticipated and the specialist quotation now received shows an increase of £1200, raising the total element cost of windows and external doors to £10 820.		2.55
(2H) *Internal doors*. The sliding hardwood doors have been replaced by side-hung softwood glazed doors, resulting in a saving of £570. The total element cost now becomes £6696.	1.21	
(3C) *Ceiling finishes*. The use of acoustic ceiling tiles to the hall has resulted in an increase of cost of £893, raising the total element cost to £9080.		1.90
(4) *Fittings*. A substantial saving has resulted from the proposal to adapt and fix second-hand fittings from the present club, instead of purchasing new fittings. The total element cost is reduced from £6495 to £1740.	10.12	
(5) *Protective installations*. The extensive fire-fighting equipment originally proposed has now been confined to four portable fire extinguishers, reducing the total element cost to £450.	3.60	
(5M) *Special installations*. As a result of the simpler burglar alarm system, the total element cost has been reduced to £1590.	3.54	
(6A) *Siteworks*. A length of 50 m of one-brick thick boundary wall has been omitted permitting a cost reduction of £4960.	10.55	
Major cost changes	£34.77	£4.45

In view of the client's expressed wish for the total cost of the project not to exceed £270 000, further design/cost investigations are proceeding with particular reference to the elements of roof, external walls, internal walls and floor finishes, as being the elements where significant cost savings may still be possible without seriously impairing the functional and aesthetic characteristics of the building.

These four elements are now investigated in detail with a view to reducing the total cost per m^2 of gross floor area to £574.46.

(2C) *Roof*. It is agreed with the architect to replace the felt-covered plywood boarding on laminated beams with felt-covered Weyroc boards supported on timber joists.

The cost of the substituted construction is

496 m^2 × £51.40 = £25 494 (£54.24/m^2)

(2E) *External walls*. Lightweight concrete blocks, 100 mm thick, are to replace the half-brick inner skin of the external walls. The saving in cost is 190 m^2 × £6.20 = £1178, and the total element cost becomes £26 984 (£57.41/m^2).

(2G) *Internal walls and partitions*. 100 mm block walls are to be substituted for half-brick walls and 75 mm block walls for 100 mm block walls.

The revised estimate of cost is

20 m^2 of one-brick wall at £39.00 =	£780
132 m^2 of 100 mm block wall at £15.35 =	£2026
100 m^2 of 75 mm block wall at £12.30 =	£1230
Total cost	£4036

This is equivalent to £8.59/m^2.

(3B) *Floor finishes*. Substitution of granolex vinyl bonded block floor tiles on cement screed for the quarry tile flooring originally proposed for the entrance hall, toilets and kitchen. The cost variation is

62 m^2 of quarry tile paving at £28.20 =	£1748
62 m^2 of granolex paving at £5.68 =	£352
Saving in cost	£1396

The revised element cost then becomes £15 750 and the cost/m^2 of gross floor area is £33.51.

A further letter to the architect summarises the current cost situation.

To: Architect
From: Quantity surveyor 8.8.94

Social Club

The cost consequences of the latest design/cost investigations are now listed.

Element	*Saving in cost/m^2 of gross floor area* £
(2C) Roof (£84.80 − £54.24)	30.56
(2E) External walls (£59.92 − £57.41)	2.51
(2G) Internal walls and partitions (£22.76 − £8.59)	14.17
(3B) Floor finishes (£36.48 − £33.51)	2.97
Total saving in cost	£50.21

These further amendments to design reduce the total cost per square metre of gross floor area to £573.72 and the total cost of the project to £269.681. This is £319 below the client's budget limit and I hope that these changes will be acceptable to the client.

CONCLUSIONS

It will be noted that all three examples have been based on the elemental cost planning approach, although it would have been quite conceivable to have developed example 2 on the comparative method, indicating a selection of choices with their appropriate costs for each of the major elements and the cost consequences of each of the choices. However, the prime aim of this chapter has been to help examination candidates in the compilation of cost plans or parts of cost plans, from sketch drawings, cost analyses of similar buildings and other relevant information. It is believed that the examples will have shown suitable methods of approach to this type of problem.

It might be argued that it is not the best practice to restrict the basis for the new cost plan to a published cost analysis of one building, with which the surveyor is not directly familiar. At the very least, a number of cost analyses should be taken and average costs computed. Limitations of space have prevented the cost examples being more comprehensive and general principles are just as well illustrated on a small contract as on a large project, and the smaller scheme has the benefit of greater simplicity and clarity.

Another aspect which may leave room for doubt in the reader's mind is that of the application of published building cost indices. There are very real dangers in applying them too rigidly and indiscriminately without regard to local circumstances. The type of building, its location and the tendering climate must each have a significant effect on price levels and all published cost data must therefore be used with the greatest discretion and be suitably adjusted, wherever necessary, to take account of local conditions. In practice the quantity surveyor will have the benefits of a vast amount of cost data of his own and published data will be largely used to form broad guidelines and for comparative purposes. Published cost data will be of particular value when dealing with special type buildings subject to the provisos already given.

An examination of BCIS cost analyses, all with prices relating to the third quarter of 1993, show wide price ranges for similar building types and this again reinforces the problems and dangers inherent in the application of even the best published cost data. The 1993 price ranges of a selection of common building types are

factories (purpose built): £123 to £1155/m^2
primary schools: £226 to £958/m^2
churches: £268 to £1084/m^2
libraries: £346 to £1140/m^2
flats: £182 to £925/m^2
houses (estates): £163 to £690/m^2
houses 'one off', detached: £186 to £2101/m^2
shops: £127 to £892/m^2
offices: £258 to £1055/m^2
sheltered housing: £208 to £837/m^2
hotels: £292 to £1115/m^2

Libraries, churches and hotels are buildings with high unit costs. These buildings have large storey heights, are constructed to high standards of quality and contain expensive fittings. It may not however be quite so readily recognised that public conveniences are also very costly buildings with high wall to floor ratios and expensive fittings and finishings. In consequence they frequently cost more than £1050/m^2 of gross floor area (1993 prices).

COST CONTROL OF ENGINEERING SERVICES

Mechanical and Electrical Services

Berryman (1971) made useful suggestions as to how the cost of heating and ventilating engineering services might be controlled during both design and post-contract stages. He admitted that the cost control functions will have to be shared and believed that the arrangements listed in table 10.5 constitute an acceptable and workable arrangement.

The cost planning of engineering services is based in the first instance on the cost analyses of existing projects. The cost information so used must be reliable and in sufficient detail to be suitable for analysis purposes, and the best source of information is often the quantity surveyor's own office. The best format for cost analyses is the elemental system in the standard form of cost analysis (SFCA) provided by the BCIS, using techniques similar to those previously described for building works. Even then a service such as air conditioning, which may form 25 per cent of the total capital value of an air conditioned building, is confined to one element (Watson, 1990).

Examination of a typical cost plan of building services, as illustrated in table 10.6, will show the disproportionate amount of expenditure between one element and another. However, the cost per

Table 10.5 Suggested procedure for cost control of engineering services in buildings

Contractor design stage	*Quantity surveyor*	*Contractor*
Pre-contract	(1) Initial budget and cost plan (2) Tender enquiries (3) Tender report (4) Cost analysis and budget reconciliation	*(1) Cost studies before design (2) Design *(3) Cost checking during design (4) Priced bills of quantities or cost sheets (5) Pre-contract variations when required
Post-contract	(5) Budget control (6) Variations or remeasurement	(6) Design and detailing (7) PC, daywork, increased or decreased costs and overtime accounts

Consultant design stage	*Consulting engineer*	*Quantity surveyor*	*Contractor*
Pre-contract	(1) Initial budget and cost plan *(2) Cost studies before design (3) Design *(4) Cost checking during design (5) Tender report	(1) Initial budget and cost plan (2) Bills of quantities when required (3) Tender enquiries (4) Tender report (5) Pre-contract variations when required (6) Cost analysis and budget reconciliation	(1) Cost sheets (if bills of quantities not required) (2) Pricing
Post-contract	(6) Post-contract design and supervision	(7) Budget control (8) Variations or remeasurement	(3) Detailing (4) PC, daywork, increased or decreased costs and overtime accounts

NOTE

(1) All post-contract duties are subject to negotiation, and consequently shared between both parties to the contract, irrespective of who is to take the initiative.

(2) The same applies to bills of quantities or cost sheets or pre-contract variations when these also are subject to negotiation.

(3) It may be helpful or necessary to delegate duties marked* to the quantity surveyor, especially where detailed measurement, analysis or integration with building costs are involved.

Source: A. Berryman, *Controlling the costs of engineering services in buildings. The Chartered Surveyor* (August 1971).

m^2 of GFA of an element is still a useful indication of cost for a new project, provided it is suitably adjusted to suit the new situation. For example, an element of one project can be of higher quality, denser distribution, more extensive in area and context, more sophisticated or more complex when compared with the equivalent element of another project. Hence it is common practice to adjust the elemental cost per m^2 of GFA by the use of *quantity factors* (Watson, 1990).

Watson (1990) has described how simple mechanical and electrical systems usually require the number of termination points to be used as a quantity factor, transportation systems can relate to the number of persons carried, heating can be based on the kW output of the heat source and ventilation on the total cubic capacity of air handled by the fans. More complex elements, such as air conditioning, may not reflect costs with precision, as too many factors combine to distort the resultant cost in quantity factor terms.

An example of the use of quantity factors applied to building services is given in table 10.7, in which two buildings of similar area and elemental cost are compared and the number of terminals or outlet points is significantly different. Hence a new project with a high density of points requires adjustment to the elemental cost in order to provide a realistic estimate.

Air conditioning can be reduced to a cost per cubic metre per second of total fan power, but this will effectively only reflect the cubic capacity of the building and the number of air changes per hour in addition to the area of the building. The

Table 10.6 Elemental cost summary: engineering services

Project: New Civic Headquarters
Area: 34 325 m²

Project No. :X532
Cost Plan No. :1
Date :05 12 92

	Element		*Cost £/m²*	*Cost £*
5A	SANITARY APPLIANCES		0.92	31 500
5B	SERVICES EQUIPMENT		29.13	1 000 000
5C	DISPOSAL INSTALLATIONS			
	Rainwater		0.60	20 500
	Soil and Waste		0.88	30 250
5D	WATER INSTALLATIONS			
	Mains Supply		3.96	136 000
	Cold Water		3.69	126 500
	Hot Water		4.49	154 000
5E	HEAT SOURCE			
	Boiler Plant		5.83	200 000
	Boiler Flues		1.46	50 000
5F	SPACE HEATING AND AIR TREATMENT			
	Heating Installation		5.94	204 000
	Chilled and Cooling Water		8.86	304 000
	Supply Ductwork		69.28	2 378 000
	Extract Ductwork		35.60	1 222 000
	Air Conditioning and Ventilation Plant		23.77	816 000
	Cooling Plant		11.89	408 000
	Automatic Controls		23.77	816 000
5G	EXTRACT VENTILATION		5.94	204 000
5H	ELECTRICAL INSTALLATIONS			
	Main Switchgear and Transformers		12.54	430 500
	Sub-Mains and distribution boards		7.01	240 500
	Power Supplies		34.49	1 184 000
	Lighting Installation		12.67	435 000
	Lighting Fittings		10.58	363 000
5I	GAS INSTALLATION		0.52	18 000
5J	LIFTS AND CONVEYORS		10.76	369 500
	PROTECTIVE INSTALLATIONS			
	Sprinkler Installation		0.87	30 000
	Hose Reels, Wet and Dry Risers		5.83	200 000
	CO2 and Foam Inlets		3.44	118 000
	Lightning Protection		0.52	18 000
5L	COMMUNICATION INSTALLATIONS			
	Fire Alarms and Detection		6.39	219 500
	Clocks		0.04	1 500
	Telephones		2.91	100 000
	TV and other communications		4.09	140 500
5M	SPECIAL INSTALLATIONS			
	Intruder Protection and Security		2.91	100 000
	Window Cleaning Equipment		10.95	376 000
6C	EXTERNAL SERVICES			
	Water Mains and Fire Mains		0.58	20 000
	Gas Mains		0.63	21 500
	Electrical Mains		2.91	100 000
	External Lighting		4.54	156 000
	PRELIMINARIES	5% of £12 742 250	18.56	637 113
	CONTINGENCIES	7.5% of £13 079 362	29.23	1 003 452
	MAIN CONTRACTORS PROFIT & ATTENDANCE	5% of £14 382 814	20.95	719 141
	GRAND TOTAL		£439.93	£15 101 955

Source: B. Watson (1992).

Table 10.7 Cold water supply – comparative costs of outlet points

Building Type A (Office Block)	
Total cost of element	£113 100
Area of building	32 500 m^2
Element cost per m^2	£3.48
Number of outlet points	887
Cost per outlet point	£127.50
Density of outlets/m^2	0.027
Building Type B (Laboratory Block)	
Total cost of element	£110 490
Area of building	31 750 m^2
Element cost per m^2	£3.48
Number of outlet points	1 232
Cost per outlet point	£89.68
Density of outlets/m^2	0.039

Source: B. Watson (1992).

factors which are not directly reflected by such a quantity factor include cooling and humidity control, sophisticated automatic controls, methods of heat and air distribution, sound attenuation requirements, the choice of fuel and type of heat source plant, water treatment, standby equipment, fire protection requirements and the relative disposition of the main plant. Watson (1990) has described how more sophisticated and detailed techniques are needed to encompass these additional factors, usually incorporating a component breakdown method as shown in tables 10.8 and 10.9.

This method effectively isolates the individual variables which affect the cost of air conditioning and deals with each one according to its particular

Table 10.8 Elemental rates for air conditioning for offices: variable air volume system

Elements	*Office block of 6 000 m^2*		*Office block of 15 000 m^2*		*Spec. ref.*
	Cost of element £	*Cost of element per m^2 of floor area* £	*Cost of element* £	*Cost of element per m^2 of floor area* £	
Boilers					
Plant and instruments	30 700	5.12	53 300	3.55	A
Flue	10 400	1.73	16 000	1.07	B
Water treatment	10 400	1.73	20 500	1.37	C
Gas installation	4 700	0.78	4 700	0.31	D
Space heating					
Distribution pipework	29 800	4.97	60 000	4.00	E
Convectors and/or radiators	5 800	0.97	14 500	0.97	F
Heating to batteries	58 800	9.80	106 200	7.08	H
Chilled water to batteries	44 400	7.40	81 400	5.43	J
Condenser cooling water					
Distribution pipework	20 900	3.48	36 000	2.40	K
Cooling plant					
Chillers	105 800	17.63	178 700	11.91	L
Cooling towers	11 800	1.97	18 000	1.20	M
Automatic controls	92 500	15.42	145 500	9.70	N
Ductwork					
Supply	449 300	74.88	1 023 400	68.23	P
Extract	95 000	15.83	214 800	14.32	P
Air conditioning plant					
Heating batteries	30 700	5.17	53 800	3.59	Q
Humidifiers & cooling batteries	54 600	9.10	94 400	6.29	Q
Fans and filters	83 700	13.95	143 000	9.53	R
Sound attenuation	40 600	6.77	99 500	6.63	S
Fire protection	23 700	3.95	43 100	2.87	T
Electrical work in connection	27 600	4.60	52 200	3.48	V
Totals	1 231 200	205.25	2 459 000	163.93	
		say 205.00		say 164.00	

Source: B. Watson (1992).

Table 10.9 Elemental rates for air conditioning for offices: induction system

Elements	*Office block of 6 000 m²*		*Office block of 15 000 m²*		*Spec. ref.*
	Cost of element £	*Cost of element per m² of floor area* £	*Cost of element* £	*Cost of element per m² of floor area* £	
Boilers					
Plant and instruments	30 700	5.12	53 300	3.55	A
Flue	10 400	1.73	16 000	1.07	B
Water treatment	10 400	1.73	20 500	1.37	C
Gas installation	4 700	0.78	4 700	0.31	D
Space heating					
Distribution pipework	29 800	4.97	60 000	4.00	E
Convectors and/or radiators	5 800	0.97	14 500	0.97	F
Induction units	110 000	18.33	294 700	19.65	G
Heating to batteries	121 300	20.22	201 600	13.44	H
Chilled water					
Distribution pipework	156 000	26.00	288 800	19.25	J
Condenser cooling water					
Distribution pipework	20 800	3.47	34 600	2.31	K
Cooling plant					
Chillers	82 200	13.70	143 100	9.54	L
Cooling towers	10 800	1.80	16 200	1.08	M
Automatic controls	102 100	17.02	182 900	12.19	N
Ductwork					
Supply	131 800	21.97	237 600	15.84	P
Extract	66 400	11.07	141 800	9.45	P
Air conditioning plant					
Heating batteries	23 100	3.85	42 000	2.80	Q
Humidifiers & cooling batteries	39 900	6.65	61 300	4.09	Q
Fans and filters	57 000	9.50	104 500	6.97	R
Sound attenuation	30 700	5.12	79 200	5.28	S
Fire protection	20 900	3.48	37 300	2.49	T
Electrical work in connection	27 300	4.55	46 200	3.08	V
Totals	1 092 100	182.03	2 080 800	138.73	
		say 182.00		say 139.00	

Source: B. Watson (1992).

characteristics. Items of plant such as heat source, chillers and air conditioning plant are identified separately. Distribution pipework and ductwork are separated so that materials standards, insulation and method of distribution can be reflected directly in the cost per m² of GFA. Methods of control together with fire protection and electrical supply components are also identified separately. Having separated each of the components, the factors affecting them can be isolated and levels of sophistication can be taken into account with a significant degree of accuracy. Tables 10.8 and 10.9 show the elemental cost breakdown of air conditioning for two office blocks of varying size and using variable air volume and induction systems for comparative purposes (Watson, 1990 and 1992).

The elemental cost breakdown should be accompanied by supplementary information defining the relevant specification standards and performance, usually in the form of *specification notes* as provided in table 10.10.

The rainwater installation provides a straightforward example, incorporating cast iron, aluminium or PVC-U pipework and fittings, with the cost per m² of GFA reflecting the appropriate specification standard. By comparison the heat source based on gas fired boilers will be cheaper in capital cost terms than oil fired or coal fired boilers; the first because of the additional require-

Table 10.10 Elemental rates for air conditioning for offices: brief specification notes

Ref.	
	Boilers
A	Plant and instruments: Three gas-fired boilers each of approximately 250 and 580 kW capacity for the two buildings respectively; together with burners, pumps, direct-mounted instruments, feed and expansion tanks. Normal standby facilities are included.
B	Flue: Mild steel insulated in boiler house, internal lining to vertical builders' stack.
C	*Water treatment*: Chemical dosage equipment.
D	*Gas installation*: Pipework internal to building, meter, solenoid valves.
	Space heating
E	Distribution pipework: Pipework from boilers to terminal equipment, all valves, fittings and supports, insulation.
F	Convector and/or radiators: Panel radiators, or natural convectors in circulation areas and staircases.
G	*Induction units*: High-velocity units suitable for four-pipe system utilising ducted fresh air.
H	*Heating to air heater batteries*: Distribution pipework to batteries, valves, fittings and supports, insulation.
J	*Chilled water to batteries and induction units*: Distribution pipework, valves, fittings and supports, insulation.
K	*Condenser cooling water*: Distribution pipework, valves, fittings and supports, insulation.
	Cooling plant
L	Chillers: Centrifugal chiller units of approximately 190 and 470 t total capacity for the two buildings respectively, including mountings and supports, insulation and pumps. Normal standby facilities are included.
M	Cooling towers: Forced or induced draught fans, roof-mounted cooling towers with supports.
N	*Automatic controls*: Pneumatic controls including motorised valves, all thermostats, control panels, actuators, interconnecting wiring and tubing
	Ductwork
P	Supply and extract: Galvanised mild steel ductwork, fittings and supports, terminal units (for VAV system), dampers, grilles and diffusers, insulation.
	Air conditioning plant
Q	Heating and cooling batteries: Humidifiers, batteries and casing and connections.
R	Fans and filters: Centrifugal and axial flow fans with casings and connections, and automatic roll type filters.
S	Sound attenuation: Silencers and duct lining (short lengths only).
T	*Fire protection*: Heat detectors, smoke detectors, gas detectors, control panel, interconnecting wiring (excluding other fire protection services not directly associated with air conditioning installation).
V	*Electrical work in connection*: Electrical supplies to control panels and mechanical plant, mechanical services distribution board.

Source: B. Watson (1992).

ment for oil tanks and the second resulting from the need for automatic stoking and ash removal facilities. Similarly the cooling plant for air conditioning can be based on refrigeration units in the plant room with cooling towers on the roof, necessitating expensive interconnecting pipework. By comparison an air cooled chiller unit will be more costly in terms of basic plant but less expensive overall (Watson, 1990).

It is important to be able to relate an increase in capital expenditure to a saving in running costs, as described in chapter 13. For example, if the capital expenditure on air conditioning plant for one system was to be increased by £10 000, the running costs would need to be reduced by at least £900 per annum, or £0.15/m^2 of GFA before such a decision would become economically viable.

When considering the financial control of engineering services contracts, it should be recognised that this cannot be restricted to capital cost alone. Cost control will not only be relevant to cost planning but also to the selection of tendering procedure, cost control during installation, running costs, maintenance costs, replacement costs and life cycle costs (Watson, 1990).

Industrial Engineering

Industrial engineering covers five broad areas – petrochemical plants, process engineering plants, off-shore platforms, power stations and coal mining. These schemes are generally much larger than building projects and generate greater problems and more claims.

With these projects, an engineer is solely responsible for the design, financial control and, in many cases, the construction. Often, however, the design responsibility rests with the contractor, who then undertakes both design and construction. Watson (1981) has described how the quantity surveyor's involvement may be limited to giving advice on the cost consequences of inadequate tender documentation with which he has no prior involvement. The areas where he is likely to be most closely involved are site evaluations of the distribution and ancillary services – a relatively insignificant proportion of total capital costs. In the case of off-shore platforms, the quantity surveyor is likely to have a greater involvement because of the extensive use of bills of quantities.

Initial estimates are often prepared by the cost engineer, partly because of the absence of historical estimating data and partly due to an initial absence of design information. For example, the proposed capacity may be the only known quantity factor, so preliminary figures are often used to predict the possible capital cost.

Traditional quantity surveying techniques are most directly relevant to the post-contract aspects of industrial engineering, where evaluation is much more straightforward and encompasses variations and full remeasurement. With materials prefabricated by another contractor, site modifications will often be evaluated on a daywork basis, although consequential costs are seldom charged to the original contractor. Constant requirements of design processes will lead to many such modifications and, with design delays, can give rise to potential claims for extension and disruption. The main objective is to ensure that the plant is ready for production, based on an optimum design, the implication of which does not unduly delay the production process. Costs unacceptable in building become a normal part of construction in industrial engineering. The role of the quantity surveyor has to be adjusted to suit these conditions (Watson, 1981).

Jackson (1981) has described how a bill of quantities for petrochemical works normally contains a closely defined preliminary section and a method of measurement which emphasises the key items, adequately subdivided to allow direct comparison with the cost control report and planning schedules. The final tender bill provides the first check on the construction elements of the budget. Subsequently, the cost engineering team will evaluate changes or remeasure sections of the work. The team will produce regular cost control reports at least monthly and these will be reviewed at regular cost control meetings. Any adverse cost trends will then be highlighted for immediate corrective action.

Settlement of disputes concerning off-shore related work can be complicated by the work usually being carried out on the fabricator's own premises. To complete the work, it will be necessary to take it out of the yard – a very expensive and time consuming operation when time is critical (Jackson, 1981).

Sufficient has been written to indicate the need for and problems associated with the cost control of industrial engineering work. This is an area where quantity surveyors have become increasingly involved, despite fierce competition from cost engineers.

11 VALUE MANAGEMENT

GENERAL PRINCIPLES

An outline introduction to this increasingly used and important technique, including its historical background, was provided in chapter 1, and this chapter aims to examine the process in greater detail to show the various approaches, the advantages to be obtained, a comparison with cost planning/cost management and some comprehensive case studies. Readers requiring a more comprehensive study of this subject are referred to Norton and McElligott (1995).

Value management is attracting considerable attention within the UK construction industry as major clients become increasingly concerned with the achievement of value for money in their construction projects. However, as stated by Green and Moss (1993), value management often means different things to different people, and there is considerable confusion between value management and value engineering. In practice the former term is favoured in the UK and the latter term is used extensively in the United States, where it is often performed by engineers with applications to manufacturing industry. There are also a variety of different approaches, occurring at different stages in the design process, and these will be examined in some detail later in the chapter.

Value management operates within an organised schedule of procedures. This enables the functional requirements and alternative solutions, with their associated costs, to be identified and developed to a strict timetable. Value management is often undertaken in the form of an intensive workshop conducted by an independent team of experienced design team professionals acting in a consultative capacity to the client. On the completion of the workshop, they produce a comprehensive report, with recommendations, for review and assessment by the client and his project design team. This procedure does not adversely affect the project design team's reponsibilities to the client. The value management team, operating in a complementary role, acts as a positive catalyst for savings and improved efficiency (Beard Dove, 1990).

In theory, value management can be undertaken at any stage of the design process. However, as a general rule, the earlier a study is undertaken the more effective it will be in providing the opportunity to rationalise design before it is so firmly established that any change will significantly increase design/planning costs. In practice the timing is often critical, although it should be stated that there can be no guarantee that overall initial capital costs will be reduced although a more efficiently designed project will almost certainly emerge.

Value Management Definitions

A number of definitions of value management have been formulated; the latest being: 'a service which maximises the functional value of a project by managing its evolution and development from concept to completion, through the comparison and audit of all decisions against a value system determined by the client or customer' (Kelly and Male, 1993).

Three other useful definitions, derived from Miles' descriptions of the fundamentals of value analysis in the United States in 1972, are as follows:

(1) ‘an organised approach to provide the necessary functions at lowest cost (whilst not affecting the quality of the product)’;

(2) ‘a structured analysis of a project by an independent consultant or person to determine the required functions of the building (product) and to consider alternative (design/construction) solutions to eliminate unnecessary cost’;

(3) ‘the search for (and elimination of) unnecessary cost; unnecessary cost being that cost which provides neither use, nor life, nor quality, nor appearance, nor customer features’.

Lawrence Miles was a purchasing engineer with GEC and was assigned the task of procuring relevant materials to expand production of turbo-superchargers for the B24 bomber from 50 to 1000 per week, at a time of severe shortage of all the necessary materials and components, and hence he was frequently unable to obtain the specified products. His declared philosophy was ‘if I cannot obtain the specified product I must obtain the alternative which performs the same function’. When alternatives were found they were tested and approved by the designer. Miles discovered that many substitutes were cheaper and were equal or better than the original product. Hence the definition of value engineering/ management that evolved from his work of ‘an organised approach to providing the necessary functions at the lowest cost’ can be applied equally well to construction work at the present time.

Reasons for the Client Commissioning Value Management Studies

Carter (1991/92) identified the following reasons why a client might wish to commission a value management study:

(1) client’s concern at the escalation of estimated costs;

(2) client’s concern at tenders received in excess of budget;

(3) client losing confidence in the design team and/or project, arising from such factors as planning delays, external factors or lack of competence;

(4) client requires an independent audit or appraisal of the project before it is submitted for sanction;

(5) client seeks to minimise capital and/or operational costs and maximise profit;

(6) client must achieve capital and/or operational savings to make a profit;

7) client wishes genuinely to seek an innovative/better solution to his project;

8) client wishes to experiment with a new technique that he has discovered;

9) a consultant recommends a new technique to the client.

ALTERNATIVE APPROACHES TO VALUE MANAGEMENT

There are a number of different approaches that can be adopted when carrying out value management with the choice often being decided by the type and nature of the project, the timing of the operation and the make up of the design team. It is customary to prepare a job plan incorporating a recognisable strategy, which normally comprises the six phases of information: creativity; evaluation; development; presentation/recommendations; and action and feedback. The various procedures are now described.

The Charette

This is undertaken after the project brief has been formulated and the design team appointed but before the actual design is commenced. The client’s representatives and the design team meet under the chairmanship of a value manager or facilitator for one or two days in order that the brief can be examined in detail and questions raised. The next stage is to generate ideas for rationalising the brief, when functional analysis of the space requirements can form a major component, and improving the project’s cost effectiveness. These ideas are then evaluated and, if accepted, are incorporated in a revised brief.

The 40 Hour Value Management Workshop/Study

This is probably the most widely accepted formal approach to value management, and is used as the basis for training of value engineers as prescribed by the Society of American Value Engineers (SAVE). It is normally undertaken at about 35 per cent of the way through the design stage which is about as late a stage as is reasonably practicable. The sketch design of the project is reviewed by an independently appointed second design team, under the chairmanship of a value management or value engineering team co-ordinator (VMTC or VETC), the composition of this team of possibly six to eight professionals reflecting the characteristics of the project under review. For example, a project involving a substantial proportion of mechanical and electrical work could create the need for four persons with these professional backgrounds to form part of the team. The workshop normally takes place near the project site, probably in a hotel or a room in the client's office. The complete drawings are sent to the VMTC/VETC for distribution to the team during the week preceding the workshop/study. During the week of the workshop/study, the team will follow strictly the stages of the job plan (Kelly and Male, 1991 and 1993).

The 40 hour study spread over five days concludes with a number of design/construction modifications which are referred to the client for endorsement and implementation. It is claimed that savings of up to 30 per cent may be achieved in the United States, but savings of this magnitude are unlikely to be obtained in the UK with the tighter cost control procedures. However, as highlighted by Carter (1992a), it can have its drawbacks as the potential exists for confrontation and the external team's proposals can be seen to be critical of the project design team and may be resisted. The short timescale may make it difficult for the external team to fully understand all aspects of the project proposals and it leaves only a restricted period of time to prepare revised designs and for them to be fully and accurately costed.

Norton (1992b) has described a value engineering study undertaken in 1992 for a bus maintenance and storage project estimated to cost $24m in New York for the City Office of Management and Budget. It is interesting to note that a specialist value engineering consultancy firm was engaged to conduct an independent 40 hour workshop following the normal job plan guidelines, and was carried out at 10 per cent design stage to give the greatest potential for savings at the earliest stages of design. As a result of the study 32 recommendations were accepted, 23 rejected and a further five underwent further study. The implemented recommendations produced savings of about 15 per cent.

One–Two Day Workshop/Study

Carter (1992a) has strongly advocated this approach as being more appropriate for use in the UK. He recommends that a two day study be held on a Friday and Monday, while a one day study can be held on any weekday. All members of the project design team should be represented including the client, facilities manager, letting agent and other relevant parties. At the begining, each team member usually makes a brief verbal presentation using drawings or other suitable material, with a maximum duration of 10 to 15 minutes.

The value manager frequently records the relevant data on flip charts, and seeks to identify major constraints, which can be physical (site, ground conditions, height, light or access), operational, statutory (company or legislative), time or cost, each having an impact on the project.

The next stage involves the preparation of a FAST diagram (Functional Analysis System Technique), which will be described and illustrated later in the chapter. The quantity surveyor/cost engineer then breaks down the cost plan (where available) over the weekend, hence the choice of Friday and Monday for the study.

The FAST diagram is then examined to identify any functions which appear to have an abnormally high cost or to identify functions which can be omitted or modified. The next step is an intensive session (brainstorming) which could reasonably be expected to generate 50 or more suggestions to modify the brief, relax the constraints or modify the design/construction proposals in order to achieve a more efficient design or technical solution to eliminate unnecessary costs.

These suggestions are reviewed as being either: (a) rejected (with reasons recorded) or (b) to be developed by the project team. The latter items are then prioritised. The value manager/engineer then compiles a comprehensive report (probably of some 40 to 50 pages), encompassing all the elements of the study and concluding with recommendations as to which items are to be developed by the project team. This report is normally issued within five to seven days or at the end of the study to the client/project sponsor for implementation.

This shortened form of study is much cheaper and quicker than the 40 hour workshop, probably costing less than £10 000 (1992 prices), and is considered to be more appropriate to the UK. Carter (1992a) has undertaken studies using this approach achieving benefit ratios between 1:30 and 1:300.

Two or Three Day Workshops

Doyle (1993) has outlined another approach to value management adopted by a joint venture of E. C. Harris and Australian Value Management and involves a planned series of highly structured think tank sessions chaired by an outside professional facilitator. The two successive workshops explore the objectives, perceptions and interpretations of the brief and address issues in a pre-emptive way.

On day one of the first workshop, arranged at the earliest possible stage, ideas which may amount to hundreds are reduced to a workable shortlist by rating their cost and functional values. On the second day, approximate cost implications are identified in groups working with the quantity surveyor and project manager. They are finally rated and prioritised for possible incorporation on the third day. After design development, a further three day workshop ensures that the project is reflecting its original aims and that cost effective solutions are being identified. It is claimed that the potential benefits using this approach are substantial and give as an example the £35m savings made on the £100m Brisbane International Airport.

The Concurrent Study

This approach uses the existing project team under the chairmanship of a value manager or facilitator. The group meets on a regular basis during the project design phases, offering maximum continuity. However, it has the disadvantage that creativity is not so evident and it may be more expensive than the 40 hour workshop (Smith, 1993).

The Package Review

This is often used in management forms of contract, wherein package reviews, consisting of a detailed appraisal of each package (or element or trade), are undertaken by the project team as an ongoing process, continuing throughout the design, procurement and construction phases. Discussions with specialist contractors and manufacturers form an important part of this process (Smith, 1993).

The Contractor's Change Proposal

This is a value management change proposal initiated by the contractor after the contract is let. Under US government contracts, the contractor is encouraged to develop value engineering (VE) proposals on a voluntary basis. The contractor then shares in any resultant savings if the VE plan is implemented (Smith, 1993). The major benefit is that it permits the contractor to be pro-active and to use his construction/engineering knowledge and expertise to improve a facility at the on-site stage. Whilst the disadvantage is that the contract may be delayed while the design team investigate the merits and viability of the proposed change. For this reason any changes tend to be relatively superficial (Kelly and Male, 1993).

Design and/or Construction Audit

This process aims to define a project's objectives, by formulating a list of the client's needs and wants, and provides a clear indication of both the cost and the worth of a project. The procedure adopted often follows that of the charette or a 40 hour workshop (Smith, 1993). Kelly and Male (1993) also describe a value engineering audit, whereby a value engineer acting on behalf of a large corporate company or government department reviews expenditure proposals submitted by subsidiary companies or regional authorities, and the procedure follows that of the normal job plan.

VALUE MANAGEMENT STRATEGY

The approach to value management (VM) can vary for each project, but it is customary to provide a job plan to establish the format to be adopted. A job plan should comprise a recognisable set of processes, as now described.

Phase 1: the information stage should cover the assembly of all relevant information appertaining to the project under review and the assimilation and analysis of this information. A cost benefit analysis of objectives should be undertaken, having regard to the client's or end user's method of calculating values, as for example through function analysis techniques and the construction of cost models and possibly FAST (function analysis system technique) diagrams, which are considered later in the chapter.

Phase 2: the creativity/speculation stage which comprises the generation of suggestions as to how the required functions can be performed or improved. Group creative techniques should be introduced such as 'synetics': the art of producing a greater end result than the sum of the individual parts.

Phase 3: the evaluation/analysis stage consists of the evaluation of ideas generated in the creativity phase, for example by collective or individual rating systems. It also entails the rejection of any unproductive, speculative ideas, of which there are inevitably a high number.

Phase 4: the development stage, where the ideas considered at the evaluation stage to have merit are examined and potential savings are costed, with consideration being given to both capital cost and the effect of operational and maintenance costs (life cycle costing). There is considerable scope for the use of cost models and computer aided calculations. Any ideas which either cost more than the original or are found to reduce quality are discarded.

Phase 5: the presentation/proposal stage, comprising the presentation of the refined ideas considered to be worth implementing, supported by drawings, calculations and costs.

Phase 6: the implementation/feedback stage, where the ideas agreed to be worthwhile are then implemented. Feedback from the sponsors of the VM exercise should ideally be passed back to the VM team to complete the learning cycle (Smith, 1993).

COMPARISON OF VALUE MANAGEMENT AND COST MANAGEMENT

Value management can be described as a service which is provided in the earlier stages of a project where the primary goal is to determine explicitly the client's needs and wants related to both cost and worth, sometimes described as judgement values, by the use of functional analysis and other problem solving techniques, which will be described later in the chapter. Whereas in cost management, the main thrust is on cost budgeting, management and control, and embraces such activities as feasibility studies, cost planning in all its aspects, the production of bills of quantities, tender evaluation, on-site measurement and valuation and the settlement of final accounts.

Value management often precedes cost management in its timing but there can be a substantial overlap between the two activities during the inception, feasibility and design stages. In both cases an early investigation is desirable while the design is in its early stages, as the longer it is delayed, the more advanced the design, the less opportunity there is for radical changes to the project, the lower the cost savings that can be made and the higher the cost of implementing them.

Cost management has been defined by Kelly and Male (1993) as a service that synthesises traditional quantity surveying skills with structured cost reduction or substitution procedures using a multi-disciplinary team. However, many quantity surveyors would disagree with this definition, argueing that it is too restrictive in defining their cost management role and that effective cost management is not dependent on a multi-disciplinary approach, valuable though it may be. Furthermore, the quantity surveyor can make an objective client project appraisal and have regard to the client's needs and wants against a background of cost.

It cannot be denied that the value management process entails a detailed methodology aimed at achieving savings in cost and/or increased value of a construction project. By approaching the problem in a well structured and organised way, an increased number of alternative solutions are likely be found, as compared with those emerging from a typical cost management approach. Furthermore, the study is usually taken proactively, as opposed to mainly reactive cost reduction investigations which are often only carried out when the project budget is exceeded (Norton, 1992a).

Functional analysis allows the division of a problem into manageable units and promotes an alternative and more comprehensive approach resulting in the consideration of more solutions. In cost management studies the main thrust is often devoted to reducing the cost of items in which savings are readily identifiable, such as the lowering of standards of finishes or reducing the quantity of expensive components and thereby reducing quality and possibly increasing maintenance cost.

Norton (1992a) believes that value management can achieve more fruitful results than cost management techniques on their own. There is no doubt that the client would receive a more comprehensive and efficient service if both systems could be used together on the same project.

VALUE MANAGEMENT TECHNIQUES

This section of the chapter examines some of the more fundamental and operationally important techniques used in value management, such as functional analysis and FAST diagrams, as it is considered that an understanding of these processes will be helpful to the reader when studying the application of value management to construction projects.

Functional Analysis

General Principles

Functional analysis is a powerful technique in the identification of the principal functional requirements of a project. In general the function of an item or system can be expressed as a concise phrase, often consisting of a verb followed by a noun, as this provides a precise and readily understandable description of the function. Useful active verbs include amplify, change, control, create, enclose, establish, improve, increase, prevent, protect, rectify, reduce, remove and support. A typical verb and noun relationship is door: v. control, n. access; and cable: v. conduct and n. current.

It should be recognised that it is not usually possible to seek alternatives to a technical solution without first identifying the functional definition. For example, light is required in a room (functional definition), and to install a component which emits light is a technical solution. A functional definition is frequently obtained by first seeking a technical solution and then defining the functional performance of that solution.

Functions can be subdivided into primary or basic and secondary. Primary functions are those without which the project would fail or the task would not be accomplished, whereas secondary functions are a characteristic of the technical solution selected for the primary function and may be non-essential, although both need identifying to fully understand the problem. Kelly and Male (1993) give the example of an electric filament lamp which satisfies the primary function of emitting light but is also accompanied by unwanted secondary functions, such as generating heat, inducing glare and looking unattractive, and these secondary functions can be resolved by further technical solutions.

Norton (1992a) has described how for function analysis purposes, most secondary functions have zero use value, but some secondary functions may be essential to the basic (primary) function, in which case they are termed required secondary functions and are allocated a value. A typical example of required secondary functions would include compliance with the Building Regulations.

Cost and worth are allocated to each function. The cost is the amount derived from the cost estimate while worth is the lowest possible cost at which the function can be performed. In practice, worth is generally derived by the value management team making an evaluation based on comparison of standards of the design component, historical cost data and/or experience. The total cost and worth of the component's functions are calculated and converted to a cost/worth ratio. Generally, when a cost/worth ratio is 2.00 or above, the component is likely to be adopted for its cost reduction effect.

Worth of secondary, non-essential functions is taken as zero. The first areas to examine for savings are those that perform secondary functions that can be reduced or deleted entirely without affecting the basic function of the component (Norton, 1992a).

Functional Analysis Applied to Construction Projects

The function of a building is to provide an environmentally controlled space suitable for its required use, and its design constitutes a technical solution to the functional requirements of the space. All products and components used in the building perform a function. Kelly and Male (1992) have described how in a functional analysis the function of each component is examined by asking the question 'what does it do?'. In a value management study the next question is likely to be 'how else can this be achieved?'. An intensive (brainstorming) session is held and other technical solutions are then generated.

Kelly and Male (1993) have subdivided functional analysis into four phases or levels, as follows:

(1) *Task*: The client perceives a problem, which may have been identified through a study of efficiency, safety, markets or profitability. Where a client sees a building as the answer to his problem, he is likely to be faced with a building procurement decision and to subsequently enter into a building contract. An alternative is to first approach a value manager who, with representatives of the client organisation, can carry out a value audit and this will help the client to decide whether the provision of a new building offers the best solution to his problem.

(2) *Spaces*: Having determined that a building is the best solution to his problem, the next stage generally involves the architect or the whole design team preparing the brief along with the client. A full performance specification of requirements may not be available from the client and it may therefore be necessary for the design team to determine the client's space requirements through the production of sketches and cost plans.

(3) *Elements*: This is the stage at which the building assumes a structural form. As Kelly and Male (1993) postulate the purpose of an element is to enclose and make comfortable the space provided, but it does not contribute to the client's requirements.

(4) *Components*: This is where the elements become part of the built form. Contact with the client at this stage is very limited since the client value system is likely to have been incorporated at previous levels. Components are chosen to satisfy the requirements of the elements in terms of surrounding and servicing space (Kelly and Male, 1993).

Norton (1992a) has adopted a different approach and believes that function determination may not always be straightforward, particularly as the basic (primary) functions of one item may be considered at different levels. For example, a building's basic function may be for a developer to create a profit while a basic function on a lower level is to enclose space. The different hierarchal tiers at which the function may be considered may be termed levels of abstraction. It is important to know at the outset of a VM study the operative levels of the function. For instance, should alternative methods to create profit be dominant or alternative methods to enclose space. In practice, the levels of abstraction are defined by factors such as the client's requirements for the study, design stage and the like. Norton (1992a) proposes that in order to assist the identification of levels of abstraction, a hierarchy may be determined based on the 'how–why?' approach.

FAST DIAGRAMS

General Principles

The FAST (Functional Analysis System Technique) evolved from the functional analysis approach to establish a hierarchy of functions and to identify the means by which they can achieve an end result or objective. The principal advantage of the method is that it breaks the overall problem down into individual and readily manageable components and permits a balanced analysis at different levels. It leads naturally to the identification of those items in the current brief or design which attract high cost for low functional value and those items of high importance coupled with low cost. As we shall see later the value of FAST diagrams can be much enhanced by adding the costs of the various activities.

The system revolves around 'how–why?' relationships in the studies by Norton (1992a) and Kelly and Male (1991). Thus they resemble a decision tree, by answering the questions 'WHY?' when reading from right to left and 'HOW?' when reading from left to right, as illustrated in figure 11.1. However, Carter (1991/92) works in the opposite direction, as shown in figure 11.2, and the author does feel that this represents a more logical approach, but both methods should give the same end result, and it could be argued that the choice of method is a matter of personal preference.

Compiling Functional Analysis (FAST) Diagrams

The following procedural notes, kindly supplied by Tim Carter of Davis Langdon Management show very clearly the means of compiling FAST diagrams for construction projects.

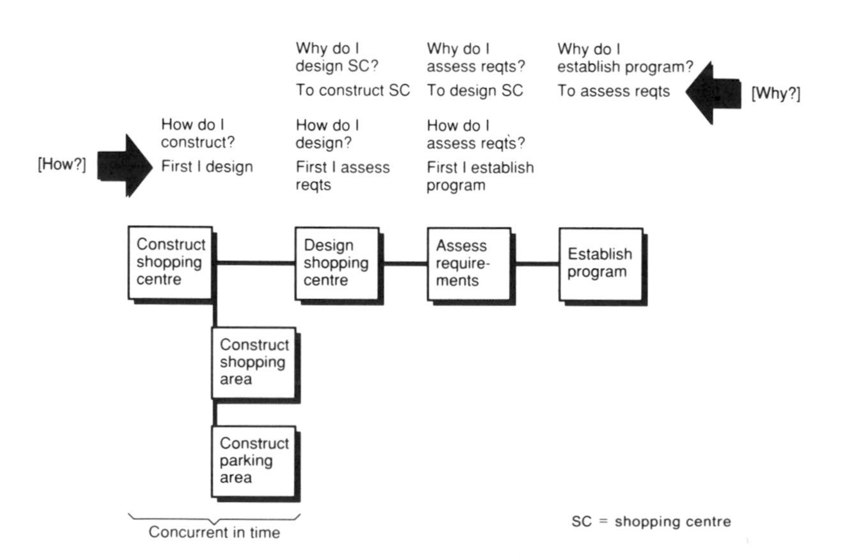

Figure 11.1 Extract from a typical FAST diagram (Source: Norton, 1992a)

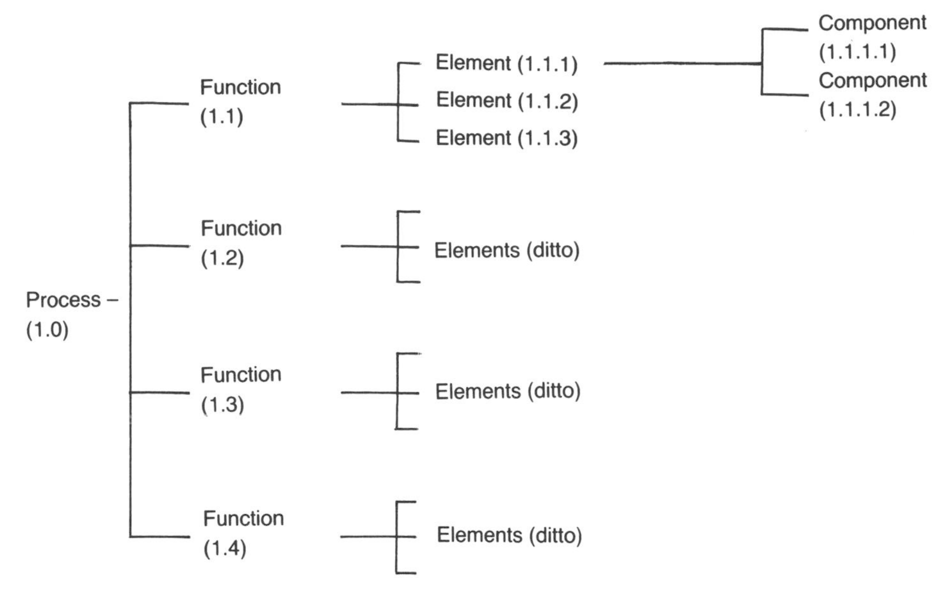

Figure 11.2 Compiling a FAST diagram

Procedure

1. Identify *key* function(s) of project.
2. Compile fast diagram, working from left to right.
WHY? → HOW?
3. Divide/subdivide functions and elements into components to *appropriate* level of detail.
4. Number each item as indicated in figure 11.2.

Typical functions (not *exclusive) identified in earlier studies*:

- Prepare site
- Provide temporary facilities
- Provide accommodation (A)
- Provide internal environment (B)
- Accommodate services (plant rooms, ducts, floor/ceiling voids, etc.)
- Enhance quality (prestige?)
- Reduce costs (operational and/or maintenance)
- Provide flexibility
- Provide for expansion
- Comply with regulations ('in house' or statutory)
- Ensure equipment reliability/availability
- Secure operations
- Safety requirements
- Enhance working conditions (provide acceptable working environment)
- Provide external environment (works)
- Provide welfare facilities } can be combined
- Circulation facilities }
- Accelerate completion.

Not all are relevant; some may be combined or be implicit in the total building function.

Typical subdivision of functions:

(A) *Provide accommodation*

- Support building

Foundations
Slab
Drainage

- Provide envelope
  - Frame
  - External walls
  - Windows/external
  - Doors
  - Roof/RWP
- Divide space
  - Upper slabs
  - Internal walls
  - Internal doors
- Finish surfaces
  - Walls
  - Floors
  - Ceilings
  - Staircases
- Fitting out accommodation
  - Reception desk
  - Signs
  - Kitchen fittings
  - Shelving, etc.

(B) *Provide internal environment*

- Disposal
  - Drainage for sprinklers
  - Vending units
- Mechanical
  - Air conditioning
  - Ventilation
  - Sprinklers
  - Vending units
  - Heating
  - Hot/cold water
  - Refuse disposal
- Primary power
  - Transformers
  - HV Switchgear
  - LV Switchgear
  - LV Cabling
- Electrical
  - Mechanical equipment
  - Lighting (general; emergency)
  - Small power
  - Cable/data highways
  - Security
  - Public address
- Lifts
  - Goods
  - Passenger
- Telecom
  - Patch panel/equipment
  - Cabling (voice/data)

1. Each item to be costed so that Item (1.0) represents total cost of budget in sanction/approved estimate.
2. Allocate costs for preliminaries/fees/inflation (if appropriate) separately – do *not* include *pro-rata* with individual items.
3. Where an item can be allocated into two or more elements, i.e.

   Raised floor can be divided into:
   (A) Provide floor finish
   (B) Provide flexibility (void below floor tiles used to accommodate cabling, etc., to allow desk/equipment configuration or layouts to be changed).

   Item (A) would be the cost of a 'basic' floor finish similar to other 'basic' floor finishes elsewhere in building and
   Item (B) would be the extra-over cost to take the combined total cost of both items to the gross cost of the raised floor, i.e.

	£/M^2
– Provide floor finish (carpet on screed)	22.00
– Provide flexibility (raised floor)	28.00
	£50.00

4. Absolute accuracy is *not* essential; it is the relative costings assigned to each function/element/component that are important.

Criteria Scoring/Alternative Analysis Matrix

Norton (1992a) has described how during the analytical phase of a value management study, the ideas generated during the creative phase are sorted into a list of feasible lower cost or energy saving alternatives. The advantages and disadvantages of each idea are listed or discussed and the ideas are subsequently ranked in order of viability. Selected viable ideas are then evaluated in detail and capital and life cycle costs estimated.

Having ascertained the cost effect of the ideas, alternatives may then be the subject of a weighted evaluation that includes a consideration of intangible factors, such as aesthetics, flexibility, reliability and the like. Tangible and intangible criteria are listed and weighted by the value management team in accordance with the client's requirements, using tools such as the criteria scoring matrix, as illustrated in figure 11.3. These weightings may then be applied to alternatives using an alternative analysis matrix or equivalent method. The alternative achieving the optimum weighted score is considered to be the most viable option and is then presented to the client during the recommendation phase (Norton, 1992a).

VALUE MANAGEMENT CASE STUDIES

Tim Carter of Davis Langdon Management kindly agreed to the inclusion of two interesting and informative value management studies being included in the book and they do very much help to bring the subject to life. The first is concerned with the site preparation for a computer centre in Northern England and the second and more detailed study encompasses a bank processing centre, which was also located in Northern England.

Computer Centre, Northern England

Carter (1992b) has described how the site for the computer centre was a very large triangular shaped area of reclaimed chemical works, bounded on one side by a small stream and on the other two sides by roads. The study was commissioned during the detailed design/tender period shortly before the site works were commenced.

The value management study occupied two days and the VM team prepared a FAST (functional cost analysis) diagram, the summary of which is shown in figure 11.4. On examination, one element which was clearly very costly was the site preparation function, which represented almost 16 per cent of the overall total investment. The detailed FAST diagram breakdown (figure

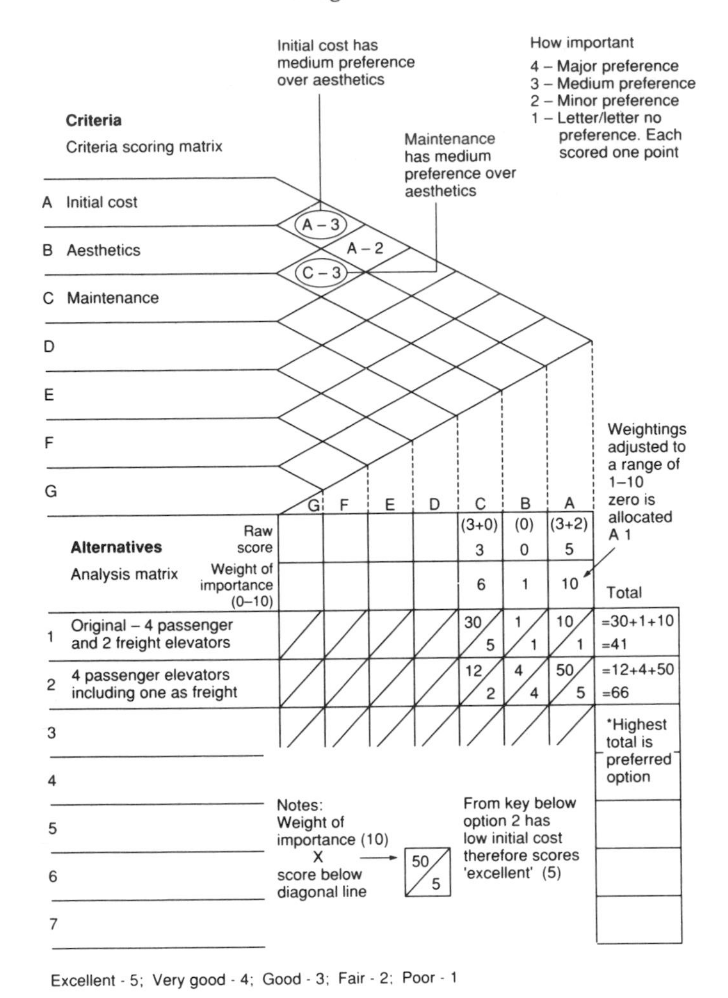

Figure 11.3 A criteria scoring matrix (Source: Norton, 1992a)

11.5) indicated a 2.00 m deep oversite fill of imported crusher-run limestone costing over £300 000. This component was considered necessary by the design team because the stream had flooded in living memory, although the exact timing was unclear, and the project site was at some risk of being inundated.

During brainstorming, the following options/ actions were suggested:

(1) Provide an earth mound levee or sheet steel piling flood bank along the stream frontage to retain the floodwater in the event of the stream flooding again.

(2) Redesign the building to provide a basement with a water-retaining structure having the ground floor above the perceived floodwater level.

(3) Check with the local meteorological records to determine the height (and date) of the earlier flood; assess the likelihood/probability of another flood occurring during the next 10–20 years and consider the 'do nothing' option if the risks were assessed to be acceptably low.

(4) Redesign the building to elevate it on piled columns with an open basement below for car parking (planned elsewhere on higher ground).

(5) Relocate the building by exchanging with

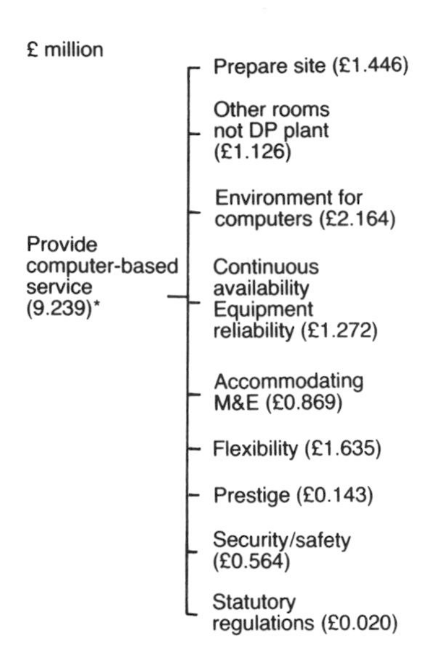

Figure 11.4 Summary of FAST diagram of Computer Centre project (Source: T. Carter, 1992b)

the car parking area (above the previous flood level, elsewhere on the site); in the event of flooding, cars could be temporarily parked above the flood level elsewhere on the site.

Regrettably, the client was unable to consider any of these options, primarily because the programme for the building was critical and any significant delay was unacceptable. Hence he elected to continue with the proposed design. However, all parties agreed that had the value management study taken place at the end of the RIBA design stage 'D' (scheme design) or, preferably, earlier, then one or more of the options listed could have been implemented. Carter (1992b) commented that this experience is all too typical and is a risk to be considered when value management studies are commissioned too late in the procurement programme.

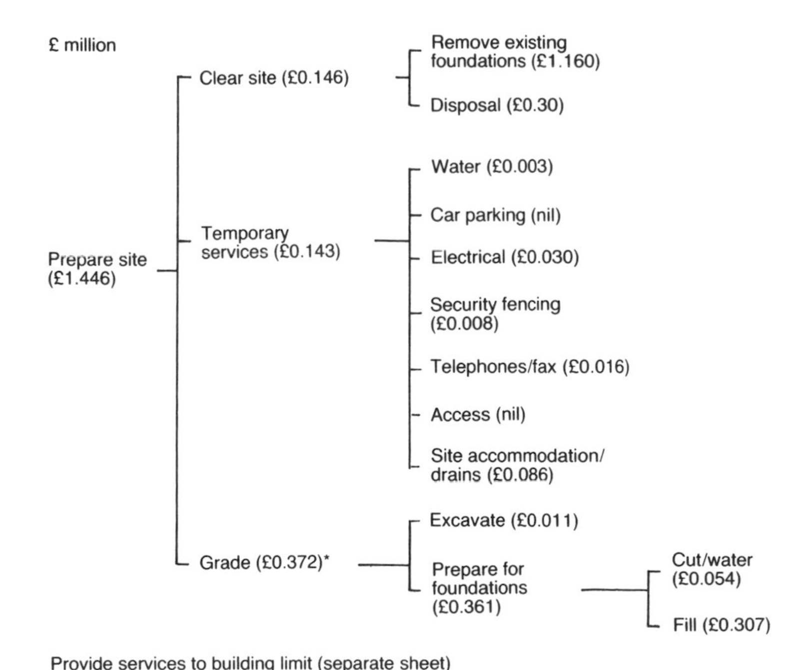

Figure 11.5 Detailed FAST diagram of site preparation work to Computer Centre project (Source: T. Carter, 1992b)

Bank Processing Centre, Northern England

This case study based on papers provided by Tim Carter of Davis Langdon Management explains the main processes and operation of value management techniques as applied to a specific project, using the two day study approach.

It starts by detailing the key elements of the client's brief and, as could be expected, these call for a high quality building which is also efficient, user friendly and secures the lowest operating costs, with an 18 month timescale.

This is followed by a schedule of the VM study team members who were present on the two allotted days (a Friday and the following Monday). It will be noted that the client is represented by the project manager, facilities manager and a director. All the design team are well represented by all the senior staff concerned with the project, as is also the appointed management contractor, as this project is being undertaken as a management contract. The meetings were chaired by a partner of the value management consultancy supported by a senior project manager.

Figure 11.6 shows a FAST (functional analysis) diagram summary showing all the functions with their estimated costs, all aimed at rationalising operations and achieving lower operating costs, at an overall cost of £8.884m. The most expensive items are providing accommodation (1.2) at £1.766m and the provision of the internal environment (1.3) at £2.034m, and these are likely to be the prime areas for investigation to reduce costs without lowering standards. A cost reduction of £0.216m is scheduled.

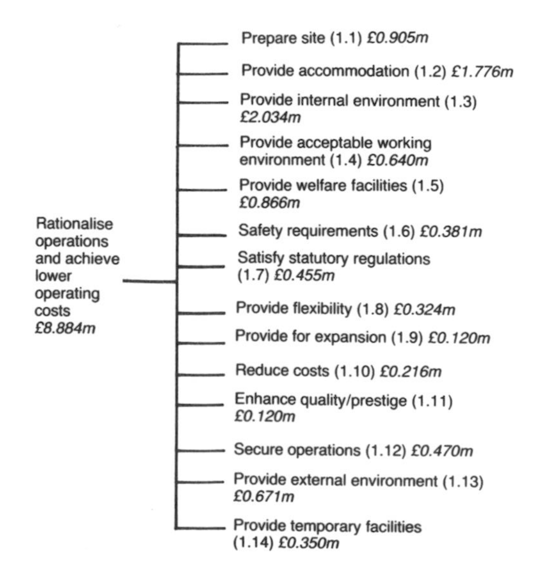

Figure 11.6 Summary of FAST diagram of Bank Processing Centre

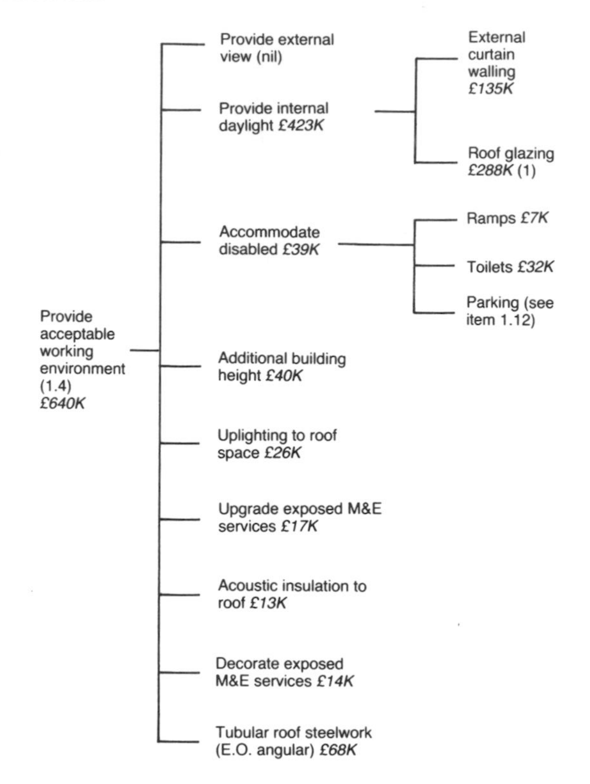

Figure 11.7 Detailed FAST diagram of provision of acceptable working environment to Bank Processing Centre

Figures 11.7 and 11.8 illustrate detailed FAST diagrams covering 'providing an acceptable working environment' and 'secure operations', each with extensive components, all individually priced. In practice, detailed FAST diagrams will be prepared for all the 14 functions.

The value management team examine all possible alternatives with the objective of meeting the required performance standards at lower cost

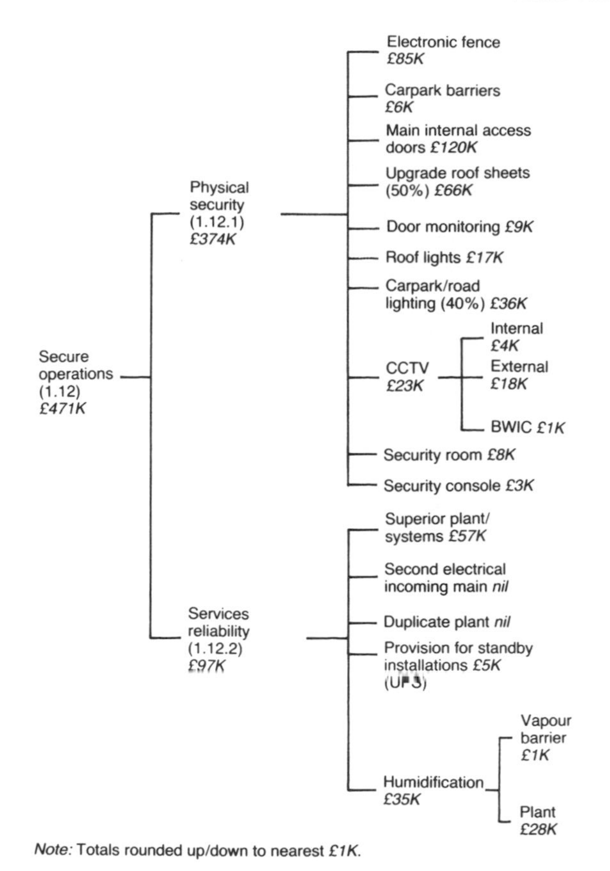

Figure 11.8 Detailed FAST diagram of secure operations to Bank Processing Centre

or improving standards for the same cost. As is normally the case, a considerable number of options were scrutinised and rejected by the management team as failing to meet the prescribed criteria, while many more were considered and were the subject of further investigation by the project team. A schedule shows these latter items with their likely cost savings, or extras where their provision is advisable to achieve the required standards and/or reduce maintenance or operating costs.

Finally, the overall cost benefit of the value management study is listed and in this particular project yielded savings of £296.1K representing an overall saving of 2.7 per cent, and taking the cost of the study at approximately £8K, a cost benefit ratio of 1:37 is obtained.

In order to obtain a feedback from all the VM study team members, a questionnaire was circulated and the results were recorded for future reference and action. It is interesting to note that the majority felt the study was both instructive and constructive, giving them a better concept of project value. There was little or no intimidation of participants by the VM chairman. There was however a feeling that the study's findings would have been identified without the VM study and that good value for money already existed. There was also a fairly strongly held view that the study was carried out too late in the design process, highlighting once again the importance of the study/workshop being held early in the design process. Finally there was strong support for the summary of cost by functions as a good test of value and the use of value management on other projects.

Some individual comments from members of a value management team on a large office and residental project in London in 1989 are the author believes worthy of note, and are now listed:

(1) It seems that the greater benefits offered to the client as a result of the two day workshop/ study were in changes resulting from re-examining and then revising the constraints on design that were previously held to apply.

(2) Not only does it give you the opportunity to produce ideas and concepts but, equally valuable, gives you the chance to re-examine the validity of earlier decisions/actions.

(3) The study should be held much earlier (say six weeks previously)

(4) Some of the suggestions accepted would not have been adopted had there not been the value manager and the workshop/study to act as catalysts. Thus the workshop's findings would not have come out anyway because most of them had been previously discussed but discounted, for whatever reason.

(5) The presence of the client at the workshop/ study was significantly useful. Not only does the client now have a better picture of what he is getting and why, but also the design team now know clearly what he wants/is willing to accept.

For these reasons alone, with no significant improvements arising from it, the workshop would have been valuable.

(6) Costing of functions is, I believe, a useful aid to bringing home the real costs of decisions taken and expressing them in terminology that both consultants and the client can more readily understand.

Bank Processing Centre – Northern England
VM Study 16–20 March 1990
Key elements of client's brief

1. Provide contingency processing operations (urgently).
2. Rationalise rented accommodation in NW England.
3. New building to be:
   - Modular
   - Single storey
   - Open plan/highly flexible
   - Utilitarian/functional
   - Energy efficient
   - Single standard
   - Expandable (100%)
   - User-friendly
   - Controlled access
   - Phase 1: 800–900 staff
     4:1 female/male
     Age 16–22
   - Comfort cooled (not A/C) for 120 W/m^2
   - Without suspended ceiling
   - Raised floors in operational areas.
4. New building to enable client to achieve/maintain *lowest* processing costs.
5. New building to project image of hi-tech/cost-effective client (i.e. as item 4 above).
6. Building *must* be operational by October 1991.

VM Study Team members (both days)*

Client	– Project manager
	– Facilities manager
	– Director
Architects	– Partner
	– Project architect
Structural	– Partner
Engineer	– Project engineer
Services	– Partner (mechanical)
Engineers	– Partner (electrical)
Management	– Contract manager
Contractor	– Commercial manager
	– Services manager
QS	– Partner
	– Associate (services/electrical)
	– Project QS (building)
	– Project QS (mechanical)
VM Chairman	– Partner
	– Senior project manager

(18 × 2 mandays)

* Friday and following Monday.

(A) *Options considered and rejected by VM team*

1. Provide site concrete batching plant
2. Reduce 600 mm concrete oversite slab
3. Bend reinforcement on site
4. Prefabricate reinforcement mesh to piled area slabs
5. Divert SW drainage into canal
6. Consider integral plant rooms
7. Re-use existing buildings for main restaurant
8. GRP/polycarbonate rooflights in lieu of glass
9. 2.00 m high block perimeter wall in lieu of metal cladding
10. Ha-ha to site perimeter in lieu of security fence
11. Client to insure all works (clause 22B)
12. Refurbish existing pumphouse; use as site offices
13. M/C to purchase temporary site offices at end of project
14. Omit tubular steelwork and use standard angles/channels.

(B) *Options considered and to be investigated by project team*

		Approx. *(saving)* *extra* £/K
1.	Bay sizes altered by 90 degrees	–
2.	PC planks to walkways in lieu of solid	7.5
3.	Renegotiate connection charges/electricity charges by NORWEB	(39)
4.	Reconsider allowance *£108K* for louvres (Item 1.3)	(20)
5.	Reconsider roof elements (materials, sizes, span, maintenance)	Range (50–100)
6.	Balance daylight/solar gain with fewer rooflights/artificial lighting	Range (8–70)
7.	Check external doors budget *£85K* (Item 1.2)	(4)
8.	High level clerestorey glazing to east wall in lieu of rooflights	–
9.	Check *£32K* budget for disabled toilet (Item 1.4)	–
10.	Catering company to provide catering equipment and rentalise and/or	–
11.	Reconsider scope of catering equipment *£319K* (Item 1.5)	(81)
12.	Reduce lightning protection and use steel frame (assess risks)	(2)
13.	Client to consider fire compartmentation policy (provision for fire wall)	(1)
14.	Check contingency of *£57K* to strengthen roof steelwork for sprinklers to be added (Item 1.8)	
15.	Verify 20% expansion provision in catering equipment	See Item 11
16.	Phase carpet installations with phased occupation	–
17.	Cheaper carpet tiling to non-walkway 'corridor' areas	–
18.	Consider bonded carpet tiling to main areas and/or 'corridors'	–
19.	Consider gravel in lieu of paving alongside canal	–
20.	Low fence or rail alongside canal as demarcation barrier?	–
21.	Consider turnstiles (card key controlled)	

	and CCTV in lieu of main reception security screen/ oscillating doors	(30)
22.	Re-assess security provision; costs v. risks	See Item 25
23.	Consider cheaper pavings around building perimeter (blocks costing *£67K* in Item 1.13)	(2)
24.	Plan landscaping one season ahead to obtain savings	–
25.	Question *£85K* 'electronic' security fence to site perimeter (Item 1.12)	46
26.	Upgrade painting specification to 'exposed' roof steelwork to reduce maintenance.	40
Net saving	£(257.5K)	

Overall cost/benefit of VM study

	Budget £/K	*VM savings* £/K
1. Construction works/MC's fee	9515	(257.5)
add		
2. Professional fees (15%)	1427	(38.6)
Estimated costs	£10942	(£296.1)

VM study yielded approx. *2.7%* savings overall
Cost of study approx. *£8K*
Cost benefit ratio = 1:37
Note: Savings would have been much higher had director not changed shortly before VM study (new director reluctant to take risks!)

Questionnaire
0 = Absolute no; 10 = Absolute yes

Questions	*Responses by value management team*														*Average*
1 Was the workshop instructive?	8	8	6	8	9	7	6	4	4	8	8	9	9	7	7.21
2 Were the two days a constructive use of time?	9	5	8	7	9	5	4	5	4	8	4	9	4	7	6.29
3 Were the participants intimidated by the outside VM chairman?	2	0	1	0	1	0	0	0	0	0	1	3	2	3	0.93
4 Full team turnout; is this an effective means of reviewing problems?	8	10	10	10	9	10	9	10	10	8	3	10	9	10	9.00
5 Do you now have a better understanding of where the money is?	7	10	8	9	8	10	5	4	2	5	2	9	8	8	6.78
6 Do you now have a better concept of project value (following the workshop)?	7	10	7	8	7	10	5	4	0	2	7	9	10	8	6.71
7 Do you now have a better understanding of the client's requirements?	8	6	–	9	7	5	1	2	0	2	5	9	5	6	5.00
8 Would the workshop's findings have come anyway?	9	9	6	5	5	9	6	5	5	10	8	6	6	3	6.21
9 Will the workshop's results prove a disruptive influence?	2	3	0	5	2	3	6	4	0	0	5	5	4	5	3.14
10 Was the workshop too late in the design process?	5	10	2	2	9	10	9	3	0	5	6	8	7	5	5.79
11 Are you concerned at the implications of the results of the workshop?	0	0	2	2	2	0	6	10	0	0	2	3	2	2	2.21
12 Did the workshop prove that good value already existed?	9	10	8	8	9	10	8	7	10	10	7	7	7	8	8.42
13 Is the summary of cost by 'functions' a good test of value?	9	8	8	8	8	8	8	10	0	5	7	8	9	7	7.36
14 Would you recommend the use of value management on other projects?	9	5	10	9	10	5	5	8	5	5	7	10	7	9	7.43

CONCLUSIONS

It has been clearly established in this chapter that value management is not merely a cost cutting exercise, it also takes account of the three-way relationship between function, cost and value.

McElligot and Norton (1995) have identified three important aspects of value improvement:

- same performance at reduced cost;
- improving performance at same cost;
- improving performance at reduced cost.

In their comprehensive and practically oriented book entitled *A Practical Guide to Value Management in Construction*, McElligot and Norton explore very fully the different job plan phases of

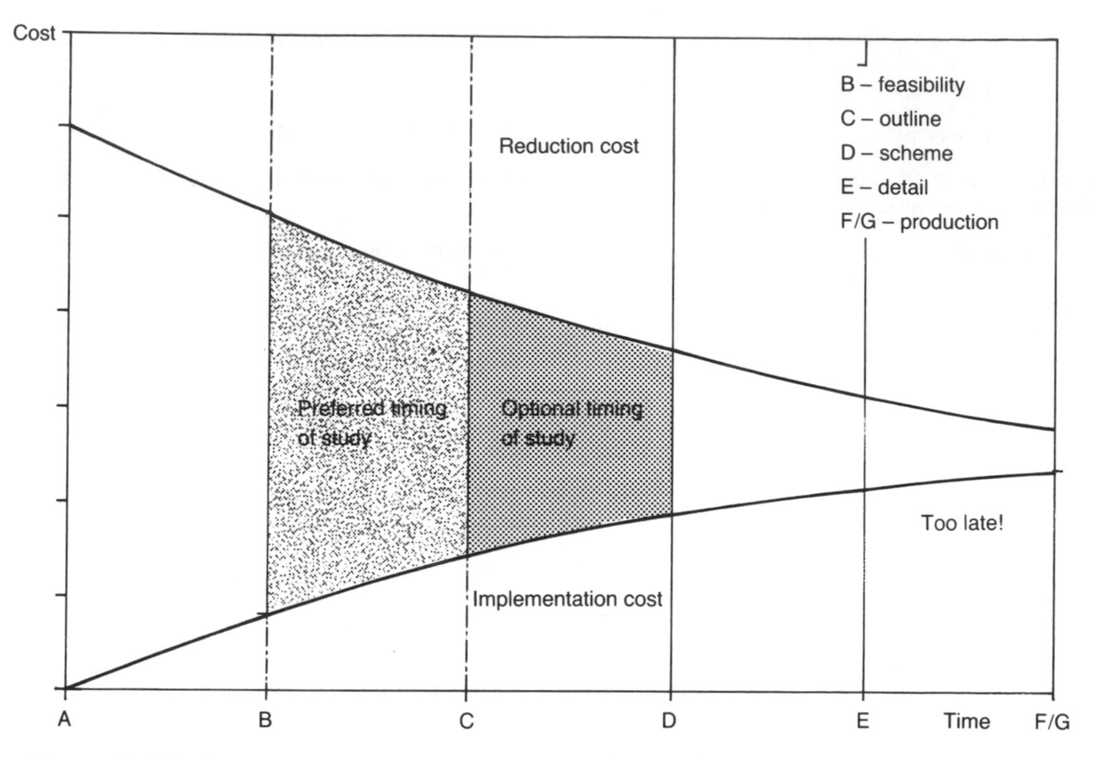

Figure 11.9 Value management: optimum timing for study (Source: T. Carter, 1991/92)

a value management study, supported by a wide range of case studies. The different techniques and approaches are examined and compared with their potential benefits, which the author has endeavoured to outline in this chapter, and the way in which value management can complement cost planning, to provide an improved service to the client.

The timing of the value management study can be critical and figure 11.9 produced by Carter (1991/92) illustrates very clearly the optimum time for conducting such a study. Currently, such studies/workshops are conducted at between 10 to 35 per cent stage of the design process, using the design team chaired by a value management team co-ordinator or by an independent value management team.

Smith (1993) emphasises that, where appropriate, consideration should be given to capital cost, life cycle costs, programme, buildability and/or optimum return on money. By adopting this approach, good value for money can be achieved. It is also important that design teams should not see value management or design/cost reviews as an attack on aesthetics or quality. Equally, alternative solutions should not be viewed as criticisms of the existing design.

Brown (1992) believes that value management represents a natural progression for the quantity surveyor in leading the search for alternative technical solutions and presenting them as evaluated and costed options.

There are doubtless many benefits to be gained by adopting value management techniques and Beard Dove have identified the following aspects as being the most important:

(1) reduce project costs;
(2) improve design efficiency;
(3) optimise value for money;
(4) concentrate design effort;
(5) advance design decisions;

(6) highlight design options for selection;
(7) improve ways to comply with the brief;
(8) afford an independent functional review.

Carter (1991/92) has critically examined both the benefits and possible disadvantages of value management, and his findings follow:

Benefits

- Examines function and cost
- Provides opportunity for options to be considered
- Seeks better technical and more cost-effective solutions
- Identifies and reviews constraints and criteria affecting the project
- Design changes can be accommodated at minimal cost (if study conducted early)
- Opportunity for in depth project review and greater understanding for all team members
- Team building
- Identifies and can eliminate unnecessary costs
- Generates greater client confidence
- Can shorten overall programme period (longer brief/design period and shorter production/construction period)
- Assists client decision making

Disadvantages

- Extra work for existing project team, which is not always reimbursed, as it is at the client's discretion
- Disruption to project team
- Can incur extra fees
- Can extend design period

Carter (1991/92) also raises the very pertinent point concerning traditional projects, as to how many where the tender comes within budget does the design team look for any savings/improved value, even though there could be a potential saving of as much as 5 to 10 per cent on many schemes.

12 VALUATION PROCESSES

This chapter is concerned with the nature of value and investment, the construction and use of valuation tables and methods of valuation. Until quite recent years these matters were considered to be solely the province of the valuation surveyor, but it has now become apparent that the quantity surveyor also needs to be familiar with some of the valuation techniques and certain of the valuation tables, in order to be able to make feasibility studies and to deal satisfactorily with future costs.

THE CONCEPT OF VALUE AND INVESTMENT

Value

The cornerstone of the economic theory of value is that an object must be scarce relative to demand to have a value. Where there is an abundance of a particular object and only limited demand for it, then the object has little or no value in an economic sense. Value constitutes a measure of the relationship between supply and demand. An increase in the value of an object is obtained either through an increase in demand or a decrease in supply. Value also measures the usefulness and scarcity of an object relative to other objects or commodities.

The degree of response of supply and demand to price changes is referred to as the elasticity of supply or demand. Where a small change in price causes a large change in demand, then the demand is elastic, but if a large change in price leaves the demand virtually unchanged, then the demand is inelastic. The elasticity of demand is very much influenced by the availability of suitable substitutes. There are also the short term and long term requirements of changes in supply to be considered.

Surveyors, being property professionals, are primarily concerned with the value of property, sometimes referred to as 'landed property', and this embraces all forms of land and buildings which may be put to a wide variety of uses. The market value of an interest in property will be the amount of money which can be obtained from a willing purchaser at a specific point in time, and is generally determined by the interaction of the forces of supply and demand. It will be appreciated that the supply of land as a whole is fixed, apart from changes due to reclamation or erosion, but the quantity of various types of property is variable, as land and buildings can be transferred from one use to another, existing buildings demolished and new ones built. The value of a specific form of property will be influenced by the amount coming on to the market at a particular time rather than the total stock in existence. It takes time to transfer one form of property to another use and to erect buildings to meet an increased demand, and so the supply of property is generally regarded as inelastic.

The demand for any particular type of property is influenced by a number of factors, such as:

(1) population changes;
(2) changes in the standard of living or in taste or fashion;
(3) changes in society;
(4) population movement;
(5) changes in social services (shops, schools, libraries, health centres and other facilities);
(6) nature of adjoining buildings/uses;
(7) changes in communications;

(8) changes in statutory requirements, such as the Town and Country Planning Acts;
(9) inflationary trends; and
(10) availability of finance.

Each property is unique, with its own specific location and characteristics and no one property is a perfect substitute for another. It is these factors which make the valuation of properties so difficult.

Investment

In a capital investment project there is an outlay of cash in return for an anticipated flow of future benefits. The consequences of capital investment extend into the future and may involve decisions as to the type and/or quality of a new building and its best location. Buildings cannot always be readily adapted to other uses, so wrong development decisions can result in heavy losses to investors. In addition, future benefits are always difficult to evaluate. When comparing alternative building solutions it is essential that total costs are used. In this situation it is necessary to compare both present and future costs on a common basis with the help of valuation tables, which will be described later in this chapter.

Property has a basic characteristic of relative durability and can be used over lengthy periods of time. It is accordingly capable of yielding an income as individuals will be prepared to make periodic payments for its use. When an investor purchases an interest in property, he is tying up a certain amount of capital in the property and will expect a reasonable return comparable with what he might have received had he invested it elsewhere. The amount of yield or rate of interest will vary with the degree of security, regularity of payment, period of investment, ease of convertibility of capital and cost of acquiring or disposing of the asset. Inflationary tendencies and taxation arrangements also have a bearing on interest rates and the relative desirability of the investment. Changes in rates of income tax, property tax, capital gains tax, and investment grants and allowances will influence interest rates. Nevertheless, interest rates on property tend, after a suitable lapse of time, to be similar to the yields of the nearest substitute in the capital market. There are, however, essential and significant differences between property and other forms of investment, as now described.

(1) There is no central market for the comparison of prices of property as with the Stock Exchange. The transfer of property by conveyance is both costly and time-consuming.
(2) It is not possible to divide property into small units like shares on the stock market, and it is therefore difficult for an investor to invest small sums in property.
(3) The management of property creates problems which do not arise with other forms of investment.
(4) The income or rate of return from property can normally only be varied at the end of comparatively long leases, whereas the income from ordinary shares can vary annually.

METHODS OF VALUATION

The main function of the valuation surveyor or valuer is to assess the value of any type of property under any set of conditions. Valuations are required for a variety of purposes – for sale, for purchase for occupation or investment, for determining auction reserves, mortgage loans, inheritance tax, or for income tax or local taxation purposes. Property values vary considerably from one district to another and so a valuer needs to have extensive experience of values in the area in which he is practising. It is a specialised function involving its own particular expertise and the quantity surveyor would be wise to consult a valuer whenever valuation of property is concerned.

A number of methods may be used to assess the market value of an interest in landed property.

Comparison Method

This method is a popular valuation technique and consists of making a direct comparison with the prices paid in the open market for other similar properties, where reasonably close substitutes are available and transactions occur quite frequently. Its prime use is for residential properties where there is likely to be a greater similarity between different properties. Difficulties do, however, frequently arise in the use of this method as it is unusual to find two entirely similar type properties – differences occur in size, amount of accommodation, quality and extent of finishings and fittings, condition of property and its situation. For instance, the price paid for one block of offices may not be a very good indicator of the value of an adjacent office building which may differ considerably in room sizes, internal layout, type of finishes and in many other ways. Furthermore, prices may vary appreciably over relatively short periods of time and so the valuer must also have regard to current trends.

The valuer generally finds it helpful to break down the property into suitable units for comparison purposes. Land can conveniently be priced per hectare or possibly per metre of frontage in the case of building land, and buildings might be reduced to the price per square metre of gross floor area (total area inside enclosing walls). It is also advisable to have regard to the underlying economic factors influencing the prices as well as the prices themselves.

Contractor's Method

The basis of this approach is that the value of the land and buildings is equivalent to the cost of erecting the buildings plus the value of the site. This is usually an unsound assumption as the value of a property is determined not by what it cost to build but by the amount which purchasers in the open market are prepared to pay for it in relation to the price the seller is prepared to take. Its main use is for buildings which rarely change hands, such as hospitals, schools, town halls and sewage treatment works, where there is little or no evidence in the form of sale prices.

When applying this method it is necessary to make allowance for depreciation in older buildings, as a building which is sixty years old cannot have the same value as a similar type of building of comparable size and construction, but using modern materials, and is only five years old. Some buildings may be excessively ornate or extravagant in their construction and finishings and the value of these properties may not necessarily be increased in proportion to the additional expense incurred. A house specially designed to meet the needs of a particular occupant may not suit the requirements of prospective purchasers. Extreme care is needed when assessing allowances for age and the Lands Tribunal prefers values based on comparable rentals (Richmond, 1994).

Residual Method

This is a valuation method which is sometimes used where the value of the property can be increased by carrying out certain works of development or redevelopment. A large house could, for instance, be profitably converted into flats when its potential will be exploited to the full. The building could be valued by taking its value after conversion and deducting the cost of conversion plus an allowance for developer's risk and profit. The residual figure will indicate the value of the property in its existing state but with a potential for development. The same method can be used for the valuation of land with potential for development as shown in chapter 15.

Reinstatement Method

This method is used to estimate the cost of rebuilding a property, probably destroyed by fire, and adding to it the value of the land on which it stands. It may be used for fire insurance purposes to calculate the annual premium.

Profits Method

This is sometimes described as the accounts method and is used where the value is largely dependent upon the earning capacity of the property, as is the case with hotels, public houses, theatres and dance halls. The usual approach is to estimate the average annual gross earnings and to deduct from them the working expenses, interest on capital and tenant's remuneration. The balance represents the amount that is available for the annual rent, which is then capitalised by an appropriate Years' Purchase, as explained later, to arrive at the capital value. It is an exceedingly indirect approach and is best checked by some other method, such as the value per cinema seat or hotel bedroom. It has, however, been found useful in rating valuations for the classes of property previously described, which require a specialist skill.

Investment Method

This method can be used where the property produces an income, as there will be a direct relationship between the income accruing and the capital value of the property. The income must show a reasonable return compatible with the interest which could be earned by investing the capital elsewhere. An example will serve to illustrate this aspect.

If an investor purchased a freehold property at £100 000 and required an eight per cent per annum rate of interest on his capital, he will only secure the required return if the net income accruing from the property is £8000 per annum (£100 000 × 8/100). When the position is reversed, it is possible to calculate the capital value from the return and required rate of interest, thus

$$£8000 \times \frac{100}{8} = £100\,000.$$

Years' Purchase

The multiplier used in the last example – 100/8 may be described as *years' purchase* (YP) or the *present value of £1 per annum*. Net income × years' purchase = capital value.

With a perpetual income, years' purchase or YP can be obtained by dividing 100 by the interest rate, thus

$$\text{YP in perpetuity} = \frac{100}{\text{rate of interest}}$$

and YP in perpetuity at 6 per cent $= \dfrac{100}{6}$

$= 16.67$

The following examples will illustrate its use.

Example 12.1. Value a freehold interest in a shop producing a net income (after deduction of all outgoings) of £7000 per annum. It can be assumed that a purchaser will require a return of six per cent on this capital.

Net income	£7000 pa
YP in perpetuity at 6 per cent (100/6)	16.67
Capital value	£116 690
Rounded off to	£116 700

Example 12.2. Value a house capable of producing a net income of £5000 per annum. This investor requires an eight per cent rate of interest.

Net income	£5 000 pa
YP in perpetuity at 8 per cent (100/8)	12.5
Capital value	£62 500

An analysis of each of these examples shows that the net income from the property represents the required rate of interest on the capital value or purchase price, for example

six per cent on £116 690

or

$$£116\,700 \times \frac{6}{100}$$

$= £7000$ pa

The figure of years' purchase (YP) varies considerably over time and with the type of property and the degree of risk involved. Thus properties involving greater risk require higher rates of interest and lower YPs. The following figures for YPs and yields are typical for the various classes of property, and vary with design, location and other factors.

Houses	– 12.5 to 9.09 YP	– 8 to 11% interest
Shops	– 20 to 11.11 YP	– 5 to 9% interest
Offices	– 16.67 to 10 YP	– 6 to 10% interest
Factories	– 12.5 to 9.09 YP	– 8 to 11% interest

Sinking Funds

Leasehold properties reduce in value throughout the duration of the lease, until finally at the termination of the lease, they cease to have any value. It is customary, therefore, for a leaseholder to provide for a sinking fund which will recoup the initial capital sum by the end of the lease. The interest on the capital and that on the sinking fund are generally at different rates (say 8 per cent and 2½ per cent). For this reason it is usual to use the dual-rate tables provided in valuation tables when valuing leasehold interests. Where a leasehold interest is sublet and the net rent so received exceeds the rent paid to the landlord, then a profit rent exists (the difference between the two). If taxation is taken into account a realistic figure might be 30 p in the £.

An example follows to show the method of valuing leasehold interests.

Example 12.3. X owns the freehold interest of a house which is let on an annual tenancy on full repairing and insuring terms at a rack rent (full rental value) of £600 pa. Y has a lease of a similar house with forty years to run at a ground rent of £600 per annum. Value the interests of X and Y.

Freehold interest of X	
Net income per annum	£6000
YP in perpetuity at eight per cent	12.5
Capital value	£75 000
Leasehold interest of Y	
Net rent received	£6000
less ground rent	600
Profit rent per annum	£5400
YP for forty years at 8½ per cent and 2½ per cent (tax 30p in the £) (figure obtained from valuation tables)	9.278
Capital value	£50 101
Rounded off to	£50 100

Note: Dual-rate tables have been used to value the leasehold interest. The remunerative rate of interest has been taken at one half per cent above the corresponding freehold rate to allow for the greater risk involved, and an accumulative interest rate of 2½ per cent for the sinking fund.

VALUATION TABLES

The quantity surveyor needs to be able to use certain valuation tables in connection with some of his cost planning calculations. Where returns are spread over a number of years, or the calculation involves both initial capital costs and annual running and maintenance costs, with possibly replacement costs at intervals throughout the life of the building, then the calculations become more complicated and valuation tables will assist in the computations. It may be desirable to obtain the *present value* (PV) of future expenditure or to convert from present and future costs to *annual equivalent* costs, and examples of both processes in life cycle costing calculations are given in chapter 13.

The best source of valuation tables is the current edition of *Parry's Valuation and Investment Tables* (Estates Gazette). Students are advised to examine a copy of these tables in order to become familiar with their general form, layout and contents. Some abridged valuation tables appear in appendices 1 to 5 at the back of this book to assist the reader in working through the various life cycle costing and other worked examples. The nature, construction and application of some of the more important valuation tables will now be described.

Amount of £1 Table

Extracts from the above table are shown in appendix 1 at the back of this book. It indicates the amount to which a sum of £1 will accumulate if invested at compound interest over a certain period of years. The table is based on the assumption that if £1 is invested for a given number of years at a specific rate of interest, at the end of the period the investor will receive his original £1 together with the compound interest which has accumulated on it. It is an important table, as it forms the basis for many of the other valuation tables.

This table is constructed in the following way.

	Amount invested at start of year	*Interest payable*	*Amount invested plus interest*	*Amount owed to investor at end of year*
Year 1	1	i	$1 + i$	$1 + i$
Year 2	$1 + i$	$(1 + i)i$	$(1 + i) + (1 + i)i$	
			$= 1 + i + i + i^2$	
			$= 1 + 2i + i^2$	$= (1 + i)^2$
Year 3	$1 + 2i + i^2$	$(1 + 2i + i^2)i$	$(1 + 2i + i^2) + (1 + 2i + i^2)i$	
			$= 1 + 2i + i^2 + i + 2i^2 + i^2$	
			$= 1 + 3i + 3i^2 + i^3$	$= (1 + i)^3$

where i = the interest payable on £1.

A relatively simple example will serve to illustrate its application.

Example 12.4. Calculate the amount to which £1 invested at 8 per cent compound interest will accumulate in 3 years.

$$i = \frac{R}{100} \text{ where } R \text{ is the rate of interest (per cent)}$$

$$i = \frac{8}{100} = 0.08$$

Amount of £1 (A) = $(1 + i)^n$ where n = number of years for which the sum is invested.

Amount of £1 in three years at eight per cent $= (1 + 0.08)^3$
$= 1.08^3$
$= 1.08 \times 1.08 \times 1.08$
$= 1.260$

A much quicker approach, particularly where a large number of years is involved, is to obtain the appropriate multiplier from the amount of £1 table, part of which is shown in appendix 1.

Capital	£1.00
Amount of £1 in three years at eight per cent	1.260 (1.2597)
Capital plus interest	£1.260

Two further examples illustrate practical applications of this particular table.

Example 12.5. To what sum will £1000 accumulate if invested for twenty years at nine per cent compound interest?

Capital (amount invested)	£1000
Amount of £1 in twenty years at nine per cent	5.604
Accumulated amount	£5604

Example 12.6. An investor pays £600 000 for a building site and it remains undeveloped for five years. Calculate the cost of the land to him at the end of this period, assuming an eight per cent rate of interest.

Purchase price	£600 000
Amount of £1 in five years at eight per cent	1.469
Total equivalent cost	£881 400

This assumes that if the investor had not purchased the building site, he would have invested the money elsewhere and obtained interest on it.

Present Value of £1 Table

This table shows the amount that must be invested now to accumulate to £1 at the end of a prescribed period at a specific rate of compound interest. Whereas the amount of £1 table commences with a capital sum of £1 and ends with £1 plus compound interest, the present value of £1 table is based on the current value of the right to receive £1 at a known future date or gives the deferred value of a future sum. The table may also be used to calculate the capital sum to be invested now to provide for a future known liability (Richmond, 1994). A part of this table is illustrated in appendix 2. Hence

$$\text{the present value of £1} = \frac{1}{\text{amount of £1}}$$

If the present value of £1 is represented by PV

$$\text{and amount of £1 by } A\text{, then } PV = \frac{1}{A}$$

$$\text{As } A = (1 + i)^n\text{, therefore } PV = \frac{1}{(1 + i)^n}$$

where n = number of years for which sum is invested.

Some examples will help to indicate the practical applications of the present value of £1 table.

Example 12.7. What capital sum must be invested today to accumulate to £80 000 in eight years' time at six per cent compound interest?

Capital in eight years' time	£80 000
Present value (PV) of £1 in eight years at six per cent	0.627
Present value (amount to be invested now)	£50 160

Example 12.8. A developer has been given the option to purchase a building site which, it is estimated, will be worth £950 000 in five years' time. The provision of public services will delay its use until that time. What sum could the developer be expected to pay now assuming an interest rate of eight per cent?

Estimated value of site in five years' time	£950 000
Present value (PV) of £1 in five years at eight per cent	0.681
Price the developer could be expected to pay now	£646 950
Rounded off to	£647 000

The sum of £647 000 can be described as the *present value* or alternatively the *deferred value* of the £950 000. This method of making allowance for the receipt of a sum at some future date is often described as deferring or discounting the sum. The present value is the reciprocal of the amount computed with the use of the first table.

Example 12.9. A building client expects to have to carry out certain alterations to a building in eight years' time and these are expected to cost £150 000. What sum should he invest now to provide sufficient funds at an operative interest rate of eight per cent?

Estimated cost in eight years' time	£150 000
PV of £1 in eight years at eight per cent	0.540
Amount to be invested now	£81 000

Amount of £1 per annum Table

The purpose of this table is to determine the sum to which a series of deposits will accrue, if invested at the end of each year at a specific rate of compound interest. The table is based on the assumption that £1 will be invested *every year* for a given number of years at a certain rate of interest.

The table is constructed on the basis of the following formula

Amount of £1 per annum after n years

$$= \frac{(1+i)^n - 1}{i} = \frac{A-1}{i}$$

where n = number of years for which sum is invested and A = amount of £1.

Where a person borrows a certain sum of money at the end of each year for a given number of years at a specified rate of interest, he may wish to know his total commitment in sums borrowed and accrued compound interest at the end of the period. Some examples will indicate the various uses to which this table can be put; part of an amount of £1 per annum table is illustrated in appendix 3.

Example 12.10. What sum will be obtained if £5000 is invested every year for ten years at seven per cent compound interest.

Annual sum	£5000
Amount of £1 pa for ten years at seven per cent	13.816
Accumulated outlay	£69 080

Example 12.11. Owing to unsatisfactory trading conditions an industrial concern ceased production for four years. Throughout this period, repair and maintenance work was carried out and this cost £25 000 per annum. Calculate the accumulated sum involved at the end of the four-year period, assuming a compound interest rate of eight per cent.

Annual repairs and maintenance	£25 000
Amount of £1 pa for four years at eight per cent	4.506
Accumulated outlay	£112 650

Annual Sinking Fund Table

This table gives the annual sum which must be invested at the *end of each year* to provide a capital sum of £1 at the end of a certain number of years at a given rate of compound interest; an extract from this table is given in appendix 4. The present value of £1 table is based on the principle that if a sum of less than £1 is invested for a certain number of years at a given rate of interest it will accumulate to £1, whereas the annual sinking fund table is based on the assumption that if an equal amount is invested *every year* at a given rate of compound interest it will accumulate to £1. The table enables the sum to be invested *each* year to be calculated. The basis of the approach is as follows

Annual sinking fund (S)

$$= \frac{1}{\text{amount of £1 per annum}}$$

$$S = \frac{i}{(1+i)^n - 1} = \frac{i}{A-1}$$

This is the reciprocal of the amount of £1 per annum table, and its use is best illustrated by practical examples.

Example 12.12. A building client is expecting to have to replace various engineering services in a building in five years' time. He wishes to know what sum he should invest at the end of each year if the rate of interest obtainable is seven per cent and the estimated cost of the replacements is £160 000.

Cost of replacements	£160 000
Sinking fund to replace £1 in five years at seven per cent	0.1739
Amount of annual sinking fund	£27 824

This total can be checked by using the amount of £1 per annum table.

Annual sinking fund payment	£27 824
Amount of £1 pa in five years at seven per cent	5.750
Capital sum to be replaced in five years	£159 988

Note: The small difference between this total and the cost of replacements stems from the restriction on the number of decimal places.

Example 12.13. A building client will need to replace a building at the end of twenty years, and the replacement cost is estimated at £1 200 000. Calculate the amount of the annual sinking fund assuming an interest rate of seven per cent is obtainable.

Cost of new building	£1 200 000
Sinking fund to replace £1 in twenty years at seven per cent	0.0244
Amount of annual sinking fund	£29 280

It is advisable to make a check as before

Annual sinking fund payment	£29 280
Amount of £1 pa in twenty years at seven per cent	40.995
Capital sum to be replaced in twenty years	£1 200 334

Present Value of £1 per annum or Years' Purchase Table

This table is used to convert a known annual sum which is received or required at the end of *each year* into an equivalent capital sum. It is an extremely useful table in life cycle costing calculations for obtaining the present lump sum value of annual payments for repairs, cleaning, lighting and heating a building throughout its effective life. Extracts from this table are given in appendix 5, and the basis of computation of the table follows.

$$\text{PV of £1 per annum} = \frac{(1+i)^n - 1}{i(1+i)^n} \text{ or } \frac{A-1}{i \times A}$$

which can be set down as

$$\frac{(1+i)^n}{i(1+i)^n} - \frac{1}{i(1+i)^n}$$

As n (the number of years) approaches perpetuity, the value of the second part of the formula becomes so small that it is really insignificant. Hence the present value of £1 per annum for perpetuity is

$$\frac{(1+i)^n}{i(1+i)^n} = \frac{1}{i}$$

An example will serve to illustrate the use of this table in cost planning.

Example 12.14. A building client wishes to compare the cost of purchasing one type of heating installation costing £65 000, having a ten-year life, with another type having the same length of life but costing £36 000 initially and £5000 per annum for servicing and replacement of certain components. It is necessary to calculate the present value of the two installations for comparison purposes. Interest is to be taken at eight per cent.

Installation A			
Total cost over ten-year period			£65 000
Installation B			
Initial cost		£36 000	
Annual replacements and servicing PV of £1 pa for ten years at eight per cent	6.710		
PV of annual costs		£33 550	
Total present value			£69 550

Installation A shows a relatively small financial advantage over installation B when all costs are reduced to present value.

There are also dual-rate years' purchase tables whereby it is possible to use a lower rate of interest for the sinking fund than for the interest on capital. The annual sinking fund permits the capital to be replaced by the time the capital value of the investment has been dissipated. A normal tax-free sinking fund rate of interest is in the order of 2½ per cent as it will be invested in a relatively risk-free asset such as gilt-edged securities. This is often referred to as an accumulative rate of interest. *Parry's Valuation Tables* also contain a series of years' purchase tables on the dual rate percentage principle with rates of interest varying from four to 25 per cent, sinking funds ranging from one to 3½ per cent, and with the part of the income used to provide the annual sinking fund instalment subject to varying rates of income tax. The figures contained in the years' purchase table in appendix 5 contain no allowance for income tax, and are restricted to a single rate of interest. Hence they provide for a sinking fund to accumulate at the same rate of interest as that which is required on the invested capital, and ignore the effect of income tax on that part of the income used to provide the annual sinking fund instalment.

Life Cycle Costing Calculations

A number of worked examples involving the use of valuation tables are provided in chapter 13. In order to evaluate both initial and future costs, it is necessary to reduce all costs to present values or to annual equivalents and this requires a knowledge of valuation tables and their general application and use.

RENTAL VALUE

Rental value forms the basis for many valuation computations and, for this reason, its main characteristics and underlying influences are now examined.

Rent and its Relationship to Value

The rental value of a property is generally considered to be the value which the average tenant is prepared to pay for its occupation. With properties other than residential the value will be influenced by the profitability of the processes undertaken. Hence, in periods of prosperity rent levels can be expected to rise. Rent is, however, usually associated with leases operating for terms of several years so that there tends to be a timelag between changes in profit levels and the modification of rental values. The term *rack rent* is often used to describe the economic rent of a property or its true or full rental value, and is the rent which a property should command in the open market.

Factors influencing the Demand for Properties and Rental Value

(1) General prosperity of the country; in times of prosperity demand for properties and their rental values will rise, whereas in periods of recession the opposite will apply.

(2) Movement of population, where areas such as south east England with increasing opportunities for employment, in periods of economic stability, cause increased demand for properties with enhanced rental values.

(3) Improved transport facilities; for instance, the provision of new roads and underground lines, open up new sites for development.

(4) Changes in the character of demand, stemming from improved living and working standards, may create a demand for new buildings and result in a lowering of rental values of older properties.

(5) Rent for commercial, industrial and agricultural properties is a proportion of profit, and so the rent that occupiers are prepared to pay is influenced by anticipated profits.

Determinants of Rental Value

(1) The rent actually paid can form a basis for computation of rental value but it may be less than true or rack rental value for one of three reasons: firstly, changes in rental values since commencement of lease; secondly, consideration may have been paid for lease by way of premium; or thirdly, a personal or business relationship exists between the lessor and lessee.

(2) Comparison of rents paid for similar pro-

perties in the district, and making allowance for long established leases, special conditions of leases and variations in such factors as size, arrangement, condition, age and location of the properties. It is advisable to use common units for purpose of comparison, such as the hectare for agricultural land and the square metre of floor area for commercial properties, with 6 m zones for shops.

(3) Assessment on the basis of profits after computation of turnover and gross profit, less outgoings and interest on capital.

(4) Computation on the basis of a certain percentage of the capital outlay.

Rent in Relation to Market Value

A person who purchases property as an investment expects a reasonable annual return for the use of his capital. The annual return or net annual income will consist of the rent received, less outgoings in the form of repairs, insurance, management expenses and the like.

Example 12.15. If an investor expects an eight per cent return on his investment and the net income of the property is £8000 pa, then he will be prepared to pay

$$£8000 \times \frac{100}{8} = £100\,000$$

Hence

$$\text{Net income} \times \frac{100}{\text{Rate of interest}} = \text{Market value}$$

This process is described as 'capitalising' the net income and the multiplier is termed 'years' purchase' or YP. Thus net annual income × years' purchase = market value.

Example 12.16. A owns two sites, No. 15 and No. 185, at either end of a shopping street. On each site he has erected almost identical shops and No. 15 lets at £22 000 pa and No. 185 at £17 000 pa, both on full repairing leases. The cost of erection of each shop is £150 000. Analyse the rents.

	No. 15	*No. 185*
Rent	£22 000	£17 000
less ten per cent return on outlay of £150 000 (construction of shop)	15 000	15 000
Ground rent	£7 000	£2 000

This indicates that No. 15 is in a superior trading position to No. 185.

Example 12.17. X owns a plot of land which he recently purchased for £65 000 and he intends to erect a house on the plot at a cost of £160 000. Estimate the true rental value. Assuming a return of eight per cent on the land,

$$\text{the rental value} = £65\,000 \times \frac{8}{100} = £5200$$

A suitable return on the building would be ten per cent,

$$\text{and the rental value} = £160\,000 \times \frac{10}{100} = £16\,000$$

Estimated true rental value £21 200

The main weakness in this calculation is that it is based on the landlord's expectations and not on what the market will necessarily be prepared to pay.

PREMIUMS

A premium is a sum of money paid by a lessee in consideration of a reduction in rent. The lessee purchases an annual profit rent and the landlord capitalises part of his future income. The primary advantages to the landlord are that he secures an immediate capital sum with restricted tax liability and the security of his annual income is increased. This may influence the landlord to grant a longer lease to the tenant and to give more readily any consents required under the lease, such as permission to alter or improve the premises.

Assessment of Premium

The lessee will not be prepared to pay, in combined premium and rent, a rental equivalent which exceeds the market value. He is entitled to expect a reduction in annual rent for the premium made up of both the interest foregone on the premium at the full leasehold rate, and a sum sufficient to recover the initial capital outlay by means of a sinking fund over the period of the lease. These two payments together constitute the annual equivalent of the premium.

Example 12.18. A lessee is taking a forty-year lease of a shop worth £15 000 pa net and will pay a premium of £30 000. What rent should he pay?

Full rental value	£15 000

Reduction of rent on account of premium
Lessee foregoes interest at, say, eight per cent on £30 000 and will need to recover £30 000 over forty years.

The annual equivalent is

$$\frac{£30\,000}{\text{YP – forty years at 8 per cent and } 2\tfrac{1}{2} \text{ per cent (tax 30p)}}$$

$$= \frac{£30\,000}{9.901} = \qquad \frac{3030}{£11\,970 \text{ pa}}$$

The landlord would stand to lose a little in excess of £6100 over the 40 year period by this arrangement, although it is likely that this loss would be more than offset by the tax relief on the premium, which would be available for investment over a forty-year period. The alternative valuations from the landlord's viewpoint follow.

No premium payable

Rent		£15 000	
YP – forty years at eight per cent		11.925	£178 875

Premium payable

Premium		£30 000	
Net income	£11 970		
YP – forty years at eight per cent	11.925	142 742	£172 742

SERVICE CHARGES

The landlords of blocks of flats and offices have to meet the cost of various outgoings from the gross income in rents obtained from the properties. The outgoings consist of rates (where not paid by tenants), repairs, maintenance of common parts of building and communal services, insurance and management. External repairs and maintenance of common parts of building could amount to ten to twelve per cent of gross rents, while internal repairs could account for a further ten per cent. Typical annual service costs in 1994 were: hot water and central heating – £160 per tap and £220 per radiator; lifts – small passenger-operated electric lift at £2700 to £3300; electric lighting – £45 per lighting point; porter, including uniform and cleaning materials – £8000 to £11 000, depending on whether porter is supplied with living accommodation; management – five per cent of total rents.

13 LIFE CYCLE COSTING

With many projects cost planning cannot be really effective unless the total costs are considered, embracing both initial and future costs. This chapter examines the concept of life cycle costing, the various approaches, problems in application and its use in practical situations. A number of related issues such as discounting future payments, lives of buildings and components, the relationship of design and maintenance and energy conservation are also considered.

CONCEPT OF LIFE CYCLE COSTING

Nature of Life Cycle Costing

The Institution of Civil Engineers (ICE, 1969) rightly emphasised the need to apply economic analysis to engineering projects in order to assess the *real* cost of using resources when establishing priorities between competing proposals. This hypothesis applies equally well to building projects, where the term real costs should encompass the initial acquisition costs and the running costs of maintaining and operating a building throughout its effective life, including refurbishment. In some cases the appraisal should extend even further to include the relative benefits accruing to owners/occupiers from alternative designs, and possibly the demolition or disposal at the end of the building's life.

The term life cycle costing is sometimes referred to as *ultimate life cost* or *total cost*, a technique of cost prediction by which the initial constructional and associated costs and the annual running and maintenance costs of a building, or part of a building, can be reduced to a common measure. This is a single sum which is the annual equivalent cost or the present value of all costs over the life of the building.

Flanagan and Norman (1983) have defined the life cycle cost of an asset as the total cost of that asset over its operating life, including the initial acquisition costs and subsequent running costs, while Hoar and Norman (1990) aptly defined the life cycle cost of an asset as the present value of the total cost of the asset over its operating life including initial capital costs, occupation costs, operating costs and the cost or benefit of the eventual disposal of the asset at the end of its life. This latter definition represents a rather fuller description of the process and is probably more helpful to the reader. Hence the life cycle costing approach is concerned with the time-stream of costs and benefits that flow throughout the life of a project, with future costs and benefits converted to present values by the use of discounting techniques, as illustrated in chapter 12 and in the life cycle costing examples which follow later in this chapter, and in this way the economic worth of an option can be assessed. The RICS in 1994 identified the phases in the wider concept of the property life cycle as land, measurement, extraction, cultivation, planning, funding, construction, agency, management, investment, refurbishment and redevelopment. Life cycle costing is employed as a design tool for the comparison of the costs of different designs, materials, components and constructional techniques. It is a valuable guide to the designer in obtaining value for money for the building client. It can also be used by property managers or developers to compare costs against the value accruing from future rents.

The term 'life cycle costing' has replaced the term 'costs in use' which was used extensively by quantity surveyors for many years following the wider acceptance of cost planning techniques. Unfortunately, the term 'costs in use' gave a wrong impression of being confined to the costs of using the building, while in fact it was used, for example, in comparing the costs of different design solutions for a new building, where the initial costs of land acquisition and erection of the building were included in the cost comparisons. Similarly when comparing the costs of alternative components, the costs in use calculations included the initial cost and maintenance and replacement costs.

The life cycle costing approach enables the way in which a building functions to be expressed in terms of the costs of repairing and renewing the fabric, finishings and fittings, of heating, lighting and servicing, and of the labour needed in operating the building. These costs can be added to the amortised initial acquisition cost of the building to give the total annual cost of providing, maintaining and operating the building. In what is probably a better and more commonly adopted approach, all the costs can be converted to the present value (PV) by discounting techniques, as described and illustrated in chapter 12. The use of this technique thus makes it possible to combine all the costs of the building and so enables the vast range of factors on which judgement is necessary to be reduced to a comparison of a single cost with the personal assessment of the value of the building. The cost implications of building designs are often wider than the effect on the initial costs. For some types of buildings the equivalent of first costs is less than the running costs, and small changes in design have a much larger impact on running costs than on first costs.

The essential role of life cycle costing is to provide a rationale for choice in circumstances where there are alternative means for achieving a given object, and where these alternatives differ not only in their initial costs but also in their subsequent operational costs. A typical example is the thermal insulation of walls which is reasonably straightforward as there is unlikely to be interference with the functions of the wall or its appearance, and all that has to be considered is the initial cost and life of the insulation and the value of the expected heat savings. Unfortunately, few building design problems are as simple as this, and the alternatives usually exhibit differences in functional efficiency and aesthetic quality. The construction of factory walls in brick or coated metal cladding is a typical illustration. Initial costs, maintenance costs, cost effects of different rates of heat loss, vulnerability to damage and appearance all have to be considered. Some may argue that aesthetic considerations are purely subjective and as such have no place in a financial appraisal. Yet there can often be financial implications; the improved appearance of a factory could result in more satisfied employees and the higher standing of the firm and its products.

Total building cost refers to all costs and expenses throughout the life of a building, irrespective of who pays them. In any economic appraisal one should not ignore the inevitable future upkeep costs necessary for a building to perform its complete function. The costs of maintenance, heating, lighting and cleaning (that is running costs) must affect the true economic worth of a building in use.

Most design decisions affect running costs as well as first costs, and what appears to be a cheaper building may in the long term be far more expensive than one with much higher initial costs. Some idea of the relationship between initial costs and running costs can be obtained from an examination of table 13.1, from which it will be seen that running costs often amount to about two-thirds of the annual equivalent of first costs. The proportions vary considerably from one building to another and so the percentage costs listed can only form a rough guide.

Land costs are relatively low with industrial buildings which are erected on lower priced land with a higher utilisation factor, whereas houses and schools have high land costs being built to a lower density on relatively highly priced land. Initial building costs represent the highest proportion of total cost with flats, and the least with industrial buildings. Heating costs of modern industrial buildings have been reduced by increased thermal insulation but are still high and schools rank low with shorter periods of use. Lighting of offices is a relatively high cost item

Table 13.1 Breakdown of typical total costs for various types of buildings

Type of annual cost	*Houses*	*High flats*	*Industrial buildings (Percentages)*	*Schools*	*Offices*
Maintenance	14	12	18	16	13
Fuel and attendance for heating and lighting	24	24	30	18	29
Initial costs (amortised)					
(a) Building	48	56	47	51	47
(b) Land and development	14	8	5	15	11
Total costs	100	100	100	100	100

stemming from the high standards of illumination that are required. On maintenance and decoration schools rank high with heavy wear and tear.

The relative importance of first and running costs respectively is influenced considerably by the financial interests of the building client. A developer building houses for sale will not usually consider running costs unless they affect the selling price. An industrialist will almost certainly be influenced by the greater tax savings obtainable for running costs as compared with first costs. An owner-occupier, on the other hand, will be concerned with the total effect of the design upon the costs of owning and operating the building. There are a number of effects of any particular design feature which cannot be costed directly, although their influence on costs may be considerable. Such factors as appearance, noise level and the level of illumination all have an influence on output, but it is difficult to evaluate their effect other than indirectly using a cost-benefit approach as described in chapter 16.

Life Cycle Cost Analysis, Management and Planning

Since the early nineteen eighties a number of terms have been used to identify different stages in life cycle costing techniques, and the following are the most widely used although opinions differ as to the sequence in which they should be implemented.

Life cycle cost analysis (LCCA) is the collection and analysis of historic data on the actual costs of occupying comparable buildings, having regard to running costs and performance. This will assist quantity surveyors in preparing life cycle cost plans for construction projects.

Life cycle cost management (LCCM) is derived from life cycle cost analysis and identifies those areas in which the costs of using the building as detailed by life cycle cost analysis can be reduced. Life cycle cost management can therefore be used to assist clients to compare building costs in a meaningful way and in assessing and controlling occupancy costs throughout the life of a building to obtain the greatest value for the client.

Life cycle cost planning (LCCP) can be considered as part of life cycle cost management and constitutes the prediction of total costs of a building, part of a building or an individual building element, taking account of initial capital costs, subsequent running costs and residual values, if appropriate, and expressing these various costs in a consistent and comparable manner by applying discounting techniques. It also includes planning the timing of work and expenditure on the building, and should also take into account such factors as the effect of performance and quality. It is desirable that the plan should be reasonably flexible and updated as necessary to encompass changing conditions, including environmental changes, particularly when formulating a full functional life plan.

Flanagan and Norman (1983) recognised that life cycle costing techniques involved the manipulation of a large amount of data. They accordingly devised a method of grouping life cycle costing activities into a hierarchal structure, as illustrated in figure 13.1 which shows how the various LCCP techniques can be dovetailed into the RIBA Plan of Work, as part of the normal cost planning process. As the design develops the initial or budget LCCP based on level 1 will be replaced by a detailed plan at level 3. The conventional cost planning sequence is shown on the left hand side of the figure. It should, however, be appreciated that LCCP can be used at any stage in the design process (Flanagan and Norman, 1983).

Terotechnology

The term terotechnology has been used extensively since the mid nineteen seventies to encompass the life cycle requirements of physical assets. It combines management, financial planning and other practices applied to physical assets to ensure economic life cycle costs. It is concerned with the specification and design for reliability and ease of maintenance of plant, machinery, equipment, buildings and structures with their installation, commissioning, maintenance, modification and replacement, and with feedback of information on design, performance and costs. It is a technology that takes into account the marketing and observance of design-maintenance – cost practice of all assets, the conservation of resources and the promotion of controlled and calculated life span of assets as against built-in or unpredictable obsolescence.

Volume and Impact of Building Maintenance Work

The significance of building maintenance work in the national economy and its broad categorisation is shown in table 13.2.

A greater awareness has developed of the need to reduce the cost of maintaining buildings. The annual expenditure on construction maintenance is equivalent to the first cost of about 1000 teacher training colleges, and a saving of one per cent on maintenance would each year equal the capital cost of something like 2000 secondary school places. Hence it is vitally important that the probable maintenance and running costs of a building should be considered at the design stage, and due attention directed towards the maintenance implications of alternative designs. A reduction in initial constructional costs often leads to higher maintenance and running costs.

Well in excess of fifty per cent of the 1993 building labour force was engaged on maintenance work covering over 43 per cent of the value of all

Table 13.2 Construction output including repairs and maintenance (Great Britain, at constant 1985 prices; £m)

Year	*Repairs and maintenance*				*All repair and maintenance work*	*Percentage of repairs and maintenance to all work*	*All work*
	Housing		*Other work*				
	Public	*Private*	*Public*	*Private*			
1988	3812	6067	3849	3489	17 217	46.6	36 959
1989	3968	6534	4000	3815	18 317	46.4	39 472
1990	4043	6340	4117	3913	18 413	45.8	40 200
1991	3479	5699	3734	3610	16 523	44.2	37 372
1992	3381	5168	3457	3403	15 410	43.4	35 535
1993 (P)	3484	4951	3269	3473	15 177	43.5	34 920

Source: DOE press release on Construction Output, 1993 (adjusted).

(P) Provisional.

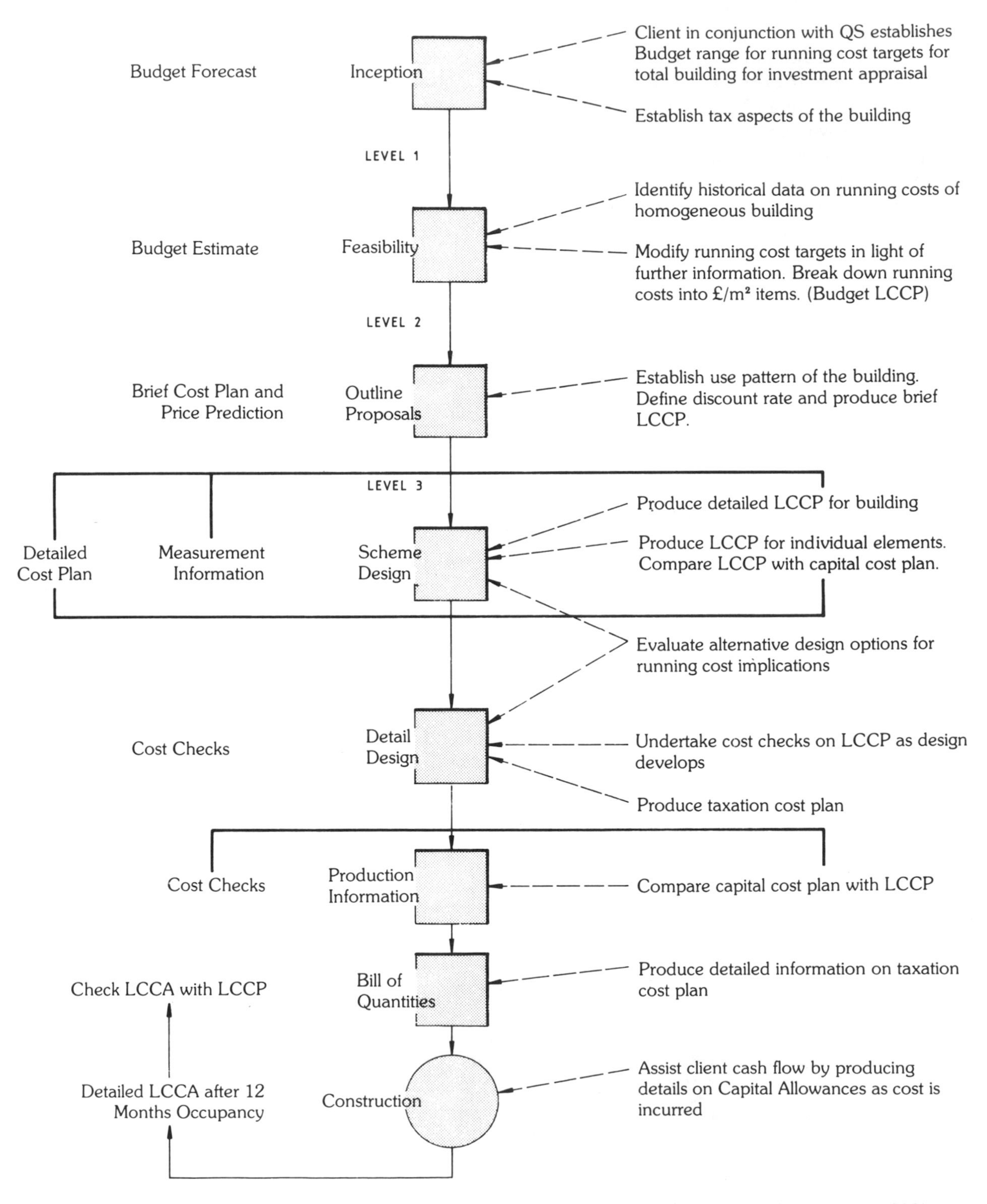

Figure 13.1 Life cycle cost and the RIBA plan of work (Source: Flanagan and Norman, 1983)

building work. On current trends, and taking into account the fact that insufficient maintenance work is currently being undertaken, it seems feasible that the proportion of the building labour force employed on maintenance work could be approaching sixty per cent by the year 2000. This imbalance must to some extent stem from the failure of earlier design teams to take full account of future costs. It will be noted that the percentage of maintenance and repair work declined progressively throughout the period 1988–92 and that the largest sector was private housing.

The magnitude and seriousness of the problem resulted in a former Minister of Public Building and Works setting up a committee on building maintenance in 1965. The terms of reference of the committee were to keep the problem of maintenance under continuous review, concentrating particularly on the relationship between design and maintenance, the dissemination of information, and the establishment of priorities for research and development. DOE subsequently established a committee on building maintenance which issued a valuable report in 1972 and this report and its recommendations were re-examined in 1988 in a conference organised by BMI on 'Building maintenance – investment for the future'. Some of the principal conclusions of this conference will be examined later in the chapter. The building maintenance cost information service (BMCIS) commenced operation in 1971 with the support of the University of Bath, the Department of the Environment and the Royal Institution of Chartered Surveyors. This service had the stated objective of helping its subscribers to derive value for money from efficient property occupancy and economic building maintenance; further details about this important service, now termed BMI and operating under the auspices of the RICS, are given later in this chapter.

Maintenance is therefore an important part of construction but has tended to lag behind new constructional work as regards modernisation and mechanisation. While output per man in the construction industry as a whole rose steadily in the nineteen eighties at between three and four per cent annually, there was little evidence of any significant increase in output per man on maintenance work. The labour-intensive nature of this section of work is largely inevitable, as many maintenance tasks can only be performed manually and demand is often dispersed and uncoordinated. Building firms engaged on maintenance work are often small in size and frequently under-capitalised. Some small contractors have been reluctant to invest in labour-saving small power tools, which can produce considerable reductions in the manhours needed for many maintenance tasks.

Value for Money

Full consideration of maintenance aspects and possible future costs at the design stage is likely to result in the building client securing better value for his money. Maintenance work often falls into two categories – the necessary and the avoidable. Avoidable maintenance work may result from faulty design or poor workmanship. The Building Research Establishment investigated the maintenance aspects of local authority housing and found that the equivalent capital costs of maintenance compared with initial costs was relatively low, except for water services (eighty-four per cent) and external painting (100 per cent). Maintenance costs of water services could be reduced substantially by protecting the services against frost damage and electrolytic action and external painting costs could be reduced by the substitution of materials which do not require painting.

A large proportion of the annual charges on a building is attributable to heating and lighting, and these costs are influenced considerably by the planning and structure of the building and the extent of thermal insulation. For example, the heat loss from a compact single-storey school with a storey height of about 2.40 m could be about thirty per cent less than from one with an irregular layout and a storey height of around 3.40 m. Furthermore, the initial costs of the building and the heating plant are likely to be about twenty per cent less.

The heating and ventilating costs of factories often amount to between one-third and two-thirds of the combined total of the running costs of the building and the annual equivalent of its initial costs, a proportion sufficiently large to justify close consideration. Roof design is an important factor and a flat roof is usually the cheapest solution from the heating viewpoint, resulting in reduced heat losses and less air to heat. The standard of insulation and amount of glazing must also be considered. It is the total costs that must be considered and the effect of each design decision should be studied with this in mind. For instance, the area of glazing affects structural, heating and lighting costs and it is advisable to adopt a form of cost analysis which shows the effect of each design on all three components. One approach is to convert the costs of each component into annual charges made up of annual equivalents of initial costs plus annual maintenance costs.

Designers could contribute significantly to a reduction in maintenance costs if they asked themselves four questions when designing each component or part of a building:

(1) How can it be reached?
(2) How can it be cleaned?
(3) How long will it last?
(4) How can it be replaced? (Seeley, 1987)

The interest of the building owner in maintenance should begin when he is considering whether to commission a new building or to purchase or rent an existing property. The client's brief for a new building is a major factor in determining the long term maintenance needs of the property. In particular a design brief should indicate performance requirements and possible changes in use, as well as the future policy for operating, cleaning and maintaining the building. The designer has a special responsibility to guide the client in these matters, but this is not always implemented effectively.

For most new buildings the design brief gives clear instructions on aspects such as space and the internal conditions but seldom states the performance or the maintenance characteristics required. This omission can result in a severe long term penalty as, for example, cleaning and redecoration can form substantial items of occupation costs and more economic designs, in terms of total cost, are likely to be achieved if these and similar factors are considered at the appropriate design stage (BMI, 1989).

Difficulties in Assessing Life Cycle Costs

There has been some reluctance on the part of many quantity surveyors to include running costs in their cost planning calculations because of the inherent difficulties in implementation. The major difficulties are now listed.

(1) The difficulty of accurately assessing the maintenance and running costs of different materials, processes and systems. There is a great scarcity of reliable historical cost data and predicting the lives of materials and components is often fraught with dangers. In these circumstances the quantity surveyor may be compelled to rely on his own knowledge of the material or component, or possibly on manufacturer's data in the case of relatively new products. Even the lives of commonly used materials like paint show surprising variations and are influenced by a whole range of factors, such as type of paint, number of coats, condition of base and extent of preparation, degree of exposure and atmospheric conditions. Owners' and occupiers' maintenance procedures also vary considerably.

(2) There are three types of payments – initial, annual and periodic, and these all have to be related to a common basis for comparison purposes. This requires a knowledge of discounted cash techniques, described later in this chapter.

(3) Tax has a bearing on maintenance costs and needs consideration, as it can reduce the impact of maintenance costs. Taxation rates and allowances are subject to considerable variation over the life of the building.

(4) The selection of suitable interest rates for calculations involving periods of up to sixty years is extremely difficult.

(5) Inflationary tendencies may not affect all costs in a uniform manner, thus distorting significantly the results of life cycle costing calcu-

lations, particularly as maintenance work has a higher labour component than new work.

(6) Where projects are to be sold as an investment on completion, the building client may show little interest in securing savings in maintenance and running costs.

(7) Where the initial funds available to the building client are severely restricted, or his interest in the project is of quite short-term duration, it is of little consequence to him to be told that he can save large sums in the future by spending more on the initial construction.

(8) Future costs can be affected by changes of taste and fashion, changing statutory requirements for buildings and the replacement of worn out components by superior updated items.

(9) The lives of different types of buildings are difficult to forecast with accuracy, as described later in the chapter.

Practical Problems Which Affect Life Cycle Costing

Some of the major practical problems which occur when carrying out a life cycle costing study were examined by Pickles (1982) and these are now discussed, as it is felt that the reader will find them both helpful and instructive.

Essentially, the main task in carrying out a LCC study is to prepare a cash flow schedule for the building including all the different user costs as they occur throughout the building's life. This requires life and maintenance profiles of components and materials to be prepared. The lives of building components can be predicted on the basis of observed rates of failure for existing buildings, although this data is rarely collected and, when it is, it often shows substantial differences in the maintenance profiles of seemingly similar buildings. The Property Services Agency (PSA) of DoE (1991) published valuable cost in use tables showing maintenance profiles and associated costs, in an attempt to quantify available data and assist in objective decision making. Probably the main weakness stems from the fact that a large proportion of the construction techniques and components in a typical modern PSA building are designed differently from those built as little as twenty years earlier. Thus collected data soon becomes out of date or is no longer applicable as new components and materials are introduced and possibly more innovative designs produced.

Hence it is evident that realistic LCC profiles are very difficult to prepare, encompassing as they do many predictions and assumptions sometimes of questionable validity. The following factors and examples will serve to illustrate the practical difficulties that can arise in their formulation:

(a) Changes in the basic prices of materials, components, energy, labour and capital are difficult to forecast with accuracy and will affect all user costs, especially the costs of fuels for heating and labour for cleaning. Over the life of a building these costs can change dramatically. Sophisticated cost models incorporating many assumptions can soon be rendered invalid by changes in basic prices, which are unlikely to be uniform across the different components.

(b) Changes in government policy, the way in which people live and work and their expectations, can also have far reaching effects on future needs and costs. Social, economic and technological changes are bound to have a significant effect on the costs incurred throughout a building's life and are all unpredictable at the time of preparing the life cycle costing plan.

(c) Emergency repairs and maintenance, arising from such factors as unforeseeable design faults or bad workmanship, which together constitute a significant proportion of maintenance costs, display a random pattern in both timing and extent which calls for a probability approach based upon statistical observation and research. While this phenomenon can be reasonably well identified by examining the failure patterns of buildings in use, the associated disruption costs can only be assessed in a very approximate form.

(d) With foreseeable maintenance work such as cleaning and redecoration, which together form a major part of total user costs, the actual decision as to the timing of the work depends to a considerable extent on management policy.

Unfortunately, many organisations take a rather disinterested view of maintenance work as highlighted in the author's research for the Paintmakers Association in 1984, and it is often performed without any records being kept. Redecoration and cleaning costs are normally planned to follow regular set cycles and therefore their estimation appears uncomplicated, but once again there is a general lack of costing data (although cleaning and painting contractors hold it in abundance), and the set cycles are not always adhered to in practice. In particular, redecoration cycles can vary significantly as shorter cycles may be introduced to meet changing tastes and fashions, to implement a new colour scheme or on an unexpected change of occupancy, while at the other extreme, work may be carried out in a piecemeal manner over a longer period to reduce disturbance to the occupants. Longer cycles can result from financial constraints leading to deferment of the repainting and increasing substantially the cost of the eventual work.

(e) Major maintenance, replacement and modernisation works are linked not only to the life of the particular building elements, but also to economic, social and technological changes, and even possible obsolescence, which is difficult to predict with any accuracy. Although some works of refurbishment can be forecast with greater accuracy than others as, for example, those to in-town enclosed shopping centres, which are often carried out at about 15 year intervals.

(f) Many variable factors contribute to the real cost of maintenance work making it very difficult to assess with accuracy. For example, varying rates of productivity arising from the scale and relative difficulty of individual maintenance and replacement tasks compared with the initial installation, the cost of renewing old work and installations, the cost of renewing and/or making good the fabric in order to carry out other works, and the cost of disturbance and disruption to the occupier. Disturbance and disruption can be very costly and can only be assessed by reference to the amount of floor space occupied when carrying out the repair/replacement work, the cost of temporarily removing furniture and equipment and associated factors, which may not truely reflect the cost to the occupier.

(g) A CLC study cannot cater satisfactorily for the way in which the building is used. Work may be carried out which is not essential for the upkeep of the building, to suit individuals or to meet changing tastes and fashions. An actual example, provided by Pickles (1982), will serve to illustrate the problem. A new departmental head requires his room carpeted in preference to the existing PVC tiling. This unexpected cost was increased since the electric underfloor heating was adversely affected, and the room required a heater for winter use; neither of these costs could have been foreseen at the design stage. The different ways in which buildings are used by the occupants could be one of the main causes of the wide variations which can occur in the maintenance and running costs of very similar type buildings.

(h) One of the main problems experienced in the evolution of both costs in use and life cycle costing has been the lack of adequate data in a suitable format and of adequate quantity to support a total cost approach to buildings. BMI, as described later in the chapter, have done much to help with their extensive assembly and circulation of actual costs and supporting data in a standard format. Unfortunately, much of the information made available from other sources is often inconsistent, and interpretation of data is made difficult as the standards of maintenance can vary widely between different buildings, efficiency of operation can vary between apparently similar tasks and is not related to cost input, and exceptionally high figures can result from past neglect arising from policy decisions to cut spending. Recording and assimilating maintenance data may also be carried out in a variety of ways causing difficulties in making comparisons.

Life Cycle Costing Terminology

It is considered advisable to define the life cycle costing terms currently in use, as problems have arisen in practice because of the varying interpretation of some of the terms. For instance some surveyors understood the term costs in use to mean the cost of using the building, rather than

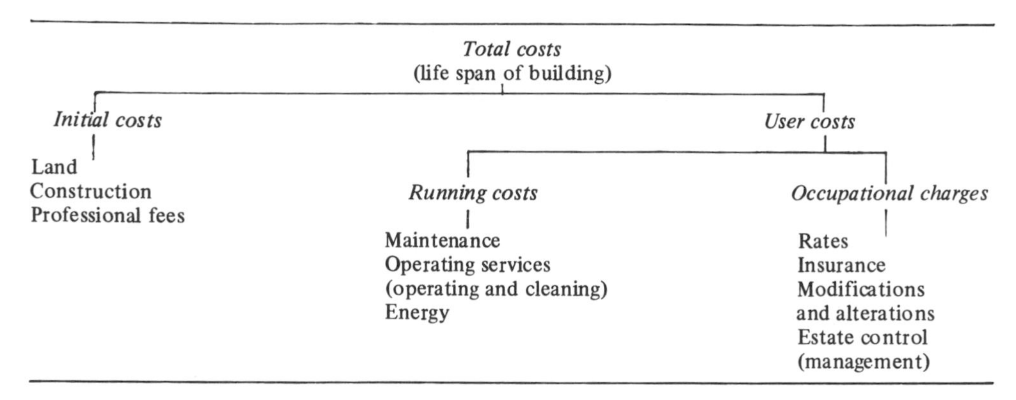

Figure 13.2 Breakdown of total costs

the intended meaning of total costs or ultimate costs, made up of both initial constructional costs and all running costs, including maintenance costs. However, as explained earlier in the chapter, this term has now been largely superseded by the much more appropriate term of life cycle costing. The construction field is particularly prone to loose interpretations, the term *maintenance* ranging from 'all construction work other than new capital works' to 'all work undertaken to keep a building in a satisfactory state of preservation'. Figure 13.2 shows a breakdown of total costs with the object of isolating the significant aspects of future costs in the design process and thus assisting in the search for suitable expressions.

Life cycle cost of an asset is the total cost of that asset over its operating life, including land acquisition costs and subsequent running costs.

Initial costs. The capital or initial expenditure on an asset when first provided.

User costs. These are synonymous with future costs and comprise both running costs and occupational charges.

Maintenance. This is defined in BS 3811:1984 as 'the combination of all technical and associated administrative actions intended to retain an item in, or restore it to, a state which can perform its required function'.

Operating services. These embrace cleaning, caretaking, operation of plant and equipment and other associated activities.

Energy. This represents the energy costs for heating, lighting, air conditioning and the like.

Modifications and alterations. These are new works required to improve or adapt an asset.

THE TECHNOLOGY OF MAINTENANCE

Maintenance seeks to preserve a building in its initial state so that it continues to serve its purpose and is an essential component in the life cycle of a building. Yet there exists a general financial climate where minimum first costs are often the only consideration, risking future maintenance problems. Cost assessment formulae and cost limits rarely take account of life cycle costing implications, and capital expenditure is all too frequently divorced from current spending. A designer of buildings could be failing in his role as adviser if he does not understand the problems connected with the maintenance and running costs of buildings and fails to apply this knowledge at the design stage. There is merit in the preparation of standard details and, where different qualities of material are involved, listing each with their effect on both capital and maintenance costs for the guidance of the client.

Building work will be a discredit to the designer

if its usefulness and convenience is permitted to fall below an acceptable standard. The definition of acceptable standards recognises three separate categories which serve to illustrate some of the complexities of maintenance.

(1) Functional performance, quality and reliability, which relate to user needs.
(2) Structural, electrical, fire and other safety aspects, for which maintenance personnel are generally responsible.
(3) The preservation of the asset and its amenities, in which the owner has the primary interest.

Expressed in another way the prime benefits of maintaining a building are to retain the value of the investment; to maintain the building in a condition in which it continues to satisfactorily fulfil its function; and to present a good appearance to the public.

TYPES OF MAINTENANCE

Figure 13.3 shows a method of classifying the main arrangements for maintenance work, with a primary classification of planned and unplanned maintenance.

The predominant characteristic of maintenance is the variety of factors that affect its incidence. These range from the initial design and cost

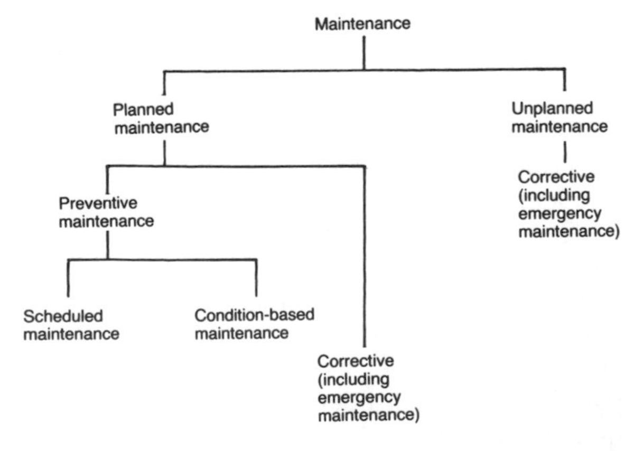

Figure 13.3 Types of maintenance (Source: BS 3811: 1984)

involving the quality of materials and workmanship, the intensity of exposure, to the efficiency of the maintenance organisation. Their interaction directly affects the durability of the buildings and their components and the resultant maintenance work. The control of maintenance, if it is to be effective, should therefore commence at the time the building is designed and continue throughout its life.

Origins of Maintenance Problems

Maintenance work is generated by a whole range of factors including weathering, corrosion, dirt, structural and thermal movement, wear, low initial expenditure, passage of time, incorrect specification, inferior design, poor detailing and damage by users. For further information on the causes and agents of deterioration and on diagnosis and investigation techniques, the reader is referred to Lee and Yuen (1993). Some of the major maintenance problems stem from the use of new materials and techniques. These require careful appraisal prior to use, but processes and products which satisfactorily withstand laboratory tests may not always be suitable in live situations. Nevertheless, the pace of technological development demands the use of such processes and products before their suitability over long periods has been adequately tested. The work of the British Board of Agrément (BBA) helps designers in their selection.

Too many post-war buildings have been built too quickly and too carelessly. Expenditure in the United Kingdom on repairs and maintenance in the next 50 to 60 years will probably equate to the cost of replacing the country's entire building stock. Following the high alumina cement scare of the nineteen seventies, identifying and investigating about 50 000 structures and rebuilding some of them cost over £70 m. The nineteen eighties revealed different, greater and even more costly problems, mostly arising from deficiencies in system building.

For example, about 50 000 dwellings were built for over 50 local authorities using the Bison wallframe system. The precast concrete panels suffered from spalling and the occupants complained of extensive water penetration. Substantial and expensive remedial work (over £20 000 per flat in the London Borough of Hillingdon and £26 700 per dwelling in Barnet in 1982) showed the defects as largely irreparable, and several authorities decided on demolition of the properties as the most economic long term solution. The repair and replacement of Britain's system-built housing could cost taxpayers over £6b.

Some architects have responded irresponsibly to the demands of their clients to cut building costs, and on occasions pay too much attention to innovative design at the expense of sound construction. The resulting absence of tried and tested prototypes means faults are not revealed before the buildings are erected in large numbers. Contractors who seek to achieve low costs and high profits also share the blame; for instance, inadequate on-site supervision has contributed to the ineffectiveness of some system building.

Bishop (1981) described how many clients, and perhaps the public at large, believed that the building industry failed to offer value for money; building projects were said to cost too much, take too long to complete and were too prone to failure, and to rapid failure at that.

A Building Research Establishment study by Freeman (1975) investigated 5000 defective buildings and contained the following findings. Faulty materials and components accounted for 12 per cent of defects and unexpected user requirements for a further 11 per cent, while faulty design contributed 58 per cent or faulty execution 37 per cent, or both (there was an overlap between the categories). As to faulty design, evidence pointed to 'a frequent failure to make use of existing authoritative design guidance' and to 'what can only be described as a careless attitude to detailed design'. This damning indictment is reinforced by a later BRE study in 1977 of 15 sites under construction, which is reputed to have found 72 000 actual and incipient errors which would have cost very little to avoid had the industry been sufficiently motivated to do so, but involved a significant proportion of the initial cost of the projects to rectify.

Designers tend to forget the client and the cost of operating and maintaining a building, notwithstanding that the cost of running and looking after a building throughout its lifetime may amount to many times its first cost. When one adds to that time lost, or used ineffectively, through poor arrangement of spaces, inadequate heating, thermal insulation and ventilating controls and the disruption caused through large scale repair operations, the cost of construction becomes of diminishing importance.

In addition, all too frequently inadequate sums are spent on maintenance work with probable cumulative effects. In 1980/81 local authority budgets were generally regarded as about 30 per cent below the level required to maintain the properties in reasonable condition. Subsequently with government pressure on local authority spending increasing, the situation worsened considerably. This will result in a greater investment in building as the maintenance backlog increases and as buildings which could have been preserved have to be replaced.

The 1991 English Housing Condition Survey found that approximately 1.5 million dwellings failed to meet the minimum standards of fitness for human habitation, which is a terrible indictment of the intolerable housing conditions under which many people are obliged to live as we approach the end of the twentieth century. Furthermore, another 1.3 million dwellings in England alone were found to be unsatisfactory in some other way; they either lacked one or more basic amenities or needed extensive repairs. The total backlog in housing repairs and maintenance in both the public and private sectors in the UK in 1994 was approaching £60b.

Some county council building surveyors asserted that in 1993/94 they were allocated less than half the budget necessary to keep their schools in a reasonable state of repair. This is supported by surveys undertaken by the Society of Chief Architects in Local Authorities (SCALA), which showed that the maintenance of school buildings had fallen well short of the level necessary to stabilise deterioration, let alone carry out much needed improvements in the older stock. Her Majesty's Inspectorate of Schools (privatised in 1993) had repeatedly reported annually on the unacceptably low condition of schools and the consequent adverse effect on the quality of teaching, and that without urgent attention to these problems the remedial cost will become prohibitive. This emphasises yet again the pressing need for the operation of an effective planned maintenance programme. Similar maintenance problems have been identified in hospitals where short term patching has become common place leading to serious difficulties in the years ahead.

Some of the principal maintenance problems will now be considered on the basis of specific elements, components or services.

Concrete. Good concrete with adequate cover to the reinforcement has the potential of almost unlimited life in the United Kingdom climate and is unlikely to be attacked by frost except in copings. However, sea water and sea spray can cause chemical attack on concrete as well as corrosion of reinforcement, and sulphate-bearing soils are a potential source of danger to concrete foundations. In these and other aggressive environments it is advisable to use special cements in well designed and compacted concrete. Particular attention must be paid to foundations constructed in shrinkable clay. Large concrete roofs require expansion joints. Serious carbonation, chemical attack on the concrete, including alkali silica reaction, cracking and spalling due to poor quality materials or workmanship, and/or corrosion of the reinforcement are all signs of distress which are indicative of a concrete of low durability (Lee and Yuen, 1993).

Brickwork. In this country frost is the chief cause of the decay of brickwork but this only occurs where bricks are frozen when saturated, principally in parapets, retaining walls and free-standing walls. There is an unfortunate tendency to omit water shedding drips and projections, resulting in unsightly surface staining and in extreme cases damp penetration. It is important to select a suitable type of brick and mortar according to the location in which they are to be used. Overloading of the brickwork can result in bulging or cracking and expansion joints are required in long lengths.

Masonry. In the United Kingdom most sandstones are immune to frost attack and limestones are rarely affected except in exposed positions such as cornices, string courses and copings to parapets. The chief cause of decay is attack by atmospheric sulphur gases. It is necessary to avoid the use of metals, such as bronze, copper and aluminium in proximity to marble and polished granite because of the danger of permanent staining of the masonry. Periodic cleaning is needed to maintain a good appearance and to remove dirt and other harmful substances.

Cladding panels. The use in modern building of prefabricated units of increasing size and complexity and framed structures with infill or cladding panels produces joints subject to relatively large movement. This has created several problems which are the subject of extensive research at the Building Research Establishment (1979a). Information is required on the satisfactory design of joints, particularly between dissimilar components, tolerance of dimensions in manufacture and changing dimensions of joints with changes in temperature and moisture. It is vital that the various sealants, mastics, putties and gaskets used for jointing shall be satisfactory under all conditions.

Roofs. Frequently otherwise durable buildings have roofs of less permanence and when preliminary estimates are reduced it is often roofs that suffer first. Flat roofs often produce serious maintenance problems due to lack of or haphazard falls, disregard of codes of practice, inadequate eaves, verge or fascia details, faulty upstands, sharp granite chippings puncturing roofing felt with possibly sodden strawboard below supported on untreated and unventilated timber, inadequate thermal insulation, heavy condensation on underside of roof slabs, ill-conceived gutters and lack of walkways.

Timber. One of the main problems with timber is caused by shrinkage which occurs because the moisture content of the wood at the time of erection was too high and it has dried out subsequently. The resultant change in moisture content can cause distortion and deflection. Unprotected external timber loses its colour and the surface becomes rough and difficult to clean. Even western red cedar, a durable species, loses durability after prolonged leaching. Fungal attack can only take place if the wood attains a moisture content of twenty per cent or more and remains wet over an appreciable period of time. There is evidence to suggest that in houses over twenty years old the incidence of some measure of attack by insects or fungi or both is greater than sixty per cent, although in some cases the extent of infection is very restricted.

Many wooden casement joints are not strong enough to withstand the racking strains imposed by wind loadings on open windows, resulting in cracking of paint films at joints which permit the entry and trapping of moisture under the paint, and that there is often excessive shrinkage on glazing putty. The common practice of storing timber for considerable periods on building sites in unsatisfactory situations prior to use, and the use of central heating for the rapid drying out of a building after the installation of interior joinery causes numerous problems in new buildings.

Metals. In most environments steel corrodes at between 0.05 and 0.10 mm per year as a result of oxidation when freely exposed in air. The corrosion risk level is severe in polluted coastal areas and in corrosive atmospheres such as swimming pools and chemical processing plant, where suitable protective coatings are usually applied. In soils the corrosion rate is similar, but pitting often takes place and at localised points much higher corrosion rates may occur. Steel reinforcement in concrete should have a minimum cover of 50 mm. Copper requires little or no maintenance as it develops a green protective patina after a number of years. Corrosion of aluminium and zinc roof coverings can take place around chimneys which emit soot particles. Unanodised aluminium and aluminium alloys are satisfactory provided the environment is not exceptionally corrosive.

The life of galvanised steel cold water cisterns depends on the nature of the water and can vary from five to thirty years. In view of the difficulty of replacing them and the damage that can occur as a result of leakage, it is advisable to paint them internally with two coats of suitable bituminous paint before installation. In districts where the water is known to be aggressive towards galvanised steel cisterns it is desirable also to fit a sacrificial magnesium anode to provide cathodic protection and to encourage protective carbonate scale deposition on any exposed parts of the metal surface. Steel members can suffer from abrasion, fatigue, loosening of connections, weld defects or impact failure.

Plastics. There is considerable confidence that suitably formulated phenolic resins, rigid polyvinylchloride (PVC), polymethyl methacrylate, glass fibre reinforced polyester, polyisobutylene and polychloroprene can withstand the effects of external weathering for periods of up to thirty years without maintenance.

Painting. Paint is normally applied in thickness little more than 0.125 mm, and often in even thinner layers, yet a great deal is expected from the coating. The selection of the most economical time to repaint, the type of paint to use and the thickness to apply are not easy as most buildings have their own peculiar features. There is scope and need for careful observation and collection of data on large numbers of buildings so that evidence can be accumulated to form the basis of a rational study of the problem.

Paint is the most vulnerable of building materials and its regular replacement constitutes the largest maintenance item resulting from weathering. Although the life of paint films has increased by twenty per cent over the last fifty years, there is considerable scope for improving painting maintenance by imposing standards of good practice. The former Greater London Council, that was probably the leading authority in this field, found that despite its known insistence on proper standards, seven per cent of unopened cans failed their standard and no less than one in three samples from painters' kettles on site were rejected, mainly because of unauthorised addition of thinners. The Council also found frequent omission of an undercoat from a four-coat paint scheme. If so much unsatisfactory painting is found by an authority that was known to impose rigid standards one wonders whether a satisfactory job is ever obtained on the average uncontrolled building site. If the former GLC methods of control could be imposed over the whole country the contribution to reducing building maintenance expenditure would probably be greater than any other single measure (Seeley, 1984b).

Services. A decision has to be made whether to expose, bury or duct services in a building. The first gives ease of maintenance and alteration but produces cleaning problems, the second seems to solve all problems until failure occurs or a major rearrangement becomes necessary, while the third is expensive, space-consuming and demands good access for inspection, day-to-day use of valves and switchgear and replacement or alteration.

Mechanical plant produces two main problems: firstly, the extent to which a number of standard units may be used for a particular purpose, for example, the use of small fans for ventilation rather than one or more larger but less standard units; and secondly, the relationship between final cost and maintenance cost of individual items and manufacturers' standard goods.

Where appropriate, tenders for such items as lifts, escalators and boilers should include the necessary maintenance costs so that these may be evaluated at the time of ordering. This is particularly desirable for buildings where no engineering maintenance staff are available on the site.

Many maintenance problems in plumbing installations stem from bad design, careless construction and improper usage. All too frequently insufficient space is provided around service installations to maintain the pipes and equipment. There is also a general deficiency in the provision of control valves, so that a large section of the installation requires shutting down before maintenance work can proceed. Stoppages in waste pipes are common occurrences and the number of stoppages could be reduced considerably by simplifying pipe layouts and ensuring an adequate gradient for waste pipes.

It is important to provide easy access for the purpose of examination, repair, replacement and operation. Many complaints are made by maintenance personnel concerning the lack of access to valves and cleaning eyes.

Most drain blockages arise from badly constructed drain joints and defects in manholes, with damage to pipes after installation and extensive deposits inside pipes also causing problems. The gradient of the pipes has little effect on the frequency of blockages. Recurring drain blockages are both inconvenient and a likely health hazard. In 1979 BRE estimated £7 m was spent annually on clearing and maintaining domestic drainage systems alone (BRE, 1979b).

Refuse chutes. Refuse disposal in high rise buildings is one of the essential services necessary to keep the buildings clean and to ensure a better environment for the occupants. Generally refuse is fed in a chute through hoppers and stored in containers housed in a storage chamber at the bottom of the chute. The containers, when full, are stored temporarily while awaiting collection vehicles. Chutes should be designed to avoid blockages, smell, noise, moisture movement, overfilling of containers and fire risk.

Lifts. At least one lift should serve all floors of high rise buildings for the movement of people and goods from floor to floor. In order to give passengers a safe, speedy and comfortable ride, the speed of the lift car requires careful consideration. For example, in Singapore a lift speed of 150 m/minute is considered satisfactory for a 30 storey block of flats. The capacity of the lift car must be adequate to accommodate all passengers at peak periods.

The ease of maintenance of lifts depends very much on the types of finishes used in the lift car and lift shaft, as well as accessibility to plant and equipment. The most practical lift door is probably plastic laminate recessed into stainless steel framing. In private flats, stainless steel is often used for aesthetic reasons although scratches are very noticeable on these surfaces. Spray painted surfaces are also commonly used but these are liable to scratches and chipping. Regular cleaning is necessary.

The lift floors are often covered with PVC sheet or tiles, which are subject to damage from hot tobacco ash. Hence waste receptacles and cigarette urns should be located close to lifts. With luxury flats, costly studded rubber sheets are often used for lift floors and these are more difficult and costly to maintain. Normally, special acoustic treatment is unnecessary for lifts. Ceilings should be of simple design, probably incorporating fluorescent lighting behind plastic panels.

Air conditioning. This has become an important feature of modern shopping complexes in the United Kingdom. Whittaker (1975) described how cost effectiveness is a good principle to adopt but there are problems as cost is borne by one party, the landlord, and effectiveness encompasses others, the tenant and the shopper, and the effectiveness has to be considered in relation to long term capital value. Because the range of requirements is so wide, designers have tended to provide a relatively simple air movement, conditioned by heating or chilling only. Hence air conditioning systems in shopping centres are much less sophisticated than in offices and other buildings. It is this form of compromise which results in many problems for the property manager, quite apart from the very large quantities of energy required to provide the air movement, and any heating or chilling. Air conditioning has also resulted in a decline in the quality of cladding materials. It will also be appreciated that no air conditioning system can cope effectively with a fully glazed building.

Cleaning. The most important and expensive single item relating to any building, after the rent and rates have been paid, is the cost of internal and external cleaning. In thirty years, or even less in some instances, the cleaning costs will have risen above the original cost of the building. It therefore follows that if buildings are designed with a view to reducing cleaning costs, the savings to the occupant can be considerable. Furthermore, it is possible for two similar buildings, of say 5000 m^2 floor area each, to vary from £24 000 to £60 000 per annum for internal and external cleaning (1994 prices).

The process of cleaning uses increasingly more sophisticated engineering equipment and is often based on complicated service schedules. The frequency of cleaning varies: floors are generally swept daily and polished weekly, windows washed monthly and flues swept every six months. Service schedules may also embrace painting for decoration and protection, cleaning of gutters and drains, and servicing lifts and central heating plant. The cost of maintaining floors varied from £1.00 to £3.00 per m^2 (1994 prices).

Toilets require cleaning every day and so the surfaces should be specified for easy cleaning. Toilet pans should be placed on walls, toilet doors should not come down to floor level, wall tiling should extend to ceiling level, tops of skirtings and corners should be rounded, and sanitary disposal units should be provided in ladies' toilets. Public entrance halls and common staircases should be surfaced with durable and easily cleansed materials. Ample storage space for cleaning equipment, water supply and sink should be provided on each floor. With tall buildings, the design should permit easy cleaning of all glazed areas.

Koh (1979) described how sophisticated cleaning techniques accompanied the growth of the commercial property market in Singapore in the nineteen sixties and seventies. Commercial cleaning encompassed many specialised services, including reconditioning of marble, terrazzo and parquet flooring, heavy duty cleaning of industrial premises, degreasing of hotel kitchen equipment, carpet shampooing, steam cleaning and exterior glass and facade cleaning of high rise buildings. Special training is required for those employed in hospital cleaning as well as computer room maintenance. When cleaning food areas, such as restaurants and supermarkets, extra care is needed since any misuse of toxic substances could be disastrous.

Building finishes are continually being improved to reduce the frequency of cleaning. As the standards of society increase, current abuse such as the indiscriminate depositing of litter and abuse of toilet facilities will hopefully decline. In the meantime a strong case could be made for more positive preventive action by the Government.

Cleaning work requires close supervision and where outside contractors are employed, the inclusion of penalty clauses covering damage to wall and floor finishes arising from the improper use of chemicals is advisable. For contracts of substantial value, a performance bond will provide an added safeguard for the client.

Consultation with cleaning staff employed in a higher education building in Nottingham emphasised the need for cleaning personnel to be consulted at the design stage in the choice of materials for fabric and finishes. Many examples could be given of where an improved design at no additional initial cost would have reduced future cleaning costs. For example, the use of slightly darker coloured floor tiles in some locations would have reduced cleaning cycles without affecting lighting levels.

Another frequent problem is that after handover the building may be used differently from the original planned use resulting in, for example, floor finishes being unsuitable for the new use and causing cleaning problems. A maintenance manual was supplied for the Nottingham building listing recommended cleaning cycles and techniques for each floor finish. Because of the large variety of floor finishes it was found necessary to depart from the schedule and rationalise the cleaning work. Furthermore the cleaning cycles must be arranged to suit the occupation of the building to cause minimum disruption.

Cleaning techniques are changing and becoming increasingly complex and are difficult to foresee in their entirety over the life of a building. For instance, cleaning cupboards in modern buildings are often too small to accommodate modern equipment.

A study of the 1993 costs of cleaning a higher education building of approximately 11 000 m^2, totalling £690/100 m^2 is shown in table 13.3.

Table 13.3 Breakdown of cleaning costs of a higher education building

Cleaning item	*Percentage cost*
Materials (cleaning fluids, towels, etc.)	4
Labour (direct labour) – teaching areas and offices daily, refreshment areas 4 times daily.	91
Laundry (cleaning curtains)	1
Equipment (replacing waste bins and cleaning equipment)	1
Window cleaning (outside contractor)	3
	100

Planning and Management of Maintenance Work

A maintenance plan should be formulated within the context of a maintenance policy, which itself has to be comprehensive, covering all types of work and all properties under maintenance. The policy should set out standards for the provision of the maintenance service in terms of amenity level (redecoration cycles), quality of work (inspections and lists of selected contractors), and day-to-day service (discretionary repairs and response times). The respective responsibilities of landlord and tenants should also be detailed.

The planning process has been subdivided into the following four stages by NFHA (1989): identifying needs; establishing priorities; developing the plan, possibly encompassing a five year period on a rolling programme; and monitoring results with feedback into the on-going plan.

Seeley (1987) has described suitable arrangements for the planning and management of maintenance work. Full records should be kept of each property stating the geographical location, age, condition, construction details by elements, details of services, floor area and cubic content, accommodation provided, current use, and any proposals for the area by the local authority which might affect the property, all based primarily on a stock condition survey. Small organisations may use card records, but most organisations will use computers for ease of recording, updating and accessing information.

It is good policy to require contractors on new projects to supply maintenance manuals giving a

Table 13.4 Maintenance manual materials schedule

Element Number – 2 (floors/ceilings)				
Item Nr	*Item*	*Location*	*Description and comments*	*Manufacturer or supplier*
2/1	Acoustic tiles	Basement access corridor	Gold bond fissured solitude tiles, 300 mm square, self-finish on Anderson's 'J-type' suspension, with white semi-gloss stove enamelled edge trim	Supplied and fixed by Anderson Construction, Twickenham
2/2	Asphalt floor	Tank room	Includes skirting 300 mm high	
2/3	Carpet tiles	Offices	'Debron' ½ metre square. Stipple range, colour 103 'brindle brown'. Arborite skirting.	Supplied and laid by Carpet Tile Co., Berkhamsted. Maker – Carpet Manufacturing Co.

Source: Seeley (1987).

Table 13.5 Maintenance manual cleaning schedule

Element Number – 2 (floors/ceilings)			*Quantity*	*Location*	*Maintenance frequency*						
Item Nr	*Item*	*Maintenance specification*			*Twice daily*	*Day*	*Week*	*One month*	*Three months*	*Six months*	*One year or longer*
2/1	Acoustic tiles	Vacuum clean with brush attachment	210 m^2	Basement access corridor							Vac. once year
2/2	Asphalt floor and skirting	Wash with clean water and wipe dry	25 m^2	Tank room						Wash	
2/3	Carpet tile	Vacuum shampoo clean with proprietary carpet shampoo and report wear	4900 m^2	Offices		Vac.			Clean		

Source: Seeley (1987).

physical record of each building as built, inspection and maintenance cycles for each element, list of specialist subcontractors and suppliers and information and instructions on maintenance for occupants. Extracts from typical maintenance manual schedules are shown in tables 13.4 and 13.5.

Inspection cycles are a vital aspect of any maintenance system. Suppliers of services will normally prescribe inspection cycles for the plant they have provided and the fabric of a building should also be inspected at regular intervals, preferably related to the endurance period of a significant component or material. External painting is generally undertaken at five-year intervals and prior to this it is good policy to carry out a thorough inspection of the property to determine the extent of necessary repair work and its probable cost. Interim inspections should be carried out more frequently, possibly at twelve-monthly intervals to detect defects which would result in progressive deterioration if left unattended until the next cyclic inspection. It is an advantage to use standard report sheets to ensure uniformity of approach. The principal challenge in devising a maintenance programme is achieving the right balance between the cost of check inspections and the resultant benefits. Gutter clearance, and checking drains and flat roofs should always be programmed.

A yearly budget prevents adequate forward planning, and ideally the budget period should extend over a number of years, possibly matching the maintenance programme based on a five-year external painting cycle. This would provide a framework for an efficient maintenance system based on detailed cyclic inspections, accurate estimates and a firm long-term budget which would enable work to be properly planned and executed. Maintenance can be performed either by directly employed labour or by private contractors. A direct labour system permits full control of operatives but entails the provision of supporting facilities such as workshops, stores and transport, and a high standard of supervision and control.

In the case of local authorities, following the passing of the Local Government Planning and Land Act 1980 and subsequent regulations made thereunder, where local authorities continue to operate substantial direct labour organisations (DLOs) they must compete for work with private contractors, keep their accounts on a trading basis and earn a specified rate of return, to ensure that they are cost effective in operation.

There are various contractual arrangements which can be used for maintenance work undertaken by private contractors.
The most common methods are:

(1) payment on the basis of time expended and materials used, plus agreed percentage additions for overheads and profit;
(2) payment on the basis of agreed measurements and an agreed schedule of prices; and
(3) a lump sum offer based on specification, bill of quantities and/or drawings.

In practice few firms apply incentives to building maintenance and a relatively small proportion of these are successful. Nevertheless, it is claimed with some justification that incentive schemes could be applied to maintenance work for which an estimate had been prepared at fairly low cost and could result in increases in output ranging up to fifty per cent as well as providing a means of labour control and a check on profitability. For example incentive bonus schemes have been applied to direct labour housing maintenance work for Middlesbrough Corporation, with work values built up from time studies and making due allowance for attendance to personal needs, recovery from fatigue according to the arduousness of the work done (resting time) and travelling time. A daily allowance is calculated to cover the drawing of stores, receiving instructions and obtaining tools. External painting work has been measured and evaluated, and the operation of financial incentives resulted in a forty per cent increase in productivity and twelve per cent reduction in costs.

Maintenance Feedback

Maintenance feedback should be an essential part of any maintenance administration. Feedback may be mainly injected into the system in the following two ways:

(1) directly to the design team; particularly information on design faults, faulty workmanship and materials failures;
(2) by general discussion within the maintenance team, when solutions to problems should be documented and passed on to all appropriate personnel.

A visual representation of feedback is illustrated in figure 13.4, and this shows some of the major stages in the operation of a maintenance scheme.

(1) management organisation of resources;
(2) work execution;
(3) appraisal of results; and
(4) corrective action through feedback to design and management teams.

To assist in the feedback of information, site defects are suitably recorded showing the symptoms, diagnosis, prognosis (projection of

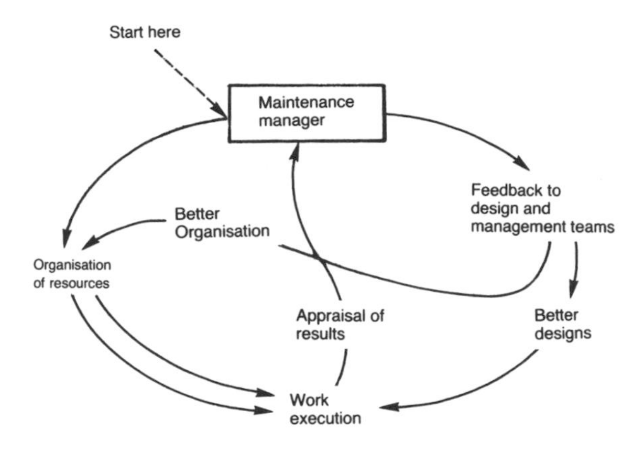

Figure 13.4 Maintenance feedback (Source: Seeley, 1987)

defect performance in time), and the agreed remedy (Seeley, 1987).

In general terms, maintenance costs rise with age; during the first twenty years costs rise fairly steeply and then settle down to a more gradual increase for the remainder of the building life, although this tendency will be distorted as replacements of components become necessary. However, the age/cost profile is not a simple one and there are a number of factors which affect the shape of the profile, such as policy and type of construction.

Some aspects of dwelling design, in terms of maintenance, are critical, such as flat roofs finished with lightweight membranes and large areas of painted woodwork. It is more difficult to pin-point many of the problems with smaller components. Government pressure on local authorities in the 1980s and by the Housing Corporation on housing associations in the 1990s to provide financial statements of maintenance expenditure has led to improved maintenance systems and, hopefully, will culminate in the production of fully documented data on component lives and costs.

Current analyses show that a high percentage of repairs to finishings and fittings result from work to doors and windows. Finishes can account for approximately 27 to 30 per cent of maintenance work, plumbing for about 25 per cent and heating and lighting for around 20 per cent. Defective ball valves were the most common plumbing problem. More attention needs to be paid to the design and treatment of external doors and windows with a view to extending their lives.

There needs to be a more analytical approach to maintenance cost feedback, which has become increasingly important with the greater use of new materials and components, otherwise their cost benefits may never be soundly evaluated (Seeley, 1987).

CURRENT AND FUTURE PAYMENTS

One of the principal difficulties in making life cycle costing calculations is that every building project involves streams of payments over a long period of time (usually the life of the building). The payments are of three main types:

(1) present payments covering the acquisition of the building site, erection of the building and professional fees;

(2) annual payments relating to minor repairs, cleaning, heating, lighting, rents and the like; and

(3) periodic payments such as external painting at five-year intervals and replacement of services or other components.

All these varying types of payments have to be converted to a common method of expression to permit a meaningful comparison to be made between alternative designs. The process is often referred to as discounting future costs and is based on the premise that if the money were not spent on the project in question it could be invested elsewhere and would be earning interest. £1000 invested today at eight per cent compound interest will accumulate to £2158.9 after 10 years (see appendix 1 – amount of £1 table). In the reverse direction the present value of £1 table (appendix 2) shows that it will be necessary to invest £463.19 now at eight per cent compound interest to accrue to £1000 in ten years time. We are often concerned with annual payments throughout the life of a building, which is commonly taken as sixty years. The present value of £1 per annum table (appendix 5) shows that an expenditure of £1 per annum throughout the sixty-year period is equivalent to a single payment of £16.1614 today taking an interest rate of six per cent. Expressed in another way, if £16.1614 were invested today at six per cent compound interest it would provide sufficient funds to be able to pay out £1 per annum for each of the sixty years. Hence it is sometimes said that it is worth spending up to an extra £16 today on initial construction if this will reduce the expenditure on maintenance work by £1 per annum throughout the sixty-year life of the building.

There are two possible approaches in making life cycle costing calculations and both will be illustrated in the worked examples used later in the chapter.

The first one is to discount all future costs at an appropriate rate of interest, often taken at five or six per cent (long term pure interest rate with no allowance for risk premium), and so to convert all payments to present value (PV) or present worth, using the valuation tables described in chapter 12.

The second one is to express all costs in the form of annual equivalents, taking into account the interest rate and annual sinking fund. A building client is entitled to interest on the capital he has invested in the project and requires a sinking fund to replace the capital when the life of the building has expired.

A simple example will serve to illustrate the discounting principle. A building is designed to last sixty years and can either be provided with a roof costing £12 000 which will last thirty years and then need replacing, or be covered with a roof costing £18 000 which will last the life of the building. It is necessary to determine which is the better proposition financially and to do this the payment in thirty years time has to be converted to its present value. The present value of a payment of £12 000 in thirty years time is found from the present value of £1 table (appendix 2) and is £12 000 × 0.17411 = £2089.32, taking an interest rate of six per cent. The calculations can be summarised as follows:

	Cost	*Present value*
Roof A		
Initial construction	£12 000	£12 000
Replacement after thirty years	£12 000	£2 089.32
Total cost	£24 000	£14 089.32
Roof B		
Initial construction	£18 000	£18 000

These calculations show roof A to be the better long-term proposition. It might be argued that it is over-simplified as, for instance, it takes no account of the cost of dismantling and removing the old roof, any temporary work that may be necessary to protect the occupants and contents of the building and any disturbance to the occupants and disruption to the use of the building in the case of roof A.

The selection of a suitable discounting interest rate is also extremely difficult. Interest rates vary considerably over time and we are concerned with exceptionally long periods. There are three ways of assessing the interest rate to be used, as shown below.

(1) The social time preference rate. This is a positive rate of interest which expresses the value persons place on having assets now, rather than at some time in the future and, adopting the kind of life tables used by insurance companies, a social time preference rate could be as low as three per cent.

(2) The rate of interest at which the Government lends and borrows, and is roughly the risk free rate of interest, possibly about five to six per cent.

(3) The opportunity cost rate of interest. This is the rate of interest which could operate if the project being evaluated were not carried out, and so freed the capital for an alternative opportunity, or more often the rate selected is the cost of borrowing less the anticipated inflation rate.

Building clients must either borrow money to finance the project or sacrifice an alternative use for their money. A realistic rate of interest in therefore either the market rate for money borrowed on the security of the building, or the average return which the building client can secure for money invested in his own business. These rates are usually considerably lower than the rate of interest often assumed when predicting the return on a business investment. The actual rate of interest to be used for cost prediction purposes will depend on such factors as the financial standing of the client and on predictions of the long-term movement of rates of interest.

A common approach for calculating a suitable discounting interest rate is to use the following formula: $r = \left\{\frac{1 + b}{1 + i}\right\} - 1$, where r = discounting rate, b = borrowing rate and i = inflation rate. Taking a borrowing rate of 8 per cent and an inflation rate of 3 per cent, the discounting rate becomes $\left\{\frac{1 + 0.08}{1 + 0.03}\right\} - 1 = 4.85$ per cent (say 5 per cent). The worked examples that follow will therefore be based on this figure, although a longer term government stock interest rate could conceivably be 9 per cent and an inflation rate of 5 per cent, in which case $r = \left\{\frac{1 + 0.09}{1 + 0.05}\right\} - 1 =$ 3.81 per cent, which could be rounded off to 4 per cent.

It is often valuable to apply a range of interest rates and then to determine the particular interest rate at which the alternative proposal ceases to be financially attractive by a process known as sensitivity analysis, thereby determining whether the initial ranking is sensitive to changes in any of the forecast elements. For example, with the previous roof cost comparison, the replacement cost of the roof after 30 years would be reduced from £2089.32 to £687.6 if the discounted interest rate had been increased from 6 to 10 per cent.

The main conclusions to be drawn are:

(1) Discounted values have far less influence when the bulk of the expenditure is in initial cost.

(2) At higher rates of interest it is often more beneficial to defer expenditure for as long as possible, while with low interest rates it is generally better to reduce future costs.

MAINTENANCE AND RUNNING COSTS

In 1970, The Royal Institute of British Architects drew attention to the need to balance capital costs against subsequent maintenance and running costs. Their report described how economies in finishes today will undoubtedly lead to inflated maintenance costs in future and that skimping external works or attempting to increase densities at the expense of environmental facilities, such as play areas, may well reduce the acceptable life of a housing estate. These aspects were further highlighted at a conference on cost control of hospitals where it was shown that the costs of servicing a hospital may be six times greater than the building costs.

Housing

Table 13.6 shows the likely relationship of the major items of maintenance costs to their initial capital costs. A survey of maintenance work on local authority houses revealed the distribution of costs shown in table 13.7. The greatest economies in maintenance can be obtained by concentrating on the most costly items, that is decorations and plumbing, by reducing the amount of external paintwork, and designing better plumbing systems well protected from frost, aggressive water and electrolytic action. Another analysis of housing maintenance work carried out by a large local authority revealed that over 12 per cent of the work resulted from faulty materials and workmanship and twenty per cent from design or

Table 13.6 Local authority traditional housing: capitalised maintenance costs as a percentage of initial costs

Item	*Percentage*
Water services	84
Sanitary appliances	22
Heating, cooking and lighting	33
Internal structure and finishes	7
Main structure	5
External services and siteworks	20
External painting	100

Table 13.7 Sources and causes of typical local authority housing maintenance costs

Item	*Percentage of total maintenance costs*	*Main causes*
Structural and cladding repairs	10	Roofs, windows and external doors
External redecoration	25	Protection of wood and metalwork
Internal repairs and renovations	25	Redecorations and minor repairs
Services, installations and sanitation	40	Ball valves, tanks, cylinders, burst and blocked pipes

specification faults. Furthermore, the cost of maintaining pre-war houses was roughly double that of post-war houses, and that of pre-1930 houses could be two to three times as high as that of post-war houses. In general, maintenance expenditure increases with the age of the house at a rate of around 6 per cent per annum excluding indexation for inflation.

In 1991 the average annual cost of repair for occupied dwellings was £1940 for private rented, £1050 for owner occupied, £820 for local authority and £710 for housing association. The high cost of repair of private rented and owner occupied dwellings largely results from over half the pre 1918 dwellings being owner occupied and a further 14 per cent were private rented, accounting for over half of the private rented sector.

Comparison of Different Buildings

The relative importance of the maintenance expenditure incurred on different parts of buildings does however vary considerably with the type of building as shown in table 13.8. Roof repairs account for between one-quarter and one-half of the expenditure on *structure*; much of that on *partitions* is spent on doors; and heating services account for between one-quarter and one-half of the *services* costs. The higher proportion of the cost on services in factories and hospitals is mainly due to the greater complexity of service arrangements in these buildings. BMI maintenance indices in 1990 showed an average breakdown of redecorations: 32 per cent; fabric: 30 per cent; and services: 38 per cent.

A study of the annual maintenance costs of schools showed that pre-war schools (average age thirty years) cost almost double those of post-war schools (average age five years) (DoE, 1972).

University Buildings

Table 13.9 shows the results of a BMI (Building Maintenance Information) study of occupancy expenditure on the main buildings of a university estate for the five year period 1985/86 to 1989/90, as it relates to building and services maintenance. BMI prepared analyses of expenditure on individual buildings in the form of property occupancy cost analyses of the type illustrated in table 13.11.

The estate comprises a main campus which, including residential accommodation, covers approximately 110 ha, a medical school which adjoins a major hospital on the periphery of the main campus, together with a second site of approximately 32 ha in area, which encompasses various buildings and playing fields. The buildings selected for detailed analysis were restricted to 32 of the main cost centres which accounted for about 75 per cent of the total maintained floor area, occupying 421 460 m^2, and in 1989/90 absorbed nearly 80 per cent of the estate expenditure on building and services maintenance. The annual maintenance expenditure over the five year period was made up of 27 to 21 per cent direct labour and materials and 73 to 79 per cent contract work; the latter appearing to be more cost effective. The total maintenance costs/100 m^2 were: 1985/86 – £708; 1986/87 – £667; 1987/88 – £748; 1988/89 – £771; and 1989/90 – £801. The total amount allocated to maintenance is related to the condition of the buildings but must also be determined by the total amount that can be drawn from the gross income of the university, particularly in the 1990s when most universities were examining ways of reducing their operating

Table 13.8 Distribution of maintenance expenditure between different buildings

Building element	Percentage of annual maintenance expenditure				
	Houses	Factories	Schools	Hospitals	Average
Structure	15	28	12	10	17
Partitions	14	–	4	6	11
Decorations	36	29	52	29	31
Fittings	–	–	3	7	5
Services	30	43	25	41	31
Other	5	–	4	7	5
Total	100	100	100	100	100

Table 13.9 University ranked average annual expenditure 1985/86 to 1989/90

Building type	Building maintenance and redecoration (£/100 m^2)	Building GFA (m^2) (approx. date of erection)	Building type	Services maintenance (£/100m^2)	Building GFA (m^2) (approx. date of erection)
Residence (1)	1387	4773 (1900)	Computer centre	698	1979 (1970)
Law library	753	4752 (1900)	Sports centre	688	7828 (1970)
Residence (4)	707	13169 (1965)	University centre	620	7902 (1960)
Staff House	682	3576 (1960)	Staff house	574	3576 (1960)
Residence (5)	671	16840 (1965)	Medical school (1)	496	21154 (1935)
Computer centre	632	1979 (1970)	Medical school (2)	473	6822 (1970)
Students' flats (1)	611	22081 (1970)	Electrical engineering	410	12232 (1965)
Chemistry dept (1)	559	4621 (1935)	Residence (1)	392	4773 (1900)
Residence (3)	452	17325 (1960)	Biology dept	371	17401 (1960)
Clinical research	396	7410 (1955)	Library	353	17424 (1955)
Sports centre	394	7828 (1970)	Residence (3)	337	17325 (1960)
Chemistry dept (2)	390	8250 (1960)	Chemistry dept	334	4621 (1935)
Biology dept	381	17401 (1960)	Residence (4)	332	13169 (1965)
Metallurgy	353	14948 (1965)	Chemistry dept (2)	331	8250 (1960)
Psychology dept	352	3463 (1900)	Clinical research	321	7410 (1955)
Arts dept (3)	324	2004 (1975)	Great hall	305	17419 (1900)
Great hall	313	17419 (1900)	Psychology dept	288	3463 (1900)
Students' flats (2)	313	3344 (1980)	Law library	276	4752 (1900)
Students' union	283	7873 (1930)	Residence (5)	273	16840 (1965)
Arts dept (2)	257	12711 (1970)	Chemical eng	263	5853 (1925)
Education dept	254	6495 (1965)	Arts dept (2)	253	12711 (1970)
Maths physics dept	249	5881 (1960)	Residence (2)	180	7181 (1890)
Medical school (1)	228	21154 (1935)	Metallurgy	175	14948 (1965)
Residence (2)	224	7181 (1890)	Education dept	170	6495 (1965)
Arts dept (1)	210	10576 (1960)	Maths physics dept	170	5881 (1960)
University centre	175	7902 (1960)	Students' flats (2)	167	3344 (1980)
Mechanical eng.	159	11971 (1950)	Arts dept (1)	160	10576 (1960)
Chemical eng.	134	5853 (1925)	Mechanical eng	150	11971 (1950)
Medical school (2)	119	6822 (1970)	Physics	142	4808 (1965)
Physics	99	4808 (1965)	Arts dept (3)	138	2004 (1975)
Library	85	17424 (1955)	Students' flats (1)	137	22081 (1970)
Electrical eng.	64	12232 (1965)	Students' union	70	7873 (1930)

Source: BMI – *University Occupancy Expenditure 2* (1991).

costs because of reductions in government funding in real terms (BMI, 1991).

The average annual expenditure over the five year period under the heads of 'building maintenance and redecoration' and 'services maintenance' are ranked for each of the 32 buildings, in descending order of cost in table 13.9. There is an amazingly wide variation in the costs of the different buildings when related to the common unit of $100\,m^2$, and the annual analyses also showed considerable variations from year to year. It was evident that apart from the general level of annual maintenance work, major upgrading work was carried out at specific points in time, often involving reroofing, replacing windows and complete redecoration and thereby producing very high rates of spend. Because the costs of such work may or may not fall within the five year period, BMI justifiably considered that it was not possible to draw firm conclusions on whether one building type was more expensive to maintain than another. Furthermore, it will be noted that the cost ranking of buildings is not necessarily related to the age or size of the building or its function, and it will be recognised that older buildings constructed of well established local materials and using well tried and proven techniques, and which in all probability have been updated and improved, may not be subject to heavy maintenance expenditure.

The BMI average annual expenditure analyses/ $100\,m^2$ for the five year period for the 8 residential buildings showed median costs of £611 for building maintenance and redecoration and £273 for services maintenance, indicating much higher building maintenance costs but lower services maintenance costs as compared with the 24 academic buildings at £313 and £320 respectively. The higher building maintenance and redecoration costs of the residential buildings will be partially explained by their much greater subdivision into smaller units with a higher proportion of fittings and possibly greater wear and tear.

The averaging of the annual expenditure in table 13.9 makes it possible to identify the buildings which have absorbed the greatest input of maintenance work in the five year period. For example, the six academic buildings with the highest expenditure on building maintenance and redecoration are the Law library, Computer centre, Chemistry department (1), Clinical research, Sports centre and Chemistry (2). Of these six buildings, two form part of the original nucleus of the university, two more were built during the major expansion in the 1960s and one forms part of the 1950s expansion of the Medical school.

The highest expenditures on services maintenance show that the main concentration has been on the social buildings and upgradings of the medical school and computer buildings, while the lowest services expenditures are recorded for the various art buildings. However, it must be recognised that the period of five years is too short to permit any definitive conclusions to be drawn on whether any particular building is likely to appear at one end of the range or the other, because of the variable incidence of major repair, refurbishment or retrofit works. Furthermore, one might reasonably expect the engineering and science buildings with their laboratories and associated services to rank high in services maintenance expenditure, whereas in fact they are spread throughout the range.

Occupancy Costs

Table 13.10 is based on a statistical analysis of building occupancy cost analyses submitted to BMCIS (now BMI) as part of a scheme which started in 1974 and has progessively been increased and the timescale extended to cover the years 1966 to 1984. The results of the study include the use of the standard deviation and coefficient of variation for the costs of cleaning, utilities, administration and overheads, as these costs usually follow a regular pattern and the coefficient of variation provides a reasonable guide to the reliability of the mean. Definitions of these terms are now provided for the benefit of the reader who is not familiar with them:

Mean: the sum of the figures divided by the number of figures, thus providing the average occupancy cost for the buildings in the survey.

Table 13.10 Average occupancy costs, £/100 m² per annum (1985 mean prices)

Building type:	*Industrial buildings*	*Offices*	*Shops*	*Homes for the aged*	*Schools*	*Universities*	*Laboratories*	*Student hostels*
Total occupancy costs:	*4103*	*4826*	*7015*	*3561*	*2470*	*3293*	*5338*	*3378*
Element costs								
1. Decorations	210	147	87	130	60	131	116	209
2. Fabric	127	209	199	66	110	131	246	155
3. Services	183	271	348	129	52	154	917	187
4. Cleaning	406	506	407	275*	520*	466	528	797
5. Utilities	1756	833	1028	895	513	846	1513	738
6. Admin. costs	416	1043	1502	310*	230*	778	942	502
7. Overheads	890	1594	1448	791	976	793	1073	792

Source: *BMCIS Study of Average Occupancy Costs* (1985).
* Estimated figures.

Standard deviation: the square root of the mean of the squares of the deviations from the mean of the sample. This figure is an indication of the dispersal of the figures within the range and around the mean. In approximate terms, assuming a normal distribution, 70 per cent of the figures will be within ± 1 standard deviation of the mean, 95 per cent will be within ± 2 standard deviations and 99 per cent within ± 3 standard deviations.

Coefficient of variation: the standard deviation expressed as a percentage of the mean. As a general rule two thirds of the costs are likely to fall within the coefficient of variation above and below the mean.

The elements listed in table 13.10 each comprise a number of sub-elements, most of which are incorporated in the example illustrated in table 13.11. It should, however, be noted that cleaning includes windows and external and internal surfaces; utilities encompasses gas, electricity, fuel oil, solid fuel, water rates, effluents and drainage charges; and administration costs embrace services attendants, laundry, porterage, security, rubbish disposal and property management.

It will be immediately apparent that the occupancy costs for the different buildings listed in table 13.10 show very wide divergences, not all of which can be readily explained. The buildings analysed in any one category will exhibit significant variations in such factors as location, size, age, construction, condition, rateable value and management policy. This leads to very wide ranges in the individual costs of elements for the same type of building and casts doubt upon the value of average occupancy costs. Hence BMI have rightly emphasised that statistical indices can do no more than measure a pattern, although this will provide a starting point for dealing with specific cases. It is always necessary to consider the particular circumstances of each case, its complexity, local conditions and other relevant factors (BMI, 1985).

As is to be expected, shops exhibit the highest total occupancy costs, followed by laboratories and offices, with schools, universities and student hostels at the bottom of the range. The high decorations costs for industrial buildings is inflated by the inclusion of some buildings where redecoration includes painting the external cladding, while some of the student hostels costs encompass either complete internal and external redecorations or complete internal redecoration, thus producing inflated figures.

The laboratory fabric costs show the very high annual cost range of £21 to £853/100 m², while offices show an even greater range and include such items as re-asphalting the roof of a 5 storey block and a similar situation applies to shops where the renewal of an asphalt roof and rooflight repairs are involved. The low costs for homes for the aged result from all the buildings being low rise and mainly of traditional construction with a fabric maintenance cost range of £2 to £179/100 m^2.

The laboratories are all research laboratories with high services costs in the total range of £99 to £4999/100 m^2. The shops comprise a high street multiple store and a university bookshop exhibiting a very wide services cost range of £393 to £778/100 m^2, but it will be appreciated that the sample is very small in this case. The seven schools in the sample exhibit a very low services maintenance cost range of £12 to £140/100 m^2.

The student hostels show the highest cleaning costs and include for complete cleaning with average annual costs ranging from £411 to £1513/100 m^2, while most of the other buildings have average annual costs in the £400 to £530 range. It is interesting to note that the highest annual cleaning cost for offices was £1787/100 m^2 and comprised government offices containing ministerial suites for which one of the management criteria was prestige.

There is a very wide range in annual utilities costs between the different buildings, from £1756 for industrial buildings and £1513 for laboratories down to £513/100 m^2 for schools. The factories are heated by oil-fired district heating systems, while two laboratories use oil-fired systems and three use district heating; the first group of laboratories show an average cost range of £3046/100 m^2 and the second group average £913/100 m^2.

The highest administrative costs are inserted against shops, but this is rather misleading as the detailed analyses show that only two shop cost analyses were received; one has a range of £1727 to £2338/100 m^2 and the other a range of £177 to £208/100 m^2. The buildings with the next highest costs are offices with an enormous range of £18 to £4068/100 m^2; the four lowest annual costs are for buildings without porterage while the highest figure includes the cost of porterage. These analyses show how necessary it is to know the local circumstances and that averages should be treated with caution.

Although the average overheads costs for offices rank as the highest for this element, the detailed analyses shows a total range for overheads costs of £78 to £5902/100 m^2. The lowest figure is for a building on an educational site in southern England which pays no rates and the highest for a building in central London. Similarly with shops there is a very wide range of costs and the highest reflects a large increase in rates. At the bottom end of the scale it is interesting to note how close the average annual overheads costs are for homes for the aged, universities and students hostels, yet in all cases the range of costs is substantial.

Occupancy Expenditure Patterns

Figure 13.5 shows the breakdown of occupancy expenditure patterns for four different types of building covering broad functional uses, based on a sample of 12 expenditure reports for at least four buildings in each category to give realistic expenditure patterns. The work was undertaken by BMI and issued in 1990.

The residential buildings comprise students' residential accommodation; the commercial buildings cover both offices and shops; the recreational buildings encompass concert halls and sports buildings; and the education facilities include university colleges, teaching blocks and research buildings.

The expenditure patterns are broadly similar across all types of buildings except that cleaning takes a significantly higher proportion of resources in residential buildings than for the other categories. Other occupancy costs show much smaller variations, but it will be noted that rather higher than normal expenditure occurs in commercial buildings on fabric maintenance and utilities, in residential buildings on decoration, in recreational buildings on administration, and in education facilities on utilities. It could, however, be argued that these latter variations are not of any great significance as there are inevitably differences in the proportions of occupancy costs between the different elements, even in the same type of building, and they are all influenced by a wide range of factors as described earlier in the

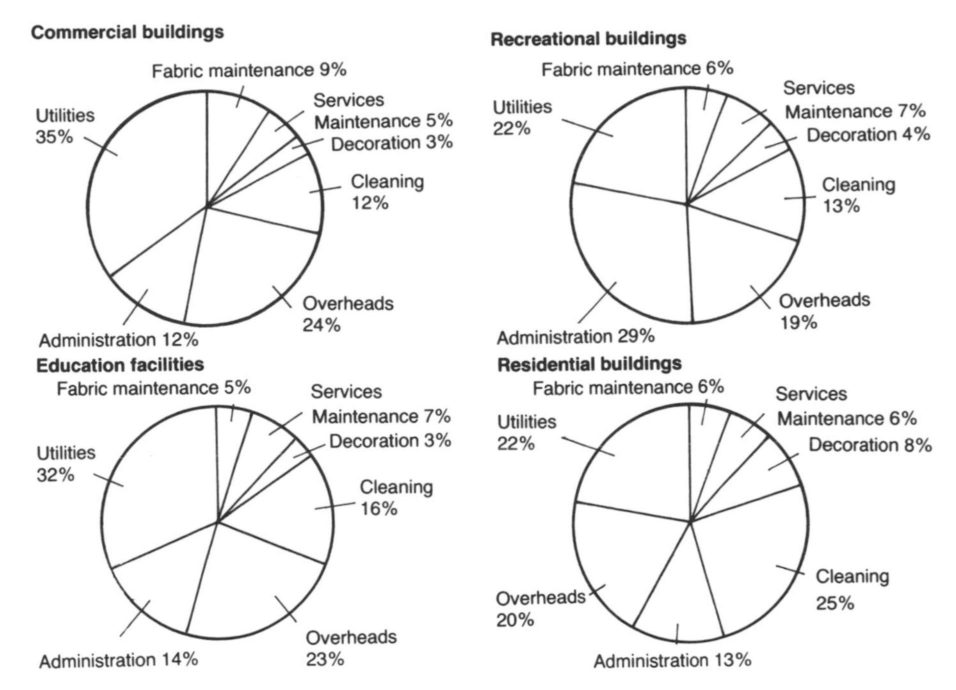

Figure 13.5 Occupancy expenditure patterns (Source: BMI – Occupancy expenditure patterns, 1990)

chapter, thus requiring all average figures to be treated as broad guidelines and to have full regard to the local circumstances.

BMI examined the occupancy expenditure patterns for student residences in universities in 1990 and computed an annual average occupancy cost of £3156/100 m^2, made up of £148 for decorations, £85 for fabric maintenance, £246 for services maintenance, £803 for cleaning, £683 for utilities, £493 for administration and £698 for overheads. Some useful unit costs were also calculated, comprising an annual occupancy cost of £425 per reading space for a university central library, £1034 per occupant for a research laboratory and £647 per student for student residences.

The 1991/92 average pattern of expenditure for seven university college buildings showed a total annual expenditure of £4286/100 m^2, while the range for the individual college buildings was £3366 to £4907/100 m^2. The average breakdown of annual occupancy costs/100 m^2 into the various elements was as follows: decorations: £81.43 (1.9 per cent); fabric maintenance: £282.88 (6.6 per cent); services maintenance: £428.60 (10.0 per cent); cleaning: £912.92 (21.3 per cent); utilities: £887.20 (20.7 per cent); administration: £1620.11 (37.8 per cent); and overheads: £72.86 (1.7 per cent).

Property Occupancy Cost Analyses

A typical property occupancy cost analysis covering a five year period is illustrated in table 13.11. It shows the costs/100 m^2 for each of the component elements or areas, subdivided into sub-elements with elemental sub-totals. It shows very clearly how occupancy costs of the same building can vary considerably from year to year depending on the incidence of major works. For example, decorations costs are minimal for the first four years under consideration as they only include touching up work, but in the fifth year full external painting occurs with a massive rise in the cost of this element. Other significant variations occur in external walls, windows and doors, other structures and internal finishes in the fabric main-

Table 13.11 Sample property occupancy cost analyses

February 1991

Financial statement for years 1985/86–1989/90 (£/100 m^2)

Gross floor area 7873 m^2

	1985/86	*1986/87*	*1987/88*	*1988/89*	*1989/90*
Decorations					
External painting	0.29	0.00	1.97	0.00	782.54
Internal painting	0.00	0.05	0.00	0.00	0.00
Decorations sub-total	0.29	0.05	1.97	0.00	782.54
Fabric					
External walls	5.03	0.30	2.65	1.13	111.89
Windows & doors	165.79	11.76	82.24	11.84	2.04
Other structure	0.92	1.97	2.97	5.07	77.36
Fittings and fixtures	0.00	0.00	0.00	0.00	0.00
Internal finishes	0.00	62.68	0.00	0.00	0.12
Joinery	0.00	0.00	0.00	0.00	4.52
Roofing	6.96	19.50	12.82	26.94	16.08
Fabric sub-total	178.70	96.21	100.68	44.98	212.01
Services					
Plumbing	5.63	0.00	0.29	0.75	0.00
Heating and ventilation	23.54	5.95	5.37	0.16	39.67
Electric power & lighting	4.42	1.37	3.10	1.07	2.53
Lifts & escalators	18.64	4.87	174.44	20.44	34.86
Other services	0.53	0.00	0.14	0.02	0.27
Services sub-total	52.76	12.19	183.34	22.44	77.33
Cleaning					
General	301.64	324.61	344.62	350.18	360.18
Local	0.00	0.00	0.00	0.00	0.00
Cleaning sub-total	301.64	324.61	344.62	350.18	360.18
Utilities					
Utilities sub-total	N/A	N/A	N/A	N/A	N/A
Administration costs					
Security	127.49	125.94	139.34	133.93	143.20
Property management	0.00	0.87	7.47	3.39	2.51
Administration sub-total	127.49	126.81	146.81	137.32	145.71
Overheads					
Property insurance	34.77	39.66	44.81	50.48	53.83
Rates	454.06	508.66	534.68	578.46	347.18
Overheads sub-total	488.83	548.32	579.49	628.94	401.01
Sub-total	1149.71	1108.19	1356.91	1183.86	1978.78
External works					
Services	45.94	35.33	60.03	42.61	70.68
Roads & grounds	29.98	80.28	34.20	48.57	89.82
External works sub-total	75.92	115.61	94.23	91.18	160.50
TOTAL	1225.63	1223.80	1451.14	1275.04	2139.28

Source: BMI – *University Occupancy Expenditure 2* (1991).

tenance element. These higher spends occur only in one or at most two years throughout the five year period, militating against the computation of a realistic average annual occupancy cost.

While wide fluctuations occur in heating and ventilation, lifts and escalators, and external works, there are a number of items where the annual costs remain reasonably constant, such as plumbing, electric power and lighting, other services, cleaning, security and overheads. However, even in these more consistent areas, costs could fluctuate widely outside the selected five year period as, for instance, electric power and lighting would show a very large increase in a certain year were rewiring of the building to be carried out. This example highlights once again the problems that occur in assessing average annual occupancy costs, unless a very long period indeed is to be investigated, such as 30 years to encompass major repairs and replacements, and it is unlikely that adequate cost records will be available for such a lengthy period.

Cleaning Costs

BMI have prepared a cleaning cost index primarily for the purpose of adjusting the cleaning section of the BMI occupancy cost analyses. It covers window cleaning, cleaning of external surfaces and internal cleaning including floors, carpets and dusting and cleaning ledges, furniture and fittings. The index attempts to model average circumstances so that it will act as a general guide to movements in cleaning costs, but it is unlikely that the make up of cleaning costs for any specific building will fit exactly the model used in the index.

The report of the National Board for Prices and Incomes on Pay and Conditions in the Contract Cleaning Trade gives the following breakdown for cleaning costs:

Wages	56%
National Insurance, etc.	9%
Materials and Equipment	7%
Administration and Overheads	18%
Profit	10%
	100%

Table 13.12 Percentage breakdown of cleaning expenditure

Building type	*Windows*	*External surfaces*	*Internal surfaces*
Factories	5	–	95
Offices	5	–	95
Shops	25	6	69
Health & welfare buildings	1	–	99
Halls, etc.	4	–	96
Education and research	3	–	97
Residential	2	–	98
Hotels	11	–	89
Average	4	–	96

Source: BMI – *Occupancy Cost Indices* (1990).

The overall weightings used in the index are as follows:

Labour	90%
Materials	6%
Plant	4%
	100%

Table 13.12 gives an average breakdown of cleaning expenditure over a range of different building types, from which it will be seen that the cleaning of internal surfaces accounts for a large proportion of the total costs. In addition, shops show a relatively high allocation of costs to cleaning windows.

Elemental Costs

Roofs. Built-up felt roofing at an initial cost similar to that of asphalt requires more maintenance and may require replacement at twenty-year intervals. Zinc sheeting may have a limited life in industrial areas.

Walls. Brickwork requires repointing at thirty to forty-year intervals. Stonework may give rise to much greater maintenance problems through its laminated structure and consequent frost damage and bad weathering. Large areas of rendering will always prove costly in maintenance.

Windows and curtain walling. Cheap timber windows are likely to prove troublesome particularly when they open up at joints. Painted hot dip galvanised steel windows are cheaper in first cost than aluminium but require periodic repainting and over the life of the building, aluminium may be competitive, but it does require frequent cleaning to prevent possible pitting.

External decoration. The cost of external painting can be reduced by using materials which do not require decorating, such as precast concrete and PVC-U rainwater goods, and self-coloured cladding panels. Large surfaces of painted wall will require redecoration at about five-year intervals and will thus be a costly maintenance item. On tall buildings roof anchors should be provided to carry painters' cradles and reduce painting costs.

Internal decoration. Emulsion paint has replaced washable distemper as it is far more hard wearing and allows the wall to dry out without damaging the paint film. Wall tiling and terrazzo are approximately 10 and 50 times more expensive respectively in initial cost than emulsion paint but require little maintenance.

Services. Service pipes should be carefully located to give protection from frost and yet be accessible for inspection and repair. PVC-U drain pipes compare very favourably with clay, particularly where long lengths are involved or ground conditions are bad.

Hot water heating systems. Low temperature hot water (LTHW) systems (82°C flow at boiler) account for most of the heating systems installed in office buildings. Both MTHW (110°C flow at boiler) and HTHW (160°C flow at boiler) heating systems are more commonly used in installations such as hospitals, prisons, and military establishments, where a central boiler house will serve high temperature primary mains to calorifiers in individual buildings. This would also apply to steam installations. The capital cost range for heating to offices was £30 to £65/m² GFA in 1993. In terms of capital costs the selection of a boiler plant and heat emitters will be the most critical (W. T. Partnership, 1993).

Table 13.13 Characteristics of heating boilers

Type of boiler	*Capacity*	*Efficiency*	*Typical 1993 cost*
Cast iron sectional	Domestic size up to 3 MW	70–76%	gas fired: £9200 oil fired: £7700 for 300 kW boiler
Modular	40–100 kW	85%	gas fired: £14 500–£18 000 for 500 kW boiler
Packaged high performance	100 kW to 3.5 MW	80–85%	gas fired: £10 500–£15 800 oil fired: £9000–£12 200 for 500 kW boiler
Condensing	12–800 kW	up to 98%	gas fired: £10 500–£12 500 for 500 kW boiler

Source: W. T. Partnership (1993).

Table 13.14 Relative cost of heat emitters

Heat emitter	*Approx. cost installed £/m² gross floor area (1993 prices)*
Steel panel radiator	7
Steel panel radiator – high output	10
Aluminium radiator – high output	10.6
Continuous natural convectors	13
Radiant panels	14.5

Source: W. T. Partnership (1993).

Table 13.13 shows the main characteristics of different heating boilers, table 13.14 the relative costs of heat emitters, and table 13.15 gives a typical elemental breakdown of a LTHW heating installation with gas fired boiler plant and continuous perimeter natural convectors for an office building of 5000 m² GFA.

Heat emitters are generally used in LTHW heating systems around the perimeter and under windows to offset down draught and consist of radiators, convectors and radiant panels. Radiators/radiant panels are used on staircases with fan convectors in entrance halls. The selection of the form of primary heating is influenced largely by the type

Table 13.15 Typical elemental breakdown of LTHW heating installation with gas-fired boiler plant and continuous perimeter natural convectors for office building of 5000 m²

	Cost/m² GFA (1993 prices)
Boiler plant 2 Nr Sectional boilers Flue, pressurisation plant and pumps Circulation pumps	4.50
Pipework, valves, fittings and insulation	14.50
Perimeter natural convectors Radiators on staircases Fan convectors in entrance	13.50
Automatic controls installation Mechanical services wiring	6.00
	£38.50
Typical cost for warmed fresh air ventilation to centre zone of deep plan office building (treated area)	£90

Source: W. T. Partnership (1993)

which will be aesthetically acceptable and unobtrusive, highlighting once again that cost is often not the only criteria to be considered. Radiators may be preferred in smaller offices whereas continuous natural convectors or radiant panels are likely to be favoured in high quality buildings. Radiant panels have the advantage of not protruding into the office and thus cause no loss of lettable floor area.

Automatic controls if required for heating alone do not need to be as sophisticated as those used with air conditioning systems. The latter need to control the temperature and humidity of the building in addition to monitoring and adjusting plant and equipment performance by means of a building management system (W. T. Partnership, 1993).

In 1992/93, Neighbourhood Energy Action quoted average annual domestic heating costs of £281 using oil, £440 with electricity (based on Economy 7 storage heating), £350 for gas and £387 using coal. However, these figures can at best form only a rough guide as confirmed by NHBC estimating the average annual running cost of heating a new home in 1993 at £750 and an older property costing at least £1150, to which VAT at 8 per cent is added as from 1994. It is very difficult to reconcile these figures and the NEA costs for gas heating are much closer to the author's average costings of around £400. The costs will be affected significantly by type of construction, amount of insulation and internal temperatures; hence there can be no truly typical figure.

Refuse disposal. Taking the annual overall cost of supplying, replacing, emptying and disposing of the contents of dustbins to flats at a base cost of 100, then the relative overall costs of other alternatives are communal containers, 110, chutes, 146 and Garchey system, 354.

THE LIVES OF BUILDINGS AND COMPONENTS

Buildings

Life cycle costing comparisons are concerned with buildings and their component parts, and the longest period over which comparisons need be made should not exceed the expected life of the building. This emphasises the need to be able to predict with reasonable accuracy how long buildings should be expected to last. One approach would be to determine the lives of a representative sample of our present stock of buildings and to assume that the average life for each class of building represents half its effective life, but there are a number of weaknesses inherent in this hypothesis. In work undertaken at the Building Research Establishment it has been generally assumed that the life expectancy of a new building is in the order of sixty years, although they often last longer, but as the annual equivalent of a capital sum is virtually constant after sixty years there is little point in assuming a longer life. It should however be borne in mind that extensive modernisation or refurbishment works will be required during the 60 year period. Other sources have assumed a life of 50 years but, in practice, the life of a building can be affected by a number of factors such as form of construction, quality of materials and components, extent of maintenance and type of use.

The period taken should be realistic, and will normally be the period over which the building is expected to earn an income or provide a service, for it is during this period that the costs will be recovered. It is necessary to distinguish between economical, functional, social and physical obsolescence of buildings. Structural or physical life is the period expiring when it ceases to be a worthwhile proposition to maintain the building because of the advanced state of deterioration whilst economic life is concerned with earning power and is the ability of the building to show an adequate return. Functional obsolescence can result from changes of use, technology or statutory requirements, such as changes in spacing standards or the need to accommodate new machinery or modern technological innovations needed to remain competitive. A new building will often need to cater for a variety of changing functions during its life, while social obsolescence can flow from changes in the needs of society in general. The structural or physical lives of buildings can be very long indeed as the continued existence of many factories constructed during the Industrial Revolution shows. It should, however, be emphasised that buildings are more usually demolished because of changes in demand rather than through becoming worn out. Optimum life is therefore determined primarily by the earning power of the building, and only secondarily by the structural stability. Changing social and economic conditions can have a considerable influence on the life of a building which can become ill-suited to present day needs and its demise may also be accelerated at times when there is a high ratio of land to building costs. Wherever possible the aim should be to extend the economic life of a building by making the structure adaptable, such as by incorporating wide floor spans and demountable partitions, and by careful management and control of the surroundings. Hence the actual physical life of a building is frequently much greater than its economic life but it is often demolished before the physical life has expired in order to permit a more profitable use of the site, or because it is found cheaper to clear and rebuild rather than to adapt the building to a change in requirements. The physical life of a building can often be extended almost indefinitely by good maintenance and suitable alterations and improvements.

There are problems in incorporating obsolescence factors into life cycle costing studies, as they are frequently subjective and difficult to predict. In view of the difficulty of predicting the needs of future generations, it would be prudent to consider carefully before providing buildings of long life with high capital and running costs and with a structure which will be difficult to adapt to changing needs.

Components

The next consideration is the life of the materials and components used in the building. For example, we may need to compare the life cycle costs of brick with precast concrete cladding, or paints as a wall finish with a plastic sheet finish, or a variety of materials for use in rainwater goods or as flat roof finishings. In each case we need to know the life of the materials, so that it will be possible to compute how many times they will need to be replaced during the life of the building, and we would also need to know what maintenance treatment will be required and how often. Unfortunately there is limited reliable information on the lives of materials and components or on maintenance frequencies, although more is becoming available through BMI and other sources.

The lives of building materials and components can be determined on the basis of observed probability of failure, but these data are rarely available and it is frequently necessary to predict the life on the basis of a knowledge of the age of early failures. Such information is often incomplete since it does not record the successes or the numbers at risk, and hence suggests a higher rate of failure than actually occurs. The best approach would probably be to prepare a tabulated list of the estimated lives of as many materials and components as possible, with the list being continually updated both as regards new materials and estimated lives. However other factors such as degree of exposure, amount of wear, amount of atmospheric pollution, and similar matters would also need recording and considering. Building Maintenance Information (BMI) provides valuable information on lives of building materials and components in varying situations and on their maintenance problems, and can greatly assist those concerned with maintenance management.

Table 13.16 Typical lives of building components

Building component	*Average life in years*
Brick walling	80
Steel profiled sheeting, PVC coated	30
Timber cladding	30
Tiled/slated roof	60
Bitument felt roof (3 layers)	20
Asphalt roof	30
Rainwater goods (PVC-U)	30
Softwood external doors	30
Softwood windows	30
Hardwood windows	50
Aluminium windows (polyester coated)	40
PVC-U windows	30
Linoleum	15
Ceramic wall tiles	50
Vinyl floor tiles	30
Cork tiles	15
Rubber flooring	25
Carpet tiles	10
Carpet (broadloom)	10
Vinyl wallpaper	10
Kitchen units (melamine faced chipboard)	20
Kitchen units (blockboard/chipboard/ plywood painted)	15
Kitchen units (good quality hardwood)	25
Wash basins, sinks, baths and WCs	25
Cold water storage tanks	25
Gas heating installation	40
Boilers (heating)	25
Calorifiers	30
Local water heaters	20
Extract fans	10
Night storage heaters	30
Electrical installations	30
Fluorescent light fittings	20
Tungsten lamp in bowl fitting	10
Air conditioning plant	10
Softwood close boarded fence	30
Concrete paths	35
Tarmacadam surfacing	15
Brick boundary walls	35

Lives of Components

Table 13.16 shows typical lives of a selection of the more common building components obtained from a variety of sources, but they will in practice vary considerably with the quality of product and workmanship in installation, degree of exposure in the case of external components, extent of maintenance and other related factors. Hence the periods listed can only form a rough guide and should not be applied indiscriminately to individual buildings, as they are likely to require adjusting for local conditions.

Many major fabric components can have a life expectancy in excess of the building's usefulness and there may often be a market for salvaged materials for restoration work. However, there are other components, particularly on the services side which have little or no residual value as, for example, raised access computer floors which are costly when new but worthless on demolition (Park, 1994).

Some of the more sophisticated components require detailed consideration with regard to maintenance and improvement during their effective lives. For example, when a lift installation has been operating for 15 to 20 years, it will almost certainly be able to improve its performance by fitting computerised controllers and new variable voltage frequency motors. Lift consultants and lift manufacturers can carry out an on-site survey of the performance of the existing life installation and, by using suitable computer programs, will be able to predict the likely improvement to be achieved by the proposed installation over that of the existing one (Hassan, 1995).

Where a group of lifts are being refurbished, the likely cost of the work may be around £100 000 (1994 prices) and the owner is more likely to have the work carried out if it can be done and paid for over a period of say five years. This has encouraged lift manufacturers to adopt a modular approach to the refurbishment of lifts incorporating the updating of part of the system usually on an annual basis. Where the building is being refurbished the owner is presented with the opportunity to bring the lift installation up to the requirements of the latest British Standards and/or Health and Safety at Work legislation, as well as improving the interior decor of the lift cars. The lift installation can amount to about 10 per cent of the capital cost of the building and, in a competitive market, it makes good financial sense to improve the effective operation and enhance the aesthetic appeal of the existing investment (Hassan, 1995).

LIFE CYCLE COST PLANS

As described earlier in the chapter, a life cycle cost plan (LCCP) identifies the total costs of the acquisition of a building or an individual building element. It takes explicit account of initial capital costs and subsequent running costs, and expresses these costs in a consistent, comparable manner by applying discounting techniques (Flanagan and Norman, 1983). The Society of Chief Quantity Surveyors in Local Government (SCQS) devised a suitable format for such a plan which enables quantity surveyors to arrange their calculations and present their estimates in an orderly way, to enable quick comparisons between projects and to identify errors or areas of the proposed design which need further critical examination from a life cycle cost point of view (SCQS, 1984).

An example of a life cycle cost plan applied to a school based on the SCQS format, using a 40 year building life and a 4 per cent discount rate and converting all costs to present value, is illustrated in appendix 7 at the back of the book. This should prove very helpful to the reader in understanding the approach to a complete project with all the basic data contained within it, together with a schedule of replacement and intermittent maintenance and one of annual maintenance. It will be noted that this example includes energy, local taxes, insurances and staffing. Further examples of LCCPs can be found in the books by Flanagan and Norman (1983 and 1989).

PRACTICAL LIFE-CYCLE COSTING EXAMPLES

A number of worked examples follow to show the application of life cycle costing techniques to design problems involving complete buildings, components and services.

Alternative Building Designs

The first two examples are designed to show the PV and annual equivalent approaches to life cycle costing calculations, for complete buildings but excluding property management costs and taxation.

Example 13.1. To find the present value of the running costs of a building with a life of sixty years, given that annual cleaning costs are £1600, annual decorations, £600, and annual repairs, £400, external painting, £4000 every five years, and a new roof will be required every thirty years at £40 000. Interest is to be taken at five per cent.

It is feasible to add together the three annual costs of cleaning, decorations and repairs, although it might possibly be argued that no decorations will be needed in the last year of the building's life. This is a little problematical as the decorations could be undertaken at the beginning of each year and the building might secure a reprieve.

PV *Cleaning, decorations and repairs* £2600 × 18.9292	£49 215.92
(PV of £1 pa for sixty years at five per cent)	
External painting £4000 × 3.37193	13 487.72
(PV of £1 at five year intervals at five per cent)*	
Roof replacement £40 000 × 0.23137	9 254.80
(PV of £1 in thirty years at five per cent)	
PV of running costs	£71 958.44

*PV of £1 in	five years at five per cent		0.78352
" "	ten years	"	0.61391
" "	fifteen years	"	0.48101
" "	twenty years	"	0.37668

PV of £1 in	twenty-five years at five per cent	0.29530
" "	thirty years "	0.23137
" "	thirty-five years "	0.18129
" "	forty years "	0.14204
" "	forty-five years "	0.11129
" "	fifty years "	0.08720
" "	fifty-five years "	0.06832
		3.37193

An alternative approach is to calculate the present values in the form of computer spreadsheets.

Example 13.2. To find the annual equivalent cost over the life of the building, with an initial construction cost of £400 000 annual costs of £16 000 for cleaning and minor repairs, quinquennial repairs of £40 000, replacement costs of £80 000 every twenty years, and demolition costs of £5800 less salvageable materials valued at £1800 at the end of 60 years. The life of the building is to be taken as sixty years and interest at five per cent (ASF at 2½ per cent).

Annual equivalent			
Building	£400 000 × 0.05735		£22 940.00
Interest (5 per cent)	0.05		
ASF for sixty years at 2½ per cent (annual sinking fund)	0.00735		
	0.05735		
Cleaning and minor repairs			£16 000.00
Larger repairs (see previous computation)	£40 000 × 3.37193	£134 877.20	
Replacements (0.37668 + 0.14204)	£80 000 × 0.51872	41 497.60	
Demolition – salvageable value (£5800 − £1800 = £4000)	£4000 × 0.05353	214.12	
To convert PV to annual equivalent multiply by interest + ASF		176 588.92	
		0.05735	
			10 127.37
Annual equivalent of life cycle costs			£49 067.37

Example 13.3. Compare the life cycle costs of the following alternative building schemes.

Scheme A. Total cost of building is £200 000 including architect's and surveyor's fees on a site costing £40 000. Annual running costs are estimated at £6000. Certain services and finishings will require replacing at a cost of £24 000 every twenty years. Other services have an estimated work life of thirty years and a replacement cost of £32 000.

Scheme B. Total cost of building is £260 00 including architect's and surveyor's fees on a site costing £40 000. Annual running costs are estimated at £4800. Certain services and finishings will require replacing at a cost of £16 000 every twenty years. Other services have an estimated working life of thirty years and a replacement cost of £20 000.

In both cases the estimated life of the building is taken as sixty years. An interest rate of five per cent and an annual sinking fund of 2½ per cent.

Note: No annual sinking fund need be applied to the site as it will still be available at the expiration of the life of the building.

Scheme A			
Cost of *site*		£40 000	
Annual equivalent in perpetuity at five per cent		0.05	
			£2 000.00
Cost of *building*		£200 000.00	
First replacement cost in twenty years	£24 000		
PV of £1 in twenty years at five per cent	0.37668		
		9 040.32	
Second replacement cost in forty years	£24 000		
PV of £1 in forty years at five per cent	0.14204		
		3 408.96	
Replacement cost in thirty years	£32 000		
PV of £1 in thirty years at five per cent	0.23137		
		£7 403.84	
PV of building and replacement costs		£219 853.12	
Annual equivalent over sixty years			
Interest at five per cent	0.05		
ASF to replace £1 in sixty years at 2½ per cent	0.00735		
		0.05735	
			12 608.58
Annual running costs			6 000.00
Life cycle costs of Scheme A			£20 608.58
Scheme B			
Cost of *site* (as calculated for Scheme A)			£2 000.00
Cost of *building*		£260 000.00	
First replacement cost in twenty years	£16 000		
PV of £1 in twenty years at five per cent	0.37668		
		6 026.88	
Second replacement cost in forty years	£16 000		
PV of £1 in forty years at five per cent	0.14204		
		2 272.64	
Replacement cost in thirty years	£20 000		
PV of £1 in thirty years at five per cent	0.23137		
		4 627.40	
PV of building and replacement costs		£272 926.92	
Annual equivalent over sixty years		0.05735	
			15 652.36
Annual running costs			4 800.00
Life cycle costs of Scheme B			£22 452.36

Scheme A is financially more favourable than scheme B, as the considerably lower initial and replacement costs in A are not offset entirely by the reduced running costs in B, after discounting future costs.

Example 13.4. A building which is to be demolished in twenty-five years time requires repainting now and will also require repainting every five years until demolition. The cost of each repainting is estimated at £1200. In ten years time £8000 is to be spent on alterations, and £600 will be spent at the end of each year on sundry repairs. What sum must be set aside now to cover the cost of all the work, assuming that the rate of interest obtainable on investment is six per cent, and ignoring the effect of taxation?

Cost of *painting*		£1200	
Present repainting	1.00000		
PV of £1 in five years at six per cent	0.74725		
" " ten years "	0.55839		
" " fifteen years "	0.41726		
" " twenty years "	0.31180		
		3.03470	
			3641.64
Cost of *alterations*		£8000	
PV of £1 in ten years at six per cent		0.55839	
			4467.12
Cost of *sundry repairs*		£600	
PV of £1 pa for twenty-four years at six per cent		12.5503	
			7530.18
Sum to be set aside			£15 638.94

It might be prudent to raise this to £16 000 to meet some of the possible future increased costs.

Example 13.5. A temporary building is to be replaced in fifteen years' time by a new building which it is estimated will then cost £240 000. What sum must be set aside at the end of each year, if the interest rate on investment (after deducting for tax) is three per cent, to accumulate to the building cost figure in fifteen years?

Cost of new building in fifteen years' time	£240 000
Sinking fund to provide £1 in fifteen years at three per cent	0.05376
Sum to be set aside each year	£12 902.40

Heating and Other Services

The examples that follow show how the heating system with the lowest initial cost can involve the heaviest long term expenditure.

In the comparison in table 13.17 a rate of interest of five per cent and a building life of sixty years have been used. Under these circumstances the installation of school 2 could be nearly fifty per cent more expensive than school 1. The breakeven point for this set of cost figures operates at an interest rate of approximately twenty per cent. A worked example will help to emphasise the need to consider running costs when comparing alternative heating schemes.

Example 13.6. The following heating schemes and costs have been submitted in connection with a proposal for a new two storey office block. The building client requires the total costs to be

Table 13.17 Comparative heating costs

		School 1		*School 2*
Capital cost of heating system		£53 800		£34 720
Annual running cost	£2 840		£6 520	
Capitalised annual running cost		53 820		123 560
Total capitalised cost (PV)		£107 620		£158 280

assessed for the two alternatives and a recommendation to be made.

The following are central heating proposals to maintain an even temperature of 17°C.

(1) Electric storage heaters: initial cost of £12 800 and estimated annual running costs of £6400.

(2) Gas-fired ducted hot air: initial cost of £24 000 and estimated annual running costs of £4400.

(1) *Electric storage heaters*

Initial cost	£12 800	
Annual equivalent over sixty years at five per cent and 2½ per cent (as example 13.2)	0.05735	734.08
Annual running costs		6400.00
Total life cycle costs with electric storage heaters		£7134.08

(2) *Gas-fired ducted hot air*

Initial cost	£24 000	
Annual equivalent over sixty years at five per cent and 2½ per cent	0.05735	
		1376.40
Annual running costs		4400.00
Total life cycle costs with gas-fired ducted hot air		£5776.40

The gas-fired ducted hot air has a nineteen per cent cost advantage over electric storage heaters when both initial and capital costs are considered. The calculations do not however include any allowance for replacement of heating equipment during the life of the building but their inclusion in this instance would not change the order of preference. Variations in the future prices of gas and electricity can also change their relative positions. The gas-fired ducted hot air system is recommended on the grounds of lower total costs and reduced loss of usable floor space.

Stone (1975) devised a method of tabulating present and future costs of alternative designs, systems and components in a way which is most meaningful and easily understood. His approach is illustrated in table 13.18.

Example 13.7. A choice is to be made between solid fuel and underfloor electric heating for a new block of flats. The cost particulars relating to installation and maintenance follow.

Table 13.18 Comparison of costs of installing and maintaining heating systems

Heating system	*Costs*	*Frequency*	*Factor*		*PV*
Solid fuel	£1400.00	initial	1.000		£1400.00
	5.60	yearly	18.929	(PV of £1 pa for sixty years)	106.00
	128.00	ten years	1.451	(PV 10 + 20 + 30 + 40 + 50)	185.73
	256.00	twenty years	0.519	(PV 20 + 40)	132.86
				Total costs	£1824.59
Electric underfloor	£800.00	initial	1.000		800.00
	64.00	ten years	1.451	(PV 10 + 20 + 30 + 40 + 50)	92.86
	112.00	fifteen years	0.824	(PV 15 + 30 + 45)	92.29
	880.00	thirty years	0.231	(PV 30)	203.28
				Total costs	£1188.43

Solid fuel
Initial costs £1400 per flat

Running costs	annual flue cleaning	£5.60
	every ten years – boiler descaling, etc.	£128
	every twenty years – replacement of boiler	£256

Electric underfloor heating
Initial costs £800 per flat

Running costs	every ten years replace thermostat	£64
	every fifteen years replace panel fire	£112
	every thirty years renew cables	£880

The calculations in table 13.18 are based on a building life of sixty years and an interest rate of five per cent. The total costs shown do not give the complete picture as they do not include energy costs. If these costs were taken into account it is possible that electric underfloor heating could lose its cost advantage.

Example 13.8. This example relates to glazing and illustrates the need to consider the cost effects of glazing on heating, including changes in running costs.

	Initial Cost of Different Glazing Systems	
	Single glazing	*Double glazing*
Glazing	£3 000	£13 800
Heating installation	38 000	32 000
Total initial costs	£41 000	£45 800

The double glazing involves an extra initial cost of £4800 (£45 800–£41 000), but offset against this will be savings in the running costs of the heating installation, which are estimated at £600 per annum. Taking a five per cent rate of interest and a sixty years' life of building, this annual sum is equivalent to £11 358 (£600 × 18.9292). The double glazing thus has an equivalent first cost advantage of £6558 (£11 358 – £4800). This calculation is somewhat oversimplified as it ignores the effects of the differing maintenance costs of the various heating and glazing systems and other related matters. A more detailed study of the costs and benefits of energy conservation is included later in this chapter.

Air Conditioning

It is possible to establish cost relationships between different combinations of heating and glazing installations. The table 13.19 shows the relative capital costs of air conditioning systems when used in conjunction with different glazing arrangements.

The relative capital costs of the different glazing systems will need to be compared with that of the air conditioning plant, to secure the optimum combination. For instance, 6 mm low emissivity glass is about 3 times more expensive than float glass. Depending on the area of fenestration, the most expensive glazing systems may not completely offset the saving in cost on air conditioning plant.

As with capital costs there are many factors which influence running costs. These include the area of the building, the proportion of glazed area, choice of single or double glazing, shading,

Table 13.19 Relative capital costs of air conditioning plant with different glazing combinations

Glazing combination	*Air conditioning capital cost*
Single sheet clear float glass, no shading	1.85 × £x
Single sheet clear float glass, light colour blinds	1.52 × £x
Body tinted glass, light colour blinds in air space, single sheet clear float glass	1.13 × £x
Low emissivity glass, air space, single sheet clear float glass	1.08 × £x
* Low emissivity glass, air space, single sheet clear float glass, light colour blinds	1.00 × £x

Note: £x is the minimum cost of air conditioning plant designed for use with the most effective glazing system.*

type of air conditioning system, choice of automatic controls, provision of standby facilities, type of heat source and orientation and disposition of the building. It is, therefore, difficult to predict running costs when the figures obtained from historical data have to be assessed against a variety of different criteria. In broad terms, if the capital expenditure on air conditioning plant were to be increased by £10 000, the running costs would normally require to be reduced by at least an average of £600 per annum before it became economically viable. This calculation is based on certain predetermined conditions.

(i) capital is raised by mortgages repayable over 20 years at a rate of interest of 8 per cent;

(ii) repayments are offset by relief on corporation tax assumed levied at 30 per cent;

(iii) air conditioning plant is regarded for tax purposes as fixtures and fittings, rather than as plant and machinery and has an assumed life of 20 years; and

(iv) running costs are calculated on the basis of a 5 per cent rate of inflation per annum.

The next example adopts a different approach to help the student in answering examination questions.

Example 13.9. A client has completed the installation of a sophisticated air conditioning system at a cost of £540 000. He has been advised that the system will require the following replacement costs:

year 2	£72 000
year 4	£24 000
year 5	£108 000
year 7	£86 000
year 9	£24 000
year 12	£240 000
year 15	£172 000

It is his intention to sell the building after the fifteenth year.

(i) What single sum would he have to deposit today in an investment yielding 4 per cent after tax has been deducted at 30 per cent, to cover these future costs?

(ii) Alternatively, how much would he have to set aside annually in equal sums to cover the costs in a similar investment?

The main complication introduced into this example is the tax to be paid on the interest received annually on the deposited sum. An additional component is needed to ensure that the desired rate of 4 per cent is a net figure. To achieve this the usual Present Value of £1 formula must be adjusted to allow for the tax liability.

Unadjusted: PV of £1 in n years $= \dfrac{1}{(1+i)^n}$

Adjusted for tax: PV of £1 in n years

$$= \frac{1}{\left\{1 + i\left(\dfrac{100 - t}{100}\right)\right\}^n}$$

where i is rate of interest required and t is rate of tax levied. Hence a larger sum is deposited where tax liability is involved to meet the future cost. For example, assuming a sum of £60 000 is required in 2 years and the relevant rate of interest is 4 per cent:

(i) Unadjusted

£60 000 × 0.925 = £55 500

(ii) Adjusted: Tax at 30 per cent

£60 000 × 0.946 = £56 760

Valuation tables are published to provide the appropriate multipliers, but if these are not available, the formula provided can be used, albeit a rather tedious process.

(i) It is unlikely that the client will wish to pay for replacements in the fifteenth year, at the time when he wishes to sell the building.

The present value of the replacement costs is calculated as follows using the Present Value of £1 table at 4 per cent (tax 30 per cent) extracted from Parry's Valuation Tables, suitably adjusted.

		£
Year 2	£72 000 × 0.946 =	68 112
Year 4	£24 000 × 0.895 =	21 480
Year 5	£108 000 × 0.872 =	94 176
Year 7	£86 000 × 0.825 =	70 950
Year 9	£24 000 × 0.780 =	18 720
Year 12	£240 000 × 0.718 =	172 320
	Total:	£445 758

(ii) This is concerned with annual investments to cover future costs instead of an initial capital sum. The concept of the annual equivalent is often used as a basis for the charging of rent but it can also be used to measure outgoings on a yearly basis.

The process involves the multiplication of the capital sum by $i + SF$, where i is the rate of interest and SF is the sinking fund or amount to be set aside annually at an appropriate rate of interest for a prescribed period. (i = 4 per cent and SF = 2½ per cent for 15 years.) Therefore,

$$£445\,758 \times (i + SF) = £445\,758 \times (0.04 + 0.056)$$
$$= £445\,758 \times 0.096$$
$$= £42\,793$$

Lift Installations

Example 13.10. An electrically operated 8 person lift installation to serve six floors is required for a new building with a planned life of 30 years. The initial cost of the lift installation is £42 000, and the running costs are made up of wiping down finishes 12 times a year at £1.60, vacuuming the floor 100 times a year at £0.12, replacing the carpet tile flooring and painting the lift car every five years at £300, replacing the installation after 20 years at a cost of £45 000 and allowing for a comprehensive maintenance contract at £920/annum (excluding the first year). Calculate the present value of the life cycle costs for the lift installation at a compound rate of interest of five per cent (1994 prices).

			£
Initial cost of lift installation			42 000.00
Annual costs			
Wiping down finishes: 12 × £1.60		£19.20	
Vacuuming floor: 100 × £0.12		12.00	
		£31.20	
PV of £1 pa for thirty years at five per cent		15.3724	479.62
Maintenance contract		£920.00	
PV of £1 pa for thirty years at five per cent	15.3724		
less first year	0.9532	14.4192	13 265.66
Replacing floor finish and repainting every five years		£300.00	
PV of £1 in 5 years at five per cent	0.78352		
PV of £1 in 10 years at five per cent	0.61391		
PV of £1 in 15 years at five per cent	0.48101		
PV of £1 in 25 years at five per cent	0.29530	2.17374	652.12
Replacing lift installation after 20 years		£45 000	
PV of £1 in twenty years at five per cent		0.37668	16 950.60
Present value of life cycle costs of lift			£73 348.00

The running costs exclude the electricity consumed which is included in the utilities budget. This example illustrates the method of including an annual maintenance contract which does not commence until the second year of the lift's life as it is incorporated in the initial installation contract. It is possible that this could occur again in the 21st year. Replacing carpet tiles and repainting can be omitted in the twentieth and thirtieth years, when the installation is demolished at the end of the building's life, although there could be some salvageable value of the equipment.

External Works

Questions often arise in practice as to the comparative long term costs of grassed areas as against various forms of paving. The following example will serve to illustrate the approach.

Example 13.11. An architect has requested advice on the comparative total costs for an area of 400 m² which is either to be paved with 50 mm precast concrete paving slabs or to be finished with 150 mm of topsoil sown with grass. Advise the architect as to the most economical proposition taking a period of sixty years and five per cent interest.

Paving slabs	£/m²
Initial cost:	
75 mm bed of ashes	2.20
50 mm slabs on mortar bed and grouting	10.20
Cost/m²	£12.40

Cost of 400 m² = 400 × £12.40 = £4960
Annual equivalent over sixty years at five per cent and 2½ per cent
= £4960 × 0.05735 = £284.46

Maintenance. Average of two days attendance per annum of one craft operative and one labourer plus allowance for materials (replacement of cracked and broken slabs).

Sixteen hours at £11.60	£185.60
Materials	60.00
Sweep: 12 times pa @ £0.10/m²	48.00
Cost/400 m²	£293.60

Topsoil sown with grass	£/m²
Initial cost:	
Cultivate ground surface and wheel, spread and level soil from spoil heap	2.00
Sow grass seed and rake soil	0.20
Fertilise ground surface	0.09
Cost/m²	£2.29

Cost of 400 m² = 400 × £2.29 = £916
Annual equivalent over sixty years at five per cent and 2½ per cent
= £916 × 0.05735 = £52.53

Maintenance. Grass will require cutting fairly frequently and receive top dressing every year to keep it in good order. Allow ten mandays for grass cutting each year.

	£/400 m²
Eighty hours at £5.40	432
Hire of machine, fuel, sharpening cutters, etc.	66
Top dressing and fertiliser	90
Remove litter 12 times pa © £0.02/m²	96
Remove leaves twice pa © £0.12/m²	96
Cost/400 m²	£780

Summary of costs

	Paving slabs	*Grass*
Initial costs	£284.46	£52.53
Maintenance	293.60	780.00
Total annual costs	£578.06	£832.53

Hence, in the long term, paving slabs are more economical than grass even though they have a much higher initial first cost. It might be possible to equate the two by cutting the grass less frequently and to reduce the frequency of removing litter and leaves, if a lower standard of maintenance were acceptable.

Components

Selection of components frequently involves a comparison of total costs as the initial costs of the various alternatives under consideration will not always indicate the best solution. Several worked examples follow to show the approach.

Example 13.12. This example is a comparison of 50 mm PVC-U, aluminium and painted cast iron rainwater goods, adopting a building life of sixty years and interest rate of five per cent. It is assumed that dismantling and erection costs of replacements will be ten per cent more than that of initial provision. The method used for tabulating the cost information for ease of appreciation follows that used by Stone (1975).

PVC-U	expenditure	£300	£330	£330
	years	0	25	50
Aluminium (polyester) coated)	expenditure	£570	£627	
	years	0	35	

(commence painting at year 15 and at 5 year intervals thereafter: £150)

Cast iron (ogee section)	expenditure	£680	£751
	years	0	40

(painting at 5 yr intervals: £150)

Material	*Installation costs*	*Renewal costs*	*Total costs*
PVC-U	£300	£330 × 0.3825 (PV 25 + 50) = £126.23	£426.23
Aluminium (polyester coated)	£570	£627 × 0.1813 (PV 35) = £113.68	
	Painting £150 × 1.5399 (PV 15, 20, 25, 30, 50, 55)	= £230.99	£914.67
Cast iron (ogee section)	£683	£751 × 0.1420 (PV 40) = £106.64	
	Painting: £150 × 3.3719 (PV 5 + 10 + 15 + 20 + 25 + 30 + 35 + 40 + 45 + 50 + 55)	= £505.79	£1295.43

Note: See example 13.1 for build up of PV rates. All gutters will require cleaning out and washing annually, but the costs will be the same (£120 pa). There could also be some minor repairs.

PVC-U gutters and downpipes show a long term cost advantage, but it must be emphasised that factors other than cost may determine the final decision. In this case it might be felt advisable to select a more expensive material than PVC-U for improved appearance. Furthermore, PVC-U can suffer damage following the thawing of heavy snow deposits on the roof and the weakening of joints by significant temperature changes.

Example 13.13. A cost comparison is made of copper, zinc, lead and polyester based bitumen felt as covering to a roof with an area of 200 m^2, taking a building life of 60 years and five per cent rate of interest.

		Life cycle costs	
Copper			
Initial cost: 200 m^2 at £48.90		£9 780	
Annual repairs: 200 × £0.12	£24		
PV of £1 pa for sixty years at five per cent	18.9292	454	
Zinc			£10 234
Initial cost: 200 m^2 at £42.60		£8 250	
Replacement in thirty years	£8 250		
Add additional dismantling and fixing costs (10 per cent)	825		
	£9 075		
PV of £1 in thirty years at five per cent	0.2314	2 100	
Annual repairs: 200 × £0.06	£12		
PV of £1 pa for sixty years at 5 per cent	18.9292	227	
			£10 577
Lead			
Initial cost: 200 m^2 at £54.50		£10 900	
Annual repairs: 200 × £0.07	£14		
PV of £1 pa for sixty years at five per cent	18.9292	265	
			£11 165
Polyester based bitumen felt (two layer)			
Initial cost: 200 m^2 at £16.80		£3 360	
Replacement every twenty years	£3 360		
Add additional dismantling and fixing costs (10 per cent)	336		
	£3 696		
PV of £1 in twenty years at five per cent 0.3767			
PV of £1 in forty years at five per cent 0.1420	0.5187	1 917	
Annual repairs: 200 × £0.60	£120		
PV of £1 pa for sixty years at five per cent	18.9292	2 272	
			£7 549

On costs alone, polyester based bitumen felt shows a definite advantage but other factors also need consideration, such as appearance and the disturbance caused by roof replacement. It would probably be advisable to allow a ten per cent margin either way on each of the total cost figures in this and other examples to make allowance for prediction errors; these will be considered later in the chapter.

Example 13.14. Advice is required on which of the following alternatives for 100 m^2 of windows in a new office building in a town centre offers the most satisfactory solution:

(1) Softwood windows to BS 644, single glazed and painted, fixed complete at an initial cost of £8600, requiring repainting every fifth year at a cost of £1160, annual cleaning and minor repairs

at £925 and replacement after thirty years at £9460.

(2) Hardwood high performance, tilt and turn, double glazed windows, costing initially £18 500 fixed complete, requiring treatment every ten years at £720, annual cleaning and minor repairs at £770 and replacement after fifty years at £320 350.

(3) Anodised aluminium, polyester coated, double glazed windows at a cost of £15 600 fixed complete, requiring painting every five years after the 15th year at £980, annual cleaning and minor repairs at a cost of £1150 and replacement after forty years at £17 160.

(4) PVC-U white double glazed windows at an initial cost of £19 900 fixed complete, requiring annual cleaning and minor repairs costing £1040 and replacement after thirty years at £21 890.

Softwood windows

Initial cost	£8 600	
Repainting every five years (less year 30): £1160 × 3.14056 (PV 5 + 10 + 15 + 20 + 25 + 35 + 40 + 45 + 50 + 55)	3 643	
Annual cleaning and minor repairs: £925 × 18.9292 (PV of £1 pa for sixty years at five per cent)	17 509	
Replacement of windows: £9460 × 0.23137 (PV of £1 in thirty years at five per cent)	2 189	£31 941

Hardwood windows

Initial cost	£18 500	
Treatment every ten years (less year 50): £720 × 1.36400 (PV 10 + 20 + 30 + 40)	982	
Annual cleaning and minor repairs: £770 × 18.9292 (PV of £1 pa for sixty years at five per cent)	14 575	
Replacement of windows: £20 350 × 0.08720 (PV of £1 in fifty years at five per cent)	1 775	£35 832

Aluminium windows

Initial cost	£20 000	
Repainting every five years from year 15: £980 × 1.63397 (PV 15 + 20 + 25 + 30 + 35 + 55)	1 601	
Annual cleaning and minor repairs: £1150 × 18.9292 (PV of £1 pa for sixty years at five per cent)	21 769	
Replacement of windows: £22 000 × 0.14204 (PV of £1 in forty years at five per cent)	3 125	£46 495

PVC-U windows

Initial cost	£19 900	
Annual cleaning and minor repairs: £1040 × 18.9292 (PV of £1 pa for sixty years at five per cent)	19 686	
Replacement of windows: £21 890 × 0.23137 (PV of £1 in thirty years at five per cent)	5 065	£44 651

The softwood windows offer the cheapest solution but not the best as the costs cover standard windows which are liable to deteriorate rapidly unless regularly and well maintained. They also are the only windows with single glazing. If value for money is to be the main criterion then the hardwood windows would top the list with good durability and appearance which could well blend in with adjoining buildings, double glazing which will halve heat loss and reduce condensation, and tilt and turn windows which permit cleaning of the windows from inside thus reducing cleaning costs. The aluminium windows are expensive and less attractive in appearance, while the PVC-U in addition to being expensive have a less predictable expected life because of lack of evidence in practice. The replacement costs have been taken as the initial costs plus ten per cent to allow for the removal of the existing windows and possible repairs to sills and reveals.

Example 13.15. Advice is required on the selection of the floor finish to the reception area of a five star hotel with an area of 300 m^2 having regard particularly to value for money, good aesthetic qualities and a minimum of maintenance and replacement work to cause least disruption to the operation of the hotel. The following six finishes have been selected for examination, taking a building life of 60 years and a rate of interest of 5 per cent.

Unit: m^2	*A* *Ceramic tiles*	*B* *Wood blocks*	*C* *Sheet vinyl*	*D* *Quality carpet*	*E* *Terrazzo tiles*	*F* *Marble*
Initial cost	28.60	41.10	12.10	36.70	43.00	70.00
Screed/bedding	6.00	8.00	8.00	8.00	6.00	6.00
Underlay				4.30		
Total initial cost	£34.60	49.10	20.10	49.00	49.00	76.00
Replacement cost*	39.60	52.10	19.60	54.50	54.00	81.00
Composite PV	0.231	0.142	0.231	1.451	0.181	0.087
PV replace cost	£9.15	7.40	4.53	79.08	9.77	7.05
Annual costs**	3.80	5.90	4.50	1.90	4.20	6.40
PV £1 pa 60 yrs	18.929	18.929	18.929	18.929	18.929	18.929
PV annual costs	£71.93	111.68	85.18	35.97	79.50	121.15
Life cycle costs (PV)	£115.68	168.18	109.81	164.05	138.27	204.20

The life expectancies of the various finishes have been taken as follows, although the sources used gave a wide range of periods.

A. ceramic tiles: 30 years; B. wood blocks: 40 years; C. sheet vinyl: 30 years; D. quality carpet: 10 years; E. terrazzo tiles: 35 years; F. marble: 50 years.

* Replacement costs have been computed from the cumulative total of removing the existing finish, temporary protection work, preparing the existing screed/bedding, and laying the replacement finish.

** Cleaning and maintenance costs have been calculated in the manner now indicated.

Finish	*Cleaning operations and costs*		*Minor repair costs*	*Total costs/m²*
A. Ceramic tiles	Sweep daily & wash weekly:	£3.30	£0.50	£3.80
B. Wood blocks	Sweep daily, wash weekly & wax polish annually:	£5.00	£0.90	£5.90
C. Sheet vinyl	Sweep daily & wash weekly:	£3.30	£1.20	£4.50
D. Quality carpet	Vacuum twice weekly & shampoo annually:	£1.10	£0.80	£1.90
E. Terrazzo tiles	Sweep daily & wash weekly:	£3.30	£0.90	£4.20
F. Marble	Sweep daily & wash weekly:	£3.30	£3.10	£6.40

The choice of finish encompasses many factors apart from the present value of the life cycle costs. For example, sheet vinyl is the cheapest but would not be suitable for the entrance to a prestige hotel, while ceramic tiles and terrazzo tiles although relatively low priced give hard, cold and unattractive finishes. Wood blocks and quality carpet (broadloom) show comparatively similar overall costs but greatly varying characteristics. Polished hardwood blocks of high quality give an attractive, hard wearing finish which deserves careful consideration, while quality carpet provides a colourful and comfortable finish but requires rather frequent renewals with consequent disruption to the busiest part of the hotel, but are less disruptive than replacing marble, tiling and wood blocks. Marble (travertine) is in a class of its own with an exceptionally attractive and impressive finish, very durable but of very high overall cost.

The client may wish to consider a wide range of characteristics, including appearance now and in the future, resistance to indentation and durability, ease of cleaning and its frequency, resource availability, extent of maintenance and remedial works, life expectancy, slip resistance, sound insulation, resistance to wear and relative comfort to the guests. It would be possible to weight each of these factors and to arrive at a cost benefit solution and it seems likely that the client may have his own personal preferences with the decision making process aided by the data provided. The main choices would seem to be between high quality carpet, hardwood blocks and marble.

Maintenance to Capital Cost Relationships

The maintenance to capital cost relationship of a large multi-storey office building with a total floor area of about 23 000 m² follows. It was restricted to a four-year occupancy period but within these limitations the figures provide useful guidelines.

Element	*Brief constructional details*	*Average annual maintenance to capital cost (percentage)*
Roof	Concrete slab and asphalt	0.100
Wall cladding	Precast concrete panels	0.026
Wall glazing	6 mm plate and 4 mm sheet glass in bronze frames (cleaning costs)	0.761

Internal partitioning	100 mm block and demountable partitions of aluminium frames with vinyl-faced chipboard	0.637
Decorations and finishes	Floor finishes mainly vinyl tiles; ceilings primarily of stove enamelled suspended modular ceiling panels and acoustic tiles; decorations mainly emulsion paint to plaster	1.254
Lifts	Five high speed lifts and one service lift	1.185
Air conditioning	Oil-fired boilers for heat load of 1 900 000 W	0.542

ENERGY CONSERVATION

Energy Audits and Surveys

Energy conservation became one of the most important factors in life cycle costing in the late 1980s and 1990s, as a means of improving long term economies and efficiency. In this context, audits and surveys of energy use aim to identify opportunities for cost effective savings, and when combined with performance monitoring and targeting methods, they provide the information needed to ensure effective energy management. An Energy Efficiency Office (EEO) study of several thousand energy surveys showed an average potential saving of about 20 per cent of each property's energy bill, with an average payback period for implementing the recommended measures of 18 months (Field, 1992).

Energy audits establish the quantity, cost and end use of each form of energy input to a building or site over a specific period. A preliminary audit can assess energy use from fuel bills and meter readings, but a site survey is required to ascertain how the energy is used and how savings can be made. A case study follows later relating to Heslington Hall, carried out by BRE's Building Research Energy Conservation Support Unit (BRECSU) which shows the value of carrying out such surveys. Readers requiring further information on the operation of energy audits and surveys are referred to the *CIBSE Applications Manual: AM5* (1991).

Instrumentation is usually required to measure room temperatures, flow temperatures, combustion gas composition and electrical loads, but the cost of such surveys can be prohibitively expensive for smaller buildings. When carrying out preliminary audits, comparison with published figures for good, fair and poor levels of energy use, as listed in table 13.21, can be used to assess the performance of different buildings (RICS, 1993a). Accountability and responsibility for energy use at all levels needs reviewing from overall financial control to individual good housekeeping, as will be illustrated later in the chapter in connection with schools.

The major aspect of most energy surveys is to determine the way in which energy is used to meet operational or comfort requirements. Hence a site survey should show the breakdown of the energy used for each of the main services in a building. For example, plotting heating energy consumption against degree days (explained later) will help to indicate the relative performance of the heating system. A wide scatter of points relative to the line of best fit will indicate that the heating is poorly controlled (EEO, 1987a). With domestic hot water, storage volumes and temperatures should be checked, and it is often worthwhile to provide local electric water heaters for summer use to avoid large standing heat losses from central boiler plant. In the case of air conditioning, a survey should identify the minimum acceptable extent to which the air should be conditioned and the most efficient way of doing it. While with lighting, new lamps, reflectors and controls can often produce major cost savings, particularly where lights are in use for long periods (Field, 1992).

Degree days indicate both the amount of time and the temperature below the base external temperature of 15.5°C, for which the national average over 20 years was 2642 degree days. Where the outside temperature exceeds 15.5°C, no heating is necessary.

A survey will identify various options for investment and energy savings. Savings are usually based on the estimated percentage of annual consumption and direct saving from a reduction in fuel cost, load, operating hours or energy loss. A financial appraisal of the economic benefits of each alternative is then needed to determine the optimum investment programme. Finally, a plan to implement the selected options should be prepared and implemented (Field, 1992).

The benefits of implementing the recommendations of an energy audit and survey relating to Heslington Hall, University of York are now illustrated. The hall is an Elizabethan building, reconstructed in Victorian times and refurbished in the 1960s to become the administrative offices of York University. An energy audit in 1979–80 revealed that the annual fuel consumption per unit area for central heating and hot water was very high. A programme of energy-saving measures was therefore adopted (EEO, 1991).

Initially, low cost measures included recommissioning the boiler, improving the time controls and relocating weekend uses to other buildings or servicing them independently, thereby reducing annual oil consumption by 30 per cent. The subsequent installation of a new gas fired boiler, independent hot water system and new controls gave a similar saving. As a result, fuel consumption dropped to just over 40 per cent of the level immediately before the energy audit, and a payback period of under two years was achieved for the energy efficiency improvements.

Further detailed examination of the use of energy surveys, the appraisal of the survey findings and the subsequent investment appraisal is provided in *Energy Appraisal of Existing Buildings* (RICS, 1993a). The RICS recommend that simple unadjusted payback periods (SPB) can be used for short paybacks (<2 years) and medium payback (2–5 years), while for long paybacks (>5 years) a discounted payback (DPB) method should be used, whereby instead of adding up all the net savings, one adds up their present values, as described and illustrated earlier in the chapter in connection with life cycle costing examples.

Table 13.20 Conversion of billed units to kWh

Energy source	*Billed units*	*To get kWh, multiply by*
Natural gas	therms	29.31
	cubic feet	0.303
Gas oil (35 sec)	litres	10.6
Light fuel oil (290 sec)	litres	11.2
Medium fuel oil (950 sec)	litres	11.3
Heavy fuel oil (3500 sec)	litres	11.4
Coal*	tonnes	7600
Anthracite*	tonnes	9200
Liquid petroleum gas (lpg)	litres	7
	tonnes	13900

Source: *RICS – Energy Appraisal of Existing Buildings* (1993).

* The calorific value of solid fuel is subject to local variation. A more accurate figure may be available from the local supplier.

Assessment of Building Energy Performance

To obtain an approximate indication of the energy performance of a building, the annual energy consumption is first determined and converted to kWh as shown in table 13.20, modify the space heating energy by multiplying by a weather correction factor, dividing degree days for the standard year (2462) by the degree days for the energy data year, suitably modified by an exposure factor for certain buildings as indicated by the RICS (1993a), add non-heating energy, such as domestic hot water and lighting, where appropriate multiply the total energy use by the hours of use factor, and estimate the floor area in m^2 and calculate the energy performance in kWh/m^2 by dividing the annual energy consumption, suitably corrected, by the floor area.

Table 13.21 assists in comparing the energy performance of the building under consideration with the total energy use of other similar buildings.

Housing

Energy Rating Systems

The National Home Energy Rating (NHER) system scores homes on a rising scale of zero to ten according to their energy efficiency. Surveys can

Table 13.21 Total energy usage in typical non-domestic buildings

Type of building	*Hrs of use pa*	*Energy usage per year (kWh/m²)*		
			Energy performance	
		good	*fair*	*poor*
Primary school (no indoor pool)	1480	<180	180–240	>240
Primary school (with pool)	1480	<230	230–310	>310
Secondary school (no pool)	1660	<190	190–240	>240
Secondary school (with pool)	2000	<250	250–310	>310
University	4250	<325	325–355	>355
Department/chain store (mechanically ventilated)		<520	520–620	>620
Supermarket/hypermarket (mechanically ventilated)		<720	720–830	>830
Small food shop – general		<510	510–580	>580
Library	2540	<200	200–280	>280
Church	3000	<90	90–170	>170
Hotel (medium size)		<310	310–420	>420
Bank or post office	2200	<180	180–240	>240
Theatre		<600	600–900	>900
Crown and county court	2400	<220	220–300	>300
Factory (small)		<230	230–300	>300
Factory (large with heat gains from manufacturing plant)		<210	210–300	>300
Warehouse (heated)		<150	150–270	>270
Clinic/health centre	2600	<36	36–46	>46

Source: *RICS. Energy Appraisal of Existing Buildings* (1993a).

be carried out from plans of new buildings or from existing dwellings of any age. Data is then processed on a microcomputer to provide a profile of fuel costs and suggestions for improving the energy efficiency of the building. NHER ratings include cooking and electricity used for lighting and other appliances, and allow for regional climatic variations.

Another energy rating system called Starpoint scores on a scale rising from one to five based on heating and hot water costs only. The NHER scheme is generally considered to be better suited for new buildings and Starpoint for existing properties. Results produced by both schemes can be compared through the government's own Standard Assessment Procedure (SAP), which is now incorporated in revised Building Regulations Approved Document L, 1995 (Joseph Rowntree Foundation, 1993).

Houses built in accordance with the 1995 Building Regulations would achieve a rating of eight on the NHER system, compared to about six if built under the 1985 Building Regulations, while the UK average rating is about four, indicating the need for extensive energy conservation work on dwellings throughout the country built before 1986. In both rating systems, the progression up the ratings is not a straight line either in terms of effort or spending, and raising the rating at the top end of the scale is more difficult and more costly than those near the bottom. Table 13.22 shows the reduction in annual fuel costs at 1992 prices with reducing NHER ratings for three different sizes of dwelling.

Table 13.22 Typical annual fuel costs for dwellings with different NHER ratings

Annual fuel costs in £ for given floor area				
	NHER	*40 m²*	*60 m²*	*80 m²*
	0	950	1340	1740
	1	710	990	1260
	2	580	790	990
	3	490	660	820
UK average rating	**4**	**430**	**570**	**700**
	5	390	500	615
	6	355	450	545
1991 Building Regs	**7**	**325**	**410**	**490**
1995 Building Regs	**8**	**300**	**370**	**440**
	9	275	330	390
	10	250	290	340

Source: *EEO home energy survey* (1992), *adjusted*.

Typical Domestic Energy Conservation Costs and Savings

A typical unimproved three bedroom house with an NHER rating of 4.4 had the following annual energy spending profile in 1993, prior to the introduction of VAT on energy costs, based on an energy audit by ECD Partnership.

heating (gas with 25 per cent peak rate electricity)	£462
water heating (gas)	£82
cooking (gas)	£30
lights and appliances	£282
standing charges	£76
TOTAL	£932

In addition the house would produce 8.5 tonnes of CO_2 per annum. The Energy Efficiency Office (EEO) estimates that heat losses in a poorly insulated house can be reduced by two-thirds in cash terms in a well insulated dwelling with a well controlled heating system. Table 13.23 shows 1992 costs of heat losses, energy conservation work and annual savings for a typical semi-detached house with gas heating and hot water, which highlight the extent of the heat losses in monetary terms in a badly insulated house and the average costs of energy conservation work and the annual savings that can accrue.

The significant benefits of higher standards of thermal insulation in new build housing were demonstrated very clearly in the study of 15 housing association new build schemes in which

Table 13.23 Typical costs of domestic heat losses, energy conservation work and annual savings

Heat losses		
	Badly insulated house (£)	*Well insulated house (£)*
Roof	*75*	*10*
Ventilation and draughts	*65*	*35*
Windows	*55*	*25*
Walls	*110*	*15*
Doors	*5*	*5*
Ground floor	*20*	*20*
Typical costs and savings		
	Capital cost (£)	*Annual saving (£)*
Hot water cylinder		
80 mm cylinder jacket and pipe lagging	10	30
Roof insulation		
100 mm professionally installed	180	65
Cavity wall insulation		
Polystyrene beads	250	300
Mineral fibre	400	95
Solid wall insulation		
Interior insulation	1600	95
Draught stripping		
Windows, doors, floors, etc. (DIY)	75	30
Double glazing		
Simple fixed secondary glazing, living room (DIY)	80	10
Living room only, more elaborate secondary glazing (DIY)	200	10
Living room only, professionally installed	550	10
Whole house, simple fixed secondary glazing	320	30
Whole house, more elaborate secondary glazing (DIY)	750	30
Whole house, professionally installed	2150	30
Reflector foil		
Fixed behind radiators on outside wall	10	5

Source: *Energy Efficiency Office Home Energy Survey.*

Table 13.24 Average additional capital cost versus fuel saving for each size of unit: gas heated (1992 prices)

	1 Bed			*2 Bed*			*3 Bed*		
	Fuel saving (£/yr)	*Capital cost (£)*	*Payback (yrs)*	*Fuel saving (£/yr)*	*Capital cost (£)*	*Payback (yrs)*	*Fuel saving (£/yr)*	*Capital cost (£)*	*Payback (yrs)*
From Building Regulations[1] to improved[2]	36	243	7	40	213	5	52	247	5
From improved to NHER = 8 (low cost)	10	41	4	12	61	5	17	70	4
From NHER = 8 to NHER = 9	25	144	6	70	564	8	94	600	6

Source: *NFHA/EEO. Affordable new low energy housing for housing associations* (1992).

[1] 'Building Regulations' is defined as Building Regulations by the trade-off method (i.e. 1985 *U*-values + double glazing).
[2] 'Improved' standard – 1991 Building Regulations fabric standards plus double glazing.

Table 13.25 Average additional capital cost versus fuel saving for each size of unit: electric systems (1992 prices)

	1 Bed			*2 Bed*			*3 Bed*		
	Fuel saving (£/yr)	*Capital cost (£)*	*Payback (yrs)*	*Fuel saving (£/yr)*	*Capt cost (£)*	*Payback (yrs)*	*Fuel saving (£/yr)*	*Capt cost (£)*	*Payback (yrs)*
From Building Regulations[1] to improved[2]	35	167	5	44	190	4	48	240	5
From improved to NHER = 8 (low cost)	51	349	7	70	439	6	85	482	6

Source: *NFHA/EEO. Affordable new low energy housing for housing associations* (1992).

[1] 'Building Regulations' is defined as Building Regulations by the trade-off method (i.e. 1985 *U*-values + double glazing).
[2] 'Improved' standard – 1991 Building Regulations fabric standards plus double glazing.

energy efficiency measures were made a priority, as described in chapter 5. The dwellings achieved significantly higher energy efficiency ratings although they were built within the operative grant structure at minimal extra capital cost. Carbon dioxide emissions were cut by a fifth and fuel costs reduced by about £21 a year by incorporating insulation measures costing less than £200 per dwelling in many cases. As a direct consequence of the survey, the Housing Corporation now includes energy efficiency in its scheme audit system.

Average fuel savings and capital costs for gas and electric schemes meeting the low cost and higher cost specifications incorporated in the above study are given in tables 13.24 and 13.25.

These new initiatives, welcome as they are, will only begin to address what is a huge national problem. The majority of UK houses achieve pitifully low NHERs of between three and five, and there are at least seven million households living in 'fuel poverty', that is, those living in the lowest 30 per cent in terms of income who do not have sufficient money to keep warm.

Some of the typical local authority housing

problems were illustrated in the Newark and Sherwood District Council energy strategy, formulated to remedy the grave deficiencies in the council's housing stock. Fifty per cent had condensation and mould problems; bedroom temperatures of less than 10°C were normal in winter; 1000 houses had no central heating, 3500 required replacement and 3000 needed improvement to their central heating systems.

Newark and Sherwood DC devised an impressive strategy costing £16.4 m over 20 years commencing in 1985. The target was 14 600 kW (heating costs per dwelling of about £255 pa at 1988 prices), to provide a minimum temperature of 21°C in the living room and 15°C elsewhere (16 hour heating day, 30 week heating season); full house central heating; controlled ventilation; insulation of cavity walls and roof spaces to 150 mm; double glazing; and energy advice. The cost per dwelling was £2 500 (only £1 130 over the pre-1985 standard modernisation cost) and the District Valuer's Assessment showed an increase of £16.4 m (the value of the total spend) (Taylor, 1993).

One of the housing associations in the national study described earlier was Sutton (Hastoe) in West Dorset, providing new build housing in Halstock, a village without a gas supply, and it showed that it was possible to make a substantial saving in power costs for a small outlay, using Economy 7 off-peak electricity, supplemented by panel heaters and an open fire for burning logs which were freely available. Automatic charge control was introduced to improve the efficiency of the storage heaters. Low energy light bulbs were fitted in the kitchen, entrance hall, stairs, bathroom and living room. The extra work, including roof, wall and floor insulation, raised the NHER rating from six to 7.9 and cost a total of £556 per dwelling, as shown in table 13.26, equivalent to a rent increase of £1.06 per week and providing estimated annual savings of £96 (£1.84 per week), with heating and hot water bills cut by almost a third, thus making good economic sense as well as producing more comfortable living conditions (Joseph Rowntree Foundation, 1993).

Table 13.26 Extra costs and savings (three bedroom house) heated by Economy 7 off-peak electricity (1992 prices)

	Contract cost of the work £	*Estimated annual saving* £
Roof insulation (200 mm)	45	6
Wall insulation (changed specification)	175	41
Floor insulation (new)	281	30
Heating controls (automatic charge control)	26	10
Low energy lights (new)	29	9
TOTAL	556	96
Estimated fuel bills	*Total bill*	*Heating/ hot water*
Before changes	607	332
After change	484	228
Percentage change	20%	31%

Source: *Joseph Rowntree Foundation. Innovations in Social Housing No. 4*, March 1993.

Schools

The Energy Efficiency Office (EEO) (1987b) advocated the use of the Normalised Performance Indicator (NPI) for calculating energy use in a school for comparison with the performance of other schools. The following procedure is outlined to arrive at the NPI:

(1) convert energy units to kWh, as detailed in table 13.20;

(2) determine the energy used for space heating;

(3) modify the space heating energy to account for weather, the weather correction factor =

$$\frac{\text{standard degree days (2462)}}{\text{degree days for energy data year}}$$

(4) modify the space heating energy to account for exposure (factors of 1.1 for sheltered locations, 1.0 for normal locations and 0.9 for exposed locations);

(5) add non-heating energy use;

(6) apply the hours of use factor – standard

hours vary from 2290 for nursery schools to 1480 for primary schools with no indoor pool, 2000 hours for secondary schools with indoor pool and 8760 for special residential schools;

(7) convert the floor area into units of square metres;

(8) calculate the NPI, which =

$$\frac{\text{corrected annual energy consumption}}{\text{floor area}}$$

This gives the amount of energy taken to heat $1\,m^2$ of the building under standard conditions.

Nottinghamshire County Council adopted a carefully considered approach to energy management in existing schools based on many years of practical experience, with the following objectives:

(1) to save money;

(2) to conserve energy whilst providing comfort conditions for the building occupants;

(3) to help protect the environment;

(4) to establish best practice methods for both existing and new buildings.

If resources are to be targeted effectively to those buildings offering the most potential for energy conservation, then an efficient monitoring and targeting system is essential. Hence the County Council aimed at identifying those buildings and to implement a range of measures which together will result in improved conditions for the occupants and financial savings for the authority. Target standards are set for all buildings in line with National Audit recommendations and regular reviews carried out to assess the potential for financial and energy savings. This assessment consists of analysing fuel bills and consumption details against such parameters as floor areas, population and degree days.

The resultant energy management measures can be summarised as follows:

Low cost measures

(1) Purchasing fuel at the most advantageous tariff.

(2) Monitoring and targeting energy use.

(3) Promoting good housekeeping; for example studies identified that 5–15 per cent of energy can be saved by switching off lights when not needed and closing doors and windows.

(4) Operating boilers at optimum efficiency.

(5) Balancing the heating system.

(6) Ensuring that the controls are set accurately.

Medium cost measures (less than 4 years payback period)

(1) Building energy management systems (BEMS) to control building energy use and comfort conditions.

(2) Electronic thermostats (tamperproof electronic thermostats installed to maintain accurate constant temperature and ensure that UK legal maximum temperature of 19°C is not exceeded; studies have shown that for each 1°C reduction of temperature, a saving of 5–8 per cent of the energy bill is made).

(3) Zone controls to take account of orientation.

(4) Draughtproofing and loft insulation.

(5) Tungsten light replacements (tungsten light replacements can be cost effective where there are long hours of use or high electric load; they range from miniature compact for tube fluorescent lamps to high pressure sodium lamps for leisure centres; there are also additional maintenance benefits because of less heat generation and longer life).

(6) Lighting controls (ranging from simple time switch devices to fully automatic mains signalling systems).

Higher cost measures (more than 4 years payback period)

A number of measures which will normally be cost effective and implemented on new buildings require a longer than normally accepted payback period for existing buildings. They could become more economically viable in the future if fuel prices escalate and costs of installing the equipment reduce. Such measures include draught lobbies, integral flat roof insulation, suspended ceilings, fluorescent lighting (low hours of use), double glazing, and waste heat recovery.

Typical annual energy costs of Nottinghamshire schools in 1993 were:

Type of school	Heating	Electricity
Secondary, with indoor pool $10\,000\,m^2$	£28.8 k	£35.2 k
Secondary, no pool $7000\,m^2$	£14 k	£21 k
Primary, with indoor pool $1800\,m^2$	£6.3 k	£7.7 k
Primary, no pool $1000\,m^2$	£2.8 k	£4.2 k

The Energy Efficiency Office (EEO, 1992) devised a wide range of energy efficiency measures for use in schools, categorised into free measures, short payback (less than 2 years), medium payback (2–5 years) and long payback (typically 5–10 years), and further classified them into those relating to the boiler room and heating system and those listed as other measures. The extensive and very informative data is reproduced in table 13.27.

EEO have defined payback periods in two ways, namely simple payback and simple rate of interest. The simple payback period is used in table 13.27 and is calculated by dividing the capital cost by the annual fuel cost savings. A more realistic approach was described earlier in the chapter involving the conversion to present values by using a discounting technique. The simple rate of return is a way of comparing the return on an investment with the cost of borrowing and the calculation consists of

$$\text{rate of return (per cent)} = \frac{\text{annual fuel cost savings} \times 100}{\text{capital cost}}$$

Office Buildings

General Background

The information in this section is based upon data obtained collected by the Building Research Energy Conservation Support Unit (BRECSU) from some 200 office buildings considered for inclusion in the EEO series of good practice case studies, and upon energy survey information for another 200 buildings.

Annual delivered energy consumption in offices can range from under 100 to over 1000 kWh/m^2 of treated floor area, costing from £4/m^2 to £40/m^2 or more at 1991 prices. The average level of fossil fuel use (gas and oil) is similar in all office types at about 200–250 kWh/m^2 at a cost of about £3/m^2. In contrast to fossil fuel use, the use of electricity varies considerably and is influenced primarily by the following factors:

(1) open plan designs, which generally use more artificial lighting;
(2) air conditioning, where the fans and pumps frequently use about twice as much electricity as refrigeration;
(3) mainframe computer rooms and their associated air conditioning (BRE, 1992).

Office Types

Average and good practice patterns of energy costs are given for the following four typical office types in figure 13.6.

Type 1: naturally ventilated, largely cellular
A fairly small, simple building with largely individual offices and possibly a few group spaces. Daylight is good and artificial lighting levels are normally less than in the other three office types and are readily controlled by individual switches located by the doors. There are few common facilities.

Type 2: Naturally ventilated, largely open plan
Mostly open plan but with some cellular offices and special areas such as conference rooms. Light levels and lighting power tend to be higher than in type 1 offices, and with deeper plans there is less daylight available. Lights also tend to be controlled in large groups and there is often more office equipment.

Type 3: Air conditioned, largely open plan
Similar in occupancy and planning to type 2 but

Table 13.27 Energy efficiency measures in schools

Payback period	*Boiler room and heating system measures*	*Other measures*
Free measures	• Eliminate unnecessary running of boilers at weekends • Eliminate holiday heating • Ensure that controls are set to provide the temperatures you want at the times you need • Ensure that the frost thermostat is set correctly • Isolate winter boiler(s) and heating circuits in summer	• Reduce domestic hot water temperatures • Isolate immersion water heaters during holidays • Use swimming pool cover where fitted • Replace 38 mm fluorescent tubes with high efficiency 26 mm tubes as the former expire (if you have switch start fittings)
Short payback (less than 2 years)	• Recommission optimiser and heating controls • Check that boiler air/fuel ratio is correct (as part of regular maintenance) • Fit boiler sequence controls • Repair leaks on distribution mains • Reduce use of supplementary electric heaters • Install, repair or replace thermostats • Insulate domestic hot water cylinder • Provide additional heating controls for individual heaters	• Reset domestic hot water thermostat and time switches and make tamperproof • Blank off unused air grilles behind radiators • Seal unused chimneys and ventilation stacks • Fit reflective foil behind radiators • Install time-switches for swimming pool circulating pumps • Replace tungsten lighting with compact fluorescent lamps
Medium payback (2–5 years)	• Install modern instruments to control boiler burners and measure flue gas composition • Discontinue night set back and install optimiser • Install thermostatic boiler controls • Improve/repair thermal insulation on the boiler • Fit an optimiser and/or compensator • Install a Building Energy Management System • Install dual fuel burners to boiler • Insulate pipework to heating system • Fit timeclocks to hot water immersion heaters • Replace central hot water boiler with point-of-use gas or off-peak electric heaters • Fit thermostatic radiator valves • Fit timeclocks to fan convector heaters • Replace on-peak electric convector heaters with off-peak storage	• Improve controls to storage radiators • Install manual swimming pool cover • Fit self-closing devices to external doors • Draught strip external doors and windows • Insulate loft spaces to current standards • Install cavity wall insulation • Install spray taps (soft water areas only) and automatic valves to showers • Install water economy equipment to WC and urinal cisterns • Install heat pump heat recovery system to swimming pool • Control run times for extract fans • Rearrange switching to light fittings
Long payback (typically 5–10 years)	• Improve zoning of the heating system • Install new condensing boiler • Replace electric (storage) heaters with gas-fired heaters	• Replace electric water heaters with gas-fired heaters • Install draught lobby to main entrance • Install double glazing • Fit secondary double glazing • Replace excessive areas of glazing with insulated infill panels • Insulate existing swimming pool roof and walls • Install automatic swimming pool cover • Install occupancy sensors to control lighting • Replace old fluorescent fittings with modern efficient fittings, e.g. high frequency fluorescent fittings

Source: *EEO. Saving energy in schools* (1992).

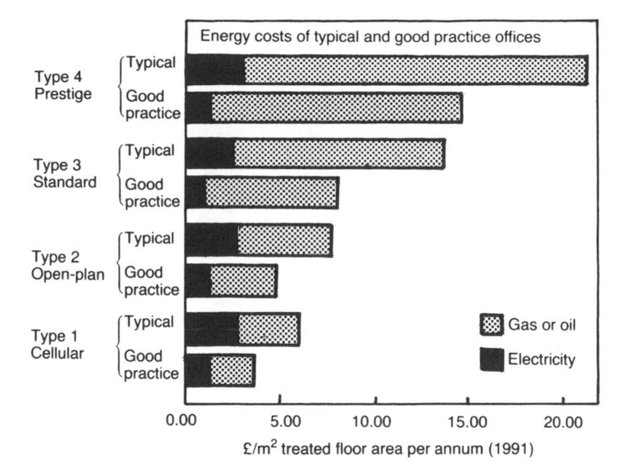

Figure 13.6 Average and good practice energy costs for four different office types (Source: BRE information paper IP 20/92)

often larger. Deeper floor plan and smaller, tinted or shaded windows reduce the availability and use of daylight still further. The air conditioning system may be either all-air (such as variable air volume) or air/water (such as induction units or fan coils).

Type 4: Prestige air conditioned
These are larger still and often comprise a national or regional head office, with a computer suite, restaurant, higher level of equipment, facilities and information technology. Hours of use are also extended because of the more diverse pattern of occupation.

Energy Use and Costs

Figure 13.6 gives the annual electricity and fossil fuel use and costs for:

- a 'typical' example near the middle of the consumption range for the national office stock as a whole, and
- a 'good current practice' building, well managed and using simple, readily available and proven technologies and design features.

The lower fossil fuel consumption in the good practice examples of all four offices is attributable to better insulation, more efficient boilers, improved control and management, and more efficient hot water systems. Lighting technology improved substantially in between the mid 1980s and mid 1990s, and it is often possible to light offices efficiently at 2.5 watts per square metre per 100 lux or less with modern fluorescent tubes, efficient reflectors, and electronic high-frequency ballasts (BRE, 1992).

Advance Factory Units

Good advice on energy efficiency in these units can be obtained from EEO *Good Practice Guide: Design Manual 61* (1993a). It was also considered helpful to the reader to refer to EEO case study 141 (1993b), relating to a single storey advance factory unit developed by The Development Board for Rural Wales (DBRW) at Ystradgynlais in Powys in 1990.

Constructional Details

The unit has a GFA of 915 m^2, comprising 802 m^2 production space, 83 m^2 and 30 m^2 plant/ancillary areas. The unit was built to a high specification with high insulation levels, double glazed windows and rooflights and draught lobby to the main entrance. The office and production areas are heated with a gas fired radiant heating system. The total cost was £333/m^2.

The office and toilet areas are heated by a low temperature hot water radiator system, fed by a balanced flue gas fired boiler located in the plant room. Optimum start is incorporated in the boiler control, and temperature control is by thermostatic radiator valves. Hot water is supplied by an electric storage water heater in the toilet area. High efficiency fluorescent light fittings are fitted in the office area with compact fluorescents in the toilets.

The production area is heated by a gas fired overhead radiant heating system, rated at 211 kW, with individual thermostatic control for each of

Table 13.28 Predicted disaggregated annual energy use for advance factory unit

2462 degree days
Occupancy: 40 h/week for 30 personnel

	Energy components	*Annual energy use*	
		kWh	*kWh/m²*
Gas	Space heating		
	Production area	56 443	70
	Office area	5 998	72
	Total gas	62 441	68
Electricity	Hot water	3 500	4
	Production lighting	10 538	13
	Office lighting	1 818	22
	Production ventilation, etc.	3 889	5
	Office equipment, etc.	1 328	16
	Total electricity	21 073	23
	Total building	83 514	91

Source: *EEO. Good practice case study 141* (1993).

the four rows of burners, and 7 day clock control. Combustion products are discharged to the outside through the plant room. Six ventilation fans installed in the walls, provide a maximum of four air changes per hour (ac/h) for summer ventilation. Three of the fans are reversible, increasing the flexibility of the system. High level trunking was provided for the installation of fluorescent light fittings by the future occupier.

Energy Use

Using the CIBSE Guide method, normalised annual energy consumption was calculated based on a standard occupancy of 8 h/day, 5 days/week and design environmental conditions of 16°C for the production area and 18°C for the office area. The projected annual energy use of 91 kWh/m^2 is well within the 'good' category (<230 kWh/m^2) for EEO performance ratings. In addition, the EEO target for annual space heating energy consumption for small, low energy factories is 85 kWh/m^2, and this was met with a projected energy use of 68 kWh/m^2 for the assumed conditions. The projected electrical energy use for lighting, hot water, ventilation and office equipment was 23 kWh/m^2, as detailed in table 13.28. The actual energy costs to May 1992, adjusted to a full year, was £7677 (£8.4/m^2), comprising gas: £1376 and electricity: £6301 for the building with a GFA of 915 m^2.

Conclusions

The occupier was impressed with the low energy consumption coupled with good comfort conditions. A comfortable environment is maintained throughout the production area. Even on cold winter days, three of the four radiant heaters suffice and the full heating potential had not so far been required.

The main conclusions drawn from the study by the EEO were as follows:

(1) Forward thinking developers will incorporate energy efficiency at the design stage to assist in meeting occupiers' quality requirements of good comfort conditions at minimal running cost.

(2) The additional capital cost is relatively small but worthwhile to increase the likelihood of early lettings.

(3) Energy efficiency can be achieved in practice without buildability problems.

Besides short payback periods being achievable, savings from energy efficiency projects in *refurbished industrial buildings*, as illustrated in table 13.29, continue for the life of the refurbishment and are largely protected against inflation through energy price increases, including VAT since 1994. The internal rate of return or net present value can as a result be favourable. Thus there is a good economic case for improving energy efficiency when refurbishing factory and warehouse buildings (BRE, 1993b).

Public Houses

EEO undertook 15 good practice case studies (nos. 44–58 inclusive) of energy efficiency meas-

Table 13.29 Savings and payback periods for energy efficiency measures in refurbished industrial buildings (1993 prices)

Activities and installations	*Energy savings*	*Annual payback (years)*
Typical savings		
Draught proofing	15–20%	1–5
Building insulation	10–15%	2–6
Boiler replacement	10–20%	1–4
Time and temperature controls	5–15%	1–5
Destratification fans	5–20%	1–3
Lighting		
Replace tungsten by fluorescent lights	40–70%	1–3
High-frequency electronic ballasts	15–20%	
Use efficient luminaire reflectors	20–50%	2–6
Install automatic lighting controls	20–50%	2–5
Localised instead of general lighting	60–80%	4–8
Specific examples		
High-speed roller shutter doors	£2 700 (14%)	2.6
External roof insulation	£100 000 (66%)	2.4
Decentralising heating system	40–60%	<2
Condensing boiler installation	10–40%	1–4
Occupancy lighting detectors	£36 000 (70%)	1.8
Time switching of lighting	£8 000 (32%)	2.1
Heat recovery for space heating	£21 000	1.3
Convert standby generators to CHP	£22 000	3.2

Source: *BRE. IP 2/93.*

ures carried out in the refurbishment of a variety of public houses. These contained some interesting data and very gratifying results from the point of view of relatively low additional capital expenditure, reduction in energy consumption and short simple payback periods. The main particulars are summarised in table 13.30. The nature and scope of the energy conservation works varies considerably as do the costs of the work and the annual energy savings. However, all 15 projects show substantial reductions in energy use/m^2 with an average approaching 30 per cent and short payback periods with an average of 1.9 years.

PREDICTION ERRORS

Very real problems arise in attempting to predict building prices or costs, either of initial constructional work or of future maintenance and running costs. As described in chapters 7 and 9, there is a marked variability in the prices quoted for the same building work, making the cost assessment of future projects extremely difficult. Again, the cost estimates are influenced by the predictions made for such factors as rates of discount, durability of materials and components, maintenance and operating costs, future relative prices, taxation and the expected life of the building. For instance, we have seen that operative interest rates are likely to fluctuate within a range of five to ten per cent or even greater, and that this can have a significant effect on predicted costs. Table 13.31 shows how prediction errors in the lives of buildings where the lives are relatively short will have an appreciable effect on annual costs. It has also been shown earlier in the chapter how variations in interest rates can alter the order of preference of alternatives where there are large differences in the proportions of initial and future costs.

Prediction errors may also arise due to changes in requirements during the life of the building, often stemming from changes in fashion or taste. Technological changes may outdate present methods and materials before they are worn out, and the likely future impact of environmental changes generally is difficult to predict. With all these sources of possible predictive errors it is advisable to state life cycle costs in rounded figures to avoid implying a degree of accuracy which cannot possibly be achieved. Ranges of figures are often more meaningful. Prediction errors can also arise from errors in measurement, unsatisfactory sampling techniques and errors in the assumptions made. Quantity surveyors would be well advised not to press unduly design solutions which show only marginal benefits on life cycle costing calculations, bearing in mind all the prediction problems outlined previously and the fact that other design considerations such as increased amenity could conceivably outweigh a small cost advantage. A quantity surveyor has to

Table 13.30 Energy efficiency in refurbished public houses

Location of public house	*Type of public house*	*Main energy conservation measures*	*Additional cost of implementation (£)*	*Annual energy saving (£)*	*Simple payback period (yrs)*	*Reduction in energy use/m²* (%)
Yew Tree, Widnes	large urban	condensing boiler; zoned heating controls; new hot water service; low energy lighting	4110	1385	3	38
Marquess of Anglesey, Covent Garden	large metropolitan	cellar heat recovery; low energy lighting; draughtstripping; energy management	1570	2330	0.66	38
White Horse, Enfield	small suburban	heating controls; new hot water service, low energy lighting; roof and pipe insulation	1473	1073	1.4	30
The Crown, Derby	large suburban	new heating zones; radiator controls; new hot water services; fan dampers; low energy lighting	3490	1310	2.7	13
The Albion, Burton on Trent	large suburban	new extension; draught lobbies; condensing boiler; heating controls; low energy lighting; roof insulation	7870	3630	2.2	16
River Wyre, Poulton Le Fylde	large rural	new hot water service; space heating controls; roof insulation	6320	1900	3.3	30
Sunshine Inn, Portsmouth	small suburban	new heating controls; low energy lighting; new hot water service	2340	1558	1.5	38
Marquis of Granby, Sompting, Worthing	large rural	electrostatic air cleaners; new hot water service; ceiling dry lining insulation; lighting controls	3670	1575	2.25	32
The Crown, Leyton-stone	large urban	new boiler; heating controls; new hot water service; low energy lighting; destratification fans	3670	2450	1.5	58
The Tree Tops, Mapperley, Nottingham	suburban	new boiler; space heating controls; new hot water service; low energy lighting; draught lobbies; roof insulation; destratification fans	3670	670	5.5	17
Cathkin Inn, Rutherglen, Glasgow	small suburban	heating controls; draughtstripping; ventilation control; pipe insulation	1300	1560	0.83	25
The Crown, Sedgley, Wolverhampton	small suburban	dry lining wall and ceiling insulation; simple controls; pipe and cylinder insulation; low energy lighting	579	711	0.83	17
The Engineer, Harpenden	small suburban	new heating controls; new hot water service; new extension; draughtstripping; low energy lighting; cellar insulation	1230	800	1.5	33
The White House, Witham, Essex	large rural	roof insulation; cellular insulation; draughtstripping; draught lobby; condensing boiler; space heating controls; new hot water service; low energy heating	2950	1425	2.1	16
Sam's Bar, Blackpool	urban	heating controls; draught lobby; air quality control; cellular insulation; loft insulation; hot water controls	2540	2180	1.2	35
Averages			3119	1637	1.90	29.07

Source: *EEO. Good practice case studies 44–58 (energy efficiency in public houses)* (1991/92).

Table 13.31 Effect of errors in predicting lives of buildings

Life x (years)	Annual equivalent percentage errors when life taken as x		
	per £100 of first cost with interest at five per cent £	instead of 40	instead of 60
20	8.02	+38	+52
30	6.51	+12	+23
40	5.83	0	+10
50	5.48	−1	+4
60	5.28	−9	0
70	5.17	−11	−2
80	5.10	−13	−3

be realistic as well as technically sound in his advice to building clients.

For large industrial projects the respective estimating accuracies have been given as rough estimates ±30 per cent, budget estimates ±15 per cent and definitive estimates ±5 per cent (Spon, 1993).

EFFECT OF TAXATION AND INSURANCE

The incidence of taxation can have a considerable effect on the design economics of buildings. For instance, with industrial buildings some relief can be obtained on the initial cost, through depreciation allowances, investment, initial and cash allowances, their actual form and impact varying from time to time. Amounts spent on maintenance and repairs, heating and lighting, and other running expenses, are classified as business expenses and are deductible from profits in the case of many types of buildings. The exact incidence of taxation varies with the circumstances of the taxpayer. Public authorities do not normally pay tax and so are not affected. Current regulations and levels of taxation tend to favour alternatives with low construction costs and high running costs. The total costs of buildings can thus be influenced considerably by the form of taxation. For example, value added tax partially offsets the tax advantage previously accruing to running costs, as new construction work is zero rated.

There is a wide variation of fiscal relief against building expenditure, ranging from the total absence of relief against investment in commercial or residential property, through the general run of investment in industrial property, to the favoured case of a building treated as plant for tax purposes and situated in a development area. The case has often been argued that whilst maintenance expenditure is wholly allowable againt liability to tax, and capital expenditure, subject to the incidence of grants and allowances, is not allowable, then a given volume of maintenance work must be less expensive to the property owner than a corresponding volume of new construction; hence building expenditure is liable to bias against new construction in favour of maintenance, even when maintenance would otherwise be uneconomic. If this is so, the demand for maintenance is increased at the expense of demand for new construction, which would put the same volume of physical resources to more productive and less labour-intensive use. Maintenance-saving investment is also stifled; the use of buildings is prolonged beyond their natural life, existing buildings are put to uneconomic uses, and the quality of the environment deteriorates.

There can however be alternative explanations. Despite every fiscal inducement, it frequently remains genuinely uneconomic to retire apparently obsolete buildings; fiscal policy has yet to realise the full development potential of land; and finally there may be an innate resistance among businessmen to investment in construction unless necessary, or for prestige.

It is also argued that fiscal considerations often have but a marginal influence on investment decisions in new building, and that the harmful effects of fiscal discrimination between new construction and maintenance may thus be less prevalent than is supposed. Furthermore, the argument described earlier is conceptually in error since money saved by building more cheaply initially would be invested elsewhere to produce at least an equal return and consequentially equal tax liability.

The design and layout of buildings may also influence rating valuations and premiums payable for fire insurance. For industrial buildings, floor space is rated according to the level of amenities provided; thus upper floors, and areas which are unheated, have low storey heights or can only carry low loads will be assessed at lower rates. However, an attempt to reduce rateable value by lowering standards may adversely affect efficiency and flexibility. Fire insurance premiums are related to the degree of fire risk and reductions in premiums may be made for design features which are likely to reduce fire spread, such as the use of non-inflammable materials or those which resist the spread of fire, and the provision of fire-fighting equipment like sprinklers. It may not pay to install sprinklers or automatic fire alarms where the annual equivalent cost of provision and maintenance is greater than the reductions in fire insurance premiums.

BMI (1989) considered that the existing fiscal system does not appear to influence the choice between spending additional capital with the intention of reducing long term maintenance costs or skimping capital expenditure at the expense of higher maintenance costs. Almost no evidence was found despite the apparently greater tax remission on maintenance costs for most commercial and industrial building owners, that such considerations had any significant effect on decisions relating to new building projects – other considerations, of technical operation, of prestige, and of the opportunity cost of capital, deciding the issues. Nevertheless, it may be expected that building owners will progressively pay greater attention to taxation considerations and the situation needs to be kept under review, to ensure that the influence of taxation does not distort the optimum economic policies relating to buildings.

Owners of commercial buildings are however concerned that Inspectors of Taxes might interpret to their detriment expenditure as between maintenance and improvements. The latter being considered as capital expenditure and therefore not allowed for tax relief. BMI (1989) has shown that there is a body of case law which would appear to substantiate these fears, leading to decisions to patch and repair when replacement by an improved system or component would be the better decision on economic grounds. It does not seem to be widely appreciated that Inspectors of Taxes generally adopt a reasonable attitude to such questions, whose interpretation is often difficult.

Capital and Revenue

Most systems of taxation distinguish between capital and revenue expenditure. Capital expenditure is money used to obtain physical assets or to permanently improve or enlarge existing assets. These assets are generally grouped into two categories: buildings; and plant and equipment.

Revenue expenditure is money spent on running a business and will include such items are rent, rates, energy costs, cleaning, insurances, management and maintenance of buildings and plant, and writing off obsolete or worn out plant. Revenue expenditure is deductable from income when calculating taxable profit.

Items of capital expenditure are not generally deductible from profits for tax purposes, although some forms of capital expenditure are eligible for capital allowances and depreciation for writing-down allowances are given for most types of capital expenditure, as provided for in the Finance Acts, but the type and amount can vary from time to time.

Capital Allowances for Buildings

These allowances are usually divided into the following two categories:

(1) capital allowances which operate in the first year;

(2) writing-down allowances which apply each year.

For example, a hotel project may be entitled to a 20 per cent initial allowance and a 4 per cent annual writing-down allowance based on the original capital cost, although it must be emphasised that such allowances may be subject to periodic alteration.

Example 13.16. A new building is being erected for use by an occupier at a cost of £3 200 000. Assuming an initial capital allowance of 20 per cent and annual writing-down allowance of 4 per cent, calculate the total tax allowance on the capital expenditure in the first year of operation. If the occupier is liable for corporation tax at the rate of 40 per cent, calculate the reduction in tax liability during the first tax year.

Capital cost of new building	£3 200 000
Initial capital allowance: 20 per cent of £3 200 000	£640 000
Writing-down allowance for first year: 4 per cent of £3 200 000	£128 000
Total capital allowance	£768 000

If corporation tax is 40 per cent, then the tax libility for the first year is:

40% of £768 000 = £307 200

Note: there will be further writing-down allowances in subsequent years.

Capital Allowances for Plant and Equipment

The question as to what constitutes plant and equipment is generally determinable by case law. Allowances can be claimed for capital expenditure on machinery, plant and equipment used for trade, profession or employment, and include such items as demountable office partitions, electrical fittings, carpets and curtains, air conditioning ducting and structural work for computer and similar installations, alteration work carried out specifically for the installation of plant or equipment, and office furniture and equipment.

The main allowance for plant and equipment is a writing-down allowance of up to 25 per cent per annum when there is sufficient tax liability to cover this amount. Different allowances may be available to companies operating in special areas.

MAINTENANCE COST RECORDS AND DATA

Form of Records

Maintenance cost records may be kept to fulfil three separate functions.

(1) Budgetary control – to produce the annual or other periodic sum which needs to be set aside to provide for maintenance and operating services.

(2) Management control – to permit the day-to-day control over maintenance expenditure.

(3) Design-cost control – to contain full information concerning causes of failures, types of failure, design faults and similar particulars. Records of roof repairs are of little use unless, for instance, they show the type of tile, quality of tile, method of laying, angle of pitch, degree of exposure and other relevant factors.

Building Maintenance Information

Building Maintenance Cost Information Service (BMCIS) commenced the distribution of building maintenance cost analyses in 1971 and was subsequently renamed Building Maintenance Information (BMI), operating under the auspices of the RICS. As the service developed it has provided valuable basic information and indicated where the greatest economies can be made. Maintenance managers and surveyors who lack adequate historic data, or who seek comparative figures against which to evaluate their performance, have access to information and guidance on cost levels and other indices through BMI. This service is essentially a co-operative system for the collection, collation and dissemination of building maintenance and other property occupancy costs, including cleaning, fuel and power. The service also provides information on the organisation of maintenance by means of published case studies and, through user reports and desk appraisals, on the maintenance consequences of design. BMI also produces interpretive information on current technical developments and problems affecting costs and relevant economic trends. Membership of the service is by subscription and is open to all organisations and individuals concerned with property ownership, management and design (BMI, 1989).

Each property occupancy cost analysis gives details of the type of building, owner/occupier, location, age, accommodation, dimensions, construction, budget procedure and maintenance organisation relating to the particular building. The analysis also breaks down the building into elements – improvements and adaptations; decoration; fabric; services; cleaning; utilities; administrative costs; overheads; and external works; and these are further subdivided into sub-elements. Against each sub-element is recorded the total cost in a stated financial year and this is converted to the cost per 100 m^2 of floor area as a unit of comparison. There are also useful energy cost analyses.

A typical BMI standard form of property occupancy cost analysis is given in table 13.32, relating to a university central library. It provides a wealth of information apart from the breakdown of costs for a particular year (1988/89), which gives the following cost allocations: decoration: 1%; fabric: 3%; services: 4%; cleaning: 16%; utilities: 22%; administration costs: 22%; and overheads: 32%. The weaknesses of a property occupancy cost analysis for a single year are apparent, as described earlier in the chapter, with only minimal work being carried on decoration, fabric and services in this particular year. Whereas the allocations for cleaning, utilities, administration costs and overheads are more realistic, as they show relatively minor variations from year to year.

The wide variations in the occupancy costs of different buildings are well illustrated in table 13.10, and BMI emphasised that statistical indices can do no more than measure a pattern, although this will provide a starting point for dealing with specific cases. It is always necessary to consider the particular circumstances of each case, the complexity, local conditions and other factors (BMI, 1985). Wide divergences can also be found when examining the occupancy costs of similar type buildings, and the reasons for the variations deserve careful scrutiny.

Building Maintenance also provides maintenance cost, cleaning cost, energy cost, labour cost, materials price, and retail prices indices. With a base of 100 in the first quarter of 1990, the appropriate indices in the first quarter of 1992 were 113.8 for general maintenance, 116.4 for cleaning and 108 for energy.

Maintenance Audit

Robertson (1983) described very effectively the nature and purpose of a maintenance audit. It forms an important part of management control which ensures that resources are obtained and used effectively and efficiently and that the organisation's objectives are accomplished. Measuring performance is now widely recognised as a fundamental part of the management control process and is a continuing activity.

A maintenance audit should desirably comprise a series of activities as are now briefly described:

- the technical audit which assesses the level of maintenance work that is achieved
- the condition audit aimed at obtaining an overall view of the condition of the organisation's estate
- the energy audit detailing the quantity and cost of energy consumed and determining whether it is being used efficiently
- the management audit which examines the management function, its policies, planning and procedures as an aid to securing best value for money
- the design audit whereby the maintenance management organisation, having implemented its own rigorous examination, can make a valuable contribution to the design process of new buildings.

GREENER BUILDINGS

Johnson (1993) has described how a building's environmental impact extends from global factors such as ozone depletion to the quality of the environment within the building. Furthermore, these impacts arise from decisions made at all stages of a building's life, including materials

Table 13.32 Typical BMI standard form of property occupancy cost analysis

BUILDING TYPE: University central library — OWNER: University

LOCATION: North-east England urban area — DATE OF ERECTION: 1966 — OCCUPIER: University

UPPER MANAGEMENT CRITERIA

Maintain building in its state within the limits of the budget allocated by the Finance Committee.

BUDGET PROCEDURE

Estimate: Overall annual budget estimate is prepared for all University buildings and grounds split into (1) Elemental Heads (2) (a) Wages and Salaries (b) Materials used and Contracting Services.
Budget: Maintenance estimate is considered along with other departmental recurrent estimates and adjusted according to allocations.
Cost control: Budget control is the responsibility of the Maintenance Officer who reviews expenditure monthly.

MAINTENANCE MANAGEMENT AND OPERATION

Responsibility: The Maintenance Officer is responsible for the maintenance of all University buildings and grounds assisted by the Supervisory Staff (Electrical, Mechanical, Buildings and Grounds), Office Manager, Secretary and Clerks.
Total estate: 220 acres including several small sites away from the main campus. 143 404 m^2 floor area Teaching and Residential accommodation.
Routine inspections: Regular visits made by Supervisors. Maintenance implemented by P.P.M. process and requisitions raised by Heads of Departments and others.
Painting frequencies: 5 year cycle externally. 1, 2, 4 & 6 years internally, depending on designated use.
Cost records and feedback: Individual jobs are cost coded according to (1) Building (2) Element, subdivided between (a) D.E.L. (b) Contract (c) P.P.M.
Directly employed labour:

Trades:	Joiners	Plumbers	Electricians	Mechanical Svs.	Gardeners
Number:	11	13	9	1	18

Incentive Schemes: None
Where the directly employed labour force is used 21% is added to basic labour rates and 22% to basic material costs to cover direct overheads.
Work done by DEL 55% and contracted out 45%
Forms of contract: For majority of work this is on a daywork basis, larger contracts use University form or RIBA Contract.
Contract supervision: By Supervisors.

BUILDING FUNCTION AND PARAMETERS

Library (book storage and display), reading, research and snack bar facilities. 465 reading spaces.
Gross floor area: 7174 m^2
Area of pitched roofs (on plan): –
Area of flat roofs (on plan): 2470 m^2
Area of external glazing: 1068 m^2
Storeys above (and including) ground floor: 4 No.
Floors below ground floor: –
Floor to ceiling height: maximum: 2.89 m

FORM OF CONSTRUCTION

Structure: Reinforced concrete columns on 22 ft grid. Precast concrete vertical external wall panels with glazing between panels. Aluminium windows in timber subframes. Concrete flat roof with roofing felt and asphalt in 2 layers. Brick and block internal partitions. Reinforced concrete floor structure.

Finishings and fittings: Felt backed vinyl floor tiles and carpet. Issue desk, shelving, etc.

Decoration: Walls and ceilings emulsion painted, eggshell in toilets. Externally, gloss paint on wood and metal, varnish on doors.

Services: Copper water services, PVC wastes. District heating system, HPHW to LPHW to perimeter convector heaters 18–21 degrees C.

Morris 10-person passenger electric lift serving 4 floors; Morris goods lift serving 2 floors, Microreader room, air conditioning unit, constant temperature room.

Note: Costs for previous years are contained in BMI *Special Report 182: University Occupancy Expenditure*.
Reference Building No. 22.

Table 13.32 (continued)

FINANCIAL STATEMENT FOR YEAR 1988/89 Gross Floor Area: 7174 m²

Element	*Total* £	*Cost per 100 m² floor area*	*Brief description of work*
0. *Improvements & adaptations*	£14 008	£195.26	Floor covering. Extra EMS. Rewire for computer
1. *Decoration*			
1.1 External decoration	–	–	
1.2 Internal decoration	1 556	21.69	
Sub-total	£1 556	£21.69	
2. *Fabric*			
2.1 External walls	1 018	14.19	
2.2 Roofs	1 234	17.20	
2.3 Other structural items	358	4.99	
2.4 Fittings and fixtures	2 410	33.59	
2.5 Internal finishes	1 253	17.47	
Sub-total	£6 273	£87.44	
3. *Services*			
3.1 Plumbing & internal drainage	524	7.30	
3.2 Heating & ventilating	2 715	37.84	
3.3 Lifts and escalators	42	0.59	
3.4 Electric power & lighting	3 519	49.05	Contains an element for P.P.M.
3.5 Other M & E services	1 556	21.69	
Sub-total	£8 356	£116.48	
4. *Cleaning*			
4.1 Windows	582	8.11	External only
4.2 External surfaces	–	–	
4.3 Internal	29 735	414.48	Includes internal window cleaning
Sub-total	£30 317	£422.60	
5. *Utilities*			
5.1 Gas	11 187	155.94	
5.2 Electricity	25 829	360.04	
5.3 Fuel oil	1 481	20.64	Allocated by area/population ratios.
5.4 Solid fuel	–	–	
5.5 Water rates	4 675	65.17	
5.6 Effluents & drainage charges	–	–	
Sub-total	£43 172	£601.78	
6. *Administrative costs*			
6.1 Services attendants	–	–	
6.2 Laundry	–	–	
6.3 Porterage	28 805	401.52	
6.4 Security	–	–	
6.5 Rubbish disposal	–	–	
6.6 Property management	15 274	212.91	Maintenance officer management only, excludes management costs of porters and cleaners.
Sub-total	£44 079	£614.43	
7. *Overheads*			
7.1 Property insurance	4 963	69.18	
7.2 Rates	58 718	818.48	
Sub-total	£63 681	£887.66	
TOTAL	£197 434	£2752.08	

External area _______ m²

Element	Total £	Cost per 100 m² floor area	Brief description of work
8. *External works*			
8.1 Repairs and decoration			
8.2 External services			
8.3 Cleaning			
8.4 Gardening			
External Works Total	£	£	

manufacture, site selection, design, construction, occupation and ultimately demolition.

In 1990 about £18b was spent on energy in buildings to provide heating, cooling, lighting and ventilation, and carbon dioxide, a major greenhouse gas is produced as a by-product. In addition air conditioning and many insulation products use chlorofluorocarbons (CFCs), another greenhouse gas which is also believed to cause ozone depletion in the upper atmosphere. The government's commitment to improve the energy efficiency of local authority housing in the 1990 environment white paper resulted in the £60 m greenhouse programme. It is estimated that schemes in the programme should reduce carbon dioxide emissions by over 50 per cent and produce energy savings of more than 40 per cent (£272 savings on total fuel costs for each household: 1993 prices) (Curtis, 1993).

Enlightened design can drastically reduce the harmful impact of buildings on the environment. Careful, creative planning will also produce a more comfortable, healthy workplace for employees. The selection of the correct materials will have a major influence on how environmentally sympathetic the building can be. In general, the closer a material or product is to nature, the more environmentally friendly it is likely to be. Another environmental issue is whether the material has been or can be recycled at the end of the building's life (Miller, 1992).

Built form has a key role to play in the eventual energy efficiency of a building. The building should optimise the balance between maximising daylight, solar heat gain and ventilation, and at the same time minimising the surface area of the walls and roof. For example, a shallow plan will permit full use to be made of natural daylighting, ventilation and solar heat gain. A deep plan may require mechanical ventilation and more artificial lighting, which will increase both the capital expenditure and the running costs. The main rooms should preferably have a southern aspect. The roof should protect the walls from excessive wetting as this causes unnecessary chilling of the building fabric. In addition, care should be taken to control the amount of heat loss from the north of the building and heat gain from the south (Miller, 1992).

Natural daylighting should be used to the full, ideally with sun screens to give shade from the summer sun and curtains, blinds, shutters or low emissivity glass to minimise heat loss on winter nights. In general lighting is best operated from zoned automatic controls that respond to daylight conditions. If fluorescent lights are used, Miller (1992) recommends that they have high frequency ballasts, to give lower running costs and reduce possible health side-effects.

The heating system should be regulated by time and thermostatic control. Lighting in certain areas should be controlled by photoelectric cells or movement sensors. Any heating or cooling system should be controlled in zones to maximise flexibility and control.

For heated buildings, the most economical fuel choice in 1993 was natural gas, which is also the least environmentally damaging, with the use of efficient condensing boilers deserving careful consideration. Oil is generally rated as second best in environmental terms, although its price does tend to fluctuate significantly. Solid fuels are placed third and electricity fourth. The reason for electricity's poor performance is that unless it is supplied from a hydroelectric plant, its generation is very inefficient and it produces the highest rate of pollution. On large developments it may be worthwhile to install gas fuelled combined heat and power (CHP) units as they are extremely efficient.

Miller (1992) describes how in order to achieve an efficient and healthy building, performance targets must be established from the earliest design stage. The consumption of energy and quality of the internal environment should be monitored and the causes of any deviations from the standards set should be investigated.

Finally, the effect of the building on the surrounding environment should be assessed. Criteria would include such factors as fire hazard, waste generation and disposal, vehicle and pedestrian traffic generation, intrusiveness of external and internal lighting, noise generation, and external landscaping around the building (Wordsworth, 1992). A precise assessment of these factors could entail carrying out cost-benefit studies of the form described in chapter 16.

FACILITIES MANAGEMENT

Nature and Scope

Facilities management (FM) is interlinked with certain activities of cost management particularly those relating to life cycle costing, and hence it was considered desirable to include basic information about this relatively new and expanding technique in this book. Readers requiring further information on this important activity are referred to *Facilities Management* by Park (1994) and Property Helpline (1994).

Facilities are generally defined as any property where people are accommodated and work, or where an organisation conducts its business, while management concerns all aspects of providing, operating, maintaining, developing and improving those facilities.

The Association of Facilities Managers has described how facilities managers typically carry responsibility for strategic planning: briefing, selecting and managing in-house and contracted resources; oversight of work, environment design, construction and fitting out of buildings, and their day-to-day operation and maintenance. In addition, they are regularly involved in property and estate management; space planning, fitting out and furnishing. They can be responsible for reprographic, communication, information technology and other services such as building maintenance, security and fire prevention, catering and transport. Hence their brief can be very wide indeed.

The RICS Facilities Management Skill Panel (1992) considered that FM embraces all the principal functions of management: planning, organising, staffing, directing, controlling and monitoring, and the bringing of all these components together in a co-ordinated way. It also postulated that FM consists of the following three distinct but interrelated areas, which can account for 30–40 per cent of the overall cost budget:

- the management of support services
- the management of property
- the management of information technology.

Watts (1992) in a keynote address at the RICS described the technique in an interesting and discerning way. ‘Effective facilities management has to provide a complete service, encompassing the whole lifestyle of buildings. Everything needs to be taken into account, from security to catering, maintenance, energy management, space planning and project management. A theme running through all these issues is sustainability and the aim of making buildings as environmentally friendly as possible. In aiming to give buildings a long, productive and efficient life, facilities management must be in tune with the spirit of the age. We are seeing a shift away from short term exploitation of natural and built resources, towards a greater emphasis on stewardship and sustainability’.

While Park (1992) believes that FM embraces the control and the most appropriate and effective use of property resources, dealing with many interrelated aspects such as space planning, space costing, asset tracking, life cycle costing, maintenance and component specifications.

Implementation of Facilities Management

Park (1992) has emphasised that it is imperative to secure the right usage of the building; wasted space, inefficient departmental interfaces together with an unattractive working environment can far outweigh the effects of a periodic rental appraisal and maintenance costs. Effective FM prolongs usefulness and therefore the effective life of property and slows down the rate and timing of decline. Buildings in this context are therefore resources of finite supply that need to be used in a controlled manner.

The operational flow chart in figure 13.7 illustrates the four phases in FM strategy, namely brief, research, implement and liaise, with the various related activities. Many of the client's requirements or conventions are cost oriented such as:

- space allocation with costs of rent, rates, energy consumption, cleaning and other related aspects
- component costs for tax accounting purposes, depreciation and replacement funding
- maintenance costs often obtained from the analysis of term contract rates to establish quarterly and annual budgets, monitor variances against cashflow and provide reports to the client's financial controller.

The research phase encompasses life cycle costing, creation of cost databases and space costing. Space planning and space costing head the list, as without the economic use of space the other functions are of limited financial value to the building user. There is an inherent need to identify and match space requirements with the buildings available. In the case of a serious mismatch, the user may require building alterations or even a new one. The space planning function seeks to match available and future planned space in the type and form needed to meet likely requirements. The space function is greatly assisted by the use of personal computer (PC) graphics and the application of computer aided design

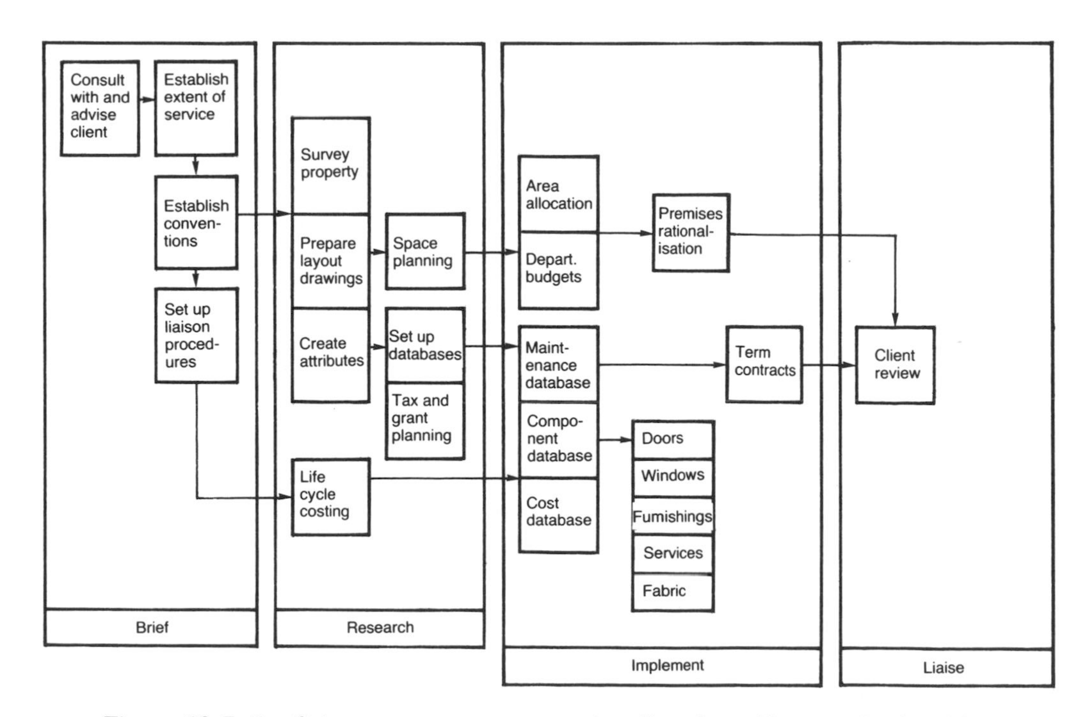

Figure 13.7 Facilities management: operation flowchart (Source: Park, 1992)

(CAD) that develops into computer aided facilities management (CAFM) linking CAD drawings with integrated databases (Park, 1994).

The operation of active space modelling within the space planning regime permits the flow of activity within the space available to be planned and applied to the design of new buildings or even the layout of existing buildings. It is important that the circulation space analysis is based upon an efficient layout that is neither over generous nor sub-standard in its allocation to occupants.

Asset tracking replaces the conventional inventory with a proactive database that locates, logs and tracks any movable asset, such as computers, emergency medical equipment and furniture. In advanced studies, assets' components and rooms can be labelled with a bar code read by a hand held data capture unit (DCU). The DCU is primed by the management computer and the operator in turn amends the data on site returning the DCU to the management computer to update the master record (Park, 1992).

All the databases are driven from a visual reference point, as it is much easier to locate a component or activity by looking at a drawing than scrutinising lists of data or reference codes. A drawing of the building is created to serve as the key. Using CAD, layers of information are stored on the drawing. These layers may be switched on and off at will and are interactive with the database, thus providing a quick and efficient referencing system (Park, 1992).

Park (1994) has described how the introduction of FM to an organisation can readily effect a ten per cent reduction in workspace costs, realistically rising to twice that amount in the medium term. FM is not confined to controlling cost, as there are several significant activities that can be managed through FM systems to assist in the smooth operation of the organisation. These activities include health and safety monitoring, maintenance and life cycle costing, specifications of components, selection of systems and software to collect, store, analyse and retrieve data, and the management of a variety of services encompassing heating, lighting, power, ventilation, telephones and communications, security systems, waste disposal, catering and staff welfare, creche provision and contracting out key support services.

Property is a finite resource that needs to repay its capital cost on an investment basis. Those buildings that serve a public need but cannot satisfy the investment return principle can only be funded with government grants or other non-commercial financial assistance. Buildings generally have a value cycle based on demand and hence changes in demand over time can result in a diminution of the building's value. The operation of FM that monitors the use of the building and adjusts it and the occupation to match current demands will slow the decline in value. Furthermore refurbishment, alteration and even change of use can prolong the life of the building. Where, however, the building has a defined life span, the facilities manager's objective will be to extract the maximum benefit from the property over its planned life without incurring excessive running expenses towards the end of the period, in accordance with good life cycle costing practice (Park, 1994).

The economic benefits of FM should show through improved productivity, better production quality and overheads control. For example, the monitoring of both space related costs like rent and rates in relation to workspace output, and the more difficult one of controlling location related costs such as energy consumption to major items of manufacturing plant and the extent of demand for office related items, such as coffee and tea vending points, photocopying, stationery and computer floppy disks, through adequate feedback and analysis. Realistic budgeting is all important to successful FM, but this must also encompass the feedback and monitoring of expenditure against the budget, be it quarterly or annually, and covering all occupancy and production costs, including maintenance, energy consumption and cleaning, with a view to achieving more efficient and economical processes. The value of maintenance and energy audits as discussed earlier in the chapter should not be overlooked.

INTELLIGENT BUILDINGS

In the 1990s there was a growing emphasis on the need for 'intelligent' buildings, referred to in the

US as 'smart' buildings. Intelligent buildings are those which satisfy the requirements of the occupants and this often entails a substantial measure of flexibility to cater for different work groups with varying needs and whose work locations or stations can also change over time.

Harvey and Ashworth (1993) have rightly emphasised that business operations are progressively more dependant on a very large investment in information technology, which facilitates the exchange of information within and between buildings as communications both internally and externally assume greater importance in many organisations. The enlarged information technology (IT) services include word processing, electronic mail, video conferencing and networks which access a wide range of business databases. Electronics contribute to a whole host of activities, including handling of the internal environment, security, regulation of access which includes lift movements, energy and lighting management and operation. In the future it is anticipated that intelligent buildings will be able to respond in a sophisticated way to changes in the external environment using highly developed materials with variable opacity and/or thermal capacity, self-cleansing properties and maintenance aspects. The Lloyds' building in the City of London is an example of a modern intelligent building, despite the objections of many of its occupants, where the services are located on the outside of the building, thus keeping the floors free of service ducting, providing flexible office layouts and minimising the cost of relocating activities.

However, it should be borne in mind that the more intelligent a building, the greater will be the difficulty in managing it. The reorganisation of heating, cooling, lighting, security and associated services will require more sophisticated systems and highly trained and skilled personnel to perform these operations, probably entailing the back-up of an expert system. Indeed, an intelligent building could be regarded as one incorporating an expert system to secure the required level of building management. BRE has developed an expert system (BREXBAS) to monitor sensor data from remote systems, apply its knowledge and reasoning capabilities to satisfactorily interpret the information and generate advice for the user. By using this system greater control can be exercised over costly building space, thus enabling it to be used more efficiently and effectively. The ultimate aim is to optimise a building's performance throughout its entire life when assessed against a range of potential requirements (Harvey and Ashworth, 1993).

14 LAND USE AND VALUE DETERMINANTS

This chapter investigates the factors which influence land use patterns and land values. It is also concerned with the whole spectrum of matters which bear upon the development of land, such as site characteristics, planning and other statutory controls, encumbrances and easements.

CHANGING LAND USE REQUIREMENTS

Land Use Patterns

Our present day basic concern is primarily the conflict between the demands of a growing economy and a fixed area of land. Great Britain has a high man/land ratio and a general pattern of increasing old and new uses although this was distorted by the 1990–93 recession.

Land is unique in that each parcel or plot has a specific location with its own particular geography. Height above sea level, slope, latitude and longitude, soil and subsoil, rainfall, sunshine, temperature, wind exposure, drainage and distance from other places – all vary from one plot to another. Some of these variables can be partially controlled by the use of other resources – capital and labour for instance, but there is no homogeneity or easy interchangeability, each site has its own peculiar characteristics. The changing pattern of urban growth is generally characterised by the broad pattern of growth, decay, rebuilding, central business and shopping core, industrial estates, and outer residential suburbs.

On the other hand, it is possible that in a twenty-five to forty-year period, scientific advances and technological change may profoundly affect patterns of urban development; and the shifting character of life styles and consumer tastes in large segments of a substantially affluent society may further complicate the problem of prediction. Land development is the consequence of many decisions and implementing actions of both a public and private nature. Priming actions often trigger secondary actions which taken together produce the total pattern of land development. Thus an industrial or commercial location decision may set in motion a whole chain of other decisions and actions, for example location decision of households, firms and institutions. Alternatively, a highway location decision, or a decision on building a new school, or a combined series of decisions of this nature can serve to prime such secondary decisions and actions. A priming action has two main characteristics – structuring effect on the distribution of land development and the timing effect in fixing the sequence of development.

The changing pattern of urban development in this country can be traced from the compact towns of the Industrial Revolution to the more open development of inter-war years assisted by the increased mobility of the urban population through improved public and private transport facilities. These developments now require extending to include the more compact, higher density residential layouts of the 1960s and 1970s and the lower rise but often compact developments of the 1980s to 1990s. Changes in competition for land arise from urbanisation and industrialisation, and at the margin of urban–industrial development, agriculture is quickly priced out of the land market. Another important implication which follows from the increasing competition for land for urban uses is the irreversibility of the results, so different from merely replacing one agricultural use by another. Thus the central problems in the economics of land use are those of location and of competition between alternative users and uses to command each particular site, although these actions may be constrained by statutory provisions.

Urban Land Use

Housing is easily the greatest urban land use, and in the New Towns of England and Wales it accounts for just over one-half of the total urban area. This is a higher percentage than in the older towns where the areas used for residential purposes generally amount to about forty-three to forty-five per cent.

Furthermore, following the recommendations of the former Ministry of Housing and Local Government (1962), the latest generation New Town proposals show a noticeable reduction in space standards. This is particularly so with housing, where areas of social housing have an average net residential density of about eight hectares per 1000 population.

There has been a continuous reduction in the area of land devoted to agricultural use but at a much reduced rate compared with the pre-war trend. During the war land was acquired by service departments at a rate exceeding 40 000 hectares per annum, and in the post-war years over one-half of this has been returned to agricultural use, and this process was still continuing in the 1980s.

Land use within an urban area can often be conveniently subdivided into four separate and distinct districts or zones.

(1) Central business district which is the optimum location for many economic activities, such as shops, offices, theatres and hotels, having maximum accessibility.

(2) Zone of transition surrounding the central business district where older buildings are being replaced.

(3) Suburban areas which are developed for residential purposes at moderate densities on cheaper land.

(4) Rural-urban fringe accommodating commuters who wish to live in rural surroundings.

In the 1980s–90s this pattern was distorted to some extent by the development by out-of-town shopping centres, DIY warehouses and business parks.

Input of Changing Use Pattern on Agriculture

Agriculture loses land in two main directions – some of its poorest land to afforestation at about 8000 hectares per annum (now reducing with the scaling down of the Forestry Commission), and some of its better land to urban uses at about 14 000 to 16 000 hectares per year. The latter rate of transfer is directly influenced by the density of the urban development, and the effectiveness of the planning authority's green belt policy.

By the year 2000 about eighty per cent of the total land surface in this country may be available for agriculture, and British agriculture will tend to adjust itself to the area of land left for its use. Increases in population and rising real incomes create demand pressures for more land at present in agricultural use, but improvements in the physical efficiency of food production economise on land area. Again if imports of food products are increased, home land can be saved for other uses. If pressures lead to a decrease in food imports, more home land will be pressed into more intensive service.

The present average increase in agricultural productivity of about 1.3 per cent per annum, coupled with some over-production, will more than offset the land likely to be needed for urban uses, provided that the choice of new urban sites is made with sensible discrimination and, in particular, that development is not concentrated to an unnecessary extent on our most highly productive farmland.

LAND USE PLANNING

Objectives

A universal interest has developed in land use planning and the determinants of land use, and there is an increasing awareness of the advisability of forward planning. The author once wrote that town and country planning, or physical planning as it is now more commonly described, is necessary to ensure that all development is coordinated with an eye to the future, and carried out in such a way to assist in producing a community environment that will advance human welfare in health, well-being and safety. Many defects are apt to arise from unplanned development, such as waste, congestion, disharmony, undesirable mixture of incompatible uses, lack of social services and unnecessary loss of good agricultural land. The broad objective of planning is to ensure that land is put to the best use from the point of view of the community, and to secure a proper balance between competing demands for land. Furthermore, the regulation and control of the use, in urban areas, of land resources within proprietary land units is necessary, in order to prevent the repetition of the undesirable mixture of land uses which has emerged over time in towns and cities. It is also necessary to consider transportation planning in tandem with land use planning, to ensure that new development is adequately served by an efficient transportation system. In the late 1980s many towns and cities had introduced park and ride schemes to encourage greater use of public transport and to ease congestion in town and city centres and subsequently the submission of a variety of mass rapid transit schemes, which required parliamentary approval, mainly aimed at bringing back electrically operated trams into urban areas, of which the most notable is Manchester.

The World Health Organisation has emphasised the importance of thinking of the metropolitan area as a coherent whole and to recognise the interplay of social, political and economic factors which must be taken into consideration. This organisation defined planning objectives as 'a model of an intended future situation' and 'a programme of action and predetermined coordination', illustrating the dynamic nature of the process and of the need to improve human conditions.

It is possible to identify certain basic principles in land use planning – the unity of the environment (fusion of town and country), comprehensiveness (controlling many activities, often conflicting) and quality of the environment. However planning authorities must accept that there is always a limit on the resources available and so they must actively encourage management and development – public and private – to accept and work for strategic goals. Various writers have emphasised the need for land use plans to take full account of social and economic factors, with the general objective of maximising social net benefit or public interest. Planning the countryside of Britain is largely a matter of conserving natural resources. The basic objective is to make full use of our minerals, soil, water and wildlife, while ensuring that we do not allow exploitation to endanger their future supply. Planning our towns, on the other hand is fundamentally concerned with the development of land resources with buildings, roads and other urban services, to accommodate changing demands.

Until the early nineteen sixties it was generally considered that land use planning objectives could be achieved by a control mechanism which operated at local planning authority level, but since then it has become accepted that many of the activities we want to influence take place over an area larger than that administered by a single authority. Hence some regional and sub-regional planning has been undertaken and there is a need for a larger degree of national planning.

In land use planning we control land use, and in controlling land use we are controlling people. Hence it is vitally important that the public are permitted to participate actively in the planning process – the subject investigated by the Skeffington Committee. The use of a series of seminars in the first stages of preparation of the plan for Milton Keynes New Town is a practical illustration of its application. Town planning frequently appears in the form of contentious or controversial public issues such as the topical debates covering pollution, conservation and traffic management.

Finally, it is interesting to examine the geographical and historical reasons for the differences between British and American patterns of urbanisation. The town planning movement in Britain had its roots in a reaction against urban growth as represented by the nineteenth-century industrial town, and it sought an orderly countryside and contained towns which would not spread across the countryside unchecked. In the United States a different attitude is adopted towards land use. The vastness of the North American continent and the tendency of its inhabitants to move on periodically from one location to the next has resulted in successive exploitation and abandonment. This has given rise to the expression 'God's own junkyard', describing many Americans' attitude to land – a resource that can be used, even squandered, with little thought for the future.

The British planning system is unusual in that it gives local planning authorities almost complete discretionary control subject to a right of appeal to the Department of the Environment. Criticisms have often been made of the extensive delays in obtaining decisions on planning applications, excessively detailed control and sometimes of ill-informed policy. The Local Government Planning and Land Act 1980 relaxed control on minor matters and introduced enterprise zones and the general policy on environment was in favour of the privatisation of resources and a minimum of state intervention.

The local planning authorities in the United Kingdom prepare structure plans which are statements of broad strategic objectives for their areas, and within this framework detailed planning is undertaken on the basis of local plans. Action area plans are prepared for areas which are to be planned and developed, redeveloped or improved in a comprehensive manner within a period of about 10 years and these provide positive guidance to developers without stifling their initiative. In the 1990s some authorities, comprising London boroughs and metropolitan authorities, adopted new style unitary development plans. Birmingham adopted such a plan in 1993, following 2½ years of consultation and a seven weeks public inquiry, and this provided the statutory framework to guide development up to year 2001.

Enterprise zones were introduced to test, as an experiment, and in a few selected areas, how far industrial and economic activity can be encouraged by the removal of certain fiscal burdens and by the elimination or streamlining of some statutory or administrative controls. The government's role becomes that of creating and maintaining the conditions necessary for expansion in private investment and production. Enterprise zones represent an attempt to create these conditions in certain areas, notably those areas of physical and economic decay where conventional government policies of increased public expenditure and public services have not succeeded in regenerating self-sustaining economic activity.

The regeneration of these declining areas is to be achieved by an amalgam of land use planning measures and fiscal provisions. Planning and development controls are simplified considerably through planning agreements, whereby deemed permissions operate for those economic activities and investments compatible with the particular circumstances of each zone.

The Town and Country Planning Act 1990 introduced the concept of the simplified planning zone (SPZ), whereby under this scheme planning permission can be granted for either some specific development or a class of development specified in the scheme, and may be unconditional or subject to conditions (Telling, 1990). The setting up of the Urban Regeneration Agency (URA) in 1992 provided the opportunity to create a better co-ordinated and more comprehensive approach to inner city problems, and this aspect is considered further in chapter 16. The URA was replaced by English Partnerships (EP) in 1993.

Economic Aspects of Land Use Planning

The economic theory which had competitively established price as its keystone, concluded that since consumer desire and producer capacity tended towards a balance at any moment in time, the most efficient allocation of resources was ever present through market forces. In certain spheres, especially in the use of land, this conclusion is suspect. The analysis is too narrow as it ignores a whole range of side effects. Furthermore, many forms of land development are relatively inflexible, as for example land developed with houses is not readily returned to agriculture, and there is no opportunity to correct the adverse effects of bad decisions.

Some writers claim that if town planning leads to higher land values than would exist without it, then a better or more efficient use of resources has been achieved, based on rent theory. This is a dubious hypothesis, as the planner may artificially reduce the supply of land for a particular kind of use and thus force up its price – putting the owner in a monopolistic situation. To ensure the best use of resources, the economic consequences of alternative courses of action should be considered, although non-economic implications frequently deserve attention. Difficulties arise in attempting to place monetary values on non-revenue producing public sector development, and the valuation of the revenue producing investment can only reflect from empirical data the imperfections of market values which are of interest in plan evaluation, and they form an inadequate basis for the latter function.

Furthermore, the fragmentation of the ownership of land may prevent it being used for the most efficient purpose. Some public urban land uses such as roads, parks, schools and sewage disposal works will be non-revenue producing, but the profitable uses often depend on the non-revenue uses. Urban planning can, by control and reorganisation of both types of use, lead to a more efficient use of urban resources. If urban planning increases the accessibility within an urban area, then this again is likely to increase efficiency and land values. Hence the economic approach to the study of urban land use patterns is an important one. The patterns that emerge are the sum total of a large number of individual appreciations of locational values. Successful planning must be based on an intimate knowledge of the economic forces at work within cities. Through planning we aim to produce a new geography, a better distribution of activity and land use related to contemporary, social and economic needs, yet fully aware that needs change more quickly than physical forms.

LAND VALUES

Land Value Determinants

While various users are in competition for sites, the sites vary considerably in their suitability for different purposes. The attributes of sites can be divided into three main groups: physical, locational, and legal consents as to use. The prices of sites are very much influenced by the use to which they can be put and can vary considerably over time. For example in the building boom in 1989, the average price of building land doubled, while in the 1990–93 recession it reduced substantially. Furthermore, residential land can be 100 times more expensive than agricultural land, as shown later in the chapter, and hence it has a very high development value. Site values are generally in the range of 20 to 25 per cent of the total cost of dwellings.

The following general model for determination of land values is useful

$$\text{land value} = \frac{\text{(aggregate gross revenues)} - \text{(total expected costs)}}{\text{capitalisation rate}}$$

Revenues are influenced by the investors' expectation of the size of the market, income spent for various urban services, urban area's competitive pull, supply of competitive urban land and prospective investment in public improvements. Expected costs are the sum of local property taxes, operating costs, interest on capital and depreciation allowances. The capitalisation rate is affected by interest rates, allowances for anticipated risk and expectations concerning capital gains.

According to land economics theory, these factors are taken into account in the property market. Users of land bid for sites in accordance with what will maximise their profits and minimise their costs. Land users in retail business and services tend to bid for space at the highest prices, and land best suited for these activities shows the highest value.

Like most other goods, the value of land is influenced considerably by the interaction of supply and demand. The supply of land is fixed although its use may change. Similarly with landed properties (buildings) the supply, particularly in the short-term, is relatively inelastic, although the supply of a particular type of building may on occasions be increased fairly quickly by conversion. In the long-term, where there is an obvious need for a certain class of property, it is possible to acquire land and erect fresh stocks of the particular building.

As a result of the 1990–93 recession, coupled with an extensive building programme and unsustainable fast growth in the late 1980s, there was a significant over-supply of offices, shops and factories because of many companies becoming bankrupt and the rental values dropped significantly. For example, in the peak of 1991 the supply of office floorspace to let in the city of London was almost three times its peak in 1976, and some observers suggest that surplus space in peripheral locations around London will not be fully occupied until the end of the century. Because of the infrequency of rent reviews, it was estimated in 1992 that over 70 per cent of central London offices were over-rented by an average of 35 per cent. In 1993 prime city rents seemed set to stabilise at their 1985 levels of over £2.8/m^2 (£30/ft^2), at under half the 1989 peak values (Fraser, 1993).

One of the most publicised examples was Olympia and York's large new office development at Canary Wharf in London's docklands where, in 1992, 40 per cent of floorspace was unlet at completion despite generous incentives being offered to tenants. The cost of the development was estimated at £1.5b, but in June 1992 its value seemed unlikely to exceed £500 m. Many large building contractors were also facing serious financial difficulties as their workload dropped substantially with tender prices cut to the bone, and they had to revalue their land banks, built up in the boom years at inflated prices, at very much reduced figures.

In practice, the market prices of the developed real properties determine the land values (the residual figure in the developer's budget, described in chapter 15, which the developer can afford to pay for the site). Land values are influenced by a variety of factors, such as accessibility and compatibility. If there is a shortage of land available for a certain use, prices will tend to rise until further land is transferred to this use, unless land use planning frustrates it.

Probably the most important factors influencing land values are:

(1) supply and demand – a limited supply of building land, or fierce demand for it will force up the price;

(2) the permitted use to which it can be put under planning regulations, of which central area uses such as shops, offices and theatres, are the most valuable;

(3) location, highly priced land often being the most accessible;

(4) physical characteristics which affect the cost of development and suitability for a given purpose, industrial areas for example needing extensive flat sites;

(5) availability of public services, such as roads, sewers, water and gas mains, electricity cables and telecommunications;

(6) form of title (freehold or leasehold) and any restrictive convenants or other encumbrances which will affect its use (as described later in the chapter); and

(7) general nature of the surrounding development and its compatability.

Values are best established by reference to the price paid for comparable properties in the market. In practice, there are few really comparable sites or buildings and there is all too frequently little information available on property transactions. Generally, auction prices are accepted as the truest indicators of market prices.

Effect of Town Planning on Land Values

According to the war-time but very authoritative Uthwatt Report (1942) the effect of town planning is to transfer land values but not to destroy them. Planning schemes frequently cause a redistribution of values, where the permitted land uses differ from the existing patterns of land use. If, for example, a town map limited the supply of land for residential purposes and permitted high densities on the outskirts and lower densities in inner districts, then the land on the outskirts would in all probability have a higher value than that nearer the centre; a movement of values has therefore occurred. Where the transfer in land values, which is caused by a redistribution of the profit uses of land, settles will partially depend on the planned redistribution of the non-profit uses, such as open spaces.

Farmland near towns often has a potential development value, described by the Uthwatt Committee as 'floating value' which increases as the likelihood of development becomes more certain. This enhanced value was the betterment of which the Land Commission required a forty per cent share under the Land Commission Act 1967 and which was abolished in 1970 and was replaced by Development Land Tax in 1976, which was abolished in 1985. The aim was to collect some share of increases in value representing the community's actions, while leaving some share to the benefit of the private individual, thus allowing the market to continue to operate. It did, however, result in a reduction in the amount of land available for development.

Successful planning should seek to understand the economic and social forces which shape our environment and assist in the allocation of land uses to meet those needs in a manner beneficial to the whole community. This involves ensuring an adequate supply of land to meet various anticipated demands. On occasions the Federation of Registered House Builders has considered the shortage of residential land to be an artificial scarcity resulting in unjustifiably high land prices. The Federation considered that the only remedy was for local planning authorities to release more land on the market, to meet significant increases in demand. At the same time the author believes that more positive steps should be taken to redevelop the large areas of derelict land in inner urban areas.

Pattern of Land Values

Commercial and similar uses are located in city centres, as they are able to pay the high land prices and secure the benefits of maximum accessibility and convenience. Hence rents serve to act as sorters and arrangers of land use patterns, and planning control alone does not decide land use. It has been suggested that the outgrowth of this market process of competitive bidding for sites among the potential users of land is an orderly pattern of land use specially organised to perform most efficiently the economic functions that characterise urban life.

Generally commercial and industrial uses can attract land away from residential uses. Competition between firms to be in the desired positions will force the land values above those of the surrounding land used for residential purposes. If all the land in a given part of the town is used for complementary purposes this is likely to enhance the land values, whereas if they are incompatible, this may lower the land values. For example, if there is a residential district well served by schools, open spaces and transport, persons will wish to live there and both property and land values will be higher than if it lacked these facilities. Developments in transport systems may also lead to changes in urban land values. For example, the extension of a bus route, or an underground railway network or the carrying out of major road improvements, may cause changes in land values in adjacent areas.

Ong (1981) undertook an interesting study into the factors which influence the general pattern of residential land values in Singapore, using multiple regression analysis. He found that proximity to schools and public transport were not critical to residential choice, but that accessibility to shopping facilities was an important factor. Environmental quality in terms of its physical and social attributes appeared a significant factor in judging the desirability of a residential neighbourhood. Size of site was not seen as a determining factor of land value in the sample of 224 transactions.

The greater the permitted residential density, the higher the price per hectare which can be obtained. For example, an increase in permitted densities from twelve to twenty dwellings per hectare would in 1993, tend to reduce site costs per dwelling by about £3000 to £5000 but could raise the land price per hectare by £50 000 to £450 000. A developer normally has to pay more per hectare for the marginal piece of land needed to complete a development site than he will for the site as a whole.

Land prices vary widely in different parts of the country. It is cheapest in parts of eastern and northern England, East Midlands, Wales and Scotland, and dearest within a 50 km radius of London (see tables 14.1 and 14.6). In consequence, even supposing a builder obtains permission to build twenty houses per hectare on land costing £450 000/hectare, it could cost him at least £35 000 a plot, after roads, drainage and other services have been provided, but before a brick is laid on the site. Shortage of residential land retards the housing programme and is a blight on the building industry. Then when land is made available for development, competition tends to force up the price which, coupled with increasing building costs, may cause the projected buildings to fail to meet the criteria for investment return or cause rents to reach levels which occupants cannot afford or are very reluctant to pay.

FACTORS INFLUENCING DEVELOPMENT

There is a wide range of factors influencing the development of a building site, from the physical characteristics of the site itself to legal restrictions, planning controls and building regulations.

Land Ownership

Legally persons hold interests in land but do not own it, as all land is held from the Crown on tenure. Nevertheless a freeholder with a freehold interest in the land is the absolute owner of the property in perpetuity, and he can do as he likes with it provided he does not contravene the law of the land or interfere with the rights of others.

A freeholder can create a lesser interest out of his absolute one, such as a leasehold interest where the leaseholder will hold the land for a limited period and subject to the payment of rent. The landlord or lessor retains an interest in the land, known as a reversion, and is entitled to receive the agreed rent. The tenant or lessee has exclusive possession of the property for the period of the lease, provided he pays the rent and observes the covenants or conditions attached to the lease. As a general rule the lessee can assign or sell his interest to another person or grant a sublease for a shorter term than his own lease. The Leasehold Reform Act 1967 permits certain tenants to purchase the freehold of the property which they occupy. With a building lease the lessee pays a ground rent for the land and undertakes to erect and maintain suitable buildings on the land. For the period of the lease, often ninety-nine years, the lessee receives an income from this possession of the buildings and land but on the termination of the lease, both buildings and land revert to the freeholder.

Site Characteristics

Each site has its own characteristics, which have an important influence on its suitability for development for a particular purpose. The main characteristics are now considered.

Soil conditions. The subsoil should have a reasonable loadbearing capacity, as poor soils create foundation problems and increase constructional costs. Clay is subject to significant changes with varying climatic conditions and can cause settlement in buildings, as described by Seeley (1995). Subsidence can also arise from former mineral workings or naturally occurring geological phenomena, such as land slipping and compressibility.

Groundwater. It is desirable that the site should be well above the highest groundwater level and free from the possibility of flooding. Working in wet conditions is difficult and a permanently wet site can give rise to unhealthy conditions for occupants, and deterioration of the buildings. It must be possible to satisfactorily dispose of surface water from low lying land.

Contours. A reasonably level site will reduce constructional costs particularly where the buildings cover large areas as with factories. Steeply sloping sites require stepped foundations and extensive earthworks, and may involve special land drainage installations as well as being inconvenient to users of the site (Seeley 1995).

Contamination. Landfill and former industrial sites are best avoided as they can involve expensive siteworks to remove potential hazards. Some of the most likely causes are the effects of acids, sulphates, heavy metals, methane, hydrogen sulphide and radon. For example, the development of Stockley Business Park in west London required the removal of three million tonnes of toxic waste which had been placed to depths exceeding 12 m over a period of 50 years, and the provision of costly measures to extract leachates and methane gas (Seeley, 1992). The cost of identifying contaminants on a site alone can be as much as £15 000/ha at 1993 prices (Vann, 1993). The implications of the Environmental Protection Act 1990 will be considered in chapter 16.

In 1994 the government, having earlier withdrawn their proposals for a contaminated land register, were seriously considering the RICS concept of a Land Quality Statement, which would form part of the planning application for any significant development and would contain assessments of contamination risk and any necessary remedial treatment. This could prove to be an efficient and reasonably economical approach and ensure adequate safeguards in the development of contaminated land, but was subsequently rejected by the government in 1994.

Obstructions. These may take various forms and all involve additional expenditure in site clearance work. Humps and hollows require levelling or filling, inconveniently sited buildings or trees need demolishing or felling, and heavy machine bases and foundations of former factories on derelict sites require costly removal, ponds require filling and possibly involve some drainage work, and ditches may need piping. Trees may also be subject to a tree preservation order, as described later in the chapter.

Services. The availability of essential services, such as sewers, water mains, electricity cables, telecommunications and possibly gas mains, of adequate capacity, is an important consideration. Private sewage works, water supply installations or electricity generating plant are expensive in both provision and operation. Culverts or sewers crossing a site may also prove costly in realignment and it may be difficult to secure adequate falls on a longer diverted route. Sites lying below sewer invert level will entail the costly provision, maintenance and operation of a pumping station in order to discharge into the public sewer. Other services crossing the site will also entail costly diversion works.

Access. Satisfactory access to the site must be available and some types of development will require good access roads leading to the site, with adequate sight lines at the junction with an existing road.

Aspect. Ideally the site should be on a gentle slope facing in a southerly direction to secure maximum sunlight and protection from the cold northerly and easterly winds. In practice the ideal site is rarely obtainable.

The site requirements for different uses vary substantially but in nearly all cases the sites have to be accessible to users, and the use of the site needs to be compatible with the uses of adjoining sites. A consideration of the requirements of two widely differing uses will serve to illustrate these points.

Site boundaries. The type and condition of site boundaries and their ownership needs identifying, together with the evaluation of any constructional problems likely to arise from buildings adjoining the site as the project progresses.

Area of site. There needs to be adequate space available on the site for the storage of plant and materials, the erection of temporary buildings and sufficient workspace.

School Site Requirements

(1) Ideally sited centrally in relation to the catchment area from which pupils will be drawn.

(2) Away from main roads and resultant noise and vibration but readily accessible to the network of distribution roads, and also bus routes where possible.

(3) Sufficient area to accommodate buildings and playing fields with ample space for probable future extensions.

(4) Preferably of regular shape and to even and gentle falls, of porous subsoil and well above highest groundwater level.

(5) Freedom from major obstructions or adverse restrictive convenants.

(6) Adequate utility services.

(7) Compatible with adjoining uses.

(8) Not excessively expensive.

(9) Planning consent forthcoming.

Factory Site Requirements

(1) Good access by road and in some cases access by water and/or rail is beneficial.

(2) Suitably located in relation to raw materials, markets and homes of the workforce.

(3) Adequate utility services.

(4) Adequate area for factory and possible future extensions together with ancillary uses such as car parking, and of regular shape, capable of subdivision into an adequate number of suitable sized plots.

(5) Reasonably level site with gradient not exceeding 1 in 20 and with subsoil of suitable load-bearing capacity above highest groundwater level.

(6) Freedom from major obstructions or adverse restrictive covenants.

(7) Compatible with adjoining uses.

(8) Not excessively expensive.

(9) Planning consent forthcoming.

Planning Controls

Planning controls stem from the operation of the Town and Country Planning Act 1990, and were administered principally by the county councils and district councils, termed local planning authorities, with the county planning authorities preparing structure plans and district planning authorities formulating local plans, unless reserved to the county planning authority and dealing directly with many matters of planning control. In the case of Greater London, the local planning authority will normally be the London borough council.

The structure and local plans are very different in both concept and presentation from the former development plans. They set out policies and proposals in written form, with any maps and diagrams being illustrative of the text rather than the basis of the plan, and are more concerned with implementation in land use and environmental terms of social and economic policies. The structure plan deals with the major planning issues for the area and sets out broad policies and proposals. Local plans elaborate on the broad policies and proposals in more detail relating them to precise areas of land and thus providing the detailed basis for both positive and regulatory planning. Structure plans require the approval of the Secretary of State, whereas local plans normally do not (Telling, 1990). Unitary development plans were also introduced, as described earlier, in certain specified areas.

Planning proposals will have direct relevance to deliberations at the brief stage as not only will they indicate the uses to which the land can be put and often permitted densities, but will also show the manner in which adjoining areas are to be developed and the location and extent of major public works, such as new roads and improvements of existing roads. A road widening may sterilise part of the site, whereas a new road may give improved means of access.

Planning permission is required for most forms of development and this is obtained by making application to the local planning authority in the prescribed manner. The local planning authority can give unconditional permission, permission subject to conditions or refuse permission altogether and there is a right of appeal from this decision to the minister. Applications can be in outline, giving brief particulars of the development in order to secure permission in principle possibly before the land is purchased, or detailed. The development has to commence within five years of the first approval. DoE estimated that there were 32 000 ha of land available with planning permission for housing in England in 1993 and that this could accommodate 805 600 dwellings.

Development is defined in the Town and Country Planning Act 1990 as 'the carrying out of building, engineering, mining or other operations in, on, over or under land, or the making of any material change in the use of any buildings or other land. The Town and Country Planning (Use Classes) Order 1987 prescribed a number of classes and where a change of use keeps within the same use class it does not constitute development and does not therefore require planning permission. A change from one use class to another, such as from an office to a shop, will require permission as the change is then material. The conversion of a single dwelling into two or more flats also constitutes development. Certain types of development are, however, expressly excluded from planning control and are deemed to be *permitted development*, and some typical examples follow:

(1) Enlargement, improvement or other alteration of a dwelling house within certain limits.
(2) Provision on land of temporary buildings and plant and the replacement of underground services.
(3) Access to certain highways.
(4) Exterior painting.
(5) Erection of gates, fences and boundary walls up to certain specified heights.

Planning proposals in the form of development plans, structure plans and local plans will normally indicate the uses to which the land may be put and the permitted maximum density to which it can be developed. In the case of residential development the unit of density will normally be the number of persons or habitable rooms permitted per hectare of site. With non-residential buildings the unit of density may be the floor space index (FSI) or plot ratio. The floor space index of a building equals

$$\frac{\text{total amount of floor space of the building (on all floors)}}{\text{total area of building site + half area of adjoining streets}}$$

whereas plot ratio is the ratio of floor area of the building to plot area, excluding any reference to adjoining roads. The floor space index has the additional merit of making allowance for the benefits of light and air accruing to a property from adjoining streets. Typical FSI guide figures for different classes of building are: shops – 1.5; offices – 2.0; wholesale warehouses – 2.25; and light industry – 1.5. The rigid application of low floor space indices or plot ratios on central area sites can result in development being uneconomical.

Readers requiring more detailed information on planning legislation and its application and implications are referred to Heap (1991) and Telling and Duxbury (1993).

Buildings of Special Interest

The Department of the Environment has compiled a list of buildings of special architectural or historic interest, and these buildings cannot be demolished, extended or altered so as to affect their character without obtaining a *listed building* consent from the local planning authority. The only exceptions are where the works are urgently needed for the safety or health of persons or the preservation of the building. Breach of these requirements or of the conditions prescribed in a consent constitutes a criminal offence. There is provision for *enforcement notices*, requiring the owner or occupier to restore the property to its former state, and for *purchase notices* whereby the owner requires the local planning authority to purchase his interest in the property because the listed building has in its present state become incapable of reasonably beneficial use. It will be appreciated that a listed building can constitute a major obstacle in a development scheme.

Tree Preservation Orders

Under the Town and Country Planning Act 1990, a local planning authority is empowered to make a tree preservation order for the preservation of a single tree, groups of trees or woodlands in the interests of amenity. Such an order prohibits the felling, lopping, uprooting, wilful damage or wilful destruction of specified trees, groups of trees or woodland without consent and may also require the replanting of woodlands felled in the course of forestry operations. Compensation may be payable in respect of loss or damage resulting from a refusal of consent or imposed conditions and orders cannot apply to trees which are dying, dead or have become dangerous. Once again, tree preservation orders can operate to the disadvantage of a developer in that they may restrict the form and extent of the development.

Conservation Areas

Under the Planning (Listed Buildings and Conservation Areas) Act 1990, it is the duty of the local planning authority to determine from time to time which parts of their area should be treated as conservation areas, on account of their special architectural or historic interest, character or appearance, which it is desirable to preserve or enhance. The local planning authority prepares a conservation area plan and, upon designation, special procedures operate for applications for planning permission, control of demolition of buildings and felling of trees, and possible stricter controls over outside advertising.

Other Statutory Requirements

Highways requirements. By virtue of the Highways Act 1980, many local authorities, particularly highway authorities, operate new street bye-laws which regulate the widths, levels and form of construction of new streets. A developer is normally required to submit plans and other details of any new streets for approval. Where the local authority believes that the new street is likely to become a main thoroughfare they may require the street to be of greater width than would normally be necessary, and if they require the street to be widened more than 6 m above the normal width, the developer must be reimbursed the extra cost. The developer is also entitled to compensation if he is required to amend his street layout to provide improved junctions with existing streets. The bye-laws prescribe minimum widths of carriageway according to the function and length of the street and whether it is open at one or both ends, and also the number and minimum widths of footpaths.

Street bye-law requirements generally conform to nationally recognised standards and the following are commonly adopted:

(1) Main estates road (spine road) in residential neighbourhood: minimum carriageway width of 7.00 m and two footpaths each with a minimum width of 3.00 m.

(2) Minor residential road: minimum carriageway width of 5.50 m and footpath to each developed frontage with a minimum width of 2.00 m.

(3) Access roads may have carriageway widths varying from 5.50 m down to 3.00 m for single lane roads with passing places.

The DoE Bulletin (1992) describes how a road and footpath network should be designed to relate to the traffic that it will have to carry and having full regard to traffic safety, access to dwellings, vehicle speeds, road junctions, visibility, footpaths, verges and parking areas. It explains how widths can be reduced, curves tightened and splays at junctions made narrower by routing non-access traffic away from dwellings. Traffic speeds can be reduced by introducing short culs de sac, carriageway offsets, chicanes and islands, small radius 90° bends, avoiding long straight stretches, road humps and in various other ways. Street parking must be reduced and wherever possible parking by residents should be kept within the curtilage of dwellings.

Highway authorities, being local authorities with highway responsibilities, often prescribe *improvement lines* alongside highways to reserve sufficient land for future road improvements. The developer is very restricted in what he is able to build between the improvement line and the street. In other cases the highway authority may prescribe a *building line* on either side of a street and it frequently consists of a line joining the fronts of existing buildings. With minor exceptions, no building work is permitted in advance of a building line. The practical effect of both improvement lines and building lines on development schemes at the design stage is to sterilise strips of land adjacent to highways. Where a developer wishes to close or divert a public right of way which is crossing his site and preventing the use of the land to best advantage, he may apply to the local authority for an order extinguishing or diverting the path. Such an order requires approval by the local planning authority and confirmation by the minister. Where a road is involved it will be necessary for the highway authority to apply to a magistrates' court or obtain an order from the minister. The ministry involved is the Department of the Environment.

Building requirements. Local authorities are responsible for ensuring that all building work is carried out to certain minimum standards of construction. Throughout England and Wales, the requirements are laid down in the operative Building Regulations. Even comparatively minor alterations and improvements require consent under Building Regulations, and the Regulations cover such aspects as fitness of materials; preparation of site and resistance to moisture; structural stability; means of escape and fire spread; conservation of fuel and power; sound insulation; stairs and ramps; access to buildings for the disabled; health and safety of people in or about the building; condensation; heat producing appliances; sanitary pipework; drainage; sanitary conveniences; and waste storage and disposal. The quantity surveyor needs to be familiar with Building Regulation requirements and their likely impact on alternative design proposals.

Plans of proposed building work have to be submitted to the appropriate local authority, as distinct from the local planning authority, for approval under Building Regulations. After approval the contractor is required to submit notices to the local authority at various stages of construction, in order that the appropriate officer to the authority (building inspector, building control officer or building surveyor) can inspect the work and ensure that it complies with the approved plans and the Building Regulations. Where contraventions occur the authority can take enforcement action.

An alternative procedure entails the submission of an initial notice to the local authority and private supervision of the works by an approved inspector.

Certain classes of building are subject to additional controls. For instance, the Factories Act 1961 prescribes certain constructional and operational requirements for factory buildings, which must affect the layout of a new factory and also its constructional and operational costs. The requirements include a minimum working area of just over 10 m^2 per employee; a normal minimum temperature of 16°C; minimum lighting and ventilation standards; minimum provision of sanitary appliances, hot and cold water supply, drinking water and storage accommodation for employees' clothes; adequate fire-fighting and warning systems and means of escape in case of fire; and suitable protective devices to lifts, hoists and openings. Other statutory provisions affecting factories include the Clean Air Acts, 1956 and 1968, as amended by the Local Government, Planning and Land Act 1980, which aim at reducing atmospheric pollution and are likely to necessitate taller factory chimneys with consequently higher costs.

The Health and Safety at Work Act 1974 created general duties on employers to ensure, so far as was reasonably practicable, the health, safety and welfare at work of employees and this can result in considerably increased constructional costs.

There are many special risks attached to work in the construction industry arising from falls from ladders, scaffolds, platforms and roofs, falling materials, use of lifting equipment or machinery, and in excavation and tunnelling work. To cater for these situations there are several sets of regulations appertaining to the construction industry, comprising Construction (General Provisions) Regulations 1961, Construction (Lifting Operations) Regulations 1961, Construction (Health and Welfare) Regulations 1966, and Construction (Working Places) Regulations 1966 (Galbraith 1991).

Other statutory provisions affecting building work include the Local Government Planning and Land Act 1980 and the Defective Premises Act 1972. The Control of Pollution Act 1974 is concerned with the deposit of waste on land, water pollution, clean air and noise abatement, now further strengthened by the Environmental Protection Act 1990.

CDM (CONDAM) regulations: new construction (design and management) regulations on safe practice, extending the responsibility for health and safety from contractors to designers and clients, came into force on 31 March 1995. The principal aim was to raise the standard of construction health and safety across the European Community by improving the co-ordination between the various parties involved at the preparation stage of a construction project, and also when work is carried out. The regulations propose the appointment by the client of one central figure for each phase of the project: a 'planning supervisor', who will assume the responsibilities of the project supervisor and who will prepare a health and safety plan, and a 'principal contractor', who will assume the responsibilities for the execution stage co-ordination.

The CDM regulations place specific duties on the design team to take into account the general principles of prevention when making decisions about design which could affect health and safety during building, maintenance or repairs. The result will inevitably impose additional costs on clients, designers and contractors. Reducing hazards by design will result in some well established design solutions having to be elaborated and possibly be made more costly or replaced by more expensive ones. Professional indemnity insurance premiums will also be affected. Indeed, this could be one of the most important events ever to affect the whole of the construction professional team.

New offices and shops have to comply with the Offices, Shops and Railway Premises Act 1963, which prescribes detailed requirements on lines similar to those contained in the Factories Act for industrial buildings. The minimum working space requirements are 3.72 m^2 per employee.

There are a number of statutes which prescribe minimum provision in relation to means of escape in case of fire and fire-fighting appliances, including the Fire Precautions Act 1971. In addition, proposals for hotels and restaurants are examined by the environmental health officer to the local authority under the Food Hygiene (General) Regulations 1970, and proposals for licensed premises need the approval of the licensing justices. Special precautions have to be taken in the construction of structures to be used for the storage of petrol and in garages to house more than twenty cars, by virtue of the Petroleum (Consolidation) Act 1928, whilst insurance companies usually prescribe requirements for the storage of large quantities of oil. The need to obtain a variety of consents from different authorities can result in a lengthening of the design period, greater probability of amendments to design and the possibility of increased costs.

ENCUMBRANCES AND EASEMENTS

On occasions land is subject to restrictions of one kind or another which adversely affect the development of the land or its enjoyment by owners and occupiers. Such restrictions can make development more costly and even sterilise parts of the site. Hence they need careful investigation in any feasibility study. The two main types of restriction are restrictive covenants and easements, but these should be considered against the background of natural rights enjoyed by occupiers of land.

Natural Rights

At common law an occupier of land has certain natural rights which impose obligations on neighbours, although some of these rights can be suspended by the granting of easements. Typical examples include the right of an occupier of land to have the support of the soil in its natural state from the land of the neighbour, although the neighbour is not made responsible for the support of a building on this land. There is also a natural right to water flowing in a defined course as a natural stream over the occupier's land, subject to the rights of other owners along its banks; these are termed *riparian owners*. In addition, water or air passing from land in one ownership to that in another should not be polluted, as an occupier of land has a natural right to pure water and air.

Restrictive Convenants

The nature and objectives of restrictive covenants vary widely; they may have been imposed on the land in previous conveyances or the freeholder of a large parcel of land may himself impose restrictions when disposing of smaller plots in order to secure satisfactory development of the whole area. Restrictive covenants impose conditions which govern the use of the land; sometimes they are so onerous as to restrict the use of the land considerably, whilst in other cases the effect is marginal.

Covenants can be of two kinds: positive and negative. A positive covenant imposes an obligation on the purchaser to do something, such as erect a building. Restrictive covenants are negative in character in that they restrict the purchaser in the manner in which he can use the land. For instance a restrictive covenant may specify that the density of residential development shall not exceed ten houses per hectare and also lay down a minimum floor area for new houses and prescribe conditions as to materials to be used or other matters affecting the external elevations, or the submission to and approval of house plans by the vendor. Restrictive covenants created since 1925 on a freehold must be registered as land charges if they are to be enforceable against all purchasers, by virtue of the Land Charges Act 1925. With building schemes where the purchaser of each plot enters into covenants with the vendor, each purchaser or his assignees can normally sue or be sued by every other purchaser for breach of covenants, provided there is a common vendor; the covenants apply to all plots and are for the benefit of all purchasers who have bought their plots on this understanding.

Where restrictive covenants are reasonable in their requirements and are framed to secure the orderly development of the area, they assist in maintaining values in that area, but where the character of the district has so changed that the restrictions are completely outdated and no longer relevant, then they are likely to retard normal development with consequent loss of value. It is accordingly essential to investigate any restrictive covenants attaching to land before its purchase is completed and to take account of the cost effect of their operation in any feasibility study.

In some areas much-needed development is thwarted by unreasonable and unacceptable restrictions imposed by restrictive covenants. For these reasons the Law of Property Act 1925, as subsequently amended by the Lands Tribunal Act 1949, the Landlord and Tenant Act 1954 and the Law of Property Act 1969, enables a person to apply to the Lands Tribunal for the modification or removal of a restriction if the changes in character of the property or the neighbourhood or other material circumstances make the restriction obsolete, or such as to impede the reasonable use of the land and do not secure any practical benefit to other persons or is contrary to the public interest, and that money will be adequate compensation for the modification or discharge. These Acts do not apply to restrictions imposed on land given gratuitously or for a nominal consideration for public purposes, nor to leaseholds of less than forty years or for recent ones of over forty years. A feasibility study must take account of any compensation payable for the modification or discharge of restrictive covenants, as it can be costly.

Section 106 Agreements

Section 106 agreements under the Town and Country Planning Act 1990 are another form of legal restriction, whereby anyone can enter into an agreement with the local planning authority which regulates the development or use of land. Normally it will be the owner or prospective purchaser of land who enters into such an agreement and typically it will require him to provide some infrastructure works, public open space or similar facility if he carries out a specified development. Sometimes a development may not be acceptable unless the developer carries out certain works, such as widening an existing public approach road to cope with the increased traffic to the development site, which could be residential, industrial, retail shopping or warehousing, business park or other use (Taylor, 1991).

Another practical example is the £230m Meadowhall out-of-town shopping and leisure centre on the outskirts of Sheffield, completed in 1990, where the developer as part of the planning permission, was required to provide extensive landscaping work, including the planting of 15 000 trees in the adjoining Don Valley linear park, which connected Meadowhall with Sheffield and formed an attractive leisure and amenity facility for the benefit of the public and the local authority. This aspect is sometimes referred to as planning gain and entails the local planning authority (LPA) requiring the developer to provide some much needed public works at his own expense in return for obtaining planning consent for a substantial development from which the developer expects to obtain a substantial return on his investment (Seeley, 1992).

A section 106 agreement runs with the land and anyone who subsequently purchases the land with the intention of carrying out the approved development will be subject to the agreement and it could represent a significant addition to the development costs.

Easements

The ownership of some properties may give the owner the right to enter another property (the *servient tenement*) and take something from it, other than water, such as sand, gravel, timber or even fish. This right is termed a *profit à prendre*. A more common right attaching to the ownership of a property is an *easement*, which the owner of the *dominant tenement* secures over another property (the servient tenement). The tenements must be in different ownerships, the right must be capable of being granted, the servient owner must not be involved in any expenditure in complying with the easement and it must not involve the removal of anything other than water from the servient tenement. A legal easement is made by a grant from the owner of the servient land or by prescription (long use of the privilege by the dominant owner under certain conditions), and is binding on all persons who occupy the servient tenement.

The more common easements relating to building development are as follows.

(1) Right of light: the right of light to a building becomes legally protected after it has been enjoyed for a period of twenty years. A right of light prevents the owner or occupier of land from erecting buildings on it which will obstruct the light passing on to the other person's land. Thus it may result in it only being possible to develop to a certain height or over a certain area.

(2) Right of support: the ownership of land carries with it the right to support from the adjoining land but not for the support of any buildings subsequently erected on it.

(3) Right of way: a private right of way may be presumed if it has been enjoyed for twenty years and becomes absolute if enjoyed for a period of forty years. Public rights of way are not easements.

(4) Right of drainage or pipe easements: the right to drain across the property of another. An easement which permits the dominant owner to lay pipes across the land of the servient owner, whether for purposes of drainage or water supply, may sterilise the strip of land through which it passes and provision will have to be made for access to and inspection of the pipes.

Land subject to easements is normally reduced in value as restrictions are placed upon the owner's full use and enjoyment of the site. As with encumbrances it is essential that their effects should be taken into account in assessing the value of the land and in any feasibility studies.

Ransom Strips

On occasions, a strip of land in another ownership is required before an adjoining site with substantial development potential can be developed. The third party owning the buffer strip of land can exercise a ransom over the site, the value of which will be subject to negotiation. The cost could amount to as much as one third of the development value of the particular site (Taylor, 1991).

MATTERS DETERMINING LAND USE AND VALUE

In concluding this chapter it might be helpful to the reader to summarise the main categories of factors which determine land use and, in consequence, land values.

(1) Location: primarily from standpoint of accessibility and compatibility with adjoining uses.
(2) Economics: with particular emphasis on supply and demand, and through this the satisfying of a need or want.
(3) Topography: with the main emphasis on the physical characteristics of sites acting as sorters of suitability for specific uses.
(4) Tenure: freehold or leasehold and if leasehold the term and conditions of the lease.
(5) Servitudes: existence or otherwise of easements such as rights of way, light and drainage, and/or restrictive covenants.
(6) Legislation: such as Town and Country Planning Acts and through them the operation of planning schemes, Building Regulations and other statutory controls.

LAND AND BUILDING VALUES

Residential Building Land Values

Table 14.1 gives average regional building land values prepared by the Valuation Office and categorised into three classes: small sites, bulk land and sites for flats or maisonettes, expressed in £/ha as at October 1993. It shows considerable variations between different regions and the various site categories. East Anglia, East Midlands and Wales have the lowest values and are significantly below the average figures for England and Wales (excluding London), while Scotland has below average values for small sites and bulk land but high values for sites for flats or maisonettes. As is to be expected both inner and outer London show much higher values in all land categories, with inner London at 3½ to 4 times average England and Wales values and outer London 2½ to 3 times. It is worth noting that average values for England and Wales (excluding London) are in the £450 000 to £555 000/ha range. In general the values of small sites and sites for flats or maisonettes are very similar but bulk land drops to about 80 per cent of the values of the other two classes. It is emphasised that all values throughout this section of the chapter are indicative as they are influenced by so many factors and are subject to so many variables.

Table 14.1 Residential building land values (October 1993) (All values are average regional values expressed in £/ha)

Region	*Small sites*	*Bulk land*	*Sites for flats or maisonettes*
East Anglia	426 000	359 000	519 000
East Midlands	456 000	373 000	438 000
Northern	496 000	438 000	440 000
North West	610 000	473 000	531 000
South East	702 000	626 000	839 000
South West	592 000	517 000	666 000
Wales	433 000	326 000	412 000
West Midlands	696 000	545 000	671 000
Yorkshire & Humberside	550 000	433 000	512 000
England & Wales (excluding London)	545 000	450 000	555 000
Inner London	2 290 000	1 590 000	1 900 000
Outer London	1 330 000	1 320 000	1 410 000
Scotland	530 000	370 000	776 000
Northern Ireland	153 000	120 000	395 000

Source: *Valuation Office – Property Market Report*, Autumn 1993.

Table 14.2 shows the changes in average housing plot prices as compared with average house prices in England and Wales over the period 1978 to 1991. The price of building plots rose progressively up to 1988 and thereafter started to fluctuate, with the largest rise between 1987 and 1988, and average house prices displayed a fairly similar pattern. Over the period under review land prices rose faster than house prices as suitable plots with planning permission became scarcer (land prices rose 7.7 times as compared with 4.5 times for houses). The ratio of plot prices as a percentage of house prices also fluctuated sig-

Table 14.2 Private sector housing land prices (at constant average density) and housing prices, 1978–91

Year	Weighted average price per plot (£)	Average price of houses (£)	Plot price as percentage of house price
1978	2 367	16 792	14.1
1979	3 395	21 455	15.8
1980	4 460	26 131	17.1
1981	4 600	27 910	16.5
1982	5 200	27 914	18.6
1983	5 900	30 943	19.1
1984	6 600	33 416	19.8
1985	8 400	36 295	23.1
1986	10 700	42 319	25.3
1987	14 086	49 435	28.5
1988	19 254	61 551	31.3
1989	18 940	72 256	26.2
1990	17 321	75 403	23.0
1991	18 163	75 119	24.0

Source: *DOE, Housing and Construction Statistics*.

nificantly, rising from 14.1 per cent in 1970 to 31.3 per cent in 1988 and then falling back to 24.0 per cent in 1991, because of the reduced building programme and the accumulation of large land banks at high prices in the boom years. In broad terms the plot accounts for about one quarter of the combined price of house and land.

Housebuilders that reduced their land banks in the recession were faced with price increases of 6 to 10 per cent in the first quarter of 1994 as they endeavoured to purchase building land to meet the increased demand for houses. The impact of land price increases coupled with constraints on the supply of housing land could pose a threat to overall recovery in this sector and, at worst, be the early signs of another inflationary cycle in the housing market. However by 1995 housing land prices began to fall again.

Office Rental Values

Table 14.3 gives regional 1993 office rental values produced by the Valuation Office, subdivided between three types of office ranging from good quality larger town centre offices to similar smaller offices and even smaller ones in converted houses of character usually just off the town centre. Those with air conditioning have been suitably indicated and are all in the highest rental groupings. In each category the highest and lowest rentals in the Valuation Office samples have been identified and listed together with their locations. Hence it will be seen that the offices in the larger or more fashionable towns and cities in each region command the highest rentals, whereas the lowest rentals occur in a very disparate array of towns. In general the offices with highest rentals in the type 2 or smaller office suites exceed those in the type 1 premises, showing their relative popularity, while those with the lowest rentals show only limited variations. The highest office rentals in inner London (Mayfair and St James) are over three times higher than those in some provincial cities such as Cambridge, Newcastle upon Tyne and Cardiff. Rather surprisingly the highest rentals in the type 3 offices are often not widely different from those in the type 1 premises and in some areas they even exceed them, highlighting the popularity of this type of accommodation for offices. The lowest rental values are often around one third of the highest and this applies across all three types of office accommodation.

By way of comparison Jones Lang Wootton (1994) compared September 1993 office rentals in 50 provincial centres and found the highest achievable rental in Birmingham at £226/m^2, closely followed by £215/m^2 at Richmond upon Thames and Uxbridge, with an average for the 50 centres of £154/m^2. The average for three Scottish centres was £179/m^2 with the highest in Edinburgh at £215/m^2. There is a reasonable correlation between these figures and those produced by the Valuation Office. Hillier Parker (1993) show average office yields rising from just under 5 per cent in 1978–85 to over 9 per cent in 1991–93.

Shop Rental Values

Table 14.4 shows October 1993 shop rental values produced by the Valuation Office separated into the three categories of prime position, good secondary position and modern out of town warehouse units, all on a regional basis expressed

Table 14.3 Office rental values (October 1993)

Region	*Type 1 (£/m²)*		*Type 2 (£/m²)*		*Type 3 (£/m²)*	
	Highest	*Lowest*	*Highest*	*Lowest*	*Highest*	*Lowest*
East Anglia	110 (Cambridge)	100 (Norwich)	120 (Peterborough)	88 (Ipswich)	120 (Cambridge)	85 (Ipswich)
East Midlands	124 (Nottingham)	65 (Lincoln)	130 (Nottingham)	63 (Leicester)	100 (Northampton)	55 (Mansfield)
Northern	105 (Newcastle upon Tyne)	50 (Gateshead)	118 (Durham)	55 (Gateshead)	100 (Durham)	45 (Barrow in Furness)
North West	185 (Manchester)	50 (Blackpool)	205 (Manchester)	50 (Blackpool)	165 (Manchester)	50 (Rochdale)
South East	*200 (Chelmsford)	65 (Eastbourne)	220 (Chelmsford)	68 (Eastbourne)	240 (Chelmsford)	60 (Milton Keynes)
South West	180 (Bristol)	75 (Truro)	160 (Bath)	70 (Weymouth)	170 (Bath)	55 (Weymouth)
Wales	110 (Cardiff)	42 (Merthyr Tydfil)	130 (Cardiff)	45 (Merthyr Tydfil)	115 (Cardiff)	53 (Swansea)
West Midlands	*240 (Birmingham)	53 (West Bromwich)	*240 (Birmingham)	59 (West Bromwich)	150 (Birmingham)	65 (West Bromwich)
Yorkshire & Humberside	180 (Leeds)	70 (Hull)	180 (Leeds)	63 (Grimsby)	140 (Leeds)	60 (Hull)
Inner London	*350 (Mayfair & St James)	140 (Lewisham)	*350 (Mayfair & St James)	140 (Lewisham)	*300 (Mayfair & St James)	110 (Lewisham)
Outer London	*230 (Uxbridge)	108 (Finchley)	*230 (Uxbridge)	130 (Ilford)	150 (Bromley)	85 (Finchley)
Scotland	210 (Edinburgh)	80 (Ayr)	210 (Edinburgh)	90 (Ayr)	180 (Edinburgh)	70 (Dundee)

Source: *Valuation Office, Property Market Report*, Autumn 1993.

Notes: The rental values are on full repairing terms for the three types of office accommodation

Type 1: Town centre location; self-contained suite over 1000 m² in office block erected in previous ten years; good standard of finish with a lift and good quality fittings to common parts; limited car parking available.

Type 2: As type 1, but suite size in range 150–400 m².

Type 3: Converted former house usually just off town centre; good quality conversion of Georgian/Victorian or similar house of character; best quality fittings throughout; self-contained suite with size range of 50–150 m²; central heating and limited car parking.

* denotes accommodation with air conditioning.

Table 14.4 Shop rental values (October 1993)

Region	Type 1 (£/m²) zoning pattern 1		Type 2 (£/m²) zoning pattern 1		Type 3 (£/m²) gross internal area	
	Highest	*Lowest*	*Highest*	*Lowest*	*Highest*	*Lowest*
East Anglia	1450 (Cambridge)	950 (Ipswich)	900 (Peterborough)	450 (Ipswich)	100 (Norwich)	70 (Ipswich)
East Midlands (few examples)	1000 (Northampton)	600 (Loughborough)	650 (Northampton)	400 (Loughborough)	90 (Nottingham)	47 (Loughborough)
Northern	1950 (Newcastle)	300 (Hexham)	950 (Newcastle)	350 (Kendal)	85 (Newcastle)	70 (Carlisle)
North West	1825 (Stockport)	485 (Crewe)	1000 (Manchester)	260 (Blackburn)	90 (Manchester)	55 (Blackpool)
South East	1650 (Milton Keynes)	700 (Chatham)	1200 (Milton Keynes)	250 (Luton)	135 (Guildford)	70 (Portsmouth)
South West	1600 (Bath)	450 (Weymouth)	875 (Bath)	225 (Weymouth)	100 (Bath)	65 (Weymouth)
Wales	1400 (Cardiff)	460 (Merthyr Tydfil)	550 (Cardiff)	285 (Merthyr Tydfil)	100 (Cardiff)	58 (Swansea)
West Midlands	1300 (Birmingham)	575 (West Bromwich)	900 (Birmingham)	350 (West Bromwich)	105 (Coventry)	53 (Hanley)
Yorkshire & Humberside	1450 (Leeds)	650 (Huddersfield)	625 (York)	375 (Bradford)	80 (Leeds)	50 (Grimsby)
Inner London (zoning pattern 2 as so few zone 1)	4000 (Kensington)	600 (Islington)	2100 (Oxford Street)	225 (Peckham)	80 (Peckham)	75 (Tower Hamlets)
Outer London	2900 (Brent Cross)	1350 (Harrow)	1500 (Brent Cross)	500 (Croydon)	120 (Croydon)	75 (Ilford)
Scotland	2075 (Edinburgh)	450 (Hamilton)	1285 (Glasgow)	220 (Dumfries)	102 (Glasgow)	62 (Dumfries)

Source: *Valuation Office, Property Market Report*, Autumn 1993.

Notes: The table indicates Zone A rental values on full repairing terms for three types of shop premises
Type 1: Prime position in principal shopping centre.
Type 2: Good secondary off peak position in principal shopping centre.
Type 3: Modern, purpose built, non food, out of town warehouse unit, *circa* 2500–5000 m^2. Edge of town location with car parking.
Zoning pattern 1: 6.10 m Zone A, 6.10 m Zone B.
Zoning pattern 2: 4.57 m Zone A, 7.26 m Zone B.
In Oxford Street, Glasgow and Edinburgh zones of 9.14 m are the norm, and the figures have been adjusted to the prescribed patterns for comparison purposes.

in £/m^2. The highest and lowest rental values in each of the regional samples have been listed and their locations inserted. The highest type 1 and type 2 rental values occur in inner London (Kensington and Oxford Street) followed by outer London (Brent Cross) and Scotland (Edinburgh and Glasgow). The regional type 1 highest rentals show a range of £1000 to £1950/m^2, while the lowest show a very wide range from £300 to £950/m^2. In like manner the type 2 rentals also show a substantial price range and the highest can be 2 to 5 or more times the lowest. While the highest and lowest type 2 rentals are always significantly lower than those for similar categories in type 1, the highest type 2 rentals often exceed the lowest type 1, and occur in the larger and more fashionable towns and cities, highlighting the importance of location as a value determinant. The out of town premises show much reduced values with their large floor areas built on much cheaper land with a total range of £47/m^2 (Loughborough) to £120/m^2 (Croydon).

By way of comparison Jones Lang Wootton (1994) in their analysis of achievable zone A shop rentals in England and Wales (excluding London) in September 1993, identified the highest rentals of £1884/m^2 in Chester and Newcastle upon Tyne. The locations with shop rentals in excess of £1600/m^2 encompassed Birmingham, Cardiff, Croydon, Kingston upon Thames, Liverpool, Manchester, Nottingham, Reading and Southampton, comprising the larger cities and the outer London ring, while the lowest was Middlesbrough at £700/m^2. Hillier Parker (1993) showed average yields on shops of under 5 per cent from 1979–85 rising to 7.5 per cent in 1992–93.

Industrial Buildings and Land Values

Industrial Buildings: Capital and Rental Values

Table 14.5 lists the regional ranges of 1993 capital and rental values from the samples selected by the Valuation Office each subdivided into four progressively larger categories of buildings. The lowest capital values are generally around 30 to 60 per cent of the highest values, whereas the same ratio for rental values is around 40 to 70 per cent. In general both capital and rental values decrease as the size of the building increases, reflecting economies of scale. Both capital and rental values are highest in outer London and the South East. The lowest capital values are in Scotland, followed by Wales, Yorkshire and Humberside, and Northern, while with rental values Northern is the lowest.

Jones Lang Wootton (1994) in their survey of 50 provincial centres in England and Wales identified Hounslow with the highest achievable industrial rent in September 1993 at £86/m^2, followed by Guildford, Hammersmith and Richmond upon Thames at £80/m^2 and the lowest group with £43/m^2 at Ipswich, Maidstone, Manchester, Newcastle upon Tyne, Nottingham and Plymouth, which shows no discernable pattern. The average rental for the 50 centres is £60/m^2 compared with the much lower figure of £45/m^2 for Scotland. In general the figures show a relatively close affinity with those produced by the Valuation Office. Hillier Parker (1993) found that average yields of industrial buildings showed a series of troughs and peaks, starting with a high of 10.5 per cent in 1974, dropping to just under 7 per cent in 1980–82, rising again to 10.5 per cent in 1986–88, falling to 8.75 per cent in 1989–90, rising to 11 per cent in 1991 and tailing off to 10.3 per cent in 1993, highlighting the difficulty of assessing future yields or returns on this type of building.

Industrial and Warehouse Land Values

Table 14.6 shows the regional average 1993 industrial land values expressed in £/ha produced by the Valuation Office. It shows how industrial land values were much higher in both inner and outer London than elsewhere in the country, followed by the South East. These are the areas where most industrial development was taking place up to the 1990–93 recession and the high demand coupled with a shortage of good potential sites pushed up the land prices. The average industrial land value in England and Wales (excluding London) was £345 000 compared with

Table 14.5 Industrial buildings: capital and rental values (October 1993)

Region	Capital values: regional ranges (£/m²)				Rental values: regional ranges (£/m²)			
	Type 1 25–75 m²	Type 2 150–200 m²	Type 3 circa 500 m²	Type 4 circa 1000 m²	Type 1 25–75 m²	Type 2 150–200 m²	Type 3 circa 500 m²	Type 4 circa 1000 m²
East Anglia	700/400	600/350	425/325	386/275	78/40	68/38	50/32	45/30
East Midlands	550/360	480/280	450/240	425/200	60/40	48/32	45/30	43/25
Northern	565/360	450/250	350/180	330/150	56/40	46/37	45/20	33/17
North West	630/290	500/255	430/180	360/165	70/40	55/35	48/25	40/23
South East	800/445	920/370	800/295	800/235	95/54	95/48	80/35	70/32
South West	700/375	600/315	500/200	400/155	70/43	62/38	56/31	46/26
Wales	480/250	440/215	360/165	320/120	60/32	55/28	45/24	40/22
West Midlands	800/360	550/325	500/250	500/200	80/42	55/38	48/30	54/28
Yorkshire & Humberside	500/250	400/200	450/130	400/100	60/32	56/28	45/24	40/22
Inner London	725/525	680/475	600/400	600/325	90/60	85/55	75/45	68/42
Outer London	900/500	900/470	950/400	900/350	100/70	95/55	95/55	90/45
Scotland	356/225	400/220	384/175	368/100	60/33	52/34	48/28	46/25

Source: *Valuation Office, Property Market Report*, Autumn 1993.

Notes: The four types are assumed to be on industrial estates, and let on FR1 terms. They are of modern construction but not of High-Tech design, and are heated by free standing heaters.

Type 1: Small starter units, 25–75 m²; steel framed, concrete block or brick construction; often built in terrace layout and let on weekly terms.

Type 2: Nursery units, 150–200 m², steel framed on concrete base, concrete block or brickwork to 2 m, with metal PVC cladding above. Eaves height 3.75–4.5 m with lined roof. Limited or no office content and common parking and loading areas.

Type 3: Industrial warehouse units, *circa* 500 m², steel frame on concrete base, concrete block or brickwork to 2 m with metal PVC covered cladding above. Eaves height 4.3–5.5 m with lined roof; 10–15 per cent office content. Detached on own site with private parking and loading facilities.

Type 4: Industrial/warehouse units, *circa* 1000 m², steel frame on concrete base, concrete block or brickwork to 2 m with metal PVC covered cladding above. Eaves height up to 7.6 m with lined roof; 10–15 per cent office content. Detached on own site with private parking and loading facilities.

Table 14.6 Industrial and warehouse land values (October 1993)

Region	Regional average values (£/ha)
East Anglia	379 000
East Midlands	294 000
Northern	123 000
North West	229 000
South East	609 000
South West	386 000
Wales	166 000
West Midlands	361 000
Yorkshire & Humberside	238 000
England & Wales (average excluding London)	345 000
Inner London	1 070 000
Outer London	1 160 000
Scotland	160 000

Source: *Valuation Office, Property Market Report*, Autumn 1993.

£160 000/ha in Scotland and around £450 000 to £550 000/ha for average residential land in England and Wales. Depressed Northern England showed the lowest industrial land prices at £123 000/ha.

Agricultural Land and Property Values

Table 14.7 shows 1993 regional agricultural land values, split between land with vacant possession and that subject to tenancy produced by the Valuation Office. As illustrated in the average England and Wales figures, dairy farming land is the most highly priced, followed by mixed, arable and hill farming. Dairy farm prices have been mainly used in analysing the average regional figures as this type of farming tends to predo-

Table 14.7 Agricultural land and property values (October 1993)

Region	*Type of farm*	*Values of agricultural land with vacant possession (£/ha)*	*Values of agricultural land subject to tenancy (£/ha)*
East Anglia	arable	4231	1556
East Midlands	dairy	6175	1975
Northern	dairy	6795	2470
North West	dairy	6773	2944
South East	mixed	4918	2020
South West	dairy	5990	2200
Wales	dairy	5805	2675
West Midlands	dairy	8520	2185
Yorkshire & Humberside	dairy	6165	1935
England & Wales	arable	4660	1710
	dairy	6600	2340
	mixed	5035	1950
	hill	2535	860
Scotland	dairy	4140	1855

Source: *Valuation Office, Property Market Report*, Autumn 1993.

Note: All values are average values for the region or country.

minate in many parts of the country. However, it was necessary to change to arable in East Anglia and mixed in the South East because of the absence of dairy farming figures in these regions. In all cases the value of the land includes the farm. Even with the same type of farming there was considerable variation in average prices in the different regions, ranging from £4140/ha in Scotland to £8520/ha in the West Midlands for dairy farming land. The values of agricultural land subject to tenancy were much less than land with vacant possession, and showed significant variations from 26 per cent in the West Midlands to 46 per cent in Wales and Scotland. It will be apparent that the price of even the most highly priced agricultural land at £8520/ha is minimal compared with its average value for residential use at between £450 000 and £550 000/ha and this illustrates the much enhanced value of land accruing from the issue of planning consent for greenfield sites for residential development.

It should be emphasised once again that all values or prices in this section of the chapter can only be indicative as the actual values of specific building sites, and capital values or rental values of buildings will vary widely according to their location, physical conditions and market forces, including the demand at a particular point in time. They are however useful in giving a general idea of their values and their relationship to one another. In practice the services of a valuer will be needed to provide a realistic valuation, possibly using some of the principles outlined in chapter 12.

15 ECONOMICS OF BUILDING DEVELOPMENT

This chapter is concerned with the characteristics of property and the basic criteria for development undertaken in both the public and private sectors, problems of land acquisition, financial considerations and sources of finance. An investigation is made of matters contained in a developer's budget and its practical application to various types of development projects.

NATURE OF PROPERTY

It seems desirable to start this chapter with a brief account of the main characteristics of property and its use as an investment medium, as these aspects are central to the building development process. Property serves several purposes, for instance it can be a factor of production, a corporate asset or an investment.

Property Characteristics

- Property market is fragmented and dispersed while that of shares (equities) and other securities such as government stocks (gilts) are highly centralised.
- Real property has the distinguishing features of being heterogeneous, indivisible and has associated management problems, which may include collecting rent, dealing with repairs, lease renewals and other related matters.
- Property is more difficult to value as there is no centralised market price and it may be difficult to ascertain a realistic value unless a comparable transaction has taken place recently.
- Its main advantages are the relative durability of property, and it normally has an element of protection against inflation with greater security to the investor.
- Property companies carry a high level of debt capital relative to equity or total capital (Isaac, 1994).

Principal Concerns about Property as an Investment Medium

- Illiquidity: normally takes three months or more to sell and buy.
- Inflexibility: necessary to purchase the whole.
- Growth of debt finance and increase in bank lending to property sector accompanied by more equity investment by institutions (although this was changing in 1994). Isaac (1994) felt this could pose a threat to the property market.
- Valuation methodology and precision are weakened by inflexible conventional methods, which entail substantial risk and are difficult to apply to the more illiquid properties, such as major shopping facilities and substantial office buildings; the problems with Queens Moat Houses valuation and other prominent disputes confirm this.
- Future of the property profession: the liberalisation of financial markets and increased importance of debt in property funding require new competencies from chartered surveyors, and the demands of the Financial Services Act for those requiring information on finance and funding will require different and greater competence in financial expertise.
- Too much emphasis is placed on the short term approach to investment and performance which is inappropriate to property financing where

long term strategy and returns are the key to successful projects (Isaac, 1994).

The RICS (1994) identified ways in which the boom and bust cycles of the commercial property market could be smoothed out. The RICS report should assist in creating a better understanding of property cycles and in developing forecasting tools that could add stability to the markets.

THE ESSENCE OF DEVELOPMENT

Every development whether it be for a public authority, industrialist or private investor, has a 'market value' – a potential worth or earning power. Even civic buildings, hospitals, churches and universities have an assessable value to the community – a cost above which it is not reasonable or feasible to build. Within certain limits of aesthetics, function and performance, the most economic development is that which shows the greatest return to the community for the minimum capital invested. This does not imply that the cheapest is the best; often the opposite is the case.

The art of phasing development to give an early return which can be used to pay for the less remunerative items is one of the objectives of a skilful developer. In this connection, an amalgamation of public and private agencies in development can be of great benefit to the community. It is also essential to integrate the planning of large scale redevelopment. For instance, the retention of an existing road layout even if only to provide a pedestrian precinct or parking facilities can economise in new construction, and the potential earning power of existing facilities should not be overlooked. Time is also important in the planning process; for maximum economy the time between capital expenditure and completion of a project should be kept to a minimum.

The private developer or industrialist will require a financial appraisal or feasibility study to determine the likely capital expenditure and probable revenue in order to arrive at the anticipated return on the money invested. Whether the project is to be financed by public or private funds, it is important to know the cost implications at the outset in order to be able to appraise the viability of the scheme. It is necessary for the developer to know the nature and extent of the proposed development, its cost and the time required to complete it. Indeed, the whole development process is becoming more sophisticated. Schemes need to be appraised from every aspect – aesthetic, fiscal and social. The long and frequently frustrating negotiations to assemble sites, obtain planning permission, barter with local authorities and secure finance, demand a truly professional approach.

Hence a developer usually wishes to know whether the investment of capital in a project will be justified by the return which he can expect to receive. It is therefore necessary to assess as accurately as possible the value of all the expected returns and benefits and to compare them with the estimated costs. One problem is to express all the benefits in monetary terms as there may well be some indirect and intangible benefits, such as more contented employees or greater prestige value, which flow from a project. The quantity surveyor is frequently called upon to make cost comparisons of different design proposals with varying capital, maintenance and running costs. The task of the quantity surveyor is to inform the developer which is the most economical scheme after taking all these costs into account.

To this end general surveyors should have a general knowledge of building costs in addition to a detailed knowledge of values, while quantity surveyors ought to have a general knowledge of values as well as a detailed knowledge of building costs. There is an evident need for the quantity surveyor to be aware that the general practice surveyor, in making his general financial appraisal, has to consider the capital cost of the works, land purchase, compensation for extinguishment of leases, bridging finance, long-term finance, rental and capital values profitability, and maintenance and other outgoings. The key factor in any successful project is generally believed to be the triangle of valuation surveyor, quantity surveyor and architect, although other professionals may also be involved. The general practice surveyor will be primarily concerned with the broad economics of development, the architect with the design of the project and the quantity surveyor with the interaction of these two aspects of development. Some of the more important matters to be resolved by the development team include:

(1) ensuring development to maximum permissible plot ratio;
(2) planning the most economical and profitable use of available floor space;
(3) implications and suitability of different methods of procuring the building;
(4) the speed of construction balanced against financial considerations, such as the cost of bridging finance and loss of rent or interest;
(5) the effect of incurring extra capital costs including expensive finishings, balanced against additional net rental value (if any); and
(6) the effect of incurring extra capital costs balanced against a consequent reduction in future maintenance and operating costs, including depreciation allowances.

DEVELOPMENT PROPERTIES

The term *development properties* refers to properties in which a developer invests capital to secure a greater and more profitable return than that previously received. Typical examples are the development of agricultural or accommodation land for residential purposes and the redevelopment of central area sites which are underdeveloped or occupied by obsolescent buildings. One method of assessing the value of sites available for development is the *residual method*, as illustrated in the following example.

Value of site when developed for the most profitable permitted use (gross development value)	£1 200 000
less	
Estimated cost of development and allowance for risk and developer's profit	£860 000
Estimated value of existing property	£340 000

Central area premises will command much higher development values than premises on the outskirts of a town, as they will produce much higher rents. The most profitable permitted use will depend upon the policy of the local planning authority and the nature of the planning proposals for the area. The advice of a valuer is vital to secure this objective. The location will also influence the decision as to the type of design and standard of finish, as a central location will justify a higher standard of design and finish than a suburban site.

BUDGETING FOR PUBLIC AND PRIVATE DEVELOPMENT

Public Finance

Central government funds are obtained from two main sources: internal and external. Internal sources are mainly confined to nationalised industries where internal cash flows provide a source of finance for the industry's investment, external sources of finance are taxes and borrowing.

In the case of local authorities, money may be obtained by grant from central government funds or may be raised by the local authority's own borrowing or local taxation. In general, local authorities meet current expenses out of local taxation and this may be augmented by the profits of successful local authority enterprises. Local authorities borrow money for capital investment by issuing securities in the ordinary market and also borrow from the Public Works Loan Board (a government agency), in addition to receiving money from central government funds and accepting money on deposit at interest. Housing associations receive grants from central government through the Housing Corporation and loans from clearing banks, building societies, merchant banks and insurance companies.

Developer's Return

The majority of building clients are seeking a financial return on the capital they invest in building projects. Nevertheless, there are some projects which are of a semi-commercial nature like local authority housing, and others which are non-commercial such as schools and churches. With the latter category of project the economic return approach will be difficult to apply and local authority housing falls between the other two classes of project with its complex political and social implications. From the quantity surveyor's point of view, it is important that the developer, in whatever category he may fall, shall prescribe the upper limit of his budget at the outset, in order that the quantity surveyor may formulate a cost plan with the object of securing value for the money expended and a realistic and advantageous distribution of costs throughout the various parts of the building. The general procedure and problems associated with budgeting for each category of development will now be considered.

Budgeting for Commercial Properties

The major financial houses have come to recognise that property, particularly commercial and industrial property, has a good investment value and generally compares favourably with other investment opportunities. Sources of finance for commercial development projects are examined later in the chapter. Investors often require, as part of the consideration for lending, a share in the profits. This has resulted in the investor showing a very real concern in the situation, design and profitability of the building. Projects must be well-conceived and efficiently executed; built to designs which are attractive, satisfy the local planning authorities and meet the needs of occupants, as well as being profitable. In large-scale developments, land assembly may have taken years involving heavy expenditure with no immediate return and, in these circumstances, the developer's judgement can no longer be merely intuitive. It must be backed by expert opinion to ensure that the investment is a secure one and will produce a return which compares well with other available investments. Assessment is not just a matter of building and land costs matched with a rent income. It concerns people and their working and shopping habits, communications and trading trends and forecasts of growth.

Modern good quality office and shop developments have, despite the problems in the 1990–93 recession, generally been in reasonably good demand for investment purposes, particularly from insurance companies and pension funds. Industrial and warehouse property has become more acceptable, although requiring a higher return, although there was substantial over-provision in the early 1990s. The one noticeable omission in the investment field is that of rented housing. Rent control, doubtless introduced with the best of motives and retained for short-term political expediency, has forced landlords to subsidise their tenants.

Financial institutions like to maintain a balance in their property portfolio between shop, office and industrial premises, including business parks and retail warehousing, but whatever type of property is considered, certain factors have to be taken into account. These factors include the return on capital invested both now and in the future, the financial security afforded by the tenant, the quality of the location and its future prospects, terms of lease and liability for insurance and repairs. Since the nineteen seventies the property investment and finance market became much more selective as to the quality of property and the need to obtain a balanced distribution as between different types of property.

Offices

A common type of commercial development project is a block of offices where the developer needs to be assured of a reasonable excess of income over expenditure. He will be laying out a substantial capital sum now in anticipation of a larger return later. The capital outlay will be for site and buildings and the return is often in the form of rents. It would be quite absurd to spend a large sum of money on a site for a proposed office block unless and until an expert financial appraisal has been carried out indicating the overall supply/demand situation and rent levels for offices in the area. This appraisal would include consideration of many related aspects, such as growth of commerce in the area, demand for all types of office space, site characteristics, transport facilities, public utility services and associated services, such as banks and financial institutions.

The rent level is influenced considerably by the quality of the accommodation, so the standard of quality must be determined at an early stage. The amount of space to be provided will depend on a number of factors, including estimated demand, site area, planning restrictions and rights of adjoining owners. The site value can be determined by the residual method, whereby building costs and developer's profit are deducted from the gross development value, and this approach will be examined in some detail later in the chapter. The quantity surveyor will prepare estimates of building costs based on the architect's preliminary designs and the standard of building needed to attract the rents set by the valuer. It is vital that the quantity surveyor's estimate is realistic, otherwise the developer may have difficulty in obtaining his desired profit margin.

In 1993/94 it could cost up to £6700 per annum to provide modern air conditioned office space for an office worker in the Strand in London, based on an average of 11.5 m^2 per person, rent and rates of around £520/m^2 and service charges of up to £80/m^2, as compared with £3160 in Grays Inn Road, based on rent and rates of about £225/m^2 and service charges of around £50/m^2. Prime office annual rents and rates varied considerably throughout the United Kingdom in 1993/94, and there was no logical perceived pattern, which was also distorted by differing regional variations over relatively short periods.

From the peak of the office letting period in 1988/89, rental values for even the best town centre locations generally dropped between 30 to 50 per cent by September 1993, and even further in central London. In addition, the extensive choice available to the relatively few prospective tenants in the market led to long rent-free periods, capital incentives and tenants' break options, thereby distorting the traditional office investment market significantly.

A Joseph Rowntree Foundation study in 1993 claimed that unwanted offices could be used to provide 60 000 flats for homeless people, viewed against a background of three million square metres of empty office space in London alone. However, not all offices are suitable for conversion and it would be a costly process.

On the office investment side there was a significant upturn in the nine months after departure from the Exchange Rate Mechanism (ERM) in September 1992, when £2b was invested in central London compared with £743 in the preceding nine months. Overseas investors, mainly from Germany, the Middle East and the Far East, exerted a strong influence and accounted for just over one half of the money invested. As Britain emerges from the recession, it is likely that the London property market could become increasingly attractive on account of the UK's unique long lease structure and the upward only rent reviews which are not available to investors elsewhere (Estates Times Review: Offices, 1993a).

In 1993/94 the majority of investors were placing increasing emphasis on initial yields, provided the property met the criteria of tenant's covenant strength and unexpired lease term. This shift had resulted from the twin effects of a scarcity of acceptable investment stock and increasing demand across the market, notably from the institutions (pension funds and insurance companies) who committed £7b to £9b to investment in commercial property in 1993.

Much publicity has been given to the problems encountered in the office developments in London Docklands. The total floorspace occupied in 1993 in London Docklands, excluding Canary Wharf, amounted to over 111 600 m^2 and there was some 148 800 m^2 of new offices available to let. Canary Wharf had 158 100 m^2 occupied by eleven major organisations, leaving 251 000 m^2 still available. In 1993/94 Docklands continued to attract service industry occupiers, banking and media companies and the education sector. The main likely factors influencing companies moving to Docklands included:

- The creation of critical mass within the Docklands area.
- The upgrading of transport facilities and road links to Docklands, and particularly the opening of the Limehouse Link.
- The announcement on the Jubilee Line extension and the improved perception of Docklands

following the increased take-up of accommodation throughout 1993.

- Docklands was one of the most competitive locations in terms of price in the UK. Overall costs which were set to fall with the next rating revaluation could be less than £160/m^2 overall over the first five years of a tenant's term. The flexible lease packages offered to tenants were also attractive (Estates Times Review: Offices, 1993a).

It would be helpful to the reader at this juncture to examine the position in 1993 with office rentals in a few selected provincial centres.

Milton Keynes. Office rents were likely to stay below £150/m^2, with little or no growth showing at review on offices let from 1988–91.

Birmingham. At Edgbaston 46 500 m^2 were available with rental levels dropping as low as £54/m^2, although a large high specification building of 5580 m^2 had an asking rent of £150/m^2.

Bristol. Rents ranged from £155 to £215/m^2 depending on location with new offices in the northern fringe showing a 15–20 per cent price advantage.

Glasgow. There were some 840 000 m^2 of empty office space available in the city centre, much of which was in older buildings, and the last stages of the 1988/89 property boom were still flowing out of the development pipeline into a very harsh environment. In 1992 ten schemes were completed in Glasgow totalling 560 000 m^2. Rental values varied between £86 and £215/m^2 depending on location, accommodation available and its quality.

Shops

The complexities surrounding schemes of large scale commercial redevelopment may not be generally appreciated. Developers must ascertain the requirements of the large space users and in many cases a large measure of pre-letting is necessary so that the development is planned around them. Furthermore, the scheme is likely to take several years to complete. Even after securing anchor covenants, because of ever-rising costs and retailer's lack of capital for expensive shopfitting, it takes time and knowhow to let all the remaining shops, with a certain loss of interest on capital in the meantime. More existing shopping streets will be pedestrianised and most modern shopping provision is in fully covered air-conditioned shopping centres, involving developers in management and promotion activities.

Apart from yield rate, inflation has caused developers and their advisers to consider future growth and the need for periodic rent reviews. Unlike offices, shop tenants of good standing have to invest capital on shop fitting and resist earlier rent reviews than seven years, as it takes time for a new branch to achieve full earning capacity. The large space users whose covenants are sought after often hold out for a fourteen-year review period, particularly if their existing branch already forms part of the development site and rehousing is necessary. This sometimes causes difficulties in obtaining finance, and there is merit in adopting a basic minimum rent plus a percentage rent based on turnover to eliminate rent reviews, but it is not likely to be favoured in this country by shop tenants. Where developers collaborate with local authorities in partnership arrangements for the redevelopment of central areas, it is essential that the developers should receive an adequate share of the return, commensurate with the work, expertise and risks involved.

A major shopping development could involve building costs of £18 to £20 m in 1994 to which soft costs of about 30 per cent give a total development cost of around £26 m. The developer is likely to be seeking an income of £2.2 m to £2.4 m per annum, based on rents of about £800 /m^2 for large stores and up to £1400/m^2 for unit shops. Developers and financial institutions are showing a strong preference for freeholds and this is not surprising when the land cost components of the total value of a property can amount to 70 per cent or more of the whole in prime High Street shopping or London offices.

There was a revival of interest in shopping

centre investment in 1993 as institutional investors re-entered the market, taking advantage of what many property specialists considered to be an under-valued sector. The purchasers were mainly property companies or the more entrepreneurial funds, investing some £900 m countrywide. They were able to take advantage of the high yield obtainable, averaging 10 per cent.

One of the more interesting developments in the market was the emergence of joint venture partnerships, where shopping experts, often with an equity stake, managed the scheme for one or more investors. A good example is the Victoria Centre, Nottingham, where Dusco's partners included ICI and PosTel. Retailers, like Boots, also stepped into the market to capitalise on their undoubted expertise.

By 1993/94 a number of the larger institutions were seeking to acquire retail centres, particularly those dominating a town centre with a capital value in excess of £30 m. Ideally the property should be freehold with a large proportion of the income well secured to multiple tenants with long leases, preferably with more than one anchor store. Centres where existing rents equated to between £550 to £1100/m^2 were preferred as rental growth prospects were all important.

There was also renewed interest in refurbishment as an effective way of bringing modern retail space on to the market. The UK has a large stock of ageing retail floorspace in need of significant capital investment to bring it up to modern standards. In the United States the average time from the opening of a shopping mall to its first refurbishment is seven years, while in the UK it is 15 years.

In 1993/94 there was a noticeable shift of power from landlords to tenants and it became necessary for the management of shopping centres and associated service costs to be more efficient and cost-effective. Shorter leases, tenants' options to break, and empty units where the landlords found themselves paying the service charge, all helped to focus attention on value for money. Cash flow difficulties also highlighted the need for accurate forecasting and efficient budget control.

A 1993 study of the rental schedule for a shopping centre in a major southern county town showed that the larger multiples were paying 50 to 70 per cent of the rent demanded of smaller, local and independent operators for similar space on a £/m^2 basis. However, until the demand from high street retailers outweighs the supply of space, developers will have to continue to provide a package of incentives to attract retailers, such as rent-free periods, shop fit contributions and capital contributions. It is also interesting to note that safety and security have become important issues in large shopping centres and, in the Victoria Centre in Nottingham, there were 19 officers and five data controllers operating round the clock security by CCTV and computerised building management systems.

On out-of-town shopping centres, Meadowhall, near Sheffield, comprising 102 300 m^2 of floorspace, 258 shopping units and six anchor retailers, scored highly on a national survey in 1993, showing a high standard of retail mix, pedestrian flow, cleaning, information feedback, public facilities, centre management and marketing.

Factories

Much of the present factory and warehouse provision is outdated and ill-suited for today's needs, whilst some of the newly erected factories are sited in unsatisfactory situations, sometimes as a result of government policies and local authority action. Industrial buildings let at lower rents than other property and gain higher yields (10–11 per cent in 1993) because of the greater risk to investors.

The Government provides a variety of incentives to industrialists who move to one of the areas for expansion in Great Britain, and those who expand or modernise an existing operation within one of these areas. These are available under the Industry Act 1972 and the Employment Act 1972. There is also provision for tax allowances on new industrial buildings and structures (first year allowance of 54 per cent and writing down allowance of 4 per cent) and for plant and machinery (first year allowance of 100 per cent).

Incentives in Urban Development Areas, for the regeneration of major areas of social and industrial decay, include regional development grants for the provision of new buildings and adapting existing buildings for new plant and machinery; and in special circumstances, selective assistance normally in the form of a grant. In some cases government factories will be available.

The Government also designated a number of Enterprise Zones in the United Kingdom, aimed at increasing economic activity by the easing of certain monetary, planning and administrative constraints. These special incentives will be additionai to whatever other benefits are available under other policies.

All industrialists are looking for a number of common essential features – a sufficient area of land for present and likely future needs, which is reasonably level and has all essential services readily available and of sufficient capacity, water supply of adequate quantity and suitable quality, good communications and access, relatively low level of local taxation and ample pool of local labour. Some industrial processes have special requirements such as clean air or specific climate conditions, and all welcome the provision of suitable living accommodation for operatives, as was supplied in new and expanding towns (Seeley, 1992).

Above all, the industrialist should be anxious to establish his plant at the 'least cost' location. He will usually have regard to market areas, ease and cost of transportation, labour position, probable operating costs and even the effects of rivals. However, it has been found in practice that it is comparatively rare for a scientific study of alternative sites to be made before production is commenced and, even when a study has been made, more than one optimum site might be indicated and the final choice may be determined by some quite trivial consideration.

The financial implications of an industrialist moving from old unsuitable premises to a new factory on an industrial estate can be conveniently considered under three heads:

(1) the provision of a new factory, complete with equipment and fittings;
(2) removal expenses and dislocation of trade;
(3) settling in period, training new staff and related matters (Seeley, 1974).

The majority of modern industrial estates are reasonably well laid out and suitably landscaped with adequate car parking space for employees and visitors. A principal failing is that many of the buildings on a typical industrial estate have been designed in isolation by different architects and there is a lack of cohesion and harmony indicating the need for co-ordination of design. Very small factories of the 'unit' or 'nest' variety, with floor areas of aound 185 to 230 m^2 have often been erected on a speculative basis, and have proved popular with new small business ventures (Seeley, 1992).

Retail Warehousing

Retail warehouses need to be located on suitable sites on main roads, in strategic locations with a suitable catchment area, and be easily accessible to the car-borne shopper. By 1990 there were in excess of 2000 retail warehouses in the UK with the DIY sector dominating the market and occupying about 55 per cent of all accommodation.

Retail warehousing showed considerable resilience during the recession of 1990–92, but by 1993 demand was rising particularly in the ranges of 900 to 2300 m^2 and 6500 m^2 upwards. B&Q had performed exceptionally well with its large catchment areas and the support of investing institutions, and was growing at the rate of about 185 000 m^2 per annum.

Growth across the retail warehouse sector has traditionally been developer led and must therefore be investor backed. The re-emergence of investors during 1993 restarted the development cycle bringing in new space. Yields below nine per cent were beginning to appear on developers' appraisals, although good, pre-funded, pre-let developments heavily outweigh speculative projects.

Business and Science Parks

There are almost as many definitions of business parks as there are schemes. They vary from two or three offices with a few flower beds to a development of 100 000 m^2 (10 ha), with a sophisticated infrastructure of roads, shops, restaurants, parking and landscaping. Debenham Tewson and Chinnocks (1990) produced the following comprehensive definition: 'A business park is a large tract of land, often in excess of 40 ha, developed in phases to a low density offering occupiers an attractive working environment and adequate parking provision . . . capable of meeting the needs of a wide range of business sectors and functions'. The most important factors were good locations and communications, prestige, environment and parking, but some lacked adequate public transport facilities.

In 1990 there were about 800 completed business park developments with a further 400 to 500 under construction or planned, providing more than 18 m square metres of usable floor space. Recessionary pressures stimulated a rethink among occupiers in 1993 and poor uptake led to a decline in owner's capital values. In 1993 rental values varied from £165/m^2 in Manchester to £130/m^2 in Warrington and Chester and £155/m^2 in Leeds. Probably one of the most prestigious and successful business parks was Stockley Park, comprising 160 ha just north of Heathrow with a 100 ha district park and public golf course. Rents varied from £235/m^2 to £280/m^2. The latter produced an 8.5 per cent yield with a capital value approaching £400 m with a floorspace of 140 000 m^2. Business parks sites could fetch prices of around £2.5 m/ha in 1993 (Estate Times, 1993b).

The revised planning draft PPG13 in 1993 aimed at cutting CO_2 emissions by reducing car travel and integrating land use planning and transportation. This may result in a reduction in out-of-town business parks in the future and a greater concentration on urban developments.

Attracting high-tech companies in the rapidly expanding computer field became a high priority in the early 1980s, and by the late 1980s Aston University and Warwick University purchased sites for the development of science parks based on the American concept, with the help of venture capital funds from banking sources. By 1989 there were 38 science parks in the UK, including three alone in Oxford and St John's College, Cambridge designed a scheme of 41 lettable units with a total lettable area of 2800 m^2 (Seeley, 1992).

Privated Housing

The majority of owner-occupiers finance their house purchase by borrowing from building societies or banks. While most house-builders can arrange ninety or ninety-five per cent mortgages over a 25 or 30 year term on their new houses, they need to be aware of the financial standing and circumstances of the kind of purchaser they aim to attract to arrive at a viable price range; for the maximum advance a building society will make is usually limited to 2½ or 3 times the gross household income. For this reason most speculative builders concentrate on low cost houses. The private housing developer has no captive market, in the form of a housing list, on which to draw but has to sell each and every one of his houses to an individual purchaser. The percentage of owner-occupied houses in Great Britain rose from 42 in 1960 to 69 in 1993.

In the nineteen seventies developers found it necessary to offer a range of house types on any one development. The range is one of price rather than plan or elevational difference within the same price range. The progression is usually in steps of around £6000 to £15 000 (1993 prices), starting with a two-bedroom terraced house and leading on to a two-bedroom flat or semi-detached bungalow, through various sizes of semi-detached houses and ending with a large three or four-bedroom detached house. The economical viability of any development assumes first priority in the design process. The developer has the choice between restricting his activities to well-tried and popular forms of traditional housing in desirable areas, and thus satisfying a demand which can be measured by instinct, experience or crude forms of market analysis, or pioneering a new form of development with all the accompanying risks. Most developers steer a middle course and as described by a North Midlands builder: 'It's a compromise in all directions. Every one of our clients would like a 'one-off' house but he can't afford it. Every builder would like to make his estate more attractive, but economics stop him.' There is always a conflict between the economies to be gained from long production runs and the desire to bring in new designs to stimulate the market. The design must also suit the requirements, tastes and pockets of purchasers whose ideas do not often keep pace with architectural and planning opinion. There is, for instance, considerable sales resistance to terraced houses although these acquired greater respectability in the nineteen seventies onwards by being described as town houses often in very accessible locations.

The main problems of the smaller house building firms are fivefold.

(1) The full extent of competition is often unknown in exact terms.

(2) Customers' exact requirements are unknown.

(3) House designs could be simplified, and they are often unnecessarily expensive.

(4) Marketing is often carried out by untrained and inexperienced persons.

(5) Builders provide a poor after-sales service and do not offer any formal maintenance contracts.

In the second quarter of 1993, the Nationwide Building Society reported an average house price in the United Kingdom of £95 371 for detached houses, £63 417 for semi-detached houses and £52 107 for terraced houses. The average price indices fell from 266 in 1989/90 to 190 in the first quarter of 1993. The average ratio of house price to earnings was 2.95 in 1993 as compared with 4.60 in 1989. Table 15.1 shows the wide variations in house prices by house type and region in the second quarter of 1993, taking the two highest and two lowest priced regions. There was a significant closing of the gap between them during the period 1982 to 1993 and the mortgage rate dropped from 13.5 to 8.00 per cent, reducing to around 7.00 per cent, or less on fixed terms, in 1994 (gross rate charged to new borrowers).

Table 15.1 House prices in the United Kingdom (second quarter 1993)

Type of property	Outer Metropolitan Area				Greater London				East Midlands				East Anglia			
	New £	Modern £	Older £	All £	New £	Modern £	Older £	All £	New £	Modern £	Older £	All £	New £	Modern £	Older £	All £
Detached house	128 365	126 815	133 777	128 723	126 446	159 833	146 455	146 209	86 441	72 433	68 048	76 264	84 294	76 218	84 295	79 403
Semi-detached house	82 249	72 546	79 465	75 386	70 336	79 297	90 939	88 188	45 602	47 564	45 871	46 854	48 700	48 247	52 674	49 973
Terraced house	61 856	61 209	59 537	60 732	62 784	72 680	76 435	75 019	44 634	39 913	37 228	38 760	42 069	39 116	42 060	41 156
Bungalows, flats and maisonettes	60 055	62 005	68 304	63 585	65 531	52 058	65 008	60 501	48 314	50 827	47 796	50 207	60 030	52 841	52 857	53 774
All	73 209	69 518	68 612	70 326	67 096	60 245	64 143	62 964	62 164	47 131	39 347	45 096	65 148	60 139	49 965	52 050

Source: Nationwide Building Society (1993).

Improvement Grants

The nature, purpose and extent of improvement grants are described in some detail in chapter 5, including renovation grants, repair and improvement of houses including care and repair services, and rehabilitation work. When considering the use of empty space above shops for residential use, the principal sources of funding can include repair and improvement grants, Historic Building Council's small grant scheme, conversion grants and town scheme grants.

The English House Condition Survey in 1991 showed that there were 1 456 000 unfit dwellings in England in 1991 compared with 1 643 000 in 1986. However, there were only 34 000 renovation grants provided in England in 1992 and on this basis it will never be possible to bring all defective properties up to a reasonable standard.

Under the 1990 renovation grant scheme a new statutory unfit standard was introduced and grants to bring the property up to standard became mandatory. However, this was also accompanied by the grants being means-tested and targeted at those on very low incomes. The 1991 survey suggests that over two-thirds of unfit dwellings could be made fit at a cost of less than £1500 per dwelling. It can also be estimated from the survey that the total cost of making dwellings fit amounted to about £3b in 1991, a manageable figure at late 1980s expenditure levels. Unfortunately, as a result of the 1990 grant changes, the average size of grant increased from £4000 in 1990 to £10 000 in 1992.

Local authority concern has centred around the extension of mandatory renovation grants because rising demand means discretionary grants are endangered, and authorities may either be unable to plan their own renewal stategies or to fulfil their statutory duty. The government had previously funded 75 per cent as specified capital grant (SCG) with local authorities contributing the remaining 25 per cent. However, in the 1993 Autumn Statement local authorities' contribution was increased to 40 per cent, to be raised from their own credit approvals, capital receipts or revenue, thereby reducing the government annual grant in England from £340 m to £260 m. People with disabilities are likely to be particularly hard hit by the proposals.

The English House Condition Survey 1991 showed the following breakdown of unfit dwellings and the average cost of remedy in different ownerships:

Type of owner	*Percentage of unfit dwellings*	*Average cost of remedy (£)*
Housing associations	6.7	2900
Owner occupiers	5.5	2700
Local authority	6.9	1600
All dwellings	7.6	3300

Budgeting for Local Authority Houses

A local authority has to balance income and expenditure relating to its housing provision but can do this in a number of different ways. The local authority has to determine the effect of each new housing scheme on its housing revenue account, and in the early nineteen nineties this way very much reduced as the local authorities' role changed from providers to enablers.

The *revenue* side of the account will contain some or all of the following items:

(1) rents for dwellings received from tenants of houses;
(2) amounts paid in housing subsidies by the government;
(3) receipts from sales of houses; and
(4) contributions by the local authority out of local taxes.

The *expenditure* side of the account will largely be made up of:

(1) loan charges on money borrowed to finance the housing scheme (cost of site, demolition, clearance, compensation for disturbed interests, buildings, siteworks, roads, sewers and other services, and professional fees);
(2) expenditure on housing management and administration;
(3) expenditure on repairs and maintenance, including renewals; and
(4) any other costs associated with housing, such as taxes and other charges.

The rents obtained from local authority houses are normally insufficient to make them self-supporting, let alone produce a profit. It is generally accepted that it is the local authority's function to provide a service and not to make a profit and this concept was clearly demonstrated in a television play *Cathy Come Home* which attracted considerable attention. Rent policy is left to each local authority to determine and the biggest single problem stems from the substantial difference between the amount paid by tenants and the economic rent required to balance the housing revenue account.

Local authorities are, however, usually concerned with the tenant's ability to pay. For many years it was national housing policy for local authorities to operate differential rent schemes under which tenants pay an economic rent, less any rebates to which they may be entitled according to their financial position. Even with differential rent schemes, the local authorities found it necessary to make some contribution from local taxes to the housing revenue account, but this practice is no longer permissible.

In the 1980s the Government particularly favoured the provision of houses for elderly persons and to replace slum clearance properties.

Most local authorities have steadily increased the rents of their older houses largely because of their inability to charge true economic rents on all new houses. It is reasonable to suppose that the time is approaching when the rent burden on older houses will have reached its limit. As local authorities developed up to their boundaries, the pressures for urban land resulted in multi-storey developments mainly in the nineteen sixties and early nineteen seventies, with consequently higher costs and social problems. These have been followed by lower rise more compact developments as described in chapter 5. An exceptionally heavy financial burden flows from the sixty-year loan period for houses, coupled with high maintenance and management costs.

Exchequer subsidies have been paid to local authorities as a result of Housing Acts, and in some cases extra subsidies were paid for expensive sites, excessive foundation costs and houses for special needs. In 1971, the government published a white paper, *Fair deal for housing* aimed at reforming the housing finance structure. The government proposals had three basic objectives: a decent home for every family at a price within its means; a fairer choice between owning and renting a house; and fairness between one citizen and another in giving and receiving help towards housing costs. In practice this has not been wholly achieved.

The Government proposed to phase out indiscriminate housing subsidies, and tenants who could afford it would move towards a fair rent based on the size, quality, location and state of repair of their dwelling. It was proposed that a rent rebate or allowance would be given to all tenants who could not afford the full rent, and it was anticipated that, in many cases, tenants would pay considerably less than the fair rent for their homes. The introduction of fair rents increased the demand for houses for purchase. In practice the fair rents scheme has generally been opposed by councils, so that the rents payable have been far below the economic rent level. Substantial increases in local authority house rents were, however, implemented in the 1980s and particularly the 1990s.

Loan charges are usually paid over a sixty-year period, although the lives of the dwellings will probably be much greater. The general order of cost of local authority dwellings is determined by design standards and cost requirements of the Department of the Environment. Local authorities are thus subject to some control on designing their houses and, in consequence, are also restricted to the extent to which they can reduce building costs, although the removal of housing cost yardsticks offered local authorities greater flexibility.

The difference between amortised annual capital costs plus expenditure on housing management and maintenance on the one hand, and revenue from rents and subsidies on the other, would have to be made up by a contribution from local taxes and reserves, if permitted. This becomes ever more distasteful as local taxes are continually being increased and other services are cut back because of lack of resources. New housing projects had to be viewed against this background and the rent and local tax resources constituted a built-in upper cost limit. Rising building costs caused many local authorities to suspend house building operations in the nineteen nineties. Local authority housing starts reached an all-time low in the nineteen nineties because of severe cuts in public expenditure.

The financial effect of a new housing project is influenced by the local authority's past history of housing work, and the proportion of house sales. Where a local authority has provided large numbers of houses over a long period, then the effect of any one new scheme will probably be no more than marginal. If substantial numbers of houses were built pre-war at very low cost, these will assist in subsidising the rents of the newer and more expensive dwellings, provided the tolerable rent limit for the older houses has not been passed. Even this cost advantage will be dissipated as the older houses are substantially modernised and refurbished at substantial cost. Each local authority's resources, commitments and problems in the housing field are quite different and the solutions reached will depend on the interplay of a number of factors, including the extent of sales of local authority houses. In the early 1990s increasing numbers of local authority houses were being transferred to housing associations.

The operation of cost planning is of considerable value in the public housing field, as the quantity surveyor can determine whether the architect's preliminary design is within the cost limits prescribed by the local authority after considering its housing revenue account. It is also important that an economical housing layout should be produced, consistent with achieving satisfactory standards of amenity, convenience and safety.

Budgeting for Housing Association Development

Rental Values

The principal aim of housing associations is to provide affordable, good quality housing to those in greatest need. Housing associations are expected by the government and the Housing Corporation to fix rents at levels affordable by those in low paid employment. There has however to be a balance between rent setting and balancing the books. The policy on rents has to take account of the capital debts on development (including interest charges), management and maintenance costs, and a sinking fund for future major works of repair and replacement. Large associations will employ expert staff and adopt sophisticated financial models to achieve these ends.

As guide lines, the National Federation of Housing Associations (NFHA) (1990) considered that 20 per cent of the tenant's income was a fair amount for rent. No matter how well managed, all associations will have some problems over tenant-associated matters, such as rent arrears, bad debts, abandoned tenancies and maintenance items.

The Budget and Cash Flow

The budget, usually annual, is the association's plan of work converted into financial terms. The budget estimates how much it will cost to achieve the targets and how much income the work will generate. It must achieve specific criteria and have regard to the constraints imposed by either Rent Surplus Fund or Revenue Deficit Grant Requirements. Within the budget there are three separate areas which should balance:

(1) capital expenditure on property development should not exceed approved advances from the lending authority and, as appropriate, the charitable or private funds or loans available;

(2) all development administrative costs should be within the Housing Association Grant (HAG) administrative allowances; even where no HAG is to be paid, the allowances provide a useful guide;

(3) the cost of managing and maintaining properties should be within the administrative allowances for management and maintenance respectively, and void and bad debts should be within the allowances for these rent losses.

Whereas the budget will show the management committee whether over a period the association can afford to undertake the desired programme of work, the cash flow forecast indicates whether at any given time it will have the cash in hand to pay for the work. Statutory grants available to housing associations are sometimes delayed so that working capital will be necessary. This is often obtained in the form of a loan from a bank or financial institution, sometimes with a guarantee from the Housing Corporation. The management committee must satisfy itself that the association is making the best use of its cash resources and minimising interest charges.

Development Procedures

Since 1989 housing associations have been letting properties on assured tenancies and setting their own rents. This resulted in the wide scale implementation of government policy that associations should use private sources of low start capital finance alongside lower, predetermined levels of public grant. These three elements: rent setting by the association; fixed and lower grant levels; and private sector low start loans; work together to transfer from the taxpayer to the housing association the financial risks attached to the development scheme. With the transfer to the association of development risk, there is greater need for scrutiny and efficiency. In 1993/94 and 1994/95 total housing association grants were progressively reduced and the government was also reducing the percentage of Housing Corporation grant, although the government still aimed to produce 150 000 social houses over three years which will be very difficult to achieve.

Other initiatives adopted by housing associations included the right to buy sales, voluntary sales, housing ownership for tenants of charitable housing associations (HOTCHA), tenant incentive scheme (TIS), improvement for sales (IFS), leasehold schemes for the elderly (LSE) and shared ownership (SO). Serious concerns remained in 1994 regarding the implication of enforced higher rents and the ability of housing associations to house those on very low incomes. The future of housing for people with special needs may be under threat and associations with little base to offer as security to the private investor may encounter particular financial difficulties.

The Housing Corporation is a government agency which has the duty of promoting, assisting and monitoring housing associations. It pays grants to associations and loans them money. Table 15.2 shows the forecast numbers of housing association approvals from 1993/97 and indicates the change in emphasis from houses for rent to houses for sale, and shows a reduction in overall completions from earlier years, coupled with a drastic reduction in local authority house building. About one half of the resources go to large associations (those with more than 2500 dwellings), with the numerous small associations with less than 500 dwellings receiving around ten per cent.

A drastic decline in rehabilitation work was also outlined by the Housing Corporation in 1993. This is very unfortunate as apart from value for money, refurbishment work should be judged against the wider social and economic benefits of urban renewal. The promotion of a design and build approach shared with administrative costs and a flexible approach to property management has much to offer. In 1985, rehabilitation work amounted to almost one half of the development programme, but by 1993 it had fallen to about 14 per cent.

Furthermore, the cuts in funding by the Housing Corporation to housing associations was in 1993 resulting in the use of more contractor-led schemes, reducing the size of dwellings, attempting to equate capital and maintenance and running costs less effectively, making some reductions in

Table 15.2 Housing association development – forecast mid range figures for approvals (1993/97)

	93/94	*94/95*	*95/96*	*96/97*
Mixed funding	35 000	25 800	21 000	22 800
Public funding	1 500	2 300	1 800	2 000
Short life	2 400	2 600	2 000	2 000
Rent	38 900	30 700	24 800	26 800
TIS	6 600	7 000	7 000	7 000
DIYSO	5 100	4 600	7 900	8 100
Conventional sale	4 500	3 600	5 300	5 400
Total incentives and sales	16 200	15 200	20 200	20 500
Total approvals	55 100	45 900	45 000	47 300

Source: Housing Corporation.

quality, and developing greenfield sites as against undertaking much needed rehabilitation. The position deteriorated in 1994–96, as forecast numbers and funds were reduced.

Budgeting for Non-commercial Projects

Probably the largest single group of non-commercial buildings are schools. It is not possible to produce a balance sheet of revenue and expenditure as there is no income received from local authority schools. It is therefore necessary to use some other criteria to establish cost standards and to ensure that local authorities receive value for money. The Department of Education and Science, established a set of cost standards for different types of school and a method of cost planning which is generally known as the elemental system and is described in chapter 7.

A similar problem arises with the building of churches. The church authorities have some idea of the size and quality of building required but the architect is rarely given a cost target. Yet the church authorities often experience considerable difficulty in raising sufficient funds and may have to resort to whist drives, garden fêtes and even direct appeals to meet the heavy cost of building new churches and of maintaining and operating existing ones. It seems evident that all church projects should be carefully cost planned and all necessary steps taken to ensure that the client obtains the best possible value for his money.

It is unsatisfactory to leave the matter entirely to the architect without any brief as to costs. Full consideration ought to be given to the actual cost of past projects analysed on the basis of common units of measure, such as places in schools and seats in churches, as well as costing on the basis of per square metre of floor area. Extensive cost records and analyses of past projects must be maintained, to enable realistic cost targets to be established from known costs set against known standards of construction and quality. On occasions the building client or developer establishes the cost target by reference to known costs of projects undertaken on his behalf in the past. Nevertheless adjustments to costs may be needed to allow for differences in building design and site and market conditions, apart from the updating aspect. In addition it should always be appreciated that cost is only one of the factors to be considered in the design of a building project, and that other factors such as appearance, quality, function and time must also receive adequate consideration.

Budgeting for Hospitals and Students' Halls of Residence

It is proposed to examine these quite disparate projects to illustrate the way in which they can be organised and funded effectively.

Hospitals

Time and cost overruns for hospital projects are commonplace with the various parties blaming each other for delays or late designs. In 1993 a leaked report from the Commons Public Accounts Committee catalogued an alleged £100 m overspend on London's Westminster and Chelsea Hospital and suggested the NHS should take legal action against those responsible for the six month construction overrun, despite the likely complexity of the project, the large number of design changes initiated by the hospital and the large reduction in the initial project period.

Large hospitals are taking on average 10 years and often as much as 15 years from initial planning to commissioning, and clinicians' needs can change radically within a few years, as medical treatment advances and hospital administrators see faster patient throughput as the bottom line. Demands from a hospital client for structural and room layout changes, often altering complex services routeing, can continue well into construction, and variation orders can run into thousands. Changes trigger conflict and contractors' claims, leading to delay and spiralling cost overruns. Unpredictable capital budgeting by the NHS has a knock-on effect on future health projects and clients and contractors suffer (Hayward, 1993).

To overcome these frequent problems, Fife Health Board opted to pioneer a fast track design and build approach, comprising a firm policy with no design changes, thus firming up its hospital outturn price from the first day and to ensure a fixed completion date for the £32 m project. The Health Board also took the rather unusual step of appointing its own project manager to oversee the scheme and an exceptionally detailed specification was prepared. Three tenderers submitted schemes and each was subject to a complex analysis based on 304 criteria covering technical, clinical and managerial suitability of the design.

For the contractor, in-house design meant contractual conflict was replaced by one of teamwork aimed at quickly solving problems. The cost to the contractor of accepting virtually all the contractual risk was reflected in the tender price. There is, however, always a danger to both sides of a fixed price contract in that it can bring design down to a cost rather than up to a requirement.

Student Halls of Residence

A research project in 1993 funded, by a consortium of housing associations under the lead of Leicester Newarke Housing Association, examined the key viability issues and considered the contribution made to student housing development by academic institutions, housing associations, funding agencies and development companies. The key viability issues were identified as: rent setting policy, location of the proposed development, funding mechanisms, quality of accommodation and affordability (Golland and Oxley, 1993).

In contrast to the new universities, many of the traditional establishments can call on subsidies in the form of land or cross-subsidy from a large existing rental income. The research analysed yields (the initial year's income expressed as a percentage of the capital outlay) and many of the established universities had returns of about five per cent. This was less than the prevailing interest rate and was a measure of the cushion offered to students in these situations.

A two tier trend in the supply of student accommodation emerged:

(1) the traditional universities with existing financial resources in locations which may lend themselves to additional revenue from vacational conferences, particularly in university towns;

(2) the new universities, sometimes aided by smaller housing organisations, neither of whom can call upon any subsidy either from existing rents or from locational advantages.

Furthermore, private sector funding institutions to safeguard their lending, may be more willing to fund accommodation on campus sites with conference potential, than to lend against schemes on inner city sites where there may be a distinct danger of void periods in the summer, but also in the case of loan default, there could be the problem of an obsolete building. The main deterrents relate to high land costs, lack of development expertise and to the inequities of passing development costs on to students in unaffordable rents. It should also be recognised that student requirements are becoming more sophisticated and hence short term, opportunistic solutions are not acceptable.

Golland and Oxley (1993) identified three options available to the small provider:

(1) exploit every opportunity at a local level which may enable a housing scheme to get off the ground, possibly working closely with local authorities to renovate redundant buildings, or identifying specific grant aid applicable to the area;

(2) develop new accommodation of a relatively modest standard, possibly at a cost of around £8000 per bedspace in 1993;

(3) incorporate a commercial use, such as shops, offices or industrial units to cross-subsidise student rents.

Refurbishment

Catt (1990) believes that the main purpose of refurbishing buildings is to add value, just as repairs delay physical obsolescence, but the author considers that it can go much further than this to preserve buildings of character and to bring them back into effective use. BRE (1990b) recognised that refurbishment work presents particular problems and a need for specialist skills amongst all the professionals involved. In practice the problems are frequently underestimated and final costs often rise unacceptably beyond original estimates.

BRE (1990c) on carrying out a detailed survey of 82 refurbishment sites in England and Wales revealed over 2000 faults which may potentially give rise to inferior performance. Of the faults discovered 57 per cent resulted from wrong design and specification and rehabilitation did not make the older house as good as new. Another disturbing feature was that the costs of rehabilitating some dwellings were typically more than twice the notional increase in the value of the dwellings.

Registered housing associations still undertake the improvement of older buildings, unfortunately in declining numbers in response to Housing Corporation requirements. They felt gratified to be able to renovate the homes long neglected by private landlords, often because rental income was insufficient to cover costs. Rehabilitation of older dwellings should not be seen as a second best choice when new build is not practicable. It is essential, if these properties are to be brought back into the useful housing stock and given a 'life' of 30 years, that a sound minimum standard of repair and improvement is established so that when families move in they will be provided with homes that give them reasonably trouble-free accommodation (NFHA, 1980).

Hidden costs often occur in rehabilitation work and they typically fall into the following categories:

- Removal of health hazards.
- The impact of current regulations, particularly in respect of fire regulations and means of escape; the requirements of fire authorities change constantly and can vary from place to place. In many cases it is wise to build in extra precautions, especially structural protection, to avoid expensive adaptions part way through a refurbishment project.
- Reconciling available headroom with the need to provide raised floors and/or suspended ceilings.
- Car parking costs can be unusually high especially when provided underground or in space which could be used for office or industrial use.
- When buildings are in conservation areas, are listed as being of architectural or historic interest or are constructed over archaeological remains, delays resulting from deliberations of conservation officers and the additional costs of remedial work may even endanger a project's feasibility.
- Environmental matters must be given full consideration and this may involve an environmental audit.

BMI has found that average rehabilitation costs when compared with average new build costs on a similar basis, show a fairly consistent relationship of around two thirds of the new build price. The only categories of building which show a significantly different relationship are flats, where average rehabilitation costs are as high as 80 per cent of the equivalent figure for new build schemes and banks, building societies and schools, where the rehabilitation costs are around 55 per cent of new build. Figure 15.1 provides an illustrated comparison of rehabilitation and new build costs at 1993 prices.

LAND ACQUISITION PROBLEMS

There are three principal methods of site acquisition available to the private developer: by private

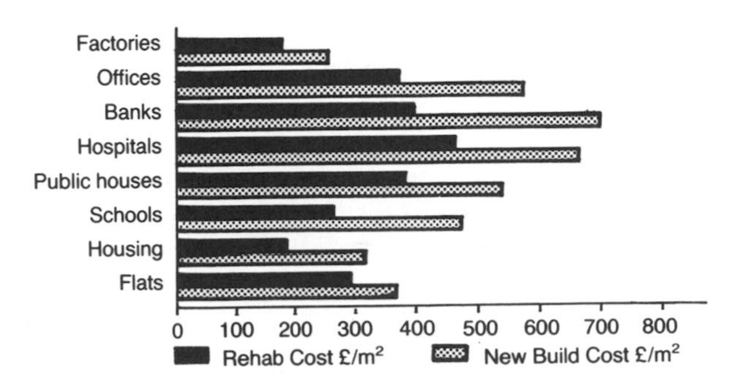

Figure 15.1 Comparison of rehab and new build costs, £/m², first quarter 1993 prices (Source: BMI – Special report 220, 1993)

treaty or negotiation, by public or private auction, and by tender. The private treaty method is favoured by the majority of developers. Next in preference come public auctions and very much lower on the list the tender system of acquisition. Public authorities also have powers of compulsory purchase.

Private developers have often suffered from a shortage of suitable land and this has contributed to the slowing down of the housing programme. In late 1970 the Secretary of State for the Environment called on planning authorities to make generous and immediate releases of land, except where this would conflict with green belt policy. He also urged wider consultations between local authorities and the building industry, particularly in areas where builders were experiencing difficulties in developing land because of the inadequacy of main services. The Minister urged local authorities to reassess the land available in relation to housing needs to ensure that there was sufficient land to meet the demand for the following five years, coupled with continuous monitoring of the balance between house building rates and the granting of planning permission. Local authorities were requested to sell land for private development wherever there was a surplus above the authorities' foreseeable requirements, and this can apply equally well to government departments. Developers have found that their only course of action to increase the amount of land available was through direct pressure on local planning authorities through the time-consuming and costly appeals procedure. In 1972 the Government set up a fund to finance land purchase by local authorities for private development; in 1975 it was advocating that local authorities establish land banks, and in the early nineteen eighties public authorities were being urged once again to dispose of surplus land in their ownership to facilitate its development. In the boom years of the late 1980s developers increased their land banks at inflated prices only to suffer in the 1990–93 recession, when land values reduced substantially and developers were compelled to reduce their holdings at much reduced prices, as building programmes declined and many completed houses remained unsold, even at advantageous prices. This factor compounded the developers' financial problems.

FINANCIAL CONSIDERATIONS

Investment and Real Property

The essential nature of an investment is the release of a capital sum in return for income to be received over a period of time. Hence every purchase of real property is an investment. The income in this case is either a cash payment or savings in rent which would otherwise be payable. Purchasers of real property can be broadly classified into two categories: those who purchase as an investment and those who purchase for occupation and use. In both cases it is advisable for the would-be purchasers of real property to consider the available alternatives of other investments and renting of property. For investment purposes the rate of interest is the cost of holding money, or the reward for parting with it. The Government operates a monetary policy through its control over interest rates and credit to influence the level of economic activity. When the bank rate is increased, the Bank of England almost invariably uses open market operations to restrict the creation of credit by the banks.

The investment aspect of real property is based on its durability and use over time. The owner can transfer the right of use to another person in return for a periodic payment termed rent. The owner is treating the transaction as an investment; he could relinquish the ownership in return for a capital sum but prefers to retain it and receive an income over time.

Needs of the Developer

Where a developer intends to build and then sell the completed development he will only require finance for a limited period. If, on the other hand, he wishes to retain the building as a permanent investment, then he will need to raise two types of finance: short-term finance to purchase land and pay the contractor, and long-term finance which can be raised either by selling an interest in the development or by borrowing against the security of the completed building (Cadman and Austin-Crowe, 1990).

Short-term money is relatively expensive to borrow because there is limited security in the land and the building under construction, and the interest rate will also be influenced by the financial status of the borrower. It is advantageous to the developer to keep his short-term borrowing to a minimum because of relatively high interest rates. In planning a development project it is important to programme expenditure so that capital remains unproductive for as short a period as possible. For instance, by phasing housebuilding, sales of the first houses completed could be used to finance construction in later phases.

Needs of the Lender

The lender of short-term finance will first consider the security of the capital which he lends and the prospects of receiving interest until the debt is repaid. He will also have regard to the risk in securing repayment of his money at the end of the period of loan. Both capital and income are at an appreciably high rate of risk. A well-established property company should not experience any great difficulty in raising short-term finance, as interest payments can be covered from income accruing from other property, but this scenario became distorted in the 1990–93 recession. The capital advanced is additionally secured by the value of the uncharged equity of the company. Small housebuilders with limited financial resources are normally able to borrow about seventy to eighty per cent or more of the value of the land and buildings.

Long-term finance is secured on completed buildings, and so the covenant of the borrower is of less importance. The margin between interest payable on capital borrowed and income arising from property will indicate the degree of security of income and consequently of capital. The borrower of money for property development is competing in the general market for his finance, so that interest rates will tend to follow the general market trend. Short-term borrowing rates will be subject to more frequent fluctuations as they will be related to the current minimum lending rate. Property as a long-term investment is likely to remain attractive since it provides, if selected with care, reasonable security of capital.

Relationship of Building Costs and Valuations

At an early stage in a development project it is necessary to decide what to build and to forecast the likely financial consequences. A valuer can advise on the optimum form of development and its ultimate value, followed by the quality of construction and finish best suited to meet the needs of the prospective market. The position of the building may be of paramount importance, whilst special architectural features are unlikely to justify higher rents in the eyes of tenants. Nevertheless there are many developers who take a pride in their projects and are willing to incorporate attractive but possibly costly features which will improve the urban scene but not necessarily increase rental values.

If sites are scarce and competition is keen, the developer will need to offer a fair price to acquire the site. The value of the site is generally computed from a residual calculation based on estimated development costs and rental values coupled with the estimated capital value of the completed project. The main headings of expenditure are usually:

(1) estimated building cost;
(2) fees payable to architect, quantity surveyor and other consultants;
(3) legal charges for letting of completed building;
(4) agent's commission on letting the accommodation;
(5) advertising costs;
(6) contingencies to cover increased building costs and unforeseen problems; and
(7) interest on finance to cover cost of money invested in the building during the construction period.

After assessing all these costs and deducting the value of the property on completion, the difference will represent the value of the site and the risk allowance on the venture. The developer is then able to make his bid for the land, which may be by way of cash, ground rent, or part ground rent and part premium.

The range of housing prices in 1993 is given in table 15.1 and site costs normally represent 25 to 30 per cent of these prices.

Possible Causes of Loss to Developer

Developers may suffer loss for a number of reasons, and some of the main ones are:

(1) payment of an exorbitant price for the land;
(2) unexpected capital expenditure stemming from such matters as problems with underground services or contaminated land, extra cost of work needed to satisfy building regulations or town planning requirements, or to comply with easements or restrictive covenants;
(3) unattractive layouts or provision of dwellings of types for which demand is limited, resulting in selling problems; and
(4) organisational weaknesses, such as inadequate or ineffective advertising, poor supervision and execution of work in the wrong sequence.

Impact of Development on the Quantity Surveyor

There is a need for an understanding by the quantity surveyor of the relationships between anticipated rents obtainable from commercial investment in building projects and the building cost limits set by these rents. For this it is necessary to know which parts of the building will be used exclusively by tenants and which parts will be in common use. Electrical fittings, telephones and, quite often, partitions are not part of the building client's expenditure. The client will wish to maximise the proportion of floor area available for use by individual tenants. The arrangement of stanchions to project outside the building can help.

The quantity surveyor needs to have a background understanding of the cost of raising finance for a building project – interest rates, amortisation periods for repayment of building loans, and their relationship to the income accruing from the finished building, whether by way of rents, manufactured goods or even school fees. In the case of public buildings such as schools and houses with tight cost limits, the quantity surveyor must ensure that he is adequately equipped with up-to-date knowledge and effective techniques to advise the client and the architect on how to build within these limits and yet achieve the best possible standards.

Quantity surveyors may be called upon to give advice to building clients on approximate cost ranges of building development before any other professional advisers have been appointed. This requires an ability to translate the client's requirements in layman's terms, such as 'a meeting hall to seat 200 people' into floor area needs and other technical requirements of the building. For instance, with the meeting hall the quantity surveyor should be able to assess the floor area and requirements for cloakrooms, toilets, fire exits, services, car parking and many other matters. It will also involve an assessment of all the professional fees, possibly including those of lawyers and estate agents, and even the cost of furniture, fixtures and fittings.

Indeed the quantity surveyor may be required to give cost advice on a whole range of development problems, such as:

(1) cost comparisons of alterations and additions to existing buildings, or of demolishing and rebuilding;
(2) cost comparisons between using a building client's direct labour force (probably his existing maintenance force) for all or part of a project, or of using independent contractors;
(3) advice on the taxation aspects of capital investment in building projects;
(4) cost comparisons between and other characteristics of different industrialised building systems or between these and traditional buildings to perform the same function; and
(5) advice on the relationships between capital

costs and subsequent running and maintenance costs of various elements of the building;

(6) advice on a value management project.

SOURCES OF DEVELOPMENT FINANCE

Corporate Finance

Retained Earnings

Fraser (1993) has described how one striking feature of property investment is the relatively low level of annual income relative to the amount of capital used, and another distinctive aspect of property companies is their relatively high level of gearing. In consequence, a high proportion of investment income is needed to pay interest on debt. In fact, for many companies the residual equity earnings are barely adequate to pay shareholders a reasonable dividend, yet alone provide for future capital expenditure.

There are two alternative methods of generating funds internally:

(1) to undertake development for sale as opposed to investment;

(2) to undertake a programme of disposals of the company's property portfolio, particularly low yielding and reversionary property. However, the main objective of most large property companies is to retain the majority of their developments as long term investments in order to keep disposals to a minimum.

Bank Borrowing

Property companies tend to rely on banks, and stock market and money market issues for much of their funding. Most property companies run up bank overdrafts from time to time as cash outflows exceed inflows over short periods. It is also customary for a company to borrow from a bank to finance short term requirements, such as interim payments to a builder carrying out a development, provided long term finance to repay the debt will be available when required.

Bank borrowing, especially from clearing banks, is normally short term and of variable interest, but this makes a company borrowing large sums vulnerable to rising interest rates and short term recall of the funds by the bank. Merchant banks and finance houses are more willing to lend for specific terms of several years, sometimes at a fixed rate of interest, but they tend to be more expensive than from clearing banks, as described later in the chapter (Fraser, 1993).

Stock Market Issues

It is customary for property companies to retain a 'narrow equity base' to maximise earnings and asset growth for existing shareholders. Fixed interest debt has in the past been a popular form of long term funding for quoted property companies, and large amounts of capital were raised by issuing *debentures* and *unsecured loan stock* in the 1950s and 1960s. Since the late 1960s the issue of fixed rate interest bonds has declined and has been largely replaced by *convertible loan stock*. In the 1980s *convertible preference shares* in addition to bonds proved to be a popular source of corporate finance.

Convertibles were also a popular form of issue in 1975 and 1991, following the two property market crashes, when property companies needed to raise long term capital to repay short term debt. Convertibles gave the investor the security of a fixed interest bond and also possibly a substantial capital gain if the company made a good recovery. *Ordinary share issues* can be an attractive source of new capital when share prices are high and dividend yields correspondingly low. With the cost of equity at a minimum when share prices are high, share issues by property companies become popular during property booms and periods of high interest rates (Fraser, 1993).

Money Market Issues

The London markets (including the Eurodollar market) became an increasingly important source of property debt in the 1980s, and provided companies with many options. They are a source of short and long term capital, zero coupon and interest bearing, at fixed and floating rates, in sterling and foreign currencies.

PROJECT FINANCE

In the financing of individual projects, the financier's capital is legally secured against the property being acquired or developed. Project finance is particularly important for smaller and unlisted property companies who do not have access to the stockmarket, whose assets may already be fully charged against previous loans, and whose financial status provides insufficient security for the financier.

Development financing has traditionally been categorised into two separate operations: short term, interim or bridging finance; and long term funding. Although Isaac (1994) has subdivided the loan periods into short term (1–2 years), medium (2–7 years) and long (more than 7 years):

(1) Short term finance is required to pay the development costs over the development period, and these encompass site purchase, payments to the building contractor and professional fees.

(2) Long term finance is required to repay the short term finance on completion of the project, and is more concerned with financing the retention of the property as a long term investment, where the developer wishes to retain ownership or some valuable interest in the completed development.

Each of these two principal forms of development financing are now examined.

Short Term

The trend in the 1970s was for insurance and pension funds to provide both short and long term finance, but the decline in institutional interest in property in the 1980s forced developers to rely on traditional short term bank finance. This conveniently coincided with a growing interest in property lending by UK and overseas banks, and resulted in a very large expansion in bank lending to property companies, accompanied by more complex and sophisticated financing arrangements.

The principal sources of short term finance are the clearing banks, merchant banks, UK branches of overseas banks and certain finance houses. Clearing banks are usually the developer's first choice, lending on a well secured base at relatively low rates, and often on a corporate rather than a project basis. Banks normally wish to limit their loan to around 65–70 per cent of development costs, with the remainder being provided by the developer. However, a larger proportion may be provided, although at higher cost, by a merchant bank or other bank specialising in property lending. Specialist lenders may fund a higher proportion of costs by providing 'mezzanine' finance, under which the bank will receive a share of the profit of the project in addition to interest on the loan. Another alternative is for the lender to provide up to 100 per cent of costs, with the top slice of the loan covered by indemnity insurance to mitigate the risk of loss in the event of the developer's default (Fraser, 1993).

Whereas banks traditionally made loans on a short term variable rate of interest, a more common approach in the early 1990s was to provide credit facilities designed to meet the specific needs of the project and the developer. The developer usually prefers finance which is not repayable until the completion of the project. This can be in the form of 'interest only' with the capital being repaid in a lump sum at maturity or 'roll-up', whereby compounded interest is paid with capital at maturity and this has the benefit of flexibility as to timing and amount. The loan term normally covers the development period, although banks will sometimes extend the period to five or seven years, possibly even to the date of the new property's first rent review.

Short term credit can be provided on a fixed interest basis, but it is more usual for the interest to be variable linked to the bank's base rate or to the London interbank offered rate (LIBOR). Agreed interest rates can vary from 0.5 per cent or less to 4 per cent above LIBOR depending on the bank's perception of the risk incurred. Exposure to changes in interest rates can be reduced by agreements for a 'cap' (maximum interest rate), 'collar' (provides upper and lower limits to the rate) or 'swap' (allows conversion from a variable to a fixed rate or vice versa) (Fraser, 1993).

In the 1980s 'non-recourse' or 'limited recourse' finance was introduced which usually involved the creation of a separate company for the sole purpose of undertaking the project. Hence collateral security was restricted to the assets of the company carrying out the project and the lender had no recourse to the parent company or to its other assets if the project failed. This arrangement restricts the developer's potential loss to the capital that he has provided for the project but increases the risk to the lender. In practice banks are unlikely to accept this arrangement without a guarantee from the parent company to limit the lender's potential loss (limited recourse).

The largest development projects often incorporate *syndicated loans*, whereby if the financial requirement is beyond that which any one bank wishes to commit to a single project, a group of banks may provide the debt collectively to spread the risk. For example, the first phase of Broadgate in the City of London was funded by a £35 m non-recourse loan from a syndicate of seven banks for a term of 25 years (Fraser, 1993).

Where a bank is proposing to lend to a property developer, it will require satisfying on the following matters:

- The developer, his financial strength, track record on previous projects and reliability, much of which can be obtained from the company's audited accounts.
- The collateral to be provided as security for the loan, normally charged against the site, fully assembled and with planning permission, and the development as it proceeds.
- The collateral to be provided as security for the loan, and the developer may be required to give personal guarantees for the recovery of the loan in the event of his bankruptcy.

The ability of the developer to repay the short term loan on completion of the project will depend on the availability of long term finance or his ability to sell the property. Hence a banker's willingness to provide short term credit may depend on whether the developer can prearrange long term finance or a commitment from an investor to buy the completed property (forward sale). Subsequently, proof of the viability of the project often rests on the ability of the developer to prelet a significant part of the development (Fraser, 1993).

Long Term Finance

The Mortgage

Long term development finance is normally raised either by mortgage or, particularly in times of credit shortage, by sale and leaseback, the latter being considered later in the chapter.

Following the Second World War, long term mortgages on commercial and industrial property were provided by insurance companies and were mainly fixed interest. In the 1990s, both fixed and variable interest mortgages were available for periods up to 25 years and on a variety of terms to suit the needs of individual projects and borrowers and can encompass the development period as well as the long term.

There are three basic methods of repayment:

(1) Equal instalment method: whereby equal amounts of capital are repaid periodically over the loan period, accompanied by declining outstanding capital and interest payments.

(2) Annuity payments: normally adopted for building society mortgages, whereby assuming no change in the level of interest rates, the combined interest and capital paid each period remains constant over the term of the loan; initially payments consist largely of interest but later of an increasing proportion of capital.

(3) Interest only: capital is repaid in a lump sum at maturity and this is usually confined to relatively short term loans.

Alternatively, repayment tranches may be spaced over the loan period; for example, a 21 year loan might have one third repayable after each of seven years, 14 years and at maturity.

The amount lent on a mortgage has normally been restricted by two criteria:

(1) the sum lent should not exceed two thirds or occasionally three quarters of the value of the property mortgaged;
(2) the net rental income from the property must exceed interest and any periodic capital repayments (Fraser, 1993).

Example 15.1 shows how these two criteria can be met and sufficient mortgage capital raised for long term funding.

Example 15.1 A prime commercial development was completed at a total cost of £4.9 m. The market value of the property was £7 m and it was fully let for a net rental income of £490 000 per annum. Mortgage finance amounting to 70 per cent of the market value was repayable on an annuity basis over 25 years at a fixed rate of interest of 6.5 per cent.

	£
Development cost (including short term finance)	4 900 000
Mortgage debt (70 per cent of £7 m)	4 900 000
Capital surplus/deficit	NIL
Net income rent	490 000 pa
Mortgage instalment	401 730 pa
Net income surplus	88 270 pa

This confirms the self-financing aspect in that:

(1) the long term mortgage was sufficient to repay all short term finance raised to pay development costs, and
(2) net rental income was more than sufficient to pay annual mortgage interest and capital repayment.

An alternative to the straight mortgage is the mortgage debenture, whereby funds are advanced against the security of a particular property, but in addition the lender has a charge over all assets of the company, repayment of the mortgage debenture is at the end of the loan period and the lender has a fixed interest investment with a substantial security but no hedge against inflation. While another variation is the convertible loan stock, mentioned earlier, wherein the lender has an option to purchase ordinary shares in the company at some future date at a fixed price in proportion to the amount of loan stock held. If the lender does not take up his option, he continues to receive interest at an agreed rate until the redemption date. The lender thereby links the advantages of a fixed interest stock with protection against inflation (Balchin *et al.*, 1994).

Sale and Leaseback

Sale and leaseback has proved to be the most important vehicle for equity sharing and, following the First World War, was regarded by developers as the principal alternative to mortgage finance, and it became particularly favourable when the government was imposing restraint on lending as part of its economic policies. Essentially, it involves the sale of the freehold or long leasehold interest by the developer to an institution in return for a long lease, with the developer subletting the property to occupying subtenants. Thus the developer continues to have the use of the property while obtaining funds to finance another development. By the 1980s rent reviews were often at five or seven year intervals with leases for 25 to 30 years, providing the investing institutions with the equity investments they required in property, often prime commercial properties in central locations with high site values.

The investing institutions were the main beneficiaries of the replacement of conventional mortgage finance by sale and leaseback. It enabled them to acquire equity investments in modern property with a reliable property company as head tenant, who collects rents from subtenants and relieves the institution of the management of the property, with the leases normally on full repairing and insuring (FRI) terms. However, the developer lost the freehold interest, much of the equity and gearing advantages. Sale and leaseback provided all necessary capital and still enabled the developer to make an adequate return, albeit a relatively small profit rent (the difference between the rent paid to the institution and that received from the tenants). Furthermore, the developer may obtain 100 per cent of the value of the asset as compared with possibly two thirds with a mortgage.

It served the needs of both parties as developers needed the institutions' finance and the institutions wished to acquire modern equity property investments. The actual terms of the agreement would be influenced by the cost and availability of finance, investment market and property market conditions, and the relative bargaining strength of the two parties. Nevertheless, equity sharing sale and leaseback arrangements can be very complex and of infinite variety, and they therefore require careful examination and analysis of the details to assess the risk and potential return to the parties.

There are two fundamental methods of sharing the investment income from the completed project namely: top/bottom-slice arrangements and side-by-side or vertical leaseback schemes, and geared and ungeared arrangements, which are considered to be outside the scope of this book. Readers requiring further information on these approaches are referred to Fraser (1993).

For the institution, the sale and leaseback transaction offers the following benefits:

(1) Periodic rent reviews provide a hedge against inflation.

(2) A leaseback received against first class shop property provides a valuable reversion on termination of the lease.

(3) Freeholds and long leaseholds are a more permanent form of investment than mortgages, which may be terminated by short notice.

(4) Ownership of property may offer greater income security than other equity investment since rent is a prior charge against a company.

(5) By dealing directly with one developer/ development company, an institution can have a stake in property without the burden of management of buildings in multi-occupation.

(6) Because of the large funds which insurance companies have to invest, a large sale and leaseback transaction of say £3 m is less trouble, both now and in the future, than investing £300 000 in ten different holdings.

(7) A geographical spread of investment, for example, through a chain of multiple shops distributed in major towns throughout the United Kingdom, reduces the risk to the investor from such factors as local unemployment.

DEVELOPER'S BUDGET/FEASIBILITY STUDY

General Philosophy and Approach

Prior to the purchase of a site for development, a developer must know what forms of development will be permitted on the site and have access to a financial appraisal or developer's budget (now often referred to as a feasibility study). To decide whether the scheme is feasible, he will require advice on a fair price for the land; the probable building costs; and the probable rent or selling price of completed building(s). The developer's budget will include the following items.

Gross development value. This is based on the estimated total annual rent accruing from the completed building less the cost of outgoings, such as maintenance, repairs and management. The net annual income is capitalised by multiplying by an appropriate years' purchase and the total so obtained is referred to as gross development value (GDV). Rental values of properties vary widely as between the same class of building in different situations and between different classes of building in similar locations. Shops or offices in the centre of a city or large town will usually command much higher rents than similar premises in suburban locations, but even the suburban premises will vary in value according to the environment (industrial, local authority housing, good class private housing, etc.), as well as with the side of the street. For example, annual rental values of suburban shops in a large provincial city might average about £250/m^2 of zone A floor area (in front 6 m deep zone or strip of shops) and about £1500/m^2 for central area shops

(1994 prices). Details of prime office rentals were given earlier in the chapter.

The valuer advises the developer on site values and future values of the development. Cost is generally associated with the supply element in the property market, while value relates to the demand side. The valuation surveyor gains his experience of values by observing and interpreting the evidence provided by the property market, in the form of sales and lettings. The valuer may often be only faintly aware of the deeper causes which influence demand. The mathematics of valuation and the use of valuation tables are incidental to the valuer's calculations, for the basic figures are likely to be obtained quite intuitively, where comparable circumstances do not exist. Hence, the approach of the quantity surveyor and valuer can be quite different. In general, the quantity surveyor possesses far more cost data than the valuer has valuation data, and he can adapt it more readily from one situation to another.

The valuation surveyor is concerned with many value/cost relationships of a project. He is particularly concerned with securing a design which will achieve maximum value in proportion to cost, and is likely to be faced with a number of fundamental issues, such as 'If a cheaper finish is used in the building what will be its effect on rental value?' A valuer usually works from rental values and any discrepancy will be magnified about fifteen times when converted into capital terms. The property market is subject to many uncertainities; changing government policy, overseas influences and new legislation can each have a significant effect on property values. Valuing is too often a matter of opinion based on experience, and often valuations of eminent valuers diverge widely. For instance, opinions conflict on the relative values of different floors of a modern office block; some valuers believe the higher floors to be the more valuable, and others the lower. Eminent valuers produced widely varying valuations of the site value of the headquarters building of the Royal Institution of Chartered Surveyors in Parliament Square, Westminster. A valuer would experience difficulty in assessing the reduction in value stemming from diminished daylighting at the centre of a large office block. These latter aspects serve to illustrate the very real difficulties which face valuers when assessing site and building values.

Cost of buildings. The cost of the proposed buildings will be assessed using one of the methods described in chapter 6, with the help of cost analyses, which were examined in chapter 9. The costs will normally be computed per square metre of floor area but the developer will generally be concerned only with the usable floor area on which a rent is payable.

Architect's and surveyor's fees. The normal allowance is ten per cent of the cost of erecting the building, but could rise to 15 per cent or more to cover consultants' fees. This covers the preparation of all contract documents and supervision of and financial arrangements for the contract, including possibly the employment of a project manager/value manager.

Legal and agency fees. An average allowance to cover legal costs on purchase of site, including stamp duty, and preparation and agreement of leases, agent's fees on letting the property and advertising costs, would be 2½ to 3 per cent of gross development value (probably about four to six per cent of building costs). The sum involved will be influenced considerably by the number of lettings.

Developer's profit. An allowance of between ten and twenty per cent of the gross development value should be made to provide a return to the developer for devoting his skill and time to the project, and for the risks that he undertakes. The risks include rising costs, falling rents and inability to lease the property on completion. The actual allowance incorporated in a developer's budget will depend on the type of development and the degree of risk involved.

Cost of finance. To purchase a building site a developer will either have to borrow money, on which interest will be payable, or use his own capital and forego the return on it. The interest paid or revenue forfeited should be charged to the development, and will cover the period from date of purchase to the time when the completed building is let or sold. Financing of the building operations will proceed throughout the contract period as payments are made to the contractor based on periodic certificates. The building cost finance is usually calculated at an agreed rate of interest on half the building cost for the full contract period or the full building cost for half the contract period assuming a constant cash flow throughout the work. A suitable annual rate of interest in 1994 was around 8 per cent, using compound interest for lengthy contract periods.

PRACTICAL APPLICATION OF DEVELOPER'S BUDGET/FEASIBILITY STUDY APPROACH

A number of worked examples follow to illustrate the application of the concept of the developer's budget to various practical development problems. It should, however, be emphasised that calculations of this type can be very complex and in practice the assistance of a valuer will normally be required.

Example 15.2. Planning consent has been given for the erection of an office block of 10 000 m^2 on a vacant building site. It is estimated that the building will produce a net income of £1 400 000 pa and will cost £680/m^2 to build. Assuming it will take eighteen months to build, determine the present market value of the site.

Gross development value		
Net income from offices	£1 400 000 pa	
YP in perpetuity at eight per cent	12.5	£17 500 000
deduct costs		
Building – 10 000 m^2 at £680/m^2	6 800 000	
add architect's, surveyor's and consultants' fees (16 per cent)	1 088 000	
	7 888 000	
One-and-a-half years building finance on half cost $\frac{£7\,888\,000}{2}$ at 11 per cent	£7 888 000	£17 500 000
	433 840	
Legal and agency fees and advertising costs (three per cent of GDV)	525 000	
Developer's profit (ten per cent of GDV)	1 750 000	
		10 596 840
	Residue =	£6 903 160

The residue represents three items: value of the land; acquisition costs (4% of value), and cost of borrowing for 2 years @ 11% pa on land value and costs.

$$\text{Let } x = \text{Land value, then } £6\,903\,160 = (x + 0.4x) \times 1.11^2$$

$$£6\,903\,160 = 1.2814x$$

$$\text{Site value} = \frac{£6\,903\,160}{1.2814} = £5\,387\,200$$

The anticipated return of the developer is usually assessed at about seven to nine per cent of development costs, and it would be useful to check the figures in this example by the anticipated return method.

Costs	£
Building, including fees	7 888 000
Building finance	433 840
Legal, agency and advertising costs	525 000
Land cost	5 387 200
Site finance and acquisition	1 515 960
Total costs	£15 750 000

The net income of £1 400 000 gives a 8.89 per cent return on the development costs of £15 750 000, and so is at the higher end of the scale. The rate of return varies considerably with fluctuations in the relationships between income, building costs and interest rates.

Example 15.3. A site is available at a purchase price of £750 000 and it is anticipated that planning permission could be obtained for a factory of 7500 m^2 or an office block of 17 500 m^2. The estimated cost of the factory including site works is £1 875 000 and the comparable cost of the office block is £9 450 000. The contract periods are assessed at one year for the factory and two years for the office block. On completion the office block is likely to have an annual rental value of £140 000 with the landlord's annual outgoings of £300 000. The normal return on office blocks in this area is seven per cent. The factory is likely to sell for £3 800 000 on completion. Legal, agency and advertising costs are likely to be £90 000 for the factory and £420 000 for the office block. Architect's and surveyor's fees are 12 per cent of construction costs and finance is at a specially advantageous interest rate of 9 per cent. Determine the form of development likely to profit the developer most.

Factory

Cost of site		£750 000
Site finance for one year at nine per cent (special rate)		£67 500
Total cost of land		£817 500
Building costs	£1 875 000	
add architect's and surveyor's fees (12 per cent)	225 000	
	2 100 000	
Building finance for one year on half cost $\frac{£2\,100\,000}{2}$ at 9 per cent	94 500	
Legal, agency and advertising costs	90 000	
		2 284 500
Cost of development, excluding developer's profit		£3 102 000
Selling price		£3 800 000
less development costs		3 102 000
Development profit (received after one year)		£698 000

Office block

Cost of site		£750 000
Site finance for two years at nine per cent		135 000
		885 000
Building costs	£9 450 000	
add architect's and surveyor's fees (12 per cent)	1 134 000	
	£10 584 000	
Building finance for two years on half cost $\frac{£10\,584\,000 \times 2}{2}$ at 9 per cent	952 560	
Legal, agency and advertising costs	420 000	
		11 956 560
Cost of development, excluding developer's profit		£12 841 560
Annual rental		£1 400 000
less outgoings		300 000
Net annual income		1 100 000
YP in perpetuity at seven per cent		14.3
Selling price		15 730 000
less development costs		12 841 560
Development profit (received after two years)		£2 888 440

The factory would be the most profitable development as it shows a return of 22.50 per cent after one year, as compared with the return of 22.49 per cent after two years for the office block. The capital expenditure on the factory project is about one quarter that involved for the office block and, if the developer is using his own capital to float the project, this will release funds for investment in other schemes. As the return on the factory development is received one year earlier, the profit of £698 000 is available for investment throughout the second year. These calculations assume that there is ample demand for both forms of development.

Example 15.4. A developer is proposing to erect a block of six lock-up shops with two floors of offices above them. The shops are likely to let at an average annual rent of £300/m^2 and the offices at £150/m^2. The net floor areas will be 500 m^2 for the shops and 900 m^2 for the offices. Circulation areas will amount to about ten per cent of total floor area and will remain under the control of the landlord, whose annual outgoings are estimated at twenty-five per cent of income. The developer requires a twelve per cent profit on the development. The freehold site is available at a purchase price of £130 000 and siteworks are estimated to cost £15 000. The rate of return in this area for similar developments is around seven per cent. The contract period is likely to be one year and finance is available at 11 per cent rate of interest. Legal, agency and advertising costs are likely to be about three per cent of gross development value. Determine the allowable building cost per m^2 of gross floor area and its feasibility.

Income from development		
Shops – £150 × 500	£75 000	
Offices – £75 × 900	67 500	
Gross income	142 500 pa	
less annual outgoings (twenty-five per cent)	35 625	
Net income from shops and offices	106 875	
YP in perpetuity at seven per cent		£106 875 14.3
Gross development value		£1 528 313
deduct costs		
Cost of site	£130 000	
Site finance – 11 per cent on £130 000 for one year	14 300	
Legal, agency and advertising costs – three per cent of £1 528 313 (GDV)	45 849	
Developers' profit (twelve per cent of cost of development = 1/10 of GDV)	152 831	
		342 980
Sum available to cover building costs, fees and finance		£1 185 333

Let x = building costs, including siteworks. Then architect's and surveyor's fees at 12 per cent = $0.12x$ and finance for one year at 11 per cent = $0.11\,\frac{(x + 0.12x)}{2}$

$$x + 0.12x + 0.062x = £1\,185\,333$$
$$1.182x = £1\,185\,333$$
$$x = \frac{£1\,185\,333}{1.182} = £1\,002\,820$$

less siteworks	15 000
Building costs	£987 820
Building area	
Shops	500 m^2
Offices	900 m^2
Net usable area	1400 m^2
add circulation area (1/9)	156 m^2
Gross floor area	1556 m^2

$$\text{Allowable building cost/m}^2 = \frac{£987\,820}{1556} = £634.85$$

An analysis of recently completed similar buildings in the district show this to be a suitable rate for the provision of shops and offices with good quality finishes and including allowance for increased costs, as it is within 1.5 per cent of the mean updated price.

Example 15.5. A developer requests advice on the value of a building site of 20 ha which he is considering purchasing. The land has outline planning permission for private residential development at a density of 120 persons per hectare. Similar building plots in the district are selling readily at £40/m^2.

It is assumed that the average size of dwelling will accommodate four persons, and the total number of building plots will be

$$20 \times \frac{120}{4} = 600$$

To arrive at the net area of the average plot, it is necessary to make allowance for roads and incidental open spaces. Assuming that roads occupy fifteen per cent of the site and incidental open spaces a further five per cent, then the effective area of the site for building purposes becomes 20 ha less twenty per cent, equalling 16 ha. The area of the average plot equals

$$\frac{16 \times 10\,000}{600} = 267\,\text{m}^2$$

as there are 10 000 m^2 in a hectare.

The value of an average building plot = 267 × 40 = £10 680 and the estimated income from the sale of all plots is 600 × £10 680 = £6 408 000.

The programme and cost of preparing the site for development have now to be considered. It is likely to take two years to construct the roads and lay sewers and a further two years before all the plots are sold. Hence the income from the sale of plots will be deferred for 2 + (2 × ½) = three years (average).

An outline layout of the development will now be prepared to determine the length of roads and sewers needed to develop the site. It is assumed that 3000 m of roads, paths and sewers at an all-in cost of £300/m will be needed, and as the constructional work is scheduled to take two years, the expenditure will be deferred an average of one year.

Income from sale of building plots 600 at £10 680	£6 408 000	
PV of £1 in three years at 8 per cent	0.7938	£5 086 670
less Development costs		
Roads and sewers – 3000 m at £300	900 000	
PV of £1 in one year (average) at 8 per cent	0.9259	
	833 310	
Legal, agency and advertising costs – four per cent of £5 086 670	203 467	
Developer's profit and risk – twelve per cent on £5 086 670	610 400	
		1 647 177
Value of land (approximately £171 975/hectare)		£3 439 493

Example 15.6. Negotiations have been proceeding with regard to a 0.6 ha site and indicate a purchase price of £80 000, inclusive of fees. A small workshop scheme of 1400 m^2 is to be constructed for rental, in conjunction with certain site development works to render the ground suitable for development, including demolition work, site levelling and ground stabilisation. A site development/construction period of nine months is anticipated. The client requires a feasibility study of the scheme.

Estimated total site development costs	£
1. Land costs: 0.6 ha @ £80 000 (including fees)	80 000

	£	
2. Construction/site development costs		
Ground levelling	70 000	
Demolition work	20 000	
Ground stabilisation	50 000	
Building costs, including access and parking	530 000	
Site investigation fees	5 000	
Design, etc. fees (8% on £670 000)	53 600	
Clerk of Works costs	7 600	
Statutory fees	9 000	745 200
Internal administrative fees (2.5% on £670 000)	16 750	16 750
3. Interest/finance charges (11% on land price during constn): £80 000 for nine months @ 11%	6 600	
On half construction costs during construction (assume constant cash flow over nine months) one half × £745 200 for nine months @ 11%	30 740	
On land, construction and other costs, prior to units let/sold: say, £862 540 for nine months @ 11%	71 160	108 500
4. Miscellaneous costs		
Agent's letting fees	8 700	
Advertising	2 000	10 700
	Total	£961 150

Costs are now related to anticipated rentals

10 units @ 47 m² = 470 m² @ £68.90 =		32 383
5 units @ 93 m² = 465 m² @ £62.40 =		29 016
2 units @ 232.5 m² = 465 m² @ £56.00 =		26 040
	Total	87 439
Less management fees @ say 3%	£2 623	
Landlord's repairs	£6 960	9 583
Estimated net rents		£77 856

Assuming total development costs of £961 150 and estimated net rentals of £77 856, the initial return on capital would be 8.1 per cent. This would require checking with the identified development objectives to determine whether the appraisal result is satisfactory and the level of return is acceptable. It is probably at the lower end of the relevant yield range which may vary between 8 and 10 per cent (English Estates *et al.*, 1986).

CHOICE BETWEEN BUILDING LEASE OR PURCHASE

When making a decision as to whether to purchase a building or lease it for say a twenty-one-year period, the alternatives must be reduced to a common time scale, the simplest approach being to take the period of the lease and to evaluate the building in twenty-one years time for the purchase calculations. However, it should be borne in mind that many building leases are now let on seven year terms. The value should be the higher of replacement cost or resale value, and where there is doubt about the estimated value it is advisable to adopt a range of values to determine how critical the figure is to the calculation. Tax allowances must be taken into account in the calculations, as purchase of the building will give rise to annual capital allowances and lease payments will be allowable against the company's taxable income. A capital allowance of around 50 per cent on the initial expenditure may be claimable, although this will be higher for small buildings and can be 100 per cent in enterprise zones. There is an annual writing down allowance of 4 per cent. With leases, the company is likely to secure a forty per cent reduction on lease payments due to tax relief, although the tax recovery will normally occur a year later. Tax rates and allowances are, however, subject to periodic review.

16 ENVIRONMENTAL ECONOMICS AND THE CONSTRUCTION INDUSTRY

In the final chapter we examine the impact of both public and private investment and of government action generally on the construction industry, together with the structure of the industry and the relationship of its output to demand and available resources. The economic aspects of urban renewal, building conservation and urban regeneration, and of new town and town development schemes are examined. Consideration is given to the philosophy and nature of cost benefit analysis and its application to a variety of environmental problems.

ENVIRONMENTAL MANAGEMENT

Reasons for Growing Importance of Environmental Matters

It was not until the late 1980s that the wider effect on the world's resources and the balance of the global environment became generally recognised, when considering the relationship of buildings to the environment (DoE, 1991), while Cadman (1990) identified 'growing environmental concern' as one of the four factors, alongside market balance, the demographic trough and European integration, that will change attitudes to planning and investment in the future. The RICS (1993b) identified three main stimuli that are likely to secure a positive reaction to environmental issues: namely, legislation, public opinion and profit.

1. *Legislation*
The nature and scope of legislation are changing rapidly as evidenced by the Report on European Environmental Legislation and its Application (Metra Martech, 1991). For example, currently permitted levels of contamination may become legislatively unacceptable as standards and expectations rise.

2. *Public opinion*
Public interest in environmental matters is growing as illustrated by the national press. A typical example was the concern expressed over the use of timber framing after it was shown in a poor light in a *World in Action* programme, with a significant effect on residential construction practices and house prices. Many green issues have been adopted by democratic governments and it is being increasingly emphasised that there are serious risks in ignoring environmental considerations.

3. *Profit*
Legislation is usually effective because non-compliance often results in a monetary fine. Public opinion is frequently sought to increase the competitive edge, obtain an enhanced image and increases in demand and perceived value (RICS, 1993b).

Environmental Definitions

BS 7750:1992 includes the following environmental definitions:

Environment as 'The surroundings and conditions in which an organisation operates, including living systems (human and others) therein. As the environmental effects . . . of the organisation may reach all parts of the world, the environment in this context extends from within the workplace to the global system'.

Environmental policy as 'A public statement of the intentions and principles of action for the organisation regarding its environmental effects, giving rise to its objectives and targets'.

Environmental management as 'Those aspects of the full management function (including planning) that develop, implement or maintain the environmental policy'.

Environmental management audit as 'A systematic evaluation to determine whether or not the environmental management system and its operation and results comply with planned arrangements, and whether or not the system is implemented effectively, and is suitable to fulfil the organisation's environmental policy and objectives'.

The RICS (1993b) formulated the following definition of *Environmental management*: 'The management of environmental factors to enable human activity to exist productively and compatibly alongside, and enhance where practical, other global and life sustaining processes, in a beneficial, sustainable and harmonious synthesis for future generations'.

Thus, where possible, the surveyor should beneficially manage the factors that affect the environment. However, where this is not possible, he should manage the client's interest to reflect the environmental constraints. Readers requiring more information on environmental management in construction are referred to Griffith (1994).

Environmental Issues

There are many environmental issues which affect the work of the surveyor, ranging from climate to contaminated land, energy and timber. The principal issues are now examined, dealing with each aspect in such depth as appears appropriate.

1. *Climate*
Climate may impinge upon land and buildings in various ways and probably the most publicised is the effect of man's actions upon the climate. For example, the misuse of energy, use of CFCs in construction and non-sustainable felling of tropical hardwood, all of which contribute to the Greenhouse effect.

Climatic changes may themselves require changes in structural performance and BRE have suggested that certain present structures may not be adequate to withstand predicted wind speeds without modification. There is sufficient evidence to demonstrate the linking of rainfall and prevailing winds to premature building failure, such as cavity wall tie corrosion and corrosion to steel framed structures. Tall structures can have an effect on the micro climate of a neighbourhood as in Canary Wharf, where five of the 450 mm square, 5 mm thick stainless steel cladding tiles were sucked off one of the two bridges that carry the Docklands Light Railway.

Motorways, industrial processes and power stations can all affect the climate in various ways, from the immediate surroundings as with motorways (noise, vibration, fumes and dust) to global effects with power stations. These factors have resulted in areas with air quality below World Health Organisation (WHO) standards. Compensation is claimable in respect of noise, dust and vibration under the Lands Compensation Act 1973.

With regard to the indoor climate, modern buildings may produce hermetically sealed or near-sealed environments, thereby reducing indoor air quality with detrimental health and occupancy conditions, ranging from Sick Building Syndrome to problems in domestic housing; legionella in domestic hot water systems and problems with asthma (RICS, 1993b).

2. *Contaminated land*
Contaminated land has become a major issue as discussed in chapter 14. In Europe an EC Directive made site owners and the producers of waste jointly responsible for cleaning up contaminated sites, while in the UK environmental legislation relating to the contamination of land and water becomes progressively more stringent. In particular, the Water Resources Act 1991 and the Environmental Protection Act 1990 (EPA) made owners of contaminated and polluted land liable for meeting the costs incurred by the local authority or the National Rivers Authority in cleaning it up.

Costs of reclaiming sites vary across the UK,

although in 1993 average figures for DoE funded reclamation projects were in the order of £200 000/ha. However, certain kinds of problem generate higher costs and the wholesale clearance of badly polluted sites can exceed six times these figures. In Holland where higher standards are implemented, costs totalling millions of pounds per hectare can arise. Even identifying contaminants on a site can cost up to £15 000/ha.

3. *Energy*

Figures produced by BRE showed that during 1991, UK buildings emitted 300 m tonnes of carbon dioxide, compared with 150 m tonnes from transport and 150 m tonnes from manufacturing. The problems relate mainly to the use of finite fossil reserves and the resultant pollution. Building products require different amounts of energy for their production. For example, timber needs to be felled, transported and shaped, while bricks and blocks need to be quarried, shaped, fired and transported. Higher thermal upgrading reduces energy consumption and its detrimental consequences for the Greenhouse effect, with consequential occupancy cost savings as illustrated in chapter 13. Other ways in which energy might be saved relate to the way in which buildings are used and processes operated, such as by changing working hours, which in its turn could lead to changed basic design and floorspace requirements. A House of Lords Report in 1991 advocated the retrofitting of the entire UK housing stock at a cost of £45b, to bring it up to European standards, but it seems, not surprisingly, to have disappeared without trace.

4. *Environmental assessment*

Environmental assessment (EA) is the process by which the environmental implications of a project are assessed. An environmental statement (ES) is the formal document which must be produced in certain cases to set out the result of the EA. Environmental information (EI) is information relating to the environmental implications of a project whether contained in the ES or in representations from statutory consultees or third parties. Proposals for sensitive developments such as quarrying, open cast mining, motorways, waste disposal sites, extensive new urban or rural developments and major out-of-town shopping schemes require detailed environmental impact assessments of their effects during and after implementation, and by 1993 approximately 1500 EAs had been prepared in the UK covering some 60 per cent of all local authorities.

Public opinion may quickly be motivated by what is seen as lack of environmental consideration, which could include such matters as loss of amenity and visual value. Commercial and environmental issues can also come into conflict in cases such as the Cardiff Bay development, where waterfront regeneration could cause the permanent loss of important estuary habitat for wildlife (RICS, 1993b).

EC Directive 85/337/EEC required member states to ensure that where projects have a significant impact on the environment, EI is obtained and considered before consent for the project is given. The main UK regulations are the Town and Country Planning (Assessment of Environmental Effects) Regulations 1988, which operate additional to the planning system by directing that, in certain cases, an ES is supplied and considered before granting planning permission. This does not cut across a planning authority's general duty to have regard to all material considerations, including environmental impact, but is an addition to that duty.

The Town and Country Planning (Development Plan) Regulations 1991 state that: 'Most policies and proposals in all types of plan will have environmental implications which should be appraised as part of the plan preparation process. Such an environmental appraisal is the process of identifying, quantifying, weighing up and reporting on the environmental and other costs and benefits . . . but . . . does not require a full environmental impact statement of the sort needed for projects likely to have serious environmental effects'.

5. *Environmental labelling*

There could be an important role for environmental labelling in property and construction, although in 1994 there seemed only limited public awareness of this process. However, environmental audits and assessments are becoming more evident in the property and construction industries and BRE have made a substantial contribution with BREEAM (Building Research Establishment Environmental Assessment Method), by assessing a building's impact on such matters as global warming, ozone depletion, acid rain, sustainable materials, recycling, local environment and indoor environment (BRE, 1990d).

The EC Environmental Council of Ministers introduced an Eco Labelling system, adopting a 'cradle to grave' approach. Among some 20 product categories, the UK long term proposals encompass light bulbs, lighting, air conditioning equipment, building materials, insulating materials and water conserving devices.

6. *Groundwater*
Problems from groundwater can arise from a variety of sources, including contaminated land, abandoned mine workings where pumping has ceased, and changes in the water table. In some cases rising groundwater results in increased construction costs as, for example, an additional £1m required to safeguard the British Library from water intrusion. A CIRIA Report (1989) stated that rising groundwater beneath London and other cities will cause severe damage to buildings and tunnels within 20 to 30 years, by the water pressure causing ground movement in the overlying clay, unless action is taken. The groundwater level under central London is rising by about a metre a year, because water is no longer being drawn from underground wells, as industry moved out of the city to other locations. Of the 150 km of the London Tube system, 40 km are within the critical zone.

7. *Health and safety*
This is likely to be one of the principal areas for concern in environmental management, particularly to prevent health and safety risks arising to the users of land and buildings and also to third parties. Safety may be affected by geographically related environmental factors such as radon gas, health clusters like meningitis and tendencies to ill health such as heart diseases. Health issues may also emanate from more location specific environmental factors such as adjoining motorways, micro waves and electricity cables, or even be building related such as dangers arising from asbestos, and legionella caused by poor maintenance of air conditioning or hot water systems. It is worth noting that BRE now believe that natural ventilation as compared with air conditioning of deep plan offices is a more viable proposition than considered hitherto. The safety of some commonly used building products is now being seriously questioned, such as cancer linked with glass fibre and some chemicals used for timber treatment, thus indicating the need for environmental advice on a wide front (RICS, 1993b).

8. *Pollution*
Pollution can take many forms ranging from lead in vehicle exhausts to chemicals in drinking water, from contaminated land to pollution from industrial processes, from dangers in demolishing existing buildings (such as asbestos) to problems arising from erecting new buildings (such as CFCs), and from pollution of external environments as with coal fired power stations to indoor environments as with Sick Building Syndrome (RICS, 1993b).

9. *Sustainability*
This is essential if the long term future of human beings and their environment is to be properly preserved. A typical example could be a development project specifying the use of timber from sustainable resources and landscaping to provide more trees than existed prior to development. The main problem may be to find a price structure that fairly reflects replacement and conservation costs. Timber is also facing stiff competition from PVC-U for windows, with the latter commanding 70 per cent of the local authority market and 30 per cent of the UK market as a whole in 1992. Another approach could be to restrict the use of greenfield sites for development by using more hilly or derelict land (RICS, 1993b).

10. *Transport*
Transport has a major environmental impact,

using large areas of land for roads and parking. Highways require additional areas to accommodate cuttings and embankments and sterilise further areas during construction, and may cut through areas of ecological interest, farm units and land drainage systems. New motorways often increase the demand by traffic as with the M25 and also increase the demand for more land for development purposes, all of which require skilful environmental management. Road pricing may be needed in certain situations to rationalise the use of roads, as done in central Singapore at peak periods.

11. *Visual aspects*
The visual environment is important to the quality of life and can be readily enhanced or reduced by man's actions. One of the easiest ways of improving the visual environment is by attractive landscaping in a variety of situations. Discerning layout and design of buildings and harmonising with adjoining developments adds immeasurably to the attractiveness of the overall scene.

12. *Historical culture and heritage*
The UK is rich in listed buildings, original frontages retained when sites are redeveloped and sites given archeological scrutiny prior to development in historic city centres like York. This heritage will always be of great value to society, as evidenced by the countless civic societies established throughout the country. Much valuable work is undertaken in this area by such bodies as the Civic Trust and English Heritage from whom expert environmental advice is readily available.

Conclusions

All surveyors have a duty to inform their clients about relevant environmental issues and to ensure that they are given adequate consideration in the decision making process. It can often be difficult to evaluate them in monetary terms and it may be necessary to use cost benefit techniques, as described later in the chapter. As more client organisations proceed along the BS 7750 environmental management route, they are likely to prefer to engage advisors who have BS 7750 certification.

Environmental Impact Assessment of a Completed Shopping Centre

It is considered useful to the reader at this stage to examine an enlightened post contract monitoring of a recently completed building project from an environmental assessment viewpoint, including identifying the relevant criteria.

The Town and Country Planning (Assessment of Environmental Effects) Regulations 1988 stated that a shopping centre was likely to have a 'significant environmental impact' if it exceeded $10\,000\,m^2$ in area, had a site larger than 5 ha, or was in close proximity to more than 500 private dwellings, although these criteria were not mandatory. Moreover, there are projects which exceed these criteria but for which no environmental impact assessments (EIAs) were carried out, and even when they were there was no guarantee that the assessments were adequate.

An EIA can be no more than a prediction of what the assessors expect to happen to the environment if the particular project is implemented, and there is no obligation for the prediction to be compared to the live situation on completion. Indeed without effective post project feedback the whole process could become quite meaningless.

In like manner the BRE Environmental Assessment Method does have some weaknesses as it is mainly concerned with the building elements, their insulation levels, use of CFCs and associated aspects. To be really effective the investigation should have a wider remit to consider, for example, whether an out-of-town shopping centre can be energy efficient if it attracts thousands of cars per day and the relative importance of energy efficiency as compared with visual acceptability in terms of environmental impact (Parker, 1993).

In a serious attempt to face up to these and many other related issues in 1993, Benoy, the designer of the recently completed £20m Waterside Shopping Centre in Lincoln, using their environmental consultancy practice, undertook a detailed evaluation of the completed project, financed by £12 000 of sponsorship including contributions from the contractor and the quantity surveyor. To put the scheme into perspective, although the total area of the shopping centre was 12 500 m^2, the derelict nature of the site led the local authority to decide that no EIA was required before planning consent was granted.

It appeared to the designer that developers and planners needed more relevant criteria and sophisticated analyses, and hence in the evaluation the local authority, local retailers and the general public were asked whether they believed that the Waterside project was really necessary. It was felt that an EIA should have to justify the need for a proposed development, although in many cases this does not occur, as if there is no proven need for a new building then all the other environmental criteria are irrelevant.

A check was made on whether the planners and developers considered alternative sites, which is a mandatory requirement on larger projects, and then proceeded to examine the impact of the construction work, which is often overlooked but which can be a major issue on large city centre projects.

Nearly all city centre developments have archaeological implications and as the Waterside project was only 5 m from the River Witham, the effect on local watercourses was considered in the evaluation. An important socio-economic issue to be investigated was the effect of the development on road traffic patterns and pedestrian flows. As the centre had no car parking provision, it was not expected to draw extra traffic from out of town but it could attract local residents from other retail areas and thereby contribute to their decay.

Another objective was to endeavour to evaluate the visual impact of the development, by asking users of the centre, English Heritage and the Royal Fine Art Commission to compare photographs of the undeveloped site with the artist's impressions that formed part of the planning application and finally the finished development. This would determine whether the artist's impressions misled the public and the planners. Other aspects considered included policies for the management of solid wastes, mainly packing and wrapping, as most shopping centre managers have little interest in solid waste management and there are normally few facilities for tenants to recycle their waste.

Ultimately, subject to Department of Trade and Industry (DTI) approval of Benoy's application for development funding, a much more comprehensive and sophisticated environmental assessment, including an elemental analysis of the building and its components would be available in 1995. The results of this investigation could have a valuable spin-off at a time when retail development and urban redevelopment were gaining momentum (Parker, 1993).

CONCEPT OF ENVIRONMENTAL ECONOMICS

The need has been identified for quantity surveyors, in a role as construction economists, to be concerned with all aspects of the construction process and the underlying forces behind the various activities. These encompass the effects of public and private investment policies and aesthetic and planning factors, all of which play a part in determining the system of economic forces which underlie the construction process. A study of economics in its broadest sense entails consideration of how people behave in their activities of producing, exchanging and consuming, and the motivating forces behind these activities.

The term 'environment' embraces constructional works and their general surroundings and thus environmental economics can be considered as the study of the forces affecting the use of assets and resources in satisfying man's need for shelter and a properly managed environment. The surveyor's skill should not be confined to the measurement of physical features or entities but should be extended to measure the forces which are at work in the deployment or changed use of the resources which shape our environment. Expressed in another way, we must deal with cause as well as effect. Hence possibly the base on which the surveying profession is founded should be changed from the land or property to the environment.

The quantity surveyor as a building/construction cost consultant or building/construction economist may be called upon to advise a building client on a number of separate but interrelated issues. The main issues are now listed, together with the broad field of knowledge and expertise that is needed to give effective advice in each case.

(1) *Why* the client should build. Requires a knowledge of economics and the ability to forecast trends in the economy, coupled with an appreciation of the client's financial and production problems and a knowledge of building costs.

(2) *Where* the client should build. Involves a knowledge of national and international markets, economics of transportation and communication, population trends, taxation benefits and building and planning legislation.

(3) *When* the client should build. Requires a knowledge of the alternative ways of financing buildings, legislation affecting capital and revenue expenditure and investment allowances.

(4) *How* the client should build. Requires expertise in contract procurement planning and administration, selection of technical advisers and a construction team, and ability to maintain financial control of the contract.

(5) *What* the client should build. Involves coordination of the specialist advisers to ensure value for money and evaluation of capital and future costs of alternative design solutions, and of completed buildings for sale or lease.

PUBLIC AND PRIVATE INVESTMENT

It is important to realise the magnitude of the sums invested annually in building work in the UK, despite the substantial cuts in funding in the public sector during the 1980s and particularly in the 1990–93 recession. Table 16.1 gives the value of British construction output from 1988 to 1993. It shows that in 1992 the total volume of new work was split between the public and private sectors in the proportions of approximately 29 and 71 per cent respectively, excluding infrastructure work which was probably shared between the two sectors with the government's strong preference for private financing of this work either wholly or partially. Maintenance and repair work accounted for some 43 per cent of the value of all building work in 1992. According to the United Nations, the output of the UK construction industry accounted for about nine per cent of the gross domestic product (GDP) in 1991, which together with Ireland and the United States were at the bottom of the international league.

The public housing sector has suffered the largest cuts and was estimated in 1993/94 to be at less than 45 per cent of the 1981 level of construction in this sector. After allowing for inflation the relative position is worsened considerably. Local authority housing starts in England and Wales in the 12 months prior to November 1993 stood at 773 compared to 43 123 in the 12 months of 1979 (Housing Minister's statement in House of Commons, 19 January 1994). Hence the public housing output in table 16.1 would appear to include the output of the housing associations, although they are not strictly classified as part of the public sector, yet they provide social housing. This view is confirmed by the DoE figures for housing starts in 1992 of 35 800 in the social sector and 119 900 in the private sector. There were also substantial transfers of local authority housing stock in 1993 with 40 404 dwellings sold to owner occupiers and 26 488 to housing associations (Minister of Housing's statement in House of Commons, 19 January 1994). The forecast figures for private housing showed only minor improvement starting from a low base and expressed in 1985 prices. Some increase in confidence among building contractors and a low level of mortgage interest should result in some improvement in 1995/96, although this is still impeded by the high level of unemployment and the large number of mortgage debts.

Building demand in the UK declined progressively quarter by quarter from 1990 to mid 1993, with building work in the private commercial sector falling by half between 1990 and 1993, because of reducing demand and excess capacity built up in the 1980s, and to a lesser extent in the private industrial sector. Hence major building contractors were endeavouring to increase their activities overseas with only limited

Table 16.1 Construction output (Great Britain): by type of work at constant (1985) prices, seasonally adjusted (£ million)*

		New housing		Other new work				All new work	Repair and maintenance				All repair and maintenance	All work
					Other new work excluding infrastructure				Housing		Other work			
	Q	Public	Private	Infra-structure	Public	Private industrial	Private commercial		Public	Private	Public	Private		
1988		752	5 465	3 006	2 486	2 394	5 640	19 743	3 812	6 067	3 849	3 489	17 217	36 959
1989		741	4 332	3 291	2 773	2 636	7 382	21 155	3 968	6 534	4 000	3 815	18 317	39 472
1990		680	3 271	3 963	3 032	2 588	8 254	21 787	4 043	6 340	4 117	3 913	18 413	40 200
1991		588	2 835	5 029	3 249	2 103	7 044	20 849	3 479	5 699	3 734	3 610	16 523	37 372
1992		986	2 896	5 218	3 300	2 009	5 716	20 125	3 381	5 168	3 457	3 403	15 410	35 535
1900	2	164	791	937	778	667	2 124	5 461	1 089	1 622	1 026	962	4 698	10 159
	3	169	793	1 043	798	595	2 063	5 460	972	1 525	1 042	974	4 513	9 973
	4	158	836	1 027	717	673	2 021	5 431	911	1 557	1 014	979	4 460	9 892
1991	1	146	663	1 261	803	546	1 901	5 319	914	1 434	957	939	4 243	9 562
	2	135	757	1 200	771	504	1 799	5 166	869	1 408	956	940	4 174	9 340
	3	144	724	1 277	743	555	1 718	5 16[illegible]	883	1 436	925	878	4 121	9 282
	4	164	691	1 292	932	499	1 626	5 20[illegible]	813	1 422	897	854	3 985	9 188
1992	1	194	702	1 320	820	513	1 593	5 14[illegible]	885	1 260	908	854	3 908	9 049
	2	230	711	1 320	845	524	1 509	5 13[illegible]	844	1 230	861	841	3 775	8 914
	3	260	752	1 262	830	506	1 367	4 97[illegible]	842	1 308	854	843	3 846	8 824
	4	301	731	1 316	806	466	1 247	4 86[illegible]	810	1 371	834	865	3 880	8 748
1993	1 R	264	701	1 572	872	469	1 124	5 00[illegible]	892	1 190	829	809	3 720	8 722
	2	294	778	1 419	862	525	1 128	5 00[illegible]	910	1 176	771	804	3 661	8 665
	3 P	294	790	1 258	882	494	1 179	4 89[illegible]	822	1 265	809	890	3 786	8 683

Source: DoE press release on Construction Output (1993).

* Output by contractors (including estimates of unrecorded output by small firms and self-employed workers) and output by public sector direct labour departments classified to construction in the 1980 Standard Industrial Classification.
(P) Provisional; (R) Revised.

success, despite the large construction programmes being implemented in some overseas countries, but increased competition was being experienced from overseas contractors, particularly those from Japan and South Korea. In 1994, bankers to Britain's largest contractors considered that there were too many firms chasing too little work and charging too little for it. They advised builders to seek mergers in an attempt to reduce the over capacity which they believed would continue until 1996.

STRUCTURE OF THE CONSTRUCTION INDUSTRY

The construction industry includes the constructors of both building and civil engineering works, and there are a considerable number of contractors who undertake both functions. Building work varies enormously in its scope and nature, from the erection of large multi-storey blocks to the execution of works of minor repair and maintenance. It is also possible to distinguish between general contractors who undertake general building work and specialist contractors who concentrate on specific trades, such as plumbing, joinery, painting, plastering and electrical work.

There has also developed 'the lump' of labour-only subcontractors who undertake specific trades for the general contractor. The main contractor usually supplies most of the equipment and all of the materials, and the self-employed operatives quote him a 'lump sum' for the work which they subsequently share between themselves. They have been heavily criticised by many building workers and the trade unions, as they are believed to manipulate national insurance payments and evade income tax and training board levy, although there has been significant tightening of the controls to prevent this, and the numbers engaged in this type of activity reduced considerably in the 1980s and 1990s. They sometimes produce poor quality work, adversely affect union structure in the industry, and their form of employment seems to offer no redress for industrial accidents. A government committee of inquiry, whilst underlining the grave disadvantages inherent in this arrangement did, nevertheless, recognise that it combined the specialist skills of the subcontractor with a psychologically good form of wage incentive. The tax loopholes were largely effectively dealt with under the Finance No. 2 Act 1975, by provisions which came into force in spring 1977.

It is possible to classify building firms according to the type of work that they undertake, as follows:

Civil engineering contractors. Those who undertake very large civil engineering projects such as motorways, bridges, dams, reservoirs and harbour works; these firms usually operate on a national or even international basis.

Building and civil engineering contractors. Those contractors who undertake large contracts in both fields, often at either national or regional level; these are medium to large-sized contractors.

General builders. These can be grouped into a number of separate categories.

(1) Large firms employing over 1000 employees operating on a national scale and undertaking the erection of a wide range of buildings from housing estates to offices, hospitals and factories.

(2) Medium-sized firms who operate at regional level and undertake all but the largest contracts.

(3) Speculative firms who specialise in building housing estates in advance of demand.

(4) Small firms who are generally restricted to the erection of single houses or small groups of dwellings, and improvements, conversions and maintenance work.

(5) Very small firms of jobbing builders who confine their activities to maintenance and repair work.

Specialist firms. Those who concentrate on a specific trade, such as roofing contractors and plasterers, or who offer a special service which may not be confined to the construction industry, such as heating and ventilating engineers and electrical contractors.

The output of the construction industry was

approaching one-half of the country's national capital formation in 1993, and almost one third of construction work occurred in the public sector.

The construction industry in the United Kingdom is very fragmented with construction firms varying widely in size as well as in the type of work they perform, as illustrated in tables 16.2 and 16.3.

Over the period 1982–91 the number of smaller construction firms increased substantially, rising by almost 65 000, despite the effects of the 1990–93 recession and the high number of bankruptcies, while the medium sized firms reduced significantly and the very largest firms stabilised. In 1992 small firms employing 24 or fewer employees accounted for 98 per cent of the total number of construction firms employing just in excess of one half of those working for private contractors and carrying out less than 40 per cent of the construction work done. By comparison the very large firms, em-

Table 16.2 Private contractors – number of firms (October each year)

	1982	*1983*	*1984*	*1985*	*1986*	*1987*	*1988*	*1989*	*1990*	*1991*	*1992*
(a) By size of firm											
1	55 498	64 585	71 386	72 896	76 946	79 354	83 484	94 218	101 223	103 169	94 452
2–3	44 872	51 370	54 533	54 405	54 223	54 712	57 878	67 189	71 498	70 452	68 486
4–7	25 249	27 489	27 081	24 171	24 455	24 838	25 639	24 984	23 403	21 664	30 395
8–13	8 630	7 129	7 241	7 164	7 067	7 074	6 156	5 869	5 362	4 981	5 240
14–24	4 994	4 949	4 922	4 582	4 520	4 485	4 306	4 212	3 935	3 429	3 574
25–34	1 733	1 684	1 604	1 519	1 394	1 507	1 467	1 478	1 420	1 186	1 146
35–59	1 682	1 603	1 529	1 480	1 502	1 520	1 471	1 458	1 305	1 100	1 148
60–79	498	541	484	441	448	456	451	450	442	382	361
80–114	411	418	406	409	369	393	406	421	432	372	317
115–299	562	563	555	512	501	507	510	530	507	431	387
300–599	152	154	144	141	130	143	138	153	150	137	103
600–1199	76	72	74	66	71	71	70	66	69	58	59
1200 and over	38	39	40	39	34	35	42	48	47	39	36
All firms	144 395	160 596	169 999	167 825	171 660	175 095	182 018	201 076	209 793	207 400	205 704

Source: DoE, Housing and Construction Statistics.

Table 16.3 Private contractors – total employment (thousands – October each year)

	1982	*1983*	*1984*	*1985*	*1986*	*1987*	*1988*	*1989*	*1990*	*1991*	*1992*
(a) By size of firm											
1	50.3	64.3	68.9	67.8	72.7	77.7	75.5	88.6	96.5	94.5	87.2
2–3	110.6	122.5	125.7	120.5	124.8	129.4	135.3	155.3	167.3	147.5	145.4
4–7	119.9	128.3	123.0	108.3	112.3	115.6	114.7	111.3	103.4	91.9	115.9
8–13	86.2	72.2	73.3	71.8	70.8	71.1	62.5	59.3	54.0	49.8	52.5
14–24	89.8	88.7	88.4	82.2	81.4	80.4	77.2	75.5	70.3	61.0	62.7
25–34	50.0	48.6	46.6	43.6	40.1	43.5	42.3	42.7	41.0	34.2	32.9
35–59	74.6	70.6	67.5	65.7	67.2	67.5	65.6	64.8	58.2	49.0	50.6
60–79	34.2	37.3	33.2	29.9	30.7	31.2	31.0	31.0	30.5	25.9	24.4
80–114	39.0	39.6	38.5	38.8	34.9	37.0	38.3	40.0	41.1	35.4	29.5
115–299	96.9	98.6	96.7	89.3	86.7	89.9	91.6	93.5	89.1	73.5	67.5
300–599	62.8	64.8	58.4	59.3	55.4	60.8	58.1	63.8	63.1	57.0	41.6
600–1199	62.7	60.7	61.1	55.2	59.2	60.0	58.3	55.3	56.7	46.0	47.8
1200 and over	111.8	110.6	109.0	108.6	99.2	97.8	113.7	129.6	125.3	100.6	84.1
All firms	988.8	1006.8	990.3	941.1	935.2	962.0	964.0	1010.6	996.7	866.6	842.2

Source: DoE, Housing and Construction Statistics.

ploying over 1200 operatives, formed the most important employers and producers with the highest output per employee. In 1992 they accounted for less than 0.02 per cent of all construction firms, yet were responsible for nearly 10 per cent of all employment and accounted for about 14 per cent of all work done.

Table 16.4 shows the spread of employees in construction work during the period 1981 to 1993 over the main categories of employer and employee. This shows that in mid 1993 the total number employed in the construction industry was in the order of 1.4 m and that this was over 400 000 less than the average for 1989. The proportions in the three employment categories in 1992 were 46 per cent with contractors, 12 per cent with public authorities and 42 per cent self-employed. These figures showed only small variations in the proportions of total construction manpower between the second quarter of 1993 and those prevailing in 1989 (contractors down 3 per cent and self-employed up 1 per cent). It is interesting to note the ratios of operatives to administrative, professional, technical and clerical staffs (APTC) with contractors and public authorities in the second quarter of 1993 – both were on 66 per cent.

There is no general optimum size of firm or organisation, the best size depending on the nature of the work, the conditions under which it needs to be carried out, and the nature of the organisation and the ability of the management. Although large firms give rise to co-ordination problems, nevertheless they secure many advantages such as increased mechanisation, fuller use of resources, economies from specialisation, improved techniques, and better financing and purchasing arrangements. On the other hand, the small firms are in an advantageous position for dealing with maintenance and repair works and small contracts for new work. The capital needs of small firms are extremely limited and they enjoy a comparatively

Table 16.4 Construction manpower: employees in employment (thousands – monthly averages of calendar months)

	Actual								
	Contractors		Public authorities[1]		All employees on register[1]	Estimated employees not on register[3]	All[4]	Self-employed[5]	All manpower
	Operatives	APTC[2]	Operatives	APTC[2]					
1981	679	230	205	104	1218	3	1221	385	1606
1982	619	223	187	95	1123	13	1136	397	1533
1983	600	214	183	88	1084	33	1116	411	1527
1984	585	214	172	85	1055	51	1106	454	1559
1985	556	213	169	84	1022	64	1086	470	1556
1986	531	212	160	82	985	66	1051	488	1538
1987	549	222	156	77	1005	53	1058	535	1592
1988	562	235	147	74	1017	69	1086	592	1677
1989	583	254	134	71	1042	67	1109	698	1806
1990	583	266	127	66	1041	76	1117	715	1832
1991	516	251	120	63	950	90	1040	657	1698
1992	437	221	112	56	825	94	919	597	1516
1993 Jan.	404	212	106	51	774	93	866	559	1425
Apr.	399	205	100	50	754	90	844	557	1401
July	382	206	94	46	728	92	820	572	1392

Source: DoE, Housing and Construction Statistics.
1. Estimates by the Department of the Environment based on quarterly returns from contractors and public authorities.
2. APTC are administrative, professional, technical and clerical staff.
3. Estimates of employees not on DoE register of firms or declassified as working proprietors.
4. This series has a different coverage from the series published by the Department of Employment. This difference has declined over the years, e.g. 105 000 in 1981; 76 000 in 1987; 18 000 in 1993.
5. Estimates based on Department of Employment's Labour Force Survey.

high ratio of liquid to fixed capital. Furthermore plant requirements are small and much of this can be hired, and materials are mainly ordered to meet current needs and purchased on credit of sufficient length to permit recovery of the cost of the materials from the client prior to paying the builders' merchant. The smaller construction firms are particularly vulnerable in periods of financial difficulty or reductions in building work, as was evidenced by the large numbers of bankruptcies of building firms during the 1990–93 recession.

The smaller firms concentrating on small projects and repair and maintenance work often restrict their area of activity to about 25 km from their base, while the very smallest repair firms may go no farther than 8 km. Much repair work needs to be carried out quickly and the small firms are better fitted to give this type of service and they also tend to undertake new work in which the larger firms are not interested.

The largest construction firms tended to grow in size in the 1980s and 1990s, often through amalgamations between contractors and property development companies. Diversification into other areas of construction was often viewed as a means of spreading risk and workload, as was vertical integration into materials production and quarrying activities like Tarmac, and combining with non-construction firms as Trafalgar House; while institutional involvement in property created links between property developers, institutions and major contractors, which assisted in maintaining continuity of work for the large contractors (Balchin *et al.*, 1994).

Labour turnover is about twice as high in construction as in industry generally, with the greatest turnover in civil engineering where labour may be employed for a single contract or even part of one. Unemployment in the construction industry can also be about twice the national average and rises even higher in the winter period, leading to the use of considerable casual labour. Because of the imperfections of the market, high unemployment can exist alongside labour shortages and frequently does in the case of skilled craft operatives.

Over 80 per cent of operatives are employed by contractors and the remainder by public authorities. Approximately half of the employees are craft operatives who usually enter through apprenticeship or other training schemes. Recruitment has suffered through the raising of the school leaving age and the industry's poor perceived public image. The reasonable clear division between craft operatives and other labour which existed in the post war years has progressively become more blurred, partly resulting from the diversity of building operations and partly from the introduction of new materials and techniques. The Construction Industry Training Board (CITB) found that the number of construction trainee operatives fell by one-third between 1980 and 1989 to less than 50 000 in 1989 (Balchin *et al.*, 1994).

In 1994 the substantial market for the UK construction industry in the European Community (EC) had still to materialise despite the estimated potential of an annual seven billion Ecus, largely resulting from the in-built restrictions in various member countries. In 1990 construction activity by British contractors in the EC amounted to less than three per cent of their total overseas construction activity.

Nevertheless, many steps have been taken towards harmonising conditions in member countries with a view to eventually removing all fiscal and social barriers, and this should help in increasing inter-community construction work. In the case of quantity surveyors, there is no equivalent profession in Europe outside the UK, although some of the basic training is included in other disciplines. Many quantity surveyors and construction cost consultants have identified the opportunities and have opened offices in a number of cities, such as Brussels and Paris.

VARIATIONS IN WORKLOAD ON THE CONSTRUCTION INDUSTRY

The construction industry covers a wide range of loosely integrated groups and organisations involved in the production, renewal, alteration, repair and maintenance of certain capital goods (building and civil engineering projects). Capital funds are required on a long-term basis to finance plant and equipment and short-term for the purchase of materials and components and hire of labour. Unlike most other production processes, production takes place essentially at the site where the product is to be used rather than in a factory. Furthermore, most constructional projects are unique with individual design, engineering and production characteristics, and are invariably of very large size and weight at fixed locations with relatively long lives.

Fluctuations in workload are essentially of a cyclical nature. The long-term fluctuations are often associated with general variations in business activity while short-term changes are generally on a seasonal basis resulting from variable weather conditions. The increased use of mechanical plant, heating apparatus, plastic-sheeted enclosures, prefabricated components and industrialised building are helping to reduce the loss in output during the winter months.

The rise and fall in building activity has in the past had a cyclical duration of about twenty years, considerably longer than the normal trade cycle, and has been described as a *building cycle*. In times of prosperity increases in income will, amongst other things, lead to increased demand for residential accommodation. This stems from a rise in marriage rates and the increase in demand for separate or improved accommodation. House prices and rents tend to rise and this encourages new contractors to undertake house building, while existing firms expand their output of new work. In addition, the higher incomes create greater demand for consumer goods which in its turn generates a need for more factory space and increased workload for the construction industry. The output of the construction industry is fairly inelastic, as it takes a considerable period of time to prepare development proposals and erect the buildings.

After about ten years the bulk of the essential demands will have been met and the downswing of building activity commences. This will be accelerated by increased building prices arising from shortages of factors of production (resources). The reducing demand will gather momentum until the major part of building activity is confined to replacement buildings and repairs and maintenance. During the downswing a surplus of buildings is likely to occur, resulting in a lowering of prices and rents, a number of construction firms will experience financial difficulties and many building operatives will leave the industry.

Up until 1940 building cycles had occurred continuously in this country over a period of 250 years and with an average duration of about twenty years. Factors influencing the length of cycle are the long production period, the sluggishness of the industry to respond to changes in demand and population changes. For instance, a high marriage rate now will result in another marriage hump in about twenty years time, with consequently higher demand for homes. There may also be an increased number of older buildings falling due for replacement. Each building cycle is unique and other factors contributing to building cycles include the internal migration of persons, and external activities such as bad harvests and wars. As will be described later in the chapter, the Government is able to exert a considerable influence on the output of construction work, and so can restrict the harmful effects of building cycles, but rarely does so.

In viewing the construction market it is convenient to divide consumers into two main categories: clients for whom construction goods are a means of production and clients for whom construction goods are an end in themselves. The second type of client requires a structure for the amenity which it will provide over its lifetime. In a modern welfare society structures are required by people collectively through the Government as a social investment. For example, a library provides an amenity which people could not afford individually and a motorway may be demanded as an addition to or improvement of the infrastructure of an economy. While the purchase of a house is a postponable capital transaction for which demand is relatively inelastic and it is influenced considerably by the age structure of the population and the state of the economy.

The demand for industrial buildings is a derived demand. The accelerator principle postulates that if the demand for a product increases then the demand for capital goods used in its production increases, but at a faster rate. Since buildings, as a part of fixed capital are likely to be a relatively small proportion of total costs, the demand for industrial buildings is likely to be relatively inelastic. Commercial buildings include offices, shops and hotels and here again the accelerator principle also applies. Technological change and changes in tastes and fashions will have their effect on demand in this market. Consumers or users of commercial buildings are often different people from those who finance the buildings. Commercial buildings are more flexible than industrial buildings and, while there are regulations which govern the comfort of office personnel, the buildings are not usually designed for specific processes.

The demand for social-type construction is determined by social policy. Public utility structures such as museums, hospitals and roads are products provided out of public funds and this is an area in which political decisions and priorities play a major part. Public sector orders form a significant but reducing proportion of the total demand on the industry. It is important, therefore, for the industry to be able to assess realistically future demands that are likely to be made upon it by public authorities.

Throughout the late nineteen seventies and early nineteen eighties, and again in the early nineteen nineties, the workload on the construction industry continued to fall. In the early 1990s the most buoyant sector was commercial building, such as supermarkets, shopping malls, out-of-town shopping centres and retail warehouses, but even there the level of demand was still relatively low because of the general lack of confidence. By 1995 a small upturn was apparent but the recovery was generally judged to be fragile. Financial institutions were anxious to find somewhere to place their funds, as there was a substantial over-supply of offices, especially in London.

The number of unemployed in the construction industry stood at around 500 000 in 1994, representing about 17 per cent of the country's unemployed and over 30 per cent of the total labour force of the industry.

This substantial rate of unemployment should result in fewer problems in finding suitable site manpower in the short term. The construction industry was, however, concerned about the long-term future supply of craft operatives in view of the reduction in apprentices and the exodus of good quality labour to other forms of employment. Materials could pose problems as manufacturers tended to become cautious about the production of fairly specialised items.

Contractors found 1993 a particularly difficult year and this resulted in very keen pricing and the submission of many ridiculously low tenders. Many large firms continued to weather the storm, partly because of the cushion of large long-term contracts and overseas work, and partly by having earlier made provision for losses by postponing purchases of plant, buying materials economically, reducing the labour force and head office staff and giving fewer bonuses to employees. However, there was also much evidence of contractors using up their accumulated profits of the late 1980s and reducing on costs to a bare minimum. Obviously this policy could not be continued indefinitely without risking the failure of firms. Medium and smaller firms were less well placed and there were many insolvencies throughout 1990–93.

It was hoped that 1994/95 would see the beginning of the end of the recession and that clients of the industry would commission work at the design stage to be ready to build when an upturn occurred, although it was generally forecast that the upturn in construction work in 1994–95 would be on a relatively small scale. Construction industry orders in the first quarter of 1994 were the highest quarterly sum for five years, which was very encouraging but it subsequently proved to be a largely one-off situation. The construction industry had experienced the worst decline in work in decades and there was a fear that its ability to cope with a sudden upturn in succeeding years could have been seriously impaired.

Considerable concern was expressed about public capital expenditure policies and the government's monetary policy in the early nineteen nineties. In 1994/95 there was little evidence of a change in government policy although there appeared some scope for modest growth.

RELATIONSHIP OF OUTPUT OF CONSTRUCTION INDUSTRY TO AVAILABLE RESOURCES

There is a need to use the country's construction resources of land, labour and capital, to best advantage, as construction is a key factor in an expanding economy and increasing demand will place a heavy strain on the country's resources. There are severe limitations on available resources and it is advisable to:

(1) continually predict demand for several years ahead, refine annual predictions and translate predictions into terms which allow resources to be assessed:
(2) continually assess resources needed and refine annual assessments;
(3) continually survey availability of resources; and
(4) compare demand with available resources and take action to match demand and resources.

It is now generally recognised that programmes need to be established for some years ahead and to be of a rolling nature (where the period of years ahead remains constant and the annual programmes are reviewed and revised periodically). Expenditure on national road building may be planned for five to ten years ahead; school building programmes are usually announced about three years ahead; and hospitals and prisons can be up to ten years ahead. However, the forward planning of public sector projects became rather disorientated during the nineteen eighties and early nineteen nineties. Many industrial firms also produce investment programmes for a number of years ahead. It frequently takes several years to increase the supply of a particular resource.

It is helpful to express the demand for resources in terms which are readily identifiable, such as cubic metres of gravel, number of bricks, tonnes of steel and number of operatives. A disadvantage of monetary values is that they do not stay constant for long periods of time. To exploit the resources economically requires accurate information on their quantity, availability and location. In like manner the possible demand must also be known – its nature, timing and geographical incidence. There is also a need to improve the prediction, assessment and survey techniques to secure effective matching of supply and demand.

A vigorous and buoyant construction industry is indicative of a strong programme of national investment: new buildings, refurbished buildings and improvements to national fixed assets are the foundation of and the first step to future economic growth and social development. Unfortunately this was far from being the case in the early 1990s.

Time lags in the construction process tend to be long and variable depending on the size and complexity of the projects. The difficulty of using the construction industry to depress or increase demand in the economy is that any change in construction activity will initially be too slow to be effective. Significant changes will only occur after a substantial period of time, perhaps when a contrary effect is desired. Thus when governments seek to stimulate economic growth, construction activity may act as a restraint to growth. Conversely, when governments seek to restrain growth, construction activity may act as a stimulant to growth.

The Cambridge Economics Group (1981) postulated that in terms of gross domestic product and gross output, works of civil engineering and housebuilding were first and second respectively, but on inflation and balance of payments, they both trailed a long way behind increases in government current spending and employment subsidies. This is because they are the options that expand the economy most and draw in the most imports. Hence the two construction options do not offer the perfect solutions to the difficulties facing Britain in recession. However, on balance the Group believed that they were the most attractive ones in that they increase employment most for the least cost to the Public Sector Borrowing Rate (PSBR) and without causing a severe drop in productivity.

EFFECT OF GOVERNMENT ACTION ON THE CONSTRUCTION INDUSTRY

The Government has a crucial role in determining demand for the construction industry's output and its growth prospects, both because public authorities buy about 30 per cent of its output and because general economic measures have a powerful influence on the demand for private housing, and industrial and commercial building. A steady, rather than wildly, fluctuating growth of demand is particularly important for the industry if it is to plan its work ahead, make sure of adequate supplies and deploy its resources effectively. Because the industry is undercapitalised it soon experiences difficulties when monetary policies are introduced to check the economy as a whole, as became very evident in the late nineteen seventies, early nineteen eighties and early nineteen nineties.

Over the years successive governments have found it necessary sometimes to restrain and at other times to stimulate the economy, and the construction industry is invariably caught up in this process. When policies of restraint operate, there is usually a reduction in the volume of public building work, particularly house building, and projects such as motorways and town centre redevelopment schemes are likely to be cut back or postponed. The adverse effects on contractors will not be immediate as contracts already in hand will normally be completed. In the longer term however the results can be serious, resulting in:

(1) unemployment of building operatives;
(2) smaller building firms being forced out of business;
(3) larger construction firms being reluctant to invest large sums in new plant and equipment or to experiment with new techniques;
(4) suppliers of materials and components being unlikely to extend or modernise their plant;
(5) recruitment of persons into the industry at all levels being made more difficult;
(6) the lack of continuity of construction work which makes for increased building costs and reduced efficiency; and
(7) the reduced resources of the industry may not be sufficient to cope with the eventual upturn and could result in overheating of the industry and inflated prices.

Every industry is affected for good or ill by the state of the national economy, and this is particularly so in an industry that is both home based and labour intensive. The construction industry's unattractive position in the late nineteen seventies, early nineteen eighties and early nineteen nineties was due largely to the poor economic condition of the country. The industry was particularly vulnerable as approximately one-half of the national capital formation was in construction and about 30 per cent of construction occurred in the public sector.

Capital investments take several years to reach fruition and often many more years to become profitable. Unfortunately politicians, who set the scene, work on a much shorter time scale than those who operate within it. There are a number of public sector capital schemes which can be postponed until the country is in better shape without causing too many social problems and little, if any, damage to the national earning capacity. But there are other public sector capital projects which are potential wealth creators, and which, if cut too sharply, will permanently damage the national ability to compete in world markets.

In 1977 the group of eight, representing the professionals, contractors' operatives and material producers was established to press the case for more construction work with Ministers and later Members of Parliament. The greatest pressure was on the need for more public spending to improve the infrastructure and to raise the efficiency of the economy through increasing building investment, and to reduce growing unemployment and the threat of developing social problems. The same problems surfaced in the 1990–93 recession. Because of the construction industry's fragmented nature, both in terms of diversity within the construction team and the large number of sites and firms scattered throughout the country, the industry has traditionally suffered from lack of cohesion and restricted political influence. The formation of the group of eight was hoped to rectify these deficiencies but its success rate was very limited and it was later abolished. The Construction Industry Council (CIC) was subsequently established, representing the main professional bodies connected with the construction industry to provide a unified forum, but it does not include contractors.

The Government's initial argument that the private sector would move in to fill the vacuum created by the cuts in public work has never been a serious proposition, because much of the construction industry's workload is either provided to users whose needs can only be catered for by the State or, as is the case with many civil engineering projects, has such long lead times as to be of little interest to private financiers who require fast returns and low risks. However, to make any headway with the Government, the industry has to convince its Treasury critics that the increase in construction employment, output and efficiency which will result from extra public spending on construction will cancel out the negative effects on public sector borrowing, interest rates, inflation and the balance of payments.

The various forms of government action and their likely repercussions on the construction industry are now considered in some detail.

The Government as a Client

The Government, either directly or through public agencies, such as local authorities, nationalised industries and other *ad hoc* bodies, purchases about 30 per cent of the output of the construction industry annually, and the performance of the industry thus plays an important role in any government programme of economic growth and social advancement. Since the second world war, the construction industry has handled many large-scale reconstruction and development schemes designed to restore and improve the infrastructure of the country. Many of these were dependent on government initiative, for instance the modernisation and electrification of railways, motorways programme, building of new and expanding towns and erection of local authority houses.

Relatively small changes in government policy may cause significant variations in the construction industry's workload, to which the industry will find difficulty in adjusting itself owing to the nature of its structure and product. Government spending on capital investment, such as schools, hospitals, roads and public housing can be curtailed with relative ease and these have the greatest impact on the work of the construction industry. A change of priorities by the Government, such as a transfer of funds from housing to motorways, creates problems for the construction industry, as these two activities are largely undertaken by different types and sizes of firm using substantially different forms of plant. Hence, it is not only the level or volume of construction activity which needs considering but also its composition. For example, substantial increases in improvement grants will result in a significant rise in the volume of work available to small builders.

Other Forms of Public Control

In chapter 14 it was shown how the Government seeks to influence the location of new industry through planning control and by giving grants, subsidies and tax incentives. New industry is therefore likely to be steered towards locations where the Government wishes to see industrial development established, such as development areas, and these may not necessarily be areas where there is a surplus of construction resources. The Government also operates controls in the public interests, of which a typical example is the Building Regulations aimed at securing minimum standards of construction in buildings, which are more exacting than those on the Continent, and this must influence costs and hence demand for building work. Another example is the Factories Act 1961 which prescribes certain minimum requirements for temperature, ventilation and lighting in factories, and these requirements must affect constructional costs.

On occasions, social and political considerations can influence the form and pattern of a construction programme. For instance in 1971, the Department of Education and Science announced that all new school building in 1973–74 was to be devoted to primary schools, and this will affect the nature, scale and location of educational building in that particular financial year. Since the second world war the form and volume of the educational building programme has varied considerably from year to year and has been very much influenced by economic, social and political factors.

Monetary Policy

The Government uses various controls, such as minimum lending rate or bank base rate, open market operations and hire purchase restrictions, to alter the level of interest rates, and to control the amount of credit available and the terms on which it can be obtained. Credit restrictions strike at contractors both directly, through banks and lending institutions, and indirectly, as through builders' suppliers who are always a ready source of trade credit and could be the mainstay of many small builders.

In the short term an increase in interest rates may not significantly affect the demand for construction work. In the long term it will result in higher prices for buildings and this may cause some developers and other building clients to refrain from investing capital in building projects, as they are no longer as profitable, and the clients themselves may experience difficulty in obtaining finance at the higher interest rates. Hence some projects are likely to be postponed or abandoned. Furthermore, contractors relying on bank overdrafts as a main source of working capital may find their loans curtailed and thus be compelled to reduce their output.

Taxation

Taxation is an important tool in a government's fiscal policy and has significant implications stemming from the alternative uses to which the money could have been put had it not been taken as tax, and the widespread distribution of the tax. Taxes are levied on both capital and income. An increase in capital tax on the value of property or on gains made from the sale of buildings or land may result in decreased demand, whilst an increase in tax levied on property income or use is likely to result in reduced demand for new buildings, unless it is offset by higher rents or profits. A practical example was the Community Land Act 1975 and the Development Land Tax Act 1976 which introduced a charge in the majority of cases whenever an owner of land realised its development value by disposing of his interest or carrying out material development. The first £50 000 of development chargeable to any person in a financial year was exempt, while the remainder was chargeable at 60 per cent. The development land tax was abolished in 1985.

Selective employment tax was introduced in 1966 as a levy on all non-manufacturing employment, to accelerate the transfer of surplus labour in service industry to manufacturing and, at the same time, to produce about £700 million per annum in additional revenue. Building was classed as an *assembly* industry and was accordingly liable for tax; this resulted in increased building costs, but subsequently the tax was removed entirely. Value added tax (VAT) was introduced in Britain in 1973. This tax operates at each stage of manufacture with the tax being levied at a flat rate per cent on the increase in value of the article accruing from the manufacturing process. New building work has a zero rating, but maintenance and repair work is subject to tax at the standard rate (17½ per cent in 1994).

Fiscal Policy

Apart from taxation, a government can influence the level of economic activity by regulating the amount of its spending. We have already seen that the level of public spending on construction work is substantial. After the first world war it became a generally accepted facet of government policy that spending on public works needed adjusting to counteract changes in spending in the private sector of the economy. It was argued that in times of deflation and rising unemployment, an expansion of public works would increase employment and investment and revive the economy (pump priming). Furthermore, public works inevitably involve construction. It seems unlikely that the construction industry could absorb a sufficient proportion of the unemployed to correct the adverse level of economic activity as many of the unemployed could not undertake construction work. Nevertheless, it appeared a worthwhile policy in the early 1990s (except to the government), with so many unemployed construction operatives and professionals, coupled with a fast decaying infrastructure and a serious lack of social housing. In 1993 the number of unemployed in the United Kingdom rose to around three million despite the falling inflation and interest rates because of the general economic recession and lack of consumer demand and confidence. The construction industry suffered severely through substantial cuts in both public and private sectors, resulting in extensive unemployment, many insolvencies and the dispersal of design teams. In direct contrast in the early 1970s, the construction industry became grossly overheated when demand far exceeded supply and building costs rose by approximately 50 per cent in one year. The substantial cuts in funding of housing associations announced for 1994/97 will have serious repercussions on the supply of social housing, the homeless situation and the workload for the house building industry, coupled with minimum housing provision by local authorities and a medium level of output of private houses.

Needs of the Industry

During the last three decades, successive governments have used both monetary and fiscal measures to alternatively restrain and stimulate the economy. This 'stop-go' policy has had serious repercussions for the construction industry, resulting in a continually fluctuating workload. It has contributed to changes in the size-structure of construction firms, restricted additional capital expenditure on plant and equipment, reduced the effective labour force, contributed to the use of labour-only subcontracting, retarded the use of new techniques and reduced efficiency. It is important to discard the concept of the industry as some form of economic regulator and to regard it as a medium for steady continuous development.

Technologically the Government has provided a lead particularly through the Building Research Establishment, and the Department of the Environment. It should, however, pay more attention to the financial needs of the industry.

EUROPEAN UNION PROPOSALS FOR GROWTH AND COMPETITIVENESS, 1993

An enlightened set of proposals to help rectify the depressing situation in 1993 emanated from the European Union (EU), on account of the economic crisis which had encompassed the whole of Europe. At their summit in Brussels in December 1993, EU Heads of State and Government approved the European Commission's (EC) White Paper on growth and competitiveness.

The paper set out a strategy aimed at reducing unemployment and creating jobs for the majority of the 17 m unemployed in Europe. It also outlined the need for non-inflationary expansion that would be outward looking, leaving room for private initiatives to develop. This growth will be backed up with a major programme of investment in infrastructure works costing the EU a total of 120 b ECU (£85b) over the period 1994 to 2000 (three times the amount of investment in the European growth initiative/recovery package agreed by the EU leaders at their summit in Edinburgh in 1992).

EU leaders also instructed the Council of Ministers and the European Parliament to prepare, in time for July 1994, the outline plans that were needed to develop rail, airport, seaport, electricity and gas infrastructures. The intention was to review the position at every end of year summit when UE leaders will assess how the action plan is progressing (RICS, 1993/94).

URBAN RENEWAL AND TOWN CENTRE REDEVELOPMENT

Basic Problems

One of the most urgent and complex problems facing many local authorities at the present time is that of urban renewal. Most towns and cities have evolved over a long period on a radial road pattern which is ill-suited for present-day traffic needs. Surrounding the pressurised inner core is often a girdle of mixed residential, commercial and industrial uses in varying stages of obsolescence, frequently termed twilight or 'blighted' areas. Comprehensive development of all these areas is essential if satisfactory layouts are to be achieved.

It is evident that nearly all the social troubles and divisions from which we suffer today stem from that fatal social division, that appalling neglect and blindness to human well-being which the land and its use should serve (Bryant, 1970).

Complete redevelopment of town centres is necessary in some towns, particularly where extensive growth is anticipated. Redevelopment is needed to overcome present deficiencies, such as bad traffic congestion, lack of parking space, inadequate loading and unloading facilities, restrictive traffic regulations, excessive fumes and noise, dangers to pedestrians, and in some places there is a definite conflict arising from the use of particular roads by different kinds of traffic. In smaller towns, pedestrianised shopping streets and precincts with rear-loading facilities, combined with a suitable network of distribution roads and adequate car parking space will often provide the best long-term solution.

Some of the conclusions contained in the Buchanan Report on Traffic in Towns in 1963 are still very relevant to town centre redevelopment today and are included here.

'The freedom with which a person can walk about and look around is a very useful guide to the civilised quality of an urban area. . . . Nothing would be more dangerous at this critical stage in planning for the new mobility offered by the motor vehicle than to underestimate its potential. . . . Even as this report is being written the opportunities are slipping past, for in many places the old obsolete street patterns are being 'frozen' by piecemeal rebuilding and will remain frozen for another half-century or longer. . . . The choice facing society is between affording, for its own convenience, a road system of some elaboration, or staggering its hours of work all round the clock to its obvious inconvenience.'

The town centre, with its concentration of people and traffic, variety of land use, intensity of building, historic and civic interest, diversity of ownership and high land and property values, is the most vital part of any town. It is the social hub of the town, the centre of local business and civic life, the entertainment and shopping centre. It must not be permitted to disintegrate under the strain of intolerable traffic conditions or the impact of haphazard development.

A number of important practical issues result from redevelopment. They entail extensive and costly diversion of underground services, many of which were routed under existing roads, which themselves are to be diverted or closed. The twilight areas are characterised by multiple-occupation, overcrowding, lack of essential facilities, neglect, dilapidation, mixed industrial and commercial uses and general environmental decay. There is both a social problem and one of physical decay, but it must be questioned whether the community can afford to allow the wholesale demolition of the houses, most of which could still have many years of life. Property owners may also be very dissatisfied with the method of delineating unfit properties in a slum clearance area, whose compensation is restricted to site value. However, urban renewal offers the opportunity to restore integrity and character to a depressed neighbourhood.

Sharp described in a gold medal address to the Royal Institution of Chartered Surveyors (1975) how past planning policies have contributed to the decay of inner areas. The promotion of decentralisation from inner cities, and making it difficult for industry and other forms of employment to survive there, were a central theme of planning policy, in the post-war years, and displayed a surprising insensitivity to their likely future problems. At local level planners were often too much concerned with amenity and too little with the positive encouragement of industry, commerce, entertainment and the whole variety of human activities which combine to produce a living community. Planners can so easily prevent development, but they seldom seem able to induce it.

The post war planning policy for inner cities

resulted in large areas of derelict land and deteriorating buildings in the 1970s to 1990s, which can be very expensive to rejuvenate. Much skilful and attractive redevelopment and rehabilitation is needed to bring life, interest and attractiveness to many of these depressed inner city areas, as was highlighted by the Civic Trust (1988). There is also a pressing need to create more ecologically rich open space in inner cities, preferably connected by 'green' networks (RICS, 1992).

Although regeneration is a complex of social and economic factors, an ICE Infrastructure Policy Group (1988) emphasised that supportive infrastructure was an essential and urgent ingredient for any initiative to suceed and to be sustained. The significant feature of urban decline is its downward spiral, whereby its characteristics reinforce each other to make decline steeper with the passage of time.

The characteristics of urban decay and deprivation vary greatly from one location to another and no single form of organisation is suited to all situations. Hence there it is desirable to exercise flexibility in the way the problems are approached and in the choice of urban development organisation, but preferably based on the local community provided with adequate powers to resolve the problems. The ICE (1988) also wisely postulated that rational and efficient development would be secured by long term indications of future capital budgets, possibly in the order of three years, with the appraisal of grant applications conducted at one level only, preferably regionally.

Housing

The poor condition of housing in inner cities is a major contributor to their accelerated decline. Private rented housing, as described in chapter 5, has steadily deteriorated, although in some inner areas, especially of London, it remains an important, albeit dilapidated source of housing provision. Much has been done by housing associations, supported by successive governments, to renovate these houses, although in the early nineteen nineties the work of the housing associations was reduced significantly by government action, particularly in relation to rehabilitation work and, in 1994, a DoE consultation paper proposed that the obligation on councils to provide permanent homes for homeless people should be replaced by offering temporary accommodation only with the private sector providing the homes. Unfortunately this is an ill-conceived, unsatisfactory and expensive approach as about one-fifth of the private rented stock is unfit for human habitation and there would be a sharp rise in the cost of housing benefit. Too many government financial decisions in the 1990s appear to be motivated by expediency and without adequate regard to the longer term consequences.

There were, in 1991, about 600 000 unfit houses in urban areas, most of which will in time be demolished. There were a further 800 000 sub-standard houses, some of which will be rehabilitated and others demolished. It would be good policy to expend more effort and money in improving the housing environment. If the housing is attractive, the population mix is a better reflection of society as a whole, as it tends to make the areas more secure and more dynamic and will assist in the attraction of industry and other uses (Baron, 1980).

The European Liaison Committee for Social Housing (CECODHAS) (1993), when commenting on corporate economic and social housing in Europe, highlighted the need for a housing policy for the citizen, the tenant and the landlord, against a backdrop of socially secured and justifiable outline conditions. The Committee believed that housing must not and cannot be surrendered solely or principally to the mechanisms of the market and those of competition, although members in EU countries all start from varied positions and even where council housing exists, the supply of adequate housing to the population is mostly no longer guaranteed. In all EU countries there is a blatant lack of housing space for certain housing groups, due to regional as well as social disparities, and the gap between poor and wealthy widening all the time with major problems of poverty and homelessness, as referred to earlier.

Financial Aspects

The main forms of finance available for the regeneration of derelict urban areas were:

- *Urban Areas* (UAs) were designated under the Inner Urban Areas Act 1978 whereby substantial government grants and loans were provided for the improvement and conversion of old industrial and commercial property for completion by the 1900s, and many local authorities benefited from these provisions.
- *Urban Development Corporations* (UDCs) were established, starting with London Docklands and Merseyside Development Corporation with the main objectives of providing adequate infrastructure, reclaiming and servicing land, renovating old buildings and developing new factories, with a sum exceeding £500m set aside each year up to the early 1990s. The main weaknesses were the lack of liaison with local authorities and the absence of associated social housing.

In 1994, England's twelve urban development corporations were planning sales of hundreds of hectares of land for construction, after the government told them to wind up operations within the following four years, after which they will hand over planning control to English Partnerships, the government's latest urban regeneration agency, or back to local authorities.

- *Enterprise Zones* where simplified planning procedures and certain financial advantages operated but these have now been terminated.
- *City Challenge* was introduced in 1991, whereby local authorities submitted bids for funds of up to £37.5m per annum for five years. It appeared to be a comprehensive and ambitious scheme involving full and effective partnership with the private sector, participation and involvement of the local community and effective arrangements for implementation and development. However, it was phased out from £245m in 1992/93 to £91m in 1994/95.

Also in 1993 in City Challenge's second year, some partnerships between local authorities and private developers were struggling to regenerate some of England's worst urban areas because of government red tape and insufficient funding. The sponsor of each individual project had to appraise the feasibility of the scheme, followed by an appraisal of the first appraisal by the City Challenge team, followed by yet another appraisal by the DoE. This procedure was far removed from the then environment secretary Michael Heseltine's original vision, when he launched City Challenge in May 1991. His aim was to cut red tape by siphoning cash from up to eight urban funds and channelling it into a single grant, for which local authorities in partnership with private backers would compete by dealing with a single funding body, thereby saving time and work (Stewart, 1993).

Furthermore, this was not new money as it was 'top sliced' from existing regeneration schemes, including City Grant, Estate Action, Urban Programme, Derelict Land Grant and Housing Association Grant. The government also wanted the public money to elicit two to four times as much private money, which proved difficult to obtain. Every Challenge area received £7.5m a year of directly related grants which had to be spent within the year or they were lost. This showed little appreciation of the nature of building projects which generally require little funding in year one, much more in the middle and less in the final year. Such a diverse programme of regeneration, new build and training needs to be flexible and this is where problems with annualisation arise because it does not provide for slippage of the programme, requiring considerable financial juggling to balance the cashflow (Stewart, 1993).

Integrated inner city policy was sought by the government through setting up the *Urban Regeneration Agency* (URA), subsequently renamed *English Partnerships* (EPs) under the Leasehold Reform, Housing and Urban Development Act 1993, whose main purposes were:

(1) to take over policy decision making from relevant ministers or Secretaries of State, such as

in DoE, Department of Employment and Department of Trade and Industry (DTI);

(2) to complement regional industrial policy of DTI and to take over the administration of the English Industrial Estates Corporation;

(3) to buy and develop inner city sites, thereby assuming the responsibility of much of DoE's Urban Programme, to award City Grants and Derelict Land Grants (DLGs) and eventually be responsible for a large slice of DoE's total budget;

(4) to administer City Challenge.

Time will tell whether this co-ordinated approach to inner city regeneration will be successful in solving the many problems of urban deprivation, in providing local authorities with a stable source of finance, formulating an effective strategy and satisfactorily encompassing the wide experience of local authorities. Early developments were far from encouraging as within three weeks of being established in November 1993, EP's budget for 1994/95 was reduced from £250m to £181m, leaving EP to secure the balance of £69m by selling assets of industrial and office buildings in a dormant property market. EP's grant was set to rise to £211m in 1995/96 and £221m in 1996/97, both well below the initial 1994/95 budget. Over and above the £69m cut, EP's resources in 1994/95 were further reduced by more than £100m because of commitments to existing schemes from the former grant regimes. In addition the budgets for Housing Action Trusts (HATs) and City Challenge were frozen in 1994/95.

Renewal Areas (RA) were areas which had a ten year life, contained between 300 and 3000 dwellings of which 75 per cent were to be privately owned, 75 per cent in poor condition and over 30 per cent of households were in receipt of housing benefit. The impact of this strategy was likely to be extremely varied because of the nature and scope of the imposed restrictions.

Eurogrants were available to assist with a variety of regeneration schemes. Unfortunately, in December 1993 regeneration schemes worth millions of pounds had been cut or delayed because European Union (EU) regional infrastructure grants earmarked for run-down regions had not been taken up by local authorities before the 31 December 1993 deadline. These included £17m of projects ranging from road improvements to business and science parks from Mid-Glamorgan County Council's 1993/94 development programme, £16m for a Europort delayed railfreight terminal in Wakefield and £2m of industrial redevelopment in Bradford. EU figures showed that of the £900m allocated for UK construction related schemes between 1988 and 1993, only £329m had been taken up and the money unclaimed by the end of December 1993 returned to EU reserves.

Local authorities blamed central government restrictions on council spending and confusion over the eligibility of private sector partners in the schemes for the low take-up of EU grants. Under EU rules, regional grants can fund up to 50 per cent of a project and the national government has to fund the balance. Local authorities claimed that government restrictions on council borrowing meant that they were unable to provide matching funds without cutting other vital services and some local authorities that had tried to raise funds through the private sector had also run into difficulties. Hence the seemingly generous Eurogrants appear to be fraught with difficulties in their implementation.

Yet in 1994, despite all these government measures, the inner cities were worse off. Unemployment was higher, investment in public and private ventures lower and deprivation more severe. The main reasons stemmed from the poor state of the national economy and the reduction of the central government financial support to local authorities in large urban areas.

In practical terms local authorities are often only able to provide the infrastructure within which the private sector can operate. The availability of finance to the public sector is so small in relation to total needs, that it must be used in the most effective way and this will normally be as a catalyst. The government financial cuts in the 1990s underlined this fact. Local authorities can provide sites, services and buildings but they must, in addition, provide the right climate for investment. They should ideally give maximum encouragement to developers, streamline the planning process and, wherever possible, plan joint schemes with them.

Developers cannot in the main retain their completed schemes as rent-producing investments and must usually sell them on completion to the institutions. Most institutions are reluctant to purchase property in unattractive locations, where it is difficult to assess future growth, and with certain types of property, such as small industrial units they are reluctant to accept the management with the risk of tenants going out of business. Finance for housing is much less of a problem, since building societies are generally prepared to support new housing in inner areas, although in the early 1990s they were also suffering from a reduced inflow of funds.

Building costs in inner city areas are higher than on greenfield sites, especially when the existing roads and services have to be renewed. The cost differential in 1994 could be as much as £6000 to £8000 for identical houses. Selling prices on the other hand could be in the range of £8000 to £15 000 less than would be obtained for comparable houses in suburban locations. The differences result in £14 000 to £23 000 less per unit being available for land and profit. However, prices vary so much in different locations that these cost relationships should only be regarded as indicative.

Basic Appraisals Objectives

Prior to formulating town centre redevelopment proposals it is essential to make an objective appraisal of the existing centre, having regard to its function, assets and deficiencies in terms of convenience and safety, usefulness by day and night as a shopping, commercial and social centre, and its civic character and architectural qualities. The objectives are primarily concerned with function (future size and purpose of town centre), layout (distribution and extent of main uses), circulation (pedestrian and vehicular movement) and character (retention and enhancement of town's individuality).

Plans for the future should aim at satisfying the human need on a modern scale by retaining or introducing some of the more picturesque uses which give character and vitality to the urban scene, such as pedestrian ways, arcades, street markets, small places or squares, landscaped spaces for rest and relaxation and facilities for amusement. The massing of buildings, their height and silhouette, and in detail the outline, textures and colours of materials, are all matters of vital importance in composing a street scene which is both attractive and harmonious.

It is desirable to carry out an economic survey and appraisal to assess town centre demand. This would embrace a physical survey of the existing shopping centre, analysis of existing traders, examination of recent transactions in sale or renting of premises, determination of market region, analysis of existing and potential turnover, examination of car parking facilities, and study of traffic flows and bypassable traffic. From the results of the surveys it will be possible to make a realistic assessment of shopping and parking needs and of the required scale and rate of development.

It is evident that the context for urban renewal and town centre redevelopment schemes will be derived from planning policies and shopping studies. The local authority needs to find out as much as possible about the demand for new shopping space, from both the public and retailers. Preliminary planning will be most useful for the preparation of the development brief. The brief will help prospective developers to assess the opportunities. Its contents should be formulated with care and will need to reflect decisions on how the developer is to be selected and to define clearly the local authority's objectives. It should, for instance, contain simple guidelines on:

(1) the amount of floor space and broad categories of uses;

(2) how the scheme fits in with policies in the structure plan, unitary development plan and any local plans;

(3) retailers' requirements and the approaches made to the local authority by groups needing accommodation;

(4) planning opportunities and constraints including development control standards; and

(5) highway and parking requirements and any light mass transit proposals and park and ride schemes.

A DoE report (1994b) was drawn up to assist landlords, local authorities and retailers in maintaining the competitiveness of UK towns. Financial resources were identified as the priority where owners and funds were not prepared to co-operate, where there might be a case for introducing the US system of raising levies in 'business improvement districts'. To provide criteria on towns' retail performance, the report advocated property values expressed as yields.

It also recommended forming management groups and a strategy of action programmes. Other recommendations included investment planning and regular public transport reviews. A simple 'health check' contained in the report focused on the attractiveness of towns, wherein the elements comprised diversity of shopping, quality of business, amenity and security.

Implementation by Partnership

Large scale town centre redevelopment schemes are often undertaken on the basis of a partnership between the local authority and the private sector. The local authority has powers of compulsory acquisition which are vital if the best pattern of development is to be achieved. Modern developments require large areas of land in order to make the best use of modern techniques of mixed development, enclosed shopping malls and sophisticated traffic arrangements. It is essential, for this reason, to be able to combine together numbers of small awkwardly shaped freehold sites of the nineteenth-century town. It is also desirable for land ownership to be in as large units as possible to balance the more profitable uses of land against unremunerative ones such as open space. The local authority also meets certain basic costs apart from land acquisition, such as site clearance and the provision of roads and services and execution of other public improvements.

The private developer often has a vital role to play with regard to availability of capital, knowledge of the market and ability to exploit commercial opportunities. To be successful the development must be in the right location with adequate and readily accessible parking space, satisfactory service access, correct amount of retail space for purchasing power of area and reasonable rent levels. An enclosed air conditioned shopping centre needs a minimum supporting population of 50 000 to 70 000, but trading in such a centre is likely to exceed that in a comparable open precinct by about twenty per cent. Many market towns do have extensive catchment areas for shopping purposes (Seeley, 1974).

It is necessary to prepare a developers' brief, probably prepared by a consultant, listing the essential feature of the scheme as a basis for competitive tendering or negotiation with developers. The brief would probably include such matters as the objectives of the scheme; introductory comments on the town and its region; planning proposals; plans for the centre; site plan; number and types of unit to be provided; special requirements such as pedestrian and vehicular access, parking provision, siting, height and architectural quality of buildings and landscaping; phasing; drawings required; disposal terms; and obligations with regard to displaced traders. It is essential that the local authority should retain the initiative in planning and guiding the redevelopment.

In disposing of land for development, the local authority usually has two principal aims in view. One is to secure the satisfactory development of the site in the interests of the public it represents, which in many cases will involve a major civic improvement and the solution of such problems as traffic congestion, car parking, lack of open space and other public amenities. The other is to obtain the maximum financial return from the developers, partly in fulfilment of the local authority's duty to the local taxpayers and partly in order to recover its own capital outlay in buying the land and constructing public services in connection with the development. These two aims tend to conflict and the solution will often involve a balance between them. It is essential to consider and reconcile the aesthetic and economic circumstances of each case.

Where a local authority has done no more than prepare a master design, it may decide to arrange a competition which offers developers opportunities for submitting alternative architectural proposals. In that event, it may prefer to fix a sale price or ground rent, and so be free to judge the competition entirely on the architectural merits of the schemes submitted. The services of a valuer are needed in the assessment of the price or rent. In other cases the local authority may wish to combine a design competition with a formula for securing the best rent obtainable. A possible procedure, in these circumstances, would be for the local authority to fix a fair ground rent and then offer the site to selected developers, in a limited architectural competition, at the pre-determined ground rent but with the proviso that the developer will be required to pay as additional ground rent, a specified percentage of any net rent derived from the development in excess of a specified return on the cost of development (after deducting the basic ground rent). The following simplified example will serve to illustrate the latter approach.

Example 16.1. A competition for a central area redevelopment scheme includes a fixed minimum ground rent of £480 000 per annum. The cost of the scheme is £24 million, the return to the developer is agreed at ten per cent and any surplus rent is to be shared equally between the landlord and the developer. Total annual rents from the buildings are estimated at £3 200 000.

Rent received (net)	£3 200 000
Return of ten per cent on £24 million	2 400 000
Surplus	800 000
less basic ground rent	480 000
Balance	£320 000
Additional ground rent = £320 000 × ½ =	£160 000

A scheme of this kind could have the initial ground rent adjusted within a short period of the commencement of the ground lease (say three years), and thereafter reviewed after possibly thirty-three and sixty-six years on a ninety-nine year lease, although many developers prefer leases for periods of up to 150 years.

Another salient feature of the partnership arrangement is the collation into one or more ownerships of all those lands which are needed to carry out the redevelopment satisfactorily, and this process has been described as *land assembly* and *site assembly*. The importance of this procedure stems from the ability to develop the area in the best interests of the community including provision for some unprofitable uses, and yet at the same time ensuring an adequate return to the developer, coupled with the availability of the local authority's powers of compulsory purchase.

The procedure for land assembly is often undertaken on the following lines.

(1) The local authority defines a comprehensive development area and prepares proposals for development often in conjunction with the developer.

(2) The developer acquires as much land as possible by agreement and the local authority can compulsorily acquire the remaining land which is essential to the scheme.

(3) The developer transfers the freehold of his land to the local authority.

(4) The local authority grants a long lease of the land to the developer often with provision for periodic revision of the ground rent. The disposal terms should take account of the enhanced site values resulting from the local authority's share of the redevelopment work and of the long term prospects. In practice, developers are normally anxious to secure a ten per cent minimum return on their capital to cover the risk element involved and the annual ground rent to the local authority may not exceed five to six per cent of the capital required for land acquisition. Furthermore, developers often seek lease periods of 125 to 150 years to enable their annual expenditure to be more favourably amortised. Obviously each case must be considered on its merits.

Financial Implications for the Local Authority

Any public body, like a private developer, must have regard to the financial aspects of renewal, although this is only one of many significant factors. The private developer makes his appraisal, and is influenced in his decisions by the return which he will receive on the capital investment. He will have regard to the cost of assembling the site, of obtaining possession, of clearance and development, and the ultimate rents to be obtained, making allowance for the period during which he will have to pay interest charges on his capital. The public authority does the same calculation but also considers its obligations, whether statutory or moral, in connection with rehousing or relocation of tenants or other people displaced by a scheme of renewal and with diversion of services. The authority also looks beyond the question of profit or loss and examines other issues such as planning, traffic, civic design, amenity and the general prosperity of an area. Many of these factors cannot be valued and in the final event the deciding issue may be the willingness of the public to meet the cost. A local authority will obtain is finance from various sources: local taxes, government grant, loans, subsidies and private capital. When interest rates are high, the burden to be carried may be heavy and frequently schemes of renewal carried out today will commit generations ahead to substantial payments.

Problems of Displaced Traders

The redevelopment work must be carefully phased and this becomes very difficult in practice, as it is most desirable to carry out the scheme as a continuous operation and this requires constant liaison between architect, surveyor, engineer and contractor. The phased programme is agreed with owners and occupiers to preserve the continuity of trading by those who are displaced and to minimise the compensation payable for disturbance. There remains the problem of the small displaced trader who frequently cannot afford to pay the rents commanded by modern shops, yet whose interests must be respected. Existing traders, both large and small, possess a wealth of experience and goodwill which can make a valuable contribution to the success of redevelopment. In particular, the small independent trader contributes variety and character to the shopping centre. It is good policy to provide small unit shops of up to 45 m^2 floor area to satisfy this need.

The Developer's Viewpoint

A prospective developer will expect the local authority's brief to go beyond the normal planning proposals and to express the desired objectives in social, financial and environmental terms. He will need to be convinced of the authority's firm commitment to the development, and have confidence in the administrative arrangements for the partnership. Ideally outline planning permission should have been granted before selection, although this is not always possible as the authority may need the developer as partner before the scheme can be promoted with some assurance of public support. His other major concern is the need to obtain a high level of cost efficiency in the broadest sense, with a satisfactory timescale for the development (RICS, 1980).

There are three main methods for the selection of a developer.

(1) *Negotiation* which is favoured by developers and involves advertisement, discussion, short-listing and selection. Its merit lies in providing mutuality of objective at an early stage and in giving the developer ample opportunity to use his own initiative. The problem is that the local authority may expose itself to criticism if it is not seen to be seeking competitive bids.

(2) *Design and Bid* arrangements are often preferred by local authorities. A number of developers are invited to submit a set of drawings supported by a financial offer. Although this method is acceptable to developers if the number of participants is restricted to three or four, they regard it as less satisfactory than negotiation. Its principal defect is that it provides a weaker basis for mutual trust and demands from developers a

degree of certainty which is premature. There is also the danger that one developer may make a higher financial offer and then wish to negotiate later arguing that certain circumstances could not have been foreseen.

(3) *Open Competition* is likely to be employed by the smaller or less experienced authorities and involves putting a scheme out to tender on a virtually unrestricted basis. It is not favoured by developers, who regard it as only suitable for cash bid tendering for a site, but not for design proposals. Once the developer has been selected, it is customary for the developer and local authority to enter into a provisional arrangement for several months. This is because there are various matters which are best decided before both sides commit themselves to a formal partnership agreement. Among these are confirmation of the programme for site assembly and of arrangements for relocating displaced homes and businesses, obtaining detailed planning approval, securing agreement with the local authority over tenant mix – which is vital to the commercial success of the centre, and negotiation with prospective key tenants. It is also important at this stage to firm up estimates of building cost and rental return and to produce an outline of the expected building programme (RICS, 1980).

The Role of the Consultant

A scheme dependent upon a partnership between a local authority and a developer is invariably complicated. The outlook, background, responsibilities, attitudes and objectives of each party are very different. The role of the consultant, normally acting for the local authority, is to combine the strengths and overcome the weaknesses of each party to the prospective partnership. Consultants can, through accumulated experience, provide detailed knowledge of shopping schemes and apply this to the authority's benefit.

For full advantage to be taken of the consultant's advice, he should be brought in at the beginning, before an initial approach is made to test the interest of the developers. This advice would be used to help the authority surmount the likely hazards inherent in this type of scheme. Some of the worst problems are now listed.

(1) An ill-prepared or ill-timed proposal where the authority has little to offer and is prejudicing its financial position;

(2) a proposal which is too ambitious, too demanding, or fails to recognise commercial feasibility;

(3) an inability by the authority to coordinate the development project – the planning, site assembly, finance, or effect the partnership and the development itself;

(4) an inadequate administrative arrangement for handling the partnership which would be unresponsive to the need for frequent and quick decisions; and

(5) a desire to exert too much control over detail – layout, design, shop representation and the like.

The development brief is the expression of the consultant's ability to balance these issues and is the basis for partnership. A good brief results from hard negotiation between local authority officers, elected members and the consultant. It must embrace all aspects of the development. The preparation of the brief brings to the fore the need for decisions on:

(1) the best basis for the developer's submissions having regard to the intended evaluation – whether financial bids alone or bids with design schemes, and if the latter what weight is to be given to the respective factors;

(2) how best to carry through negotiations, their timing, length and the ability to terminate them;

(3) how best to deal with finance, such as the funding of an authority's capital shortfall on acquisitions and accommodation cost;

(4) the timing of selection and contractual arrangements with developers on compulsory purchase orders, planning permission and related matters; and

(5) where the developer is likely to be a retailer-developer, his suitability as a developer and problems of rent review (RICS, 1980).

The Investor's Viewpoint

The major institutional investor views town centre schemes in very much the same light as a developer, indeed increasingly it is the developer. The investment of large sums in inner city and major shopping developments involves risks for which an appropriate return is expected. Building costs are a major cause of concern to institutions and developers alike, and a large part of the risk in undertaking a town centre scheme lies in uncertainty over how far costs are going to increase during the development period. It follows that financial calculations need to be very cautious and precise, although in times of fluctuating inflation and variable building market conditions increasing reliance has to be placed on experience and intuition. The services of a quantity surveyor for cost planning aspects are extremely important.

Mainly because of its cost effects, timing is of vital concern to the institutional investor. For example, one city authority involved in a successful town centre partnership kept a sub-committee on call at 24 hours' notice throughout the planning and development period specifically to deal quickly with any problem.

Where a multi-million pounds investment is being made institutions expect a long term of years for the lease. In terms of capital assets this means little difference to the freeholder. The freeholder should expect a modest share of future growth, even if no share exists at the start of the lease.

The relationship between the institution and the local authority is most important. The crucial decisions during the negotiating period are, before documentation, on the form of the partnership agreement, the specification of the development and the minimum asking rents. After documentation, decisions may be needed on variations in the building contract and on the leasing programme (RICS, 1980).

Detailed Working Arrangements

Once the parties to the scheme have agreed terms, the legal documents have to be prepared. Typically these consist of a development agreement and a lease granted to the developer after completion of the project. Important terms which tend to be controversial are:

(1) What is the developer to build? There are basically three stages in the preparation of the drawings: the sketch plans, those contract drawings which are finalised before building starts, and the further working drawings and variations prepared during the construction period. There are several potential difficulties. Either reasonably detailed drawings must be finalised before the development agreement is entered into or the agreement must contain a mechanism for those details to be approved by the local authority later.

(2) What control is the local authority to have over how the developer builds? The success of the scheme will depend very much on the correct choice of architects, quantity surveyors, engineers and building contractors. The local authority will normally therefore require the developer to use an agreed professional team and main contractor, or have the right to approve the choices made by the developer if they have not already been chosen. During the development period the developer has a very important project management role to perform and a good developer can make a significant contribution to controlling costs and to the quality and speed of construction.

(3) The forfeiture provisions in the development agreement. If the developer commits a breach of its obligations under the development agreement, the local authority will have the usual common law remedies, including the right to claim damages. The authority may seek an express right to forfeit the agreement without compensation if the developer goes into liquidation or commits a breach. If the developer wished to raise finance on the security of its interest in the development some modification to the strict forfeiture clause will be required by the funder.

(4) The fire insurance provisions in the lease. It is extremely difficult to assess the correct amount of the fire insurance cover for a large shopping centre. The lease normally requires the amount of the insurance cover to be agreed between the parties or determined by an expert or arbitrator in the event of disagreement.

(5) The management of the centre after com-

pletion. The continued success of the centre will depend very much on the quality of the management. The lease may include the right for the local authority to employ independent managing agents to take over the managing role if the developer is in serious breach of his management obligations, with the fees of these agents being met by the developer.

(6) The rent payable under the lease. In the case of partnership schemes it has become usual for the net income to be shared between the parties as opposed to the developer paying a fixed ground rent, subject to periodic review. At its simplest the rent payable is an amount equal to a percentage, agreed at the outset, of the net income. Additional refinements might include, for example, provision for the percentage to be fixed later according to an agreed formula based on the actual costs of the development (RICS, 1980).

Many feel that a lease is not the ideal basis for a partnership agreement. An alternative is a co-ownership scheme, which provides for the freehold or leasehold interest in the property to be conveyed to a custodian trustee who holds it upon trust for sale for the co-owners as tenants in common. This arrangement has been used for a number of major office schemes.

Town Centre Redevelopment Proposals

Proper segregation of town centre operations is essential for safety, maximum concentration of use and amenity and comfort. Furthermore, there are a variety of ways in which this can be undertaken, such as horizontal or vertical segregation of vehicles and pedestrians, and segregation of commercial vehicles and/or public transport from other classes of vehicle.

One of the biggest difficulties in converting existing streets into pedestrian ways is the provision of rear access to 'land-locked' shops to provide adequate unloading space off the highway. Another difficulty stems from the varying depth of shops.

An interesting approach to town centre redevelopment was carried out at Basingstoke where, to ensure that shopping will be safe and attractive to the townspeople and visitors, the shopping areas in both the new and existing sections of the town are free of vehicular traffic. In the new area the shopping platform is above service traffic and the main car parks are immediately alongside in multi-storey decks. The original proposals showed layers of car parking under the shopping deck but the later alternative was claimed to provide greater convenience for handling goods, easier pedestrian access from the sides and an architectural character of design more in harmony with the scale and character of the existing town. In the existing shopping centre it is intended that the shopping streets shall be reserved for pedestrians only, this being achieved by traffic management measures or by the longer term possibility of providing rear service access and car parking at the rear of the premises.

Many large enclosed air-conditioned shopping centres were provided in city centres in the nineteen seventies and eighties and this form of development is continuing. Typical examples are the successful Victoria and Broadmarsh shopping centres at Nottingham, with extensive car parking provision, located on either side of the existing open traditional city centre around the old market square which has since been partially pedestrianised. Both of these shopping centres were the subject of costly refurbishment in the late 1980s, which is indicative of the large measure of success that they have achieved.

NEW AND EXPANDING TOWNS

General Background

The post-war era saw the emergence of new public development on a scale hitherto unrealised. The most outstanding and significant developments have been those of new and expanding towns. New towns were provided by government appointed development corporations under the provisions of the New Towns Act 1946, with the declared objective of establishing 'self-contained and balanced communities for work and living'. Their principal aim was to relieve congestion in the larger cities, particularly London, and eight of the first generation new towns formed a ring around London at a distance of about 32 to 48 km from it. A new town development corporation obtained its working capital from the Treasury and it was empowered to acquire, hold, manage and dispose of land and other property; carry out building and other operations; provide water, gas, electricity, sewerage and other services; and carry on any business or undertaking for the purposes of a new town. The first new towns to be designated had design populations of 50 000 to 60 000, but later new towns had larger populations culminating in Milton Keynes with a population of 250 000, based on a square grid road system enclosing 1 km^2 squares.

The dispersal policy which promoted the new town schemes was later extended to the expansion of widely dispersed smaller country towns by virtue of the Town Development Act 1952. Under this Act the large city authority concluded an agreement with a much smaller provincial authority for the transfer of population and industry. The large authority was termed the *exporting authority* and the smaller authority was called the *receiving authority*. The Government could make contributions towards some of the development expenses of a receiving authority, in connection with housing, water supply and main drainage. In addition, the exporting authority, and sometimes the county council for the receiving area, also gave financial assistance. In some of the later schemes the county council was a party to the agreement. There were two main types of town expansion scheme:

(1) agency schemes, where the exporting authority carried out the bulk of the development work, and was subsequently reimbursed by the receiving authority when the work became revenue-producing; and

(2) nomination schemes, in which the receiving authority carried out all the development work but received contributions from the exporting authority.

Apart from helping to relieve congestion in large cities, town development schemes infused new life into small static or declining country towns which often had considerable history, character and charm, but at a relatively high cost. Expansion provided more and varied opportunities for employment and improved public and other services. Industrialists could build, purchase or rent factories on new industrial estates and houses were available for employees within a reasonable distance of the factories. The biggest problems have been attracting sufficient suitable industries and the synchronisation of the housing and industrial construction programmes, whilst the provision of social services and amenities barely kept pace with development in the early stages. Economic and human factors dominated many town development activities. First and foremost, it was vital that the selected site possessed a real potential for development and that a sound and diversified industrial structure was secured. It is doubtful whether small-scale expansion schemes could ever be sound economic ventures, and the later proposals aim at doubling the size of much larger towns like Peterborough, Northampton and Ipswich using development corporations. The practical difficulties of implementing town development schemes have been described in *Planned Expansion of Country Towns* (Seeley, 1974).

In the late nineteen seventies the Government began to turn its attention towards the problems of declining population and employment in the inner urban areas of major cities, with the result that the new towns programme was curtailed and proposals were initiated in 1978 for the winding up of outstanding schemes under the Town Development Act 1952. This transfer of emphasis from the new towns to the inner urban areas was reaffirmed in the Local Government, Planning and Land Act 1980, which encouraged the sale of new town assets in the market and gave the Secretary of State power to make orders reducing the designated areas of new towns with a view to expediting their completion, and these have since been implemented. The last of the third generation new towns, Milton Keynes, was scheduled for handover to local authorities and the disbanding of the Development Corporation in 1992.

Development Costs

A new town development corporation was a unique kind of developer, being dependent on the Government for finance and with no other powers of borrowing. Indeed, it started operations with no assets and in the early years had to borrow to meet revenue expenditure. Unlike a private developer, a development corporation undertook many different kinds of development, many of which were non-remunerative in themselves; and in relating cost and yield it was concerned with the new town as a single financial unit and not with individual sections of it. Industrial and commercial development was sold or rented and the cost of trading undertakings met out of charges. A development corporation also provided non-remunerative services like sewage treatment works, roads and open spaces which benefit all townspeople and would normally be provided by local authorities. Apart from any local authority contributions the costs of these services were met from the yield accruing from other services with a financial return.

In the case of an expanding town, the local authority was faced with a large annual programme of high cost houses. These costs were recovered in one or more of the methods listed below, none of which was really satisfactory.

(1) To spread the increased costs over all local authority houses whereby all the tenants of older dwellings were continually being subjected to rent increases.

(2) To apply a rent surcharge to all new houses to cover additional costs resulting in the payment of differential rents for the same type of dwelling in different parts of the town.

(3) To cover the increased housing costs by increased local taxes so spreading the additional costs over all local taxpayers and avoiding large rent increases on local authority dwellings.

It is apparent that questions of town size may be particularly relevant when considering the costs of expanding existing communities. Studies were made to assess the relative costs of projected expansion schemes at Ipswich, Peterborough and Worcester, and comparable data was published for the expansion of Basingstoke. These studies pointed to certain characteristics that gave rise to differences in total cost, such as the higher cost of town centre development at Peterborough, because of higher existing values of property to be redeveloped, and higher road costs owing to problems of developing a more difficult site.

Furthermore, theses studies indicated that within certain brackets of expansion the cost per head of additional population decreased as the scale of expansion increased, so that larger expansions were generally more economical than smaller ones. It was also relevant that the expenditure involved in carrying out an expansion would normally improve local conditions of traffic congestion and improve amenities for the town as a whole, all of which would enhance the effective return on the capital investment involved.

Town expansion was also discussed at the 1965 Annual Conference of the Royal Institution of Chartered Surveyors, in the wider context of whether there was any method of relating efficiency to the total size of a town and at what size the closest economic relationship between population and cost might be achieved. It was generally agreed that this might be a fruitless investigation because of the impossibility of ascertaining reliable economic data and of evaluating the many other factors, such as amenity and social and cultural advantages, that are relevant to the size of a community. This Conference served to emphasise how little is known about many of the fundamental influences on town design.

On the basis of limited historical published data, expanding town schemes were considerably more expensive per head of population than new town projects. They also indicated reducing costs per head of additional population with a larger scale of expansion. Unfortunately, different sources were producing differing costs and arriving at conflicting conclusions. It was very difficult to obtain accurate information on all the costs of expanding a town – the services were provided by a variety of organisations both public and private, and all the constituent costs were continually changing but not at a uniform rate, and this gave rise to difficulties in making comparisons (Seeley, 1974).

BUILDING CONSERVATION AND URBAN REGENERATION

Conservation Areas: General Background

Many parts of built up areas of the UK have been designated as Conservation Areas, and in 1993 there were no less than 7000 such areas designated in England (4 per cent of the built fabric). They are described in the Town and Country Planning Act 1990, section 277 as 'areas of special architectural or historic interest, the character or appearance of which it is desirable to preserve or enhance'.

Taylor (1991) has described how apart from the obvious central areas of historic towns and cities, conservation areas sometimes include substantial areas of mundane buildings with little architectural or civic design interest, and undeveloped areas on the outskirts of towns and villages. For example, the Hertford Conservation Area encompasses much of the town and includes numerous post-war buildings of little, if any, architectural design interest.

A RTPI Report in 1993 concluded that too many conservation areas lacked proper management and funding and that many local authorities failed to develop their aims for conservation areas, or to communicate these aims to local residents, businesses and developers.

Conservation and Re-use of Redundant Buildings

DoE (1987) have described how in the 1980s rehabilitating old houses and the re-use of old industrial and commercial buildings, instead of demolishing them and redeveloping the sites, became widespread. There are numerous large old buildings in the UK which are no longer needed for their original purposes, especially in the older towns and cities. They frequently have more character than the modern buildings which would replace them, but if left empty they soon decay and blight the area around them, and it is often cheaper to adapt an existing building than to erect a new one although it may create more problems. In the late 1980s and early 1990s, interest grew in the conversion and re-use of old buildings, not only amongst those who wish to conserve our heritage, but also those who wish to bring back economic life into run-down areas or just to carry out their own innovative projects, as in the Lace Market area of Nottingham. This form of adaptive re-use has an important role to play in urban regeneration.

DoE has supported many conversion schemes through the Urban Programme (UP), and in 1990 the EC, which had already provided substantial financial assistance for regeneration schemes in the UK, introduced a working paper which, among other environmental matters, recommended the international listing of historic buildings and towns and other conservation initiatives, which aimed to achieve real improvements in the quality of the urban environment within the community through guidelines and legislation. The DoE Handbook on re-using redundant buildings (1987) aptly explains what is involved in devising and implementing successful re-use schemes and provides practical information on the organisational and management aspects. It also contains 14 case studies describing how different organisations from the public, private and voluntary sectors carried out successful conversion schemes in a wide variety of circumstances, and embraces the conversion of old factories, warehouses, mills, railway stations and other buildings. There was a wide range of new uses to which 400 converted buildings in the UK had been put in the mid-1980s, encompassing workspaces (44.5%), leisure/retail (32.0%), housing (9.25%) and mixed use (14.25%). In conservation areas, particular emphasis needs to be placed on scale and the use of local materials.

There are several informative and well illustrated books covering the rehabilitation and re-use of old buildings and some of them are briefly described. *Saving Old Buildings* (Cantaouzino and Brandt, 1980) provides numerous examples of the conservation of buildings of character in many countries and their re-use for a variety of purposes, including cultural, commercial and housing, and the adaptation of churches, railway stations, small rural buildings and large country houses and castles. *The Housing Rehabilitation Handbood* (Benson *et al.*, 1980) gives guidance on basic repairs, improvements and conversions and provides case studies on the rehabilitation of housing for the disabled, single persons and the elderly. Highfield (1987) emphasises the potential value of the vast stock of old buildings in the UK and estimates that the cost of rehabilitating an existing building could amount to 50 to 80% of the cost of new construction, although it is likely to have a shorter life span.

Registered housing associations provide many thousands of good homes every year in the UK by improving older neglected dwellings, many of them in long blighted inner city areas. They provide accommodation for rent to those least able to compete in the housing market and also take nominations from local authority housing lists. Housing associations set out to be caring landlords who deal with repairs and tenants' other problems sympathetically, efficiently and effectively.

In the UK, all government departments and the Property Services Agency (PSA) are required to conform to the government's policies for the conservation and improvement of the environment when carrying out their responsibilities, and to meet the costs involved. Government departments should set an example of the highest order in the care of historic buildings and protected areas, and good procedural guidelines are contained in *The Conservation Handbook* (PSA, 1988).

Listed Buildings

A Nottinghamshire County Council survey in 1993 revealed that one in twenty of the county's listed buildings was at risk and many were in danger of collapse. Of the 6721 listed buildings in the county, many of which were in private ownership, 388 were considered to be at risk. The longer buildings are left empty and neglected, the more at risk they become. The main causes for the deteriorating situation were identified as neglect, lack of knowledge of traditional maintenance methods and a failure to appreciate the irretrievable loss of the county's unique buildings.

The rate of decline of some listed properties had raised fears about the unscrupulous development of their sites. Some developers could allow buildings to fall into disrepair in the hope that they may be demolished, leaving the sites available for redevelopment.

A research report published by the RICS/ English Heritage (EH) in 1993 showed that between 1980 and 1992, listed buildings had surprisingly out-performed other categories of building investments. The number of listed office buildings in institutional ownership was small (less than ten per cent of all office investments) and each unit was of a size and total value which was typically less than one third that of similarly located modern stock. Nonetheless, across the whole of the UK and also within the major concentrations of office ownerships within central London, listed buildings appeared to have held their own in terms of overall performance.

Moreover, listed buildings appeared to have attracted just as much occupier demand as other categories and achieved rates of rental value growth which are as good as, or better than, those of other categories of offices. The results of the report (RICS/EH, 1993) establish a consistency of return which should encourage investors to consider investment in listed buildings.

A Town Conservation Case Study

The Civic Trust Regeneration Unit has been involved in the preparation of numerous urban strategies on behalf of local authorities throughout the UK in recent years, each with a strong emphasis on conservation aspects. Readers wishing to pursue this topic in more detail are referred to Just and Williams *Urban Regeneration: A Practical Guide* (1996). I have selected the small Welsh town of Llanidloes because of its great character and history and the very discerning way in which the project was approached by the Civic Trust for Wales and the Civic Trust Regeneration

Unit (1992), with full regard to the conservation implications.

Conservation Policies

Llanidloes is a thriving town situated at the crossroads of Wales with a population approaching 3000, which has developed from a simple thirteenth century town plan. Much of the quality of the town stems from the historic street pattern and its attractive setting, combined with the quality of the buildings, their variety of purpose and age, and the mixture of styles and materials which together create a town of outstanding vernacular quality. Furthermore, Llanidloes is made unique by the preservation of so many building features and details from so many periods, including (but largely unspoilt by) the twentieth century. Indeed, the history of the town can be read in its townscape (Civic Trust, 1992).

At the heart of the strategy for Llanidloes are the policies for the conservation of the rich heritage, including the preservation of specific buildings and general street scenes, the enhancement of the River Severn skirting the town which forms a major but neglected asset, and the proper management of the surrounding countryside for public enjoyment.

Implementation of a Town Scheme

Conservation area status, which embraces much of the town and which the Civic Trust recommended should be extended, provides the local authority with additional planning powers to control new developments and to preserve the existing buildings. It was considered that Llanidloes would benefit from a more positive Town Scheme to assist in the effective conservation of the town's architectural heritage by providing grant aid to owners. Town Scheme grants (up to 50 per cent of the cost of eligible works) would be offered from a fund provided on an annual basis by Welsh Historic Monuments (Cadw) and the local authorities to help with the cost of restoring and repairing buildings of architectural or historic merit. Eligible works include structural repairs to the external fabric of the building; redecoration arising from structural repairs; and fees for professional advisors. In England,Town Scheme grants are also available for internal structural repairs and it was hoped that this policy could operate in Llanidloes.

The Civic Trust considered that a Town Scheme was essential to the overall strategy to conserve Llanidloes' special character, complement design guidelines and development control policies as part of a 'carrot and stick' approach.

'Use of care and repair' grant scheme

Not every property within the conservation area will need Town Scheme grant. Furthermore, it takes several years for a Town Scheme to have a significant visible impact on the townscape as properties are restored and refurbished. If the total fund annually amounted to say £25 000, this would only benefit five to ten properties per annum.

Moreover, damage to the quality of the built environment is not restricted to disrepair. It can and does often arise from the ill-considered replacement of traditional features such as windows and doors by modern, inappropriate fittings. The cumulative effect of such replacement is to erode local quality and to undermine the value of the careful restoration work applied to selected buildings. It therefore seems appropriate to encourage householders to follow good practice and maintain original features in addition to enforcing planning restrictions.

The Civic Trust therefore wisely proposed that the District Council establish a 'care and repair' scheme as operated by a number of local authorities, sometimes in collaboration with housing associations, as in the case of Newark and Sherwood District Council and Nottingham Community Housing Association. Grants are made available to cover up to 40 per cent of the costs of repair or replacement of windows and doors to properties within the conservation area, provided that the work is carried out to the satisfaction of the local authority's staff. The Civic Trust rightly believed it preferable to repair rather than replace wherever possible, and a modest fund of say

£5000 per annum would help owners considerably and target properties unlikely to be eligible for Town Scheme grant.

Design guidelines
Even with strong planning controls and a positive grant scheme for repairing buildings, problems can still arise. A new shopfront that is in discord with its neighbours; a quantity of new PVCU windows replacing traditional timber sashes; a traffic scheme requiring a batch of new traffic signs – all can mar local character just as much as straightforward neglect and decay. Hence there exists a vital need to establish clear design guidelines against which all new development can be judged. The Civic Trust described how good design is relatively easy to illustrate but extremely difficult to define.

It is generally accepted that buildings should be in harmony and respect their surroundings in the manner described by HRH the Prince of Wales in *A Vision of Britain* (1989). However directly examples of good design are provided, there is a danger that these examples will be widely used producing a new and dull uniformity. Design guidelines comprise basic principles illustrated with examples of good and bad practice and do not attempt to stifle innovation. The preservation of fine buildings and other features is essential in retaining our heritage and can be very worthwhile.

Design guidelines incorporate the following objectives;

- *Retention of existing buildings and features* as part of a strategy of sustainable development to maintain and recycle old buildings where possible.
- *Simplicity and restraint* are major facets of new schemes with particular attention paid to scale, massing, roof line, materials and finish.
- *Attention to detail*, particularly when replacing shopfronts and domestic doors and windows. Replacing shopfronts entails modest fascias, retention of stall risers, avoidance of large areas of plate glass, use of internal security shutters, and traditional sun blinds as opposed to plastic hood canopies. Signwriting should be on wood or glass, not plastic lettering or black illuminated signs. Domestic replacement windows should be in softwood or sustainable hardwood and PVCU windows will generally not be accepted in conservation areas.
- *Integrating new buildings into their surroundings* and for new industrial estates and housing developments, this requires careful selection of roofing and the use of landscaping to soften the impact of the buildings or conceal them.
- *Aiming for quality* with regard to spaces beside and between buildings. In the town centre this entails the quality of surfacing materials (paving, kerbs, edging, channels and road surfaces) and the design of street furniture, such as street lighting, litter bins, seats and flower baskets, while in car parks, open spaces and residential areas, skilful landscaped treatment of the spaces is important, together with the integration of the space with its surroundings through careful edge treatment, walling, railings or hedges (Civic Trust, 1992).

PRIVATE FINANCE PROJECTS

In 1992 the government launched a private finance initiative aimed at involving the private sector more fully in financing and managing projects and services which have traditionally been the responsibility of government. The term 'private finance' is possibly rather misleading as it amounts to more than just the provision of private finance. The government wished to use private sector management and expertise to secure improved public services and better value for the taxpayer. The concept was to be applied not only to infrastructure projects such as railways or roads, but also to the provision of specific services such as patient care or prisons.

There are, broadly, three types of private finance projects.

(1) *Financially freestanding projects* where the private sector contractor undertakes the project on the basis that costs will be recovered by charging the users as with toll roads or bridges.

(2) *Joint ventures* which consist of projects to

which both the public and private sectors contribute but where the private sector has overall control. Private sector partners will normally be chosen through competition and the government's contribution should be clearly defined and limited, with the allocation of risk also clearly defined and agreed in advance.

(3) *Services sold to the public sector* usually where a significant part of the cost is capital expenditure, such as the provision of prison places by the private sector designing, constructing, managing and financing new prisons. The public sector purchaser will need to be assured that better value for money will be obtained than by any alternative (Treasury, 1994).

Figure 16.1 gives examples of private finance projects and shows their main characteristics, wide scope and large costs involved.

COST BENEFIT ANALYSIS

Nature of Technique

Cost benefit analysis has its origins in a paper presented by a French economist, Dupuit, in 1884 on the utility of public works. The technique was further developed in the United States, where its sphere of operation was extended into many aspects of society including river and harbour projects and flood control schemes. Cost benefit analysis aims at setting out the factors which need to be taken into account in making economic choices. Most of the choices to which it has been applied involve investment projects and decisions – whether or not a particular project is worthwhile financially, which is the best of several alternative projects, or even when to undertake a particular project. The aim is generally to maximise the present values of all benefits less that of all costs, subject to specified restraints. Cost benefit analysis has been defined as 'a technique of use in either investment appraisal or the review of the operation of a service for analysing and measuring the costs and benefits to the community of adopting specified courses of action and for examining the incidence of these costs and benefits between different sections of the community'.

It has the basic objective of identifying and measuring the costs and benefits which stem from either the investment of monies or the operation of a service, but in particular it is concerned with examining not only those costs and benefits which have a direct impact on the providing authority but also those which are of an external nature and accrue to other persons. Furthermore, the costs and benefits to be measured are those which accrue throughout the life of the project.

The principal criteria to be determined are:

(1) Which costs and benefits are to be included?
(2) How are they to be valued?
(3) At what interest rate are they to be discounted?
(4) What are the relevant constraints?

A suitable methodology for use in a cost benefit study is as follows:

(1) define the problems to be studied;
(2) identify the alternative courses of action;
(3) identify the costs and benefits, both to the providing authority and to external parties;
(4) evaluate the costs and benefits; and
(5) draw conclusions as to the alternative to be adopted.

Enumeration of Costs and Benefits

Some of the more commonly used expressions are defined and described.

Social costs may be defined as the sum total of costs involved as the result of an economic action. *Private costs* are those which affect the decisions of the performers; hence production costs include those of labour, materials, land and capital. There may also be *external costs*, for instance damage to buildings or decline of property values through smoke emanating from a factory; these costs are not met by the industrialist.

Similar effects may occur on the other side of the equation – benefits are reflected in the amount

Second Severn Crossing
Work was well under way on the privately financed construction of a second road bridge across the Severn Estuary. Costing about £300 million, the private sector company had a maximum 30-year concession to operate the crossing in order to recover its costs.

The Manchester Metrolink
This tramway system opened in 1992 and runs from Bury to Altrincham passing through Manchester city centre. The private sector company involved designed, built and now operates the tram system and has a concession to do so for up to 15 years. The cost of the project was £140 million.

West Coast Main Line
A competition will be held in 1994 to select a private sector consortium to modernise the main railway line from London to Glasgow and maintain it for a defined period. The cost of the project will be up to £600 million. The successful private sector company will receive service payments linked to the performance of the services using the line.

Channel Tunnel Rail Link
The Government launched a competition early in 1994 for a private sector partner to design, construct and operate this rail link. Tenderers were invited to bid for the amount of government contribution they required and also for how the risks can be shared. The estimated cost was £2–3 billion.

The Heathrow Express
Construction of this £300 million link between Heathrow and Paddington Station had begun in 1994. It was a joint venture between the BAA and British Rail. Costs will be recouped through passenger fares.

Docklands Light Railway, Lewisham extension
This is a £100 million project to extend the DLR south to Lewisham. It will be designed, built, financed and maintained by the private sector.

Aintree HNS Trust
The private sector will build and operate a 100-bed patient hotel, conference, catering and other facilities. The NHS Trust would buy a proportion of the facilities offered.

Nursing home accommodation in Eastbourne
The private sector will build and run nursing home accommodation in and around Eastbourne. The contract will be initially for 10 years.

New prisons and secure training centres
The private sector will provide custodial services which it is proposed will include the finance, design, construction, management and maintenance of six new prisons.

New and improved roads
The Government intended to introduce contracts under which the private sector will design, build, finance and operate roads. The private sector operators will receive payments from the Government in relation to the level of traffic using these roads. The first contract should be let in 1995.

Figure 16.1 Examples of private finance projects (Source: HM Treasury, Economic Briefing No. 6, Feb. 1994)

paid by consumers for goods produced; but in addition, favourable externalities might also accrue to society. An example could be a dam which in addition to generating electricity for sale in the market, gives flood protection to others for which they may not pay.

Where there are strong relationships on either the supply or the demand side, allowance must be made for these in cost benefit calculations. Thus where an authority responsible for a long stretch of river constructs a dam at a point upstream, this will affect the water level and hence the operation of existing or potential dams downstream. The construction of a fast motorway which in itself speeds up traffic and reduces accidents, may lead to more congestion or more accidents on feeder roads if they are left unimproved.

Externalities

These are the costs and benefits which accrue to bodies other than the one sponsoring a project. The promoters of public investment projects should take into account the external effects of their actions in so far as they alter the physical production possibilities of other producers or the satisfactions that consumers can obtain from given resources; they should not take *side-effects* into account if the sole effect is through the prices of products or factors.

An example of an external effect to be taken into account would be the construction of a reservoir by the upstream authority of a river basin which results in more dredging by a downstream authority. An example of a side-effect is where the improvement of a road leads to a greater profitability of the garages and restaurants on that road and the employment of more labour by them and higher rent payments. Any net difference in profitability and any net rise in rents and land values are simply a reflection of the benefits of more journeys being undertaken, and it would be double counting if these were included too.

Constraints

Projects are frequently subject to a variety of constraints or restricting factors and these are now categorised.

Physical constraints. The most common is the production function which relates the physical inputs and outputs of a project. Where a choice is involved between different projects or concerning the size or timing of a particular project, external physical restraints may also be relevant.

Legal constraints. Restrictions such as rights of access and time needed for public inquiries may be encountered.

Administrative constraints. These might possibly limit the size of the project.

Uncertainties. These may result from possible unreliability of estimates on future trends.

Distributional and budgetary constraints. For instance, tolls on a motorway will affect the volume of traffic and may influence the width of the carriageways.

Applications of Cost Benefit Analysis

Cost benefit analysis techniques have been applied to a wide range of projects.

In this country one of the first cost benefit studies was that conducted on the economic assessment of the London – Birmingham motorway (M1). This attempted to provide the economic justification for the expenditure of large sums of public money on motorway construction by showing the benefits which would flow from their development. It was restricted inasmuch as it was concerned with the benefits directly attributable to the construction of motorways and did not take account of the effect these new roads would have on neighbouring communities.

A further important study was undertaken on the construction of the Victoria Underground line in London. This study took place after the decision to construct the line had been taken and attempted to show the benefits which would accrue to different sectors of the population when it became operational. It is a particularly useful study for the way it illustrates the difficulty of placing measurements on certain intangible items, such as time savings during leisure hours, and also for the way it emphasises the importance of exercising extreme care in deciding the cut-off points in a practical situation.

The action of public authorities often has a ripple effect – the costs and benefits spread out from the centre and become more diffuse and difficult to measure as they become more remote from the direct action taken by the authority. A good example is that of a local authority providing a housing estate which affects the tenants, shopkeepers and others who provide them with services. It could also affect servicing arangements in neighbouring communities and a decision has to be taken as to the extent to which attempts are made to measure these indirect effects. Determining cut-off points is often one of the most difficult problems because if it is too tightly circumscribed major effects will be omitted from the study.

Lichfield evolved a methodology known as the *planning balance sheet* by which he applied cost benefit analysis to a wide range of town and regional planning problems. A brief account of the methodology of analysis will explain its distinctive features, although this has not been applied very extensively in practice.

An initial step was to enumerate the sectors of the community which were affected by the alternative proposals, treating them on the one hand as producers and operators of the investment to be made in the new project and on the other hand as consumers of the goods and services arising from that project. Then for each sector the question was asked: 'What would be the difference in costs and in benefits which would accrue under the respective schemes under examination?'

The costs and benefits consisted of all those which were of relevance to the planning decision. Thus they included those which were direct as between the parties to the transaction and those which were indirect and came within the conventional definition of social costs and benefits; it included those which related to real resources and those which were transfers; it included those which could not be measured as well as those which could. Thus it was possible to evolve and summarise a set of social accounts for each sector of the community showing the differences in costs and benefits which would accrue to them under the alternative plans. This final summary of social accounts did not produce the decision itself any more than any other economic calculus, but it could form the basis for the judgement leading to the decision. In some studies the judgement that ought to be made was apparent; in others the issues were more finely balanced.

Application to Development Decisions

Cost benefit studies suitably refined could make a useful contribution in arriving at decisions on a wide range of development problems, of which the following are typical examples.

(1) Should a new town programme be carried out as a small number of large developments or a larger number of more modest size?

(2) Should a new town be built on a 'greenfield' site or be grafted on to an existing town of substantial size?

(3) What funds should be allocated to redeveloping a central area?

(4) How should the choice be made between improving old houses and clearing the site and building anew?

(5) What economic criteria are there for choosing which areas of old housing to improve when improvement is the preferred course rather than replacement?

The general principles of choice between replacement and improvement of old housing is one of the more obvious applications but even here there is the major practical problem of evaluating the difference in the standard of accommodation provided by new houses or flats and old houses improved. Scoring the accommodation according to the presence or absence of specific features would be one way, albeit not very precise. Another approach would be to use free-market rents (fair rents) as a basis for evaluating the difference in standards, but there are rarely sufficient to do this. Furthermore, as incomes rise occupiers are likely to become more exacting in their requirements and attach increased importance to features which do not exist in the improved older dwelling.

Similarly town development is a complex process with far-reaching physical, social and economic consequences. Its impact varies as between public and private developers and the community at large, and so a method is needed for determining the consequences of decisions in this field from the various points of view. Thus the need arises for evaluating alternative solutions to the same problem and alternative ways of investing the same resources, with due regard to appearance, amenity and costs. The term amenity as used in this context includes the factors of safety and convenience. It is much easier to reach rational decisions when the consequences of alternative solutions have been quantified and due consideration must be given to all social and economic benefits. Both costs and benefits are spread over long periods of time involving difficult future predictions and discounting, some costs and benefits are indirect and are not easily costed and some intangible benefits become a matter of opinion.

A cost benefit study of the provision of a tunnel or bridge to cross the Thames at Thamesmead indicated a close balance between the quantified factors (traffic and housing benefits being roughly equal to the higher cost of a tunnel). The unquantified environmental and delay factors favoured a tunnel, so that the final decision would have to be based on subjective matters, always provided that the extra money for the tunnel was available.

Application to Building Proposals

There is little evidence of cost benefit analysis being applied to building projects to any great extent and difficulties can arise if subjective evaluations of benefits are included as, for example, when comparing different floor finishes. One useful example was a study carried out by the Department of the Environment and published in 1971, to compare the costs and benefits of planned open offices with air conditioning, and traditional cellular offices without air conditioning. In the study it was assumed that the open offices would be planned to certain essential standards of space, lighting, acoustic treatment, layout and furnishing for functional reasons. Differences were calculated on a cost per capita basis, and discounted to a net present value (NPV) in the year of building at eight per cent, and also at three per cent and sixteen per cent. It was assumed that the life of the buildings would be sixty years and 1966 prices were used throughout. An allowance was made for a relative increase in the cost of labour-intensive industries in calculating maintenance and cleaning costs. The following costs and benefits were quantified.

Costs. Capital costs of building, mechanical and electrical depreciation and running costs, and initial provision, maintenance and renewal of furniture.

Staff benefits. Increased productivity due to teamwork and better communications (one per cent staff salaries and overheads); increased productivity from air conditioning (1.5 per cent staff salaries, etc.), less sick leave because of better working conditions (two per cent of sick leave); saving on messenger services (two messengers per 100 staff); increased productivity from better lighting (one per cent staff salaries, etc.); and better recruiting and staff satisfaction (0.1 per cent of non-career grade staff salaries).

Building benefits. Greater flexibility – avoid remodelling internally after forty years (at cost of ten per cent of original cost of building), more economical use of space (valued at 0.1 m^2/head rising to 0.5 m^2/head during forty years before remodelling), no removal of partitions (valued at one per cent of partitions moved each year); low maintenance materials – reduction in replacement costs and decoration and in staff disturbance; and reduction in cleaning costs (owing to easier and fewer surfaces and building kept cleaner through air conditioning).

The study concluded that the higher costs of open plan arrangements with air conditioning were more than offset by the additional benefits at interest rates of eight per cent and three per cent; at sixteen per cent the cellular offices were slightly better. At eight per cent interest air conditioned planned open offices were calculated to give a capitalised saving of about £260 per head compared with traditional non-air conditioned cellular offices and £200 per head compared with air conditioned cellular offices, at 1966 prices. Unfortunately no updated costs have been published and doubts have been expressed about some of the assumptions made in the study.

Conclusions

Wide divergences of view have been expressed about the role and usefulness of cost benefit analysis. Certainly problems arise from uncertainties about the consequences of various courses of action and the difficulties of measuring the costs and benefits. Diverse types of benefit, avoidance of double counting, dealing with externalities and choice of discount rates pose a formidable range of problems.

On the other hand, is there a better alternative? The situation was well summed up in the Resources for Tomorrow Conference at Montreal in 1961: 'There was general agreement that benefit cost analysis is a basically useful tool in project evaluation. While it has certain limitations and is sometimes difficult to apply, it is, nevertheless, an objective approach to the selection of projects.' It was emphasised that benefit cost analysis (American terminology) should be regarded only as a tool to be used in the decision-making process but not as a substitute for that process.

An important advantage of cost benefit study is that it compels those responsible to quantify costs and benefits as far as possible, rather than resting content with vague qualitative judgements or personal hunches. Furthermore, quantification and evaluation of benefits, however rough, does give some indication of the charges which consumers are willing to pay. Its limitation are clearly shown in the Roskill Commission's report on the third London airport, where after costing all the intangibles relating to the four sites and arriving at total figures in excess of £4000 million, the cheapest site at Cublington was only five per cent less than the most expensive at Foulness. This is hardly a large enough margin to be conclusive in making a choicc.

Even where 'shadow prices' cannot be computed for all the main costs and benefits, cost benefit analysis may still prove useful to the decision maker. In some cases, the valued and unvalued factors may both point in the same direction. If not, and if there is only one unvalued factor, such as noise, it is possible to calculate the value which the decision maker would have to put on this to justify a particular decision. Where there are several unvalued costs or benefits, the problem becomes more complex, but it may still be possible to show that such projects are worse than others on all counts; for instance, one transport strategy might be more expensive, save less travelling time, and have worse distributional effects on the environment than another. In some cases it might be more meaningful to assign limiting values (maxima and mimina) to factors that cannot be valued more precisely.

The technique of cost benefit analysis is still in an early stage of development and its application to building economics requires many more realistic case studies in order to develop and refine the process, so that it can become a positive aid in the decision-making process. The best results are obtained from cost benefit analysis when it is used for choosing between a limited range of alternatives with few intangibles, because of the difficulties of definition and evaluation (Balchin *et al.*, 1994), as was highlighted when attempting to apply it to the resiting of Covent Garden Market.

APPENDIX 1: AMOUNT OF £1 TABLE

Years	*2*	*2½*	*3*	*3½*	*4*	*4½*	*Rate per cent* *5*
1	1.0200	1.0250	1.0300	1.0350	1.0400	1.0450	1.0500
2	1.0403	1.0506	1.0608	1.0712	1.0815	1.0920	1.1024
3	1.0612	1.0768	1.0927	1.1087	1.1248	1.1411	1.1576
4	1.0824	1.1038	1.1255	1.1475	1.1698	1.1925	1.2155
5	1.1040	1.1314	1.1592	1.1876	1.2166	1.2461	1.2762
6	1.1261	1.1596	1.1940	1.2292	1.2653	1.3022	1.3400
7	1.1486	1.1886	1.2298	1.2722	1.3159	1.3608	1.4071
8	1.1716	1.2184	1.2667	1.3168	1.3685	1.4221	1.4774
9	1.1950	1.2488	1.3047	1.3628	1.4233	1.4860	1.5513
10	1.2189	1.2800	1.3439	1.4105	1.4802	1.5529	1.6288
11	1.2433	1.3120	1.3842	1.4599	1.5394	1.6228	1.7103
12	1.2682	1.3448	1.4257	1.5110	1.6010	1.6958	1.7958
13	1.2936	1.3785	1.4685	1.5639	1.6650	1.7721	1.8856
14	1.3194	1.4129	1.5125	1.6186	1.7316	1.8519	1.9799
15	1.3458	1.4482	1.5579	1.6753	1.8009	1.9352	2.0789
16	1.3727	1.4845	1.6047	1.7339	1.8729	2.0223	2.1828
17	1.4002	1.5216	1.6528	1.7946	1.9479	2.1133	2.2920
18	1.4282	1.5596	1.7024	1.8574	2.0258	2.2084	2.4066
19	1.4568	1.5986	1.7535	1.9225	2.1068	2.3078	2.5269
20	1.4859	1.6386	1.8061	1.9897	2.1911	2.4117	2.6532
21	1.5156	1.6795	1.8602	2.0594	2.2787	2.5202	2.7859
22	1.5459	1.7215	1.9161	2.1315	2.3699	2.6336	2.9252
23	1.5768	1.7646	1.9735	2.2061	2.4647	2.7521	3.0715
24	1.6084	1.8087	2.0327	2.2833	2.5633	2.8760	3.2250
25	1.6406	1.8539	2.0937	2.3632	2.6658	3.0054	3.3863
26	1.6734	1.9002	2.1565	2.4459	2.7724	3.1406	3.5556
27	1.7068	1.9477	2.2212	2.5315	2.8833	3.2820	3.7834
28	1.7410	1.9964	2.2879	2.6201	2.9987	3.4296	3.9201
29	1.7758	2.0464	2.3565	2.7118	3.1180	3.5840	4.1161
30	1.8113	2.0975	2.4272	2.8067	3.2433	3.7453	4.3219
35	1.9998	2.3732	2.8138	3.3335	3.9460	4.6673	5.5160
40	2.2080	2.6850	3.2620	3.9592	4.8010	5.8163	7.0399
45	2.4378	3.0379	3.7815	4.7023	5.8411	7.2482	8.9850
50	2.6915	3.4371	4.3839	5.5849	7.1066	9.0326	11.4673
55	2.9717	3.8887	5.0821	6.6331	8.6463	11.2563	14.6356
60	3.2810	4.3997	5.8916	7.8780	10.5196	14.0274	18.6791
65	3.6225	4.9779	6.8299	9.3566	12.7987	17.4807	23.8398
70	3.9995	5.6321	7.9178	11.1128	15.5716	21.7841	30.4264
75	4.4158	6.3722	9.1789	13.1985	18.9452	27.1469	38.8326
80	4.8754	7.2095	10.6408	15.6757	23.0497	33.8300	49.5614
85	5.3828	8.1569	12.3357	18.6178	28.0436	42.1584	63.2543
90	5.9431	9.2288	14.3004	22.1121	34.1193	52.5371	80.7303
95	6.5616	10.4416	16.5781	26.2623	41.5113	65.4707	103.6346
100	7.2446	11.8137	19.2186	31.1914	50.5049	81.5885	131.5012

compound interest 5½	6	6½	7	7½	8	9	10
1.0550	1.0600	1.0650	1.0700	1.0750	1.0800	1.0900	1.1000
1.1130	1.1235	1.1342	1.1448	1.1556	1.1663	1.1880	1.2099
1.1742	1.1910	1.2079	1.2250	1.2422	1.2597	1.2950	1.3309
1.2388	1.2624	1.2864	1.3107	1.3354	1.3604	1.4115	1.4640
1.3069	1.3382	1.3700	1.4025	1.4356	1.4693	1.5386	1.6105
1.3788	1.4185	1.4591	1.5007	1.5433	1.5868	1.6771	1.7715
1.4546	1.5036	1.5589	1.6057	1.6590	1.7138	1.8280	1.9487
1.5346	1.5938	1.6549	1.7181	1.7834	1.8509	1.9925	2.1435
1.6190	1.6894	1.7625	1.8384	1.9172	1.9990	2.1718	2.3579
1.7081	1.7908	1.8771	1.9671	2.0610	2.1581	2.3673	2.5937
1.8020	1.8982	1.9991	2.1048	2.2156	2.3316	2.5804	2.8531
1.9012	2.0121	2.1290	2.2521	2.3817	2.5189	2.8126	3.1384
2.0057	2.1329	2.2674	2.4098	2.5604	2.7196	3.0658	3.4522
2.1160	2.2609	2.4148	2.5785	2.7524	2.9371	3.3417	3.7974
2.2324	2.3965	2.5718	2.7590	2.9588	3.1721	3.6424	4.1772
2.3552	2.5403	2.7390	2.9521	3.1807	3.4259	3.9703	4.5949
2.4848	2.6927	2.9170	3.1588	3.4193	3.7000	4.3276	5.0544
2.6214	2.8543	3.1066	3.3799	3.6758	3.9960	4.7171	5.5599
2.7656	3.0255	3.3085	3.6165	3.9514	4.3157	5.1416	6.1159
2.9177	3.2071	3.5236	3.8696	4.2478	4.6609	5.6044	6.7274
3.0782	3.3995	3.7526	4.1405	4.5664	5.0338	6.1088	7.4002
3.2475	3.6035	3.9966	4.4304	4.9089	5.4365	6.6586	8.1402
3.4261	3.8197	4.2563	4.7405	5.2770	5.8714	7.2578	8.9543
3.6145	4.0489	4.5330	5.0723	5.6728	6.3411	7.9110	9.8497
3.8133	4.2918	4.8276	5.4274	6.0983	6.8484	8.6230	10.8347
4.0231	4.5493	5.1414	5.8073	6.5557	7.3963	9.3991	11.9181
4.2444	4.8223	5.4756	6.2138	7.0473	7.9880	10.2450	13.1099
4.4778	5.1116	5.8316	6.6488	7.5759	8.6271	11.1671	14.4209
4.7241	5.4183	6.2106	7.1142	8.1441	9.3172	12.1721	15.8630
4.9839	5.7434	6.6143	7.6122	8.7549	10.0626	13.2676	17.4494
6.5138	7.6860	9.0622	10.6765	12.5688	14.7853	20.4139	28.1024
8.5133	10.2857	12.4160	14.9744	18.0442	21.7245	31.4094	45.2592
11.1265	13.7646	17.0110	21.0024	25.9048	31.9204	48.3272	72.8904
14.5419	18.4201	23.3066	29.4570	37.1897	46.9016	74.3575	117.3908
19.0057	24.6503	31.9321	41.3149	53.3906	68.9138	114.4082	189.0591
24.8397	32.9876	43.7498	57.9464	76.6492	101.2570	176.0312	304.4816
32.4645	44.1449	59.9410	81.2728	110.0398	148.7798	270.8459	490.3706
42.4299	59.0759	82.1244	113.9893	157.9764	218.6063	416.7300	789.7468
55.4542	79.0569	112.5176	159.8760	226.7956	321.2045	641.1908	1 271.8952
72.4764	105.7959	154.1589	224.2343	325.5945	471.9547	986.5515	2 048.4000
94.7237	141.5788	211.2110	314.5002	467.4330	693.4564	1 517.9319	3 298.9687
123.8002	189.4645	289.3774	441.1029	671.0605	1 018.9149	2 335.5264	5 313.0221
161.8019	253.5462	396.4721	618.6696	963.3942	1 497.1203	3 593.4969	8 556.6753
211.4686	339.3020	543.2012	867.7162	1 383.0771	2 199.7612	5 529.0406	13 780.6110

APPENDIX 2: PRESENT VALUE OF £1 TABLE

Years	*2*	*2½*	*3*	*3½*	*4*	*4½*	*Rate per cent* 5
1	0.98039	0.97560	0.97087	0.96618	0.96153	0.95693	0.95238
2	0.96116	0.95181	0.94259	0.93351	0.92455	0.91573	0.90702
3	0.94232	0.92859	0.91514	0.90194	0.88899	0.87629	0.86383
4	0.92384	0.90595	0.88848	0.87144	0.85480	0.83856	0.82270
5	0.90573	0.88385	0.86260	0.84197	0.82192	0.80245	0.78352
6	0.88797	0.86229	0.83748	0.81350	0.79031	0.76789	0.74621
7	0.87056	0.84126	0.81309	0.78599	0.75991	0.73482	0.71068
8	0.85349	0.82074	0.78940	0.75941	0.73069	0.70318	0.67683
9	0.83675	0.80072	0.76641	0.73373	0.70258	0.67290	0.64460
10	0.82034	0.78119	0.74409	0.70891	0.67556	0.64392	0.61391
11	0.80426	0.76214	0.72242	0.68494	0.64958	0.61619	0.58467
12	0.78849	0.74355	0.70137	0.66178	0.62459	0.58966	0.55683
13	0.77303	0.72542	0.68095	0.63940	0.60057	0.56427	0.53032
14	0.75787	0.70772	0.66111	0.61778	0.57747	0.53997	0.50506
15	0.74301	0.69046	0.64186	0.59689	0.55526	0.51672	0.48101
16	0.72844	0.67362	0.62316	0.57670	0.53390	0.49446	0.45811
17	0.71416	0.65719	0.60501	0.55720	0.51337	0.47317	0.43629
18	0.70015	0.64116	0.58739	0.53836	0.49362	0.45280	0.41552
19	0.68643	0.62552	0.57028	0.52015	0.47464	0.43330	0.39573
20	0.67297	0.61027	0.55367	0.50256	0.45638	0.41464	0.37688
21	0.65977	0.59538	0.53754	0.48557	0.43383	0.39678	0.35894
22	0.64683	0.58086	0.52189	0.46915	0.42195	0.37970	0.34184
23	0.63415	0.56669	0.50669	0.45328	0.40572	0.36335	0.32557
24	0.62172	0.55287	0.49193	0.43795	0.39012	0.34770	0.31006
25	0.60953	0.53939	0.47760	0.42314	0.37511	0.33273	0.29530
26	0.59757	0.52623	0.46369	0.40883	0.36068	0.31840	0.28124
27	0.58586	0.51339	0.45018	0.39501	0.34681	0.30469	0.26784
28	0.57437	0.50087	0.43707	0.38165	0.33347	0.29157	0.25509
29	0.56311	0.48866	0.42434	0.36874	0.32065	0.27901	0.24294
30	0.55207	0.47674	0.41198	0.35627	0.30831	0.26700	0.23137
35	0.50002	0.42137	0.35538	0.29997	0.25341	0.21425	0.18129
40	0.45289	0.37243	0.30655	0.25257	0.20828	0.17192	0.14204
45	0.41019	0.32917	0.26443	0.21265	0.17119	0.13796	0.11129
50	0.37152	0.29094	0.22810	0.17905	0.14071	0.11070	0.08720
55	0.33650	0.25715	0.19676	0.15075	0.11565	0.08883	0.06832
60	0.30478	0.22728	0.16973	0.12693	0.09506	0.07128	0.05353
65	0.27605	0.20088	0.14641	0.10687	0.07813	0.05720	0.04194
70	0.25002	0.17755	0.12629	0.08998	0.06421	0.04590	0.03286
75	0.22645	0.15693	0.10894	0.07576	0.05278	0.03683	0.02575
80	0.20510	0.13870	0.09397	0.06379	0.04338	0.02955	0.02017
85	0.18577	0.12259	0.08106	0.05371	0.03565	0.02372	0.01580
90	0.16826	0.10835	0.06992	0.04522	0.02930	0.01903	0.01238
95	0.15239	0.09577	0.06032	0.03807	0.02408	0.01527	0.00970
100	0.13803	0.08464	0.05203	0.03206	0.01980	0.01225	0.00760

compound interest 5½	6	6½	7	7½	8	9	10
0.94786	0.94339	0.93896	0.93457	0.93023	0.92592	0.91743	0.90909
0.89845	0.88999	0.88165	0.87343	0.86533	0.85733	0.84168	0.82644
0.85161	0.83961	0.82784	0.81629	0.80496	0.79383	0.77218	0.75131
0.80721	0.79209	0.77732	0.76289	0.74880	0.73502	0.70842	0.68301
0.76513	0.74725	0.72988	0.71208	0.69655	0.68058	0.64993	0.62092
0.72524	0.70496	0.68933	0.66634	0.64796	0.63016	0.59626	0.56447
0.68743	0.66505	0.64350	0.62274	0.60275	0.58349	0.54703	0.51315
0.65159	0.62741	0.60423	0.58200	0.55070	0.54026	0.50186	0.46650
0.61762	0.59189	0.56735	0.54393	0.52158	0.50024	0.46042	0.42409
0.58543	0.55839	0.53272	0.50834	0.48519	0.46319	0.42241	0.38554
0.55491	0.52678	0.50021	0.47509	0.45134	0.42888	0.38753	0.35049
0.52598	0.49696	0.46968	0.44401	0.41985	0.39711	0.35553	0.31863
0.49856	0.46883	0.44101	0.41496	0.39056	0.36769	0.32617	0.28966
0.47256	0.44230	0.41410	0.38781	0.36331	0.34046	0.29924	0.26333
0.44793	0.41726	0.38882	0.36244	0.33796	0.31524	0.27453	0.23939
0.42458	0.39364	0.36509	0.33873	0.31438	0.29189	0.25186	0.21762
0.40244	0.37136	0.34281	0.31657	0.29245	0.27026	0.23107	0.19784
0.38146	0.35034	0.32188	0.29586	0.27204	0.25024	0.21199	0.17985
0.36157	0.33051	0.30224	0.27650	0.25306	0.23171	0.19448	0.16350
0.34272	0.31180	0.28379	0.25841	0.23541	0.21454	0.17843	0.14864
0.32486	0.29415	0.26647	0.24151	0.21898	0.19865	0.16369	0.13513
0.30792	0.27750	0.25021	0.22571	0.20371	0.18394	0.15018	0.12284
0.29187	0.26179	0.23494	0.21094	0.18949	0.17031	0.13778	0.11167
0.27665	0.24697	0.22060	0.19714	0.17627	0.15769	0.12640	0.10152
0.26223	0.23299	0.20713	0.18424	0.16397	0.14601	0.11596	0.09229
0.24856	0.21981	0.19449	0.17219	0.15253	0.13520	0.10639	0.08390
0.23560	0.20736	0.18262	0.16093	0.14189	0.12518	0.09760	0.07627
0.22332	0.19563	0.17147	0.15040	0.13199	0.11591	0.08954	0.06934
0.21167	0.18455	0.16101	0.14056	0.12278	0.10732	0.08215	0.06303
0.20064	0.17411	0.15118	0.13136	0.11422	0.09937	0.07537	0.05730
0.15351	0.13010	0.11034	0.09366	0.07956	0.06763	0.04898	0.03558
0.11746	0.09722	0.08054	0.06678	0.05541	0.04603	0.03183	0.02209
0.08987	0.07265	0.05878	0.04761	0.03860	0.03132	0.02069	0.01371
0.06876	0.05428	0.04290	0.03394	0.02688	0.02132	0.01344	0.00851
0.05261	0.04056	0.03131	0.02420	0.01872	0.01451	0.00874	0.00528
0.04025	0.03031	0.02285	0.01725	0.01304	0.00987	0.00568	0.00328
0.03080	0.02265	0.01668	0.01230	0.00908	0.00672	0.00369	0.00203
0.02356	0.01692	0.01217	0.00877	0.00633	0.00457	0.00239	0.00126
0.01803	0.01264	0.00888	0.00525	0.00440	0.00311	0.00155	0.00078
0.01379	0.00945	0.00648	0.00445	0.00307	0.00211	0.00101	0.00048
0.01055	0.00706	0.00473	0.00317	0.00213	0.00144	0.00065	0.00030
0.00807	0.00527	0.00345	0.00226	0.00149	0.00098	0.00042	0.00018
0.00618	0.00394	0.00252	0.00161	0.00103	0.00066	0.00027	0.00011
0.00472	0.00294	0.00184	0.00115	0.00072	0.00045	0.00018	0.00007

APPENDIX 3: AMOUNT OF £1 PER ANNUM TABLE

							Rate per cent
Years	2	$2^1/_2$	3	$3^1/_2$	4	$4^1/_2$	5
1	1.000	1.000	1.000	1.000	1.000	1.000	1.200
2	2.019	2.024	2.029	2.034	2.039	2.044	1.049
3	3.060	3.075	3.090	3.106	3.121	3.137	3.152
4	4.121	4.152	4.183	4.214	4.246	4.278	4.310
5	5.204	5.256	5.309	5.362	5.416	5.470	5.525
6	6.308	6.387	6.468	6.550	6.632	6.716	6.801
7	7.434	7.547	7.662	7.779	7.898	8.019	8.142
8	8.582	8.736	8.892	9.051	9.214	9.380	9.549
9	9.754	9.954	10.159	10.368	10.582	10.802	11.026
10	10.949	10.203	11.463	11.731	12.006	12.288	12.577
11	12.168	12.483	12.807	13.141	13.486	13.841	14.206
12	13.412	13.795	14.192	14.601	15.025	15.464	15.917
13	14.680	15.140	15.617	16.113	16.626	17.159	17.712
14	15.973	16.518	17.086	17.676	18.291	18.932	19.598
15	17.293	17.931	18.598	19.295	20.023	20.784	21.578
16	18.639	19.380	20.156	20.971	21.824	22.719	23.657
17	20.012	20.864	21.761	22.705	23.697	24.741	25.840
18	21.412	22.386	23.414	24.499	25.645	26.855	28.132
19	22.840	23.946	25.116	26.357	27.671	29.063	30.539
20	24.297	25.544	26.870	28.279	29.778	31.371	33.065
21	25.783	27.183	28.676	30.269	31.969	33.783	35.719
22	27.298	28.862	30.536	32.328	34.247	36.303	38.505
23	28.844	30.584	32.452	34.460	36.617	38.937	41.430
24	30.421	32.349	34.426	36.666	39.082	41.689	44.501
25	32.030	34.157	36.459	38.949	41.645	44.565	47.727
26	33.670	36.011	38.553	41.313	44.311	47.570	51.113
27	35.344	37.911	40.709	43.759	47.084	50.711	54.669
28	37.051	39.859	42.930	46.290	49.967	53.993	58.402
29	38.792	41.856	45.218	48.910	52.966	57.423	62.322
30	40.568	43.902	47.575	51.622	56.083	61.007	66.438
35	49.994	54.928	60.462	66.674	73.652	81.496	90.320
40	60.401	67.402	75.401	84.550	95.025	107.030	120.799
45	71.892	81.516	92.719	105.781	121.029	138.849	159.700
50	84.579	97.484	112.796	130.997	152.667	178.503	209.347
55	98.586	115.550	136.071	160.946	191.159	227.917	272.712
60	114.051	135.991	163.053	196.516	237.990	289.497	353.583
65	131.126	159.118	194.332	238.762	294.968	366.237	456.797
70	149.977	185.284	230.594	288.937	364.290	461.869	588.528
75	170.791	214.888	272.630	348.529	448.631	581.044	756.653
80	193.771	248.382	321.362	419.306	551.244	729.557	971.228
85	219.143	286.278	377.856	503.367	676.090	914.632	1 245.086
90	247.156	329.154	443.348	603.204	827.983	1 145.268	1 594.607
95	278.084	377.664	519.271	721.780	1 012.784	1 432.684	2 040.693
100	312.232	432.548	607.287	862.611	1 237.623	1 790.855	2 610.025

compound interest $5\frac{1}{2}$	6	$6\frac{1}{2}$	7	$7\frac{1}{2}$	8	9	10
1.000	1.000	1.000	1.000	1.000	1.000	1.000	1.000
2.054	2.059	2.064	2.069	2.074	2.079	2.089	2.099
3.168	3.183	3.199	3.214	3.230	3.246	3.278	3.309
4.342	4.374	4.407	4.439	4.472	4.506	4.573	4.640
5.581	5.637	5.693	5.750	5.808	5.866	5.984	6.105
6.888	6.975	7.063	7.153	7.244	7.335	7.523	7.715
8.266	8.393	8.522	8.654	8.787	8.922	9.200	9.487
9.721	9.897	10.076	10.259	10.446	10.636	11.028	11.435
11.256	11.491	11.731	11.977	12.229	12.487	13.021	13.579
12.875	13.180	13.494	13.816	14.147	14.486	15.192	15.937
14.583	14.971	15.371	15.783	16.208	16.645	17.560	18.531
16.385	16.869	17.370	17.888	18.423	18.977	20.140	21.384
18.286	18.882	19.499	20.140	20.805	21.495	22.953	24.522
20.292	21.015	21.767	22.550	23.365	24.214	26.019	27.974
22.408	23.275	24.182	25.129	26.118	27.152	29.360	31.772
24.641	25.672	26.754	27.888	29.077	30.324	33.003	35.949
26.996	28.212	29.493	30.840	32.258	33.750	36.973	40.544
29.481	30.905	32.410	33.999	35.677	37.450	41.301	45.599
32.102	33.759	35.516	37.378	39.358	41.446	46.018	51.159
34.863	36.785	38.825	40.995	43.304	45.761	51.160	57.274
37.786	39.992	42.348	44.865	47.552	50.422	56.764	64.002
40.864	43.392	46.101	49.005	52.118	55.456	62.873	71.402
44.111	46.995	50.098	53.436	57.027	60.893	69.531	79.543
47.537	50.815	54.354	58.176	62.304	66.764	76.789	88.497
51.152	54.864	58.887	63.249	67.977	73.105	84.700	98.347
54.965	59.156	63.715	68.676	74.070	79.954	93.323	109.181
58.989	63.705	68.856	74.483	80.631	87.350	102.723	121.099
63.233	68.528	74.332	80.697	87.679	95.338	112.968	134.209
67.711	73.639	80.164	87.346	95.255	103.965	124.135	148.630
72.435	79.058	86.374	94.460	103.399	113.283	136.307	164.494
100.251	111.434	124.034	138.236	154.251	172.316	215.710	271.024
136.605	154.761	175.631	199.635	227.256	259.056	337.882	442.592
184.119	212.743	246.324	285.749	332.064	386.505	525.858	718.904
246.217	290.335	343.179	406.528	482.529	573.770	815.083	1163.908
327.377	394.172	475.879	575.928	698.542	848.923	1260.091	1880.591
433.450	533.128	657.689	813.520	1008.656	1253.213	1944.791	3034.816
572.083	719.082	906.785	1146.755	1453.865	1847.247	2998.288	4893.706
753.271	967.932	1248.068	1614.134	2093.019	2720.079	4619.223	7887.468
990.076	1300.948	1715.655	2269.657	3010.608	4002.556	7113.232	12708.952
1299.571	1746.599	2356.290	3189.062	4327.926	5886.934	10950.572	20474.000
1704.068	2342.981	3234.016	4478.575	6219.107	8655.705	16854.798	32979.687
2232.730	3141.075	4436.576	6287.185	8934.140	12723.936	25939.182	53120.221
2932.730	4209.103	6084.187	8823.852	12831.922	18701.503	39916.632	85556.753
3826.702	5638.367	8341.557	12381.661	18427.694	27484.515	61422.673	137796.110

APPENDIX 4: ANNUAL SINKING FUND TABLE

Years	2	2½	3	3½	4	4½	*Rate per cent* 5
1	1.00000	1.00000	1.00000	1.00000	1.00000	1.00000	1.00000
2	0.49505	0.49382	0.49261	0.49140	0.49019	0.48899	0.48780
3	0.32675	0.32513	0.32353	0.32195	0.32034	0.31877	0.31720
4	0.24262	0.24081	0.23902	0.23725	0.23549	0.23374	0.23201
5	0.19215	0.19024	0.18835	0.18648	0.18462	0.18279	0.18097
6	0.15852	0.15654	0.15459	0.15266	0.15076	0.14867	0.14701
7	0.13451	0.13249	0.13050	0.12894	0.12660	0.12470	0.12281
8	0.11650	0.11446	0.11245	0.11047	0.10852	0.10660	0.10472
9	0.10251	0.10045	0.09843	0.09644	0.09449	0.09257	0.09069
10	0.09132	0.08925	0.08723	0.08524	0.08329	0.08137	0.07950
11	0.08217	0.08010	0.07807	0.07609	0.07414	0.07224	0.07038
12	0.07455	0.07248	0.07046	0.06848	0.06655	0.06466	0.06282
13	0.06811	0.06604	0.06402	0.06206	0.06014	0.05827	0.05645
14	0.06260	0.06053	0.05852	0.05657	0.05466	0.05292	0.05102
15	0.05782	0.05576	0.05376	0.05182	0.04994	0.04811	0.04634
16	0.05565	0.05159	0.04961	0.04768	0.04582	0.04401	0.04226
17	0.04996	0.04792	0.04595	0.04404	0.04219	0.04041	0.03869
18	0.04670	0.04467	0.04270	0.04081	0.03899	0.03723	0.03554
19	0.04378	0.04176	0.03981	0.03794	0.03613	0.03440	0.03274
20	0.04115	0.03914	0.03721	0.03536	0.03358	0.03187	0.03024
21	0.03878	0.03678	0.03487	0.03303	0.03128	0.02960	0.02799
22	0.03663	0.03464	0.03274	0.03093	0.02919	0.02754	0.02597
23	0.03460	0.03269	0.03081	0.02901	0.02730	0.02568	0.02413
24	0.03287	0.03091	0.02904	0.02727	0.02558	0.02398	0.02247
25	0.03122	0.02927	0.02742	0.02567	0.02401	0.02243	0.02095
26	0.02969	0.02776	0.02593	0.02420	0.02256	0.02102	0.01956
27	0.02829	0.02637	0.02456	0.02285	0.02123	0.01971	0.01829
28	0.02698	0.02508	0.02329	0.02160	0.02001	0.01852	0.01712
29	0.02577	0.02389	0.02211	0.02044	0.01887	0.01741	0.01604
30	0.02464	0.02277	0.02101	0.01937	0.01783	0.01639	0.01505
35	0.02000	0.01820	0.01653	0.01499	0.01357	0.01227	0.01107
40	0.01655	0.01483	0.01326	0.01182	0.01052	0.00934	0.00827
45	0.01390	0.01226	0.01078	0.00945	0.00826	0.00720	0.00626
50	0.01182	0.01025	0.00886	0.00763	0.00655	0.00560	0.00477
55	0.01014	0.00865	0.00734	0.00621	0.00523	0.00438	0.00366
60	0.00876	0.00735	0.00613	0.00508	0.00420	0.00345	0.00282
65	0.00762	0.00628	0.00514	0.00418	0.00339	0.00273	0.00218
70	0.00666	0.00539	0.00433	0.00346	0.00274	0.00216	0.00169
75	0.00585	0.00465	0.00366	0.00286	0.00222	0.00172	0.00132
80	0.00516	0.00402	0.00311	0.00238	0.00181	0.00137	0.00102
85	0.00456	0.00349	0.00264	0.00198	0.00147	0.00109	0.00080
90	0.00404	0.00303	0.00225	0.00165	0.00120	0.00087	0.00062
95	0.00359	0.00264	0.00192	0.00138	0.00098	0.00069	0.00049
100	0.00320	0.00231	0.00164	0.00115	0.00080	0.00055	0.00038

compound interest							
5½	*6*	*6½*	*7*	*7½*	*8*	*9*	*10*
1.00000	1.00000	1.00000	1.00000	1.00000	1.00000	1.00000	1.00000
0.48661	0.48543	0.48426	0.48309	0.48192	0.48076	0.47846	0.47619
0.31565	0.31410	0.31257	0.31105	0.30953	0.30803	0.30505	0.30211
0.23029	0.22859	0.22690	0.22522	0.22356	0.22192	0.21866	0.21547
0.17917	0.17739	0.17563	0.17389	0.17216	0.17045	0.16709	0.16379
0.14517	0.14336	0.14156	0.13979	0.13804	0.13631	0.13291	0.12960
0.12096	0.11913	0.11733	0.11555	0.11380	0.11207	0.10869	0.10540
0.10286	0.10103	0.09923	0.09746	0.09572	0.09401	0.09067	0.08744
0.08883	0.08702	0.08523	0.08348	0.08176	0.08007	0.07679	0.07364
0.07766	0.07586	0.07410	0.07237	0.07068	0.06902	0.06582	0.06274
0.06857	0.06679	0.06505	0.06335	0.06169	0.06007	0.05694	0.05396
0.06102	0.05927	0.05756	0.05590	0.05427	0.05269	0.04965	0.04676
0.05468	0.05296	0.05128	0.04965	0.04806	0.04652	0.04356	0.04077
0.04927	0.04758	0.04594	0.04434	0.04279	0.04129	0.03843	0.03574
0.04462	0.04296	0.04135	0.03979	0.03828	0.03682	0.03405	0.03147
0.04058	0.03895	0.03737	0.03585	0.03439	0.03297	0.03029	0.02781
0.03704	0.03544	0.03390	0.03242	0.03100	0.02962	0.02704	0.02466
0.03391	0.03235	0.03085	0.02941	0.02802	0.02670	0.02421	0.02193
0.03115	0.02962	0.02815	0.02675	0.02541	0.02412	0.02173	0.01954
0.02867	0.02718	0.02575	0.02439	0.02309	0.02185	0.01954	0.01745
0.02646	0.02500	0.02361	0.02228	0.02102	0.01983	0.01761	0.01562
0.02447	0.02304	0.02169	0.02040	0.01918	0.01803	0.01590	0.01400
0.02266	0.02127	0.01996	0.01871	0.01753	0.01642	0.01438	0.01257
0.02103	0.01967	0.01839	0.01718	0.01605	0.01497	0.01302	0.01129
0.01954	0.01822	0.01698	0.01581	0.01471	0.01367	0.01180	0.01016
0.01819	0.01690	0.01569	0.01456	0.01349	0.01250	0.01071	0.00915
0.01695	0.01569	0.01452	0.01342	0.01240	0.01144	0.00973	0.00825
0.01581	0.01459	0.01345	0.01239	0.01140	0.01048	0.00885	0.00745
0.01476	0.01357	0.01247	0.01144	0.01049	0.00961	0.00805	0.00672
0.01380	0.01264	0.01157	0.01058	0.00967	0.00882	0.00733	0.00607
0.00997	0.00897	0.00806	0.00723	0.00648	0.00580	0.00463	0.00368
0.00732	0.00646	0.00569	0.00500	0.00440	0.00386	0.00295	0.00225
0.00543	0.00470	0.00405	0.00349	0.00301	0.00258	0.00190	0.00139
0.00406	0.00344	0.00291	0.00245	0.00207	0.00174	0.00122	0.00085
0.00305	0.00253	0.00210	0.00173	0.00143	0.00117	0.00079	0.00053
0.00230	0.00187	0.00152	0.00122	0.00099	0.00079	0.00051	0.00032
0.00174	0.00139	0.00110	0.00087	0.00068	0.00054	0.00033	0.00020
0.00132	0.00103	0.00080	0.00061	0.00047	0.00036	0.00021	0.00012
0.00101	0.00076	0.00058	0.00044	0.00033	0.00024	0.00014	0.00007
0.00076	0.00057	0.00042	0.00031	0.00023	0.00016	0.00009	0.00004
0.00058	0.00042	0.00030	0.00022	0.00016	0.00011	0.00005	0.00003
0.00044	0.00031	0.00022	0.00015	0.00011	0.00007	0.00003	0.00001
0.00034	0.00023	0.00016	0.00011	0.00007	0.00005	0.00002	0.00001
0.00026	0.00017	0.00011	0.00008	0.00005	0.00003	0.00001	0.00000

APPENDIX 5: PRESENT VALUE OF £1 PER ANNUM OR YEARS' PURCHASE TABLE

Years	2	2½	3	3½	4	4½	*Rate per cent* 5
1	0.9803	0.9756	0.9708	0.9661	0.9615	0.9569	0.9523
2	1.9415	1.9274	1.9134	1.8996	1.8860	1.8726	1.8594
3	2.8838	2.8560	2.8286	2.8016	2.7750	2.7489	2.7232
4	3.8077	3.7619	3.7170	3.6730	3.6296	3.5875	3.5459
5	4.7134	4.6458	4.5797	4.5150	4.4518	4.3899	4.3294
6	5.6014	5.5081	5.4171	5.3285	5.2421	5.1578	5.0756
7	6.4719	6.3493	6.2302	6.1145	6.0020	5.8927	5.7863
8	7.3254	7.1701	7.0196	6.8739	6.7827	6.5958	6.4632
9	8.1622	7.9708	7.7861	7.6076	7.4353	7.2687	7.1078
10	8.9825	8.7520	8.5302	8.3166	8.1108	7.9127	7.7217
11	9.7868	9.5142	9.2326	9.0015	8.7604	8.5289	8.3064
12	10.5753	10.2577	9.9540	9.6633	9.3850	9.1185	8.8632
13	11.3483	10.9831	10.6349	10.3027	9.9856	9.6828	9.3935
14	12.1062	11.6909	11.2960	10.9205	10.5631	10.2228	9.8996
15	12.8492	12.3813	11.9379	11.5174	11.1183	10.7395	10.3796
16	13.5777	13.0550	12.5611	12.0941	11.6522	11.2340	10.8377
17	14.2918	13.7121	13.1661	12.6513	12.1656	11.7071	11.2740
18	15.9920	14.3533	13.7535	13.1896	12.6592	12.1599	11.6895
19	15.6784	14.9788	14.3237	13.7098	13.1339	12.5932	12.0853
20	16.3514	15.5891	14.8774	14.2124	13.5903	13.0079	12.4622
21	17.0112	16.1845	15.4150	14.6979	14.0291	13.4047	12.8211
22	17.6580	16.7654	15.9369	15.1671	14.4511	13.7844	13.1630
23	18.2922	17.3321	16.4436	15.6204	14.8568	14.1477	13.4885
24	18.9139	17.8849	16.9355	16.0583	15.2469	14.4954	13.7986
25	19.5234	18.4243	17.4131	16.4815	15.6220	14.8282	14.0939
26	20.1210	18.9506	17.8768	16.8903	15.9827	15.1466	14.3751
27	20.7068	19.4640	18.3270	17.2853	16.3295	15.4513	14.6430
28	21.2812	19.9648	18.7641	17.6670	16.6630	15.7428	14.8981
29	21.8443	20.4535	19.1884	18.0357	16.9837	16.0218	15.1410
30	22.3964	20.9302	19.6004	18.3920	17.2920	16.2888	15.3724
35	24.9986	23.1451	21.4872	20.0006	18.6646	17.4610	16.3741
40	27.3554	25.1027	23.1147	21.3550	19.7927	18.4015	17.1590
45	29.4901	26.8330	24.5187	22.4954	20.7200	19.1563	17.7740
50	31.4236	28.3623	25.7297	23.4556	21.4821	19.7620	18.2559
55	33.1747	29.7139	26.7744	24.2640	22.1086	20.2480	18.6334
60	34.7608	30.9086	27.6755	24.9447	22.6234	20.6380	18.9292
65	36.1974	31.9645	28.4528	25.5178	23.0466	20.9509	19.1610
70	37.4986	32.8978	29.1234	26.0003	23.3945	21.2021	19.3426
75	38.6771	33.7227	29.7018	26.4066	23.6804	21.4036	19.4849
80	39.7445	34.4518	30.2007	26.7487	23.9153	21.5653	19.5964
85	40.7112	35.0962	30.6311	27.0368	24.1085	21.6951	19.6838
90	41.5869	35.6657	31.0024	27.2793	24.2672	21.7992	19.7522
95	42.3800	36.1691	31.3226	27.4835	24.3977	21.8827	19.8058
100	43.0983	36.6141	31.5989	27.6554	24.5049	21.9498	19.8479

compound interest

5½	6	6½	7	7½	8	9	10
0.9478	0.9433	0.9389	0.9345	0.9302	0.9259	0.9174	0.9090
1.8463	1.8333	1.8206	1.8080	1.7956	1.7832	1.7591	1.7355
2.6979	2.6730	2.6484	2.6243	2.6003	2.5770	2.5312	2.4868
3.5051	3.4651	3.4257	3.3872	3.3493	3.3121	3.2397	3.1698
4.2702	4.2123	4.1556	4.1001	4.0458	3.9927	3.8896	3.7907
4.9955	4.9178	4.8410	4.7665	4.6938	4.6228	4.4859	4.3552
5.6829	5.5823	5.4845	5.3892	5.2966	5.2063	5.0329	4.8684
6.3345	6.2097	6.0887	5.9712	5.8573	5.7466	5.5348	5.3349
6.9521	6.8016	6.6561	6.5152	6.3788	6.2468	5.9952	5.7590
7.5376	7.3600	7.1888	7.0235	6.8640	6.7100	6.4176	6.1445
8.0925	7.8868	7.6890	7.4986	7.3154	7.1389	6.8051	6.4950
8.6185	8.3838	8.1587	7.9426	7.7352	7.5360	7.1607	6.8136
9.1176	8.8526	8.5997	8.3576	8.1258	7.9037	7.4869	7.1033
9.5896	9.2949	9.0138	8.7454	8.4891	8.2442	7.7861	7.3666
10.0375	9.7122	9.4026	9.1079	8.8271	8.5594	8.0606	7.6060
10.4621	10.1058	9.7677	9.4465	9.1415	8.8513	8.3125	7.8237
10.8646	10.4772	10.1105	9.7632	9.4339	9.1216	8.5436	8.0215
11.2460	10.8276	10.4324	10.0590	9.7060	9.3718	8.7556	8.2014
11.6076	11.1581	10.7347	10.3355	9.9590	9.6035	8.9501	8.3649
11.9503	11.4690	11.0185	10.5940	10.1944	9.8181	9.1285	8.5135
12.2752	11.7640	11.2849	10.8355	10.4134	10.0168	9.2922	8.6486
12.5831	12.0415	11.5351	11.0612	10.6171	10.2007	9.4424	8.7715
12.8750	12.3033	11.7701	11.2721	10.8066	10.3710	9.5802	8.8832
13.1516	12.5503	11.9907	11.4693	10.9829	10.5287	9.7066	8.9847
13.4139	12.7833	12.1978	11.6535	11.1459	10.6747	9.8225	9.0770
13.6824	13.0031	12.3923	11.8257	11.2994	10.8099	9.9289	9.1609
13.8980	13.2105	12.5749	11.9867	11.4413	10.9351	10.0265	9.2372
14.1214	13.4061	12.7464	12.1371	11.5733	11.0510	10.1161	9.3065
14.3331	13.5907	12.9074	12.2776	11.6961	11.1584	10.1982	9.3696
14.5837	13.7648	13.0586	12.4090	11.8103	11.2577	10.2736	9.4296
15.3905	14.4982	13.6869	12.9476	12.2725	11.6545	10.5668	9.6441
16.0461	15.0462	14.1455	13.3317	12.5944	11.9246	10.7573	9.7790
16.5477	15.4558	14.4802	13.6055	12.8186	12.1084	10.8811	9.8628
16.9315	15.7618	14.7246	13.8007	12.9748	12.2334	10.9616	9.9148
17.2251	15.9905	14.9028	13.9399	13.0836	12.3186	11.0139	9.9471
17.4498	16.1614	15.0329	14.0391	13.1593	12.3765	11.0479	9.9671
17.6217	16.2891	15.1279	14.1099	13.2121	12.4159	11.0700	9.9796
17.7533	16.3845	15.1972	14.1603	13.2489	12.4428	11.0844	9.9873
17.8539	16.4558	15.2478	14.1963	13.2745	12.4610	11.0937	9.9921
17.9309	16.5091	15.2848	14.2220	13.2923	12.4735	11.0998	9.9951
17.9898	16.5489	15.3117	14.2402	13.3048	12.4819	11.1037	9.9969
18.0348	16.5786	15.3314	14.2533	13.3134	12.4877	11.1063	9.9981
18.0694	16.6009	15.3458	14.2626	13.3194	12.4916	11.1080	9.9988
18.0958	16.6175	15.3562	14.2692	13.3236	12.4943	11.1091	9.9992

APPENDIX 6: METRIC CONVERSION TABLE

Length

1 in = 25.44 mm (approximately 25 mm) $\left(\text{then } \frac{\text{mm}}{100} \times 4 = \text{inches}\right)$
1 ft = 304.8 mm (approximately 300 mm)
1 yd = 0.914 m (approximately 910 mm)
1 mile = 1.609 km (approximately $1\frac{3}{5}$ km)
1 m = 3.281 ft = 1.094 yd (approximately 1.1 yd)
(10 m = 11 yd approximately)
1 km = 0.621 mile ($\frac{5}{8}$ mile approximately)

Area

1 ft^2 = 0.093 m^2
1 yd^2 = 0.836 m^2
1 acre = 0.405 ha (1 ha or hectare = 10 000 m^2)
1 $mile^2$ = 2.590 km^2
1 m^2 = 10.764 ft^2 = 1.196 yd^2 (approximately 1.2 yd^2)
1 ha = 2.471 acres (approximately $2\frac{1}{2}$ acres)
1 km^2 = 0.386 $mile^2$

Volume

1 ft^3 = 0.028 m^3
1 yd^2 = 0.765 m^3
1 m^3 = 35.315 ft^3 = 1.308 yd^3 (approximately 1.3 yd^3)
1 ft^3 = 28.32 litres (1000 litres = 1 m^3)
1 gal = 4.546 litres
1 litre = 0.220 gal (approximately $4\frac{1}{2}$ litres to the gallon)

Mass

1 lb = 0.454 kg (kilogramme)
1 cwt = 50.80 kg (approximately 50 kg)
1 ton = 1.016 tonnes (1 tonne = 1000 kg = 0.984 ton)
1 kg = 2.205 lb (approximately $2\frac{1}{5}$ lb)

Density

1 lb/ft^3 = 16.019 kg/m^3
1 kg/m^3 = 0.062 lb/ft^3

Velocity

1 ft/s = 0.305 m/s
1 mile/h = 1.609 km/h

Energy

1 therm = 105.506 MJ (megajoules)
1 Btu = 1.055 kJ (kilojoules)

Thermal conductivity $1\ \text{Btu/ft}^2\text{h}^\circ\text{F} = 5.678\ \text{W/m}^{2\circ}\text{C}$ (where W = watt)

Temperature

$x^\circ\text{F} = \frac{5}{9}(x - 32)^\circ\text{C}$
$x^\circ\text{C} = \frac{9}{5}x + 32^\circ\text{F}$
0°C = 32°F (freezing)
5°C = 41°F
10°C = 50°F (rather cold)
15°C = 59°F
20°C = 68°F (quite warm)
25°C = 77°F
30°C = 86°F (very hot)

Pressure

$1\ \text{lbf/in}^2 = 0.0069\ \text{N/mm}^2 = 6894.8\ \text{N/m}^2$
$(1\ \text{MN/m}^2 = 1\ \text{N/mm}^2)$
$1\ \text{lbf/ft}^2 = 47.88\ \text{N/m}^2$ (newtons/square metre)
$1\ \text{tonf/in}^2 = 15.44\ \text{MN/m}^2$ (meganewtons/square metre)
$1\ \text{tonf/ft}^2 = 107.3\ \text{kN/m}^2$ (kilonewtons/square metre)

For speedy but approximate conversion:

$$\text{lbf/ft}^2 = \frac{\text{kN/m}^2}{20}, \text{ hence } 40\ \text{lbf/ft}^2 = 2\ \text{kN/m}^2$$

$$\text{and tonf/ft}^2 = \text{kN/m}^2 \times 10, \text{ hence } 2\ \text{tonf/ft}^2 = 20\ \text{kN/m}^2$$

Floor loadings

office floors – general usage: $50\ \text{lbf/ft}^2 = 2.50\ \text{kN/m}^2$
office floors – data/processing equipment: $70\ \text{lbf/ft}^2 = 3.50\ \text{kN/m}^2$
factory floors: $100\ \text{lbf/ft}^2 = 5.00\ \text{kN/m}^2$

Safe bearing capacity of soil

$1\ \text{tonf/ft}^2 = 107.25\ \text{kN/m}^2$
$2\ \text{tonf/ft}^2 = 214.50\ \text{kN/m}^2$
$4\ \text{tonf/ft}^2 = 429.00\ \text{kN/m}^2$

Stresses in concrete

$100\ \text{lbf/in}^2 = 0.70\ \text{MN/m}^2$
$1000\ \text{lbf/in}^2 = 7.00\ \text{MN/m}^2$
$3000\ \text{lbf/in}^2 = 21.00\ \text{MN/m}^2$
$6000\ \text{lbf/in}^2 = 41.00\ \text{MN/m}^2$

Costs

$£1/\text{m}^2 = £0.092/\text{ft}^2$
$£1/\text{ft}^2 = £10.764/\text{m}^2$ (approximately $£11/\text{m}^2$)
$£2.50/\text{ft}^2 = £27/\text{m}^2$
$£5/\text{ft}^2 = £54/\text{m}^2$
$£7.50/\text{ft}^2 = £81/\text{m}^2$
$£10/\text{ft}^2 = £108/\text{m}^2$
$£50/\text{ft}^2 = £538/\text{m}^2$

APPENDIX 7: LIFE CYCLE COST PLAN

DATE: 27 April 1995

PROJECT INFORMATION
PROJECT: Oaktown South Primary School
LOCATION: Oaktown, Kenshire.

QUANTITY SURVEYOR: I.B. Able FRICS

PRICE BASE DATE: June 1993

TIME HORIZON: 40 years

DISCOUNT RATE: 4 %

EXCEPTIONS:

OCCUPANCY: 1600 hours per annum

BRIEF DESCRIPTION: Clasp Mark 5, single storey pitched roof, profiled cladding, grassed and hard play areas.

Gross Floor Area 1029 M² Functional Unit 240 pupils

SUMMARY OF LIFE CYCLE COSTS

ITEM	*TARGET COSTS (NOT DISCOUNTED) £*		*NET PRESENT VALUE £*	*NET PRESENT VALUE PER m² OF GROSS FLOOR AREA £*	*NET PRESENT VALUE PER FUNCTIONAL UNIT £*
	Total to Time Horizon	Total Per Functional Unit per annum			
1.1 CAPITAL – LAND	25 200	2.63	25 200	24.49	105.00
1.2 CAPITAL – BUILDING	478 067	49.80	478 067	464.59	1991.95
2. RENTS/LEASES					
3. OPERATIONS COSTS	891 893	92.91	441 331	428.89	1838.88
4.1 MAINTENANCE – INTERMITTENT	221 611	23.08	82 373	80.05	343.23
4.2 MAINTENANCE – ANNUAL	163 360	17.01	80 835	78.55	336.81
5. OTHER					
6. SALVAGE & RESIDUALS					
TOTALS	1 760 131	185.43	1 107 806	1076.57	4615.87
ANNUAL EQUIVALENT VALUES			55 969	54.39	233.20
FULL YEAR EFFECT	£54 695				

ITEM	Target Costs £		Net Present Value £	Net Present Value per m² of Gross Floor Area £
	Total to time horizon	Per m² of Gross Internal Floor Area per annum		
1. CAPITAL				
1.1. Land	25 200	0.61	25 200	24.49
TOTAL LAND N.B. 1. All costs include professional fees where applicable. 2. All prices are firm price equivalent.	25 200	0.61	25 200	24.49
1.2. BUILDING				
1. Building	424 045	10.31	424 045	412.09
2. External Works				
3. Furniture & Equipment	54 022	1.31	54 022	52.49
TOTAL BUILDING N.B. 1. All costs include professional fees where applicable. 2. All prices are firm price equivalent.	478 067	11.61	478 067	464.58
2. RENTS/LEASES				
1. Rents				
2. Leases				
TOTAL RENT				
3. OPERATIONS				
3.1. ENERGY				
1. Heating } 2. Hot Water }	86 933	2.11	43 016	41.80
3. Lighting } 4. Power }	72 000	1.74	35 627	34.63
5. Cooking				
6. Other				
TOTAL OF GROUP	158 933	3.85	78 643	76.43
3.2. RATES/INSURANCES				
1. General Rate	214 667	5.21	106 223	103.23
2. Sewerage	34 720	0.84	17 180	16.69
3. Water	6 933	0.18	3 431	3.33
4. Insurances	26 667	0.65	13 196	12.83
5. Refuse				
6. Other				
TOTAL OF GROUP	282 987	6.88	140 030	136.08
3.3. STAFFING				
1. Caretaking	423 307	10.28	209 463	203.56
2. Cleaning & Equipment	26 667	0.65	13 196	12.83
3. Gardening				
4. Catering				
5. Laundering				
6. Other				
TOTAL OF GROUP	449 974	10.93	222.659	216.39
TOTAL – OPERATIONS COSTS	891 894	21.66	441 332	428.90
4. MAINTENANCE – INTERMITTENT AND ANNUAL				
4.1. MAINTENANCE – INTERMITTENT				
4.1.1. Building Elements	19 485	0.48	4 553	4.43
1. External Walls	3 431	0.08	1 287	1.25
2. Roof	35 423	0.87	14 597	14.19

ITEM	Target Costs £		Net Present Value £	Net Present Value per m² of Gross Floor Area £
	Total to time horizon	Per m² of Gross Internal Floor Area per annum		
MAINTENANCE – INTERMITTENT (*contd*)				
3. Fittings, Fixtures and Furniture				
4. Internal Finishes	12 836	0.31	5 003	4.87
5. Other				
TOTAL OF GROUP	71 175	1.74	25 440	24.74
4.1.2. Services Elements				
1. Sanitary, Hot and Cold Water and Internal Drainage	2 512	0.07	835	0.81
2. Heating and Ventilation	41 124	1.00	13 251	12.88
3. Lifts and Escalators				
4. Lighting Installation	28 869	0.71	10 829	10.52
5. Kitchen Equipment				
6. Other	15 464	0.37	6 373	6.20
TOTAL OF GROUP	87 969	2.15	31 288	30.41
4.1.3. Decoration				
1. External Decoration	7 072	0.17	3 231	3.13
2. Internal Decoration	24 008	0.59	11 347	11.03
TOTAL OF GROUP	31 080	0.76	14 578	14.16
4.1.4. EXTERNAL WORKS				
1. Paths and Pavings	28 127	0.68	10 063	9.77
2. Land and Property				
3. Fencing and Walls	3 260	0.08	1 005	0.97
4. Other				
TOTAL OF GROUP	31 387	0.76	11 068	10.74
TOTAL MAINTENANCE – INTERMITTENT N.B. 1. All costs include professional fees where applicable.	221 611	5.41	82 374	80.05
4.2. MAINTENANCE – ANNUAL				
1. Building Elements	26 667	0.65	13 196	12.83
2. Services Elements } 3. Heating and Hot Water, etc. }	37 333	0.91	18 473	17.95
4. Lighting	16 000	0.39	7 917	7.69
5. Lifts and Escalators				
6. External Works	83 360	2.03	41 248	40.08
7. Other				
TOTAL MAINTENANCE – ANNUAL N.B. All costs include professional fees where applicable.	163 360	3.98	80 834	78.55
5. OTHER BUILDING RELATED COSTS				
1. Hierarchy Costs				
2. Transport				
3. Other				
TOTAL OTHER COSTS				
6. SALVAGE AND RESIDUALS				
1. Demolition				
2. Resale Value				
3. Other				
TOTAL SALVAGE & RESIDUALS				

OAKTOWN SOUTH PRIMARY SCHOOL, KENSHIRE – SCHEDULE OF REPLACEMENT AND INTERMITTENT MAINTENANCE

Element	*Quantity*	*Cost*	*Replacement Interval years*	*Proportion Replaced*	*Yr*	*Present Value Factor*	*Present Value £*
Roof							
Repairs of gutters, linings, etc.	283 m	£16.16/m	25	75%		0.3751	1 287
External Walls							
Repointing of cladding	364 m²	£5.48/m²	20	100%		0.4564	910
Replacing cladding	364 m²	£64.07/m²	35	75%		0.2534	4 432
External Redecoration	–	£884 total	5	100%	5	0.8219	
					10	0.6756	
					15	0.5553	
					20	0.4564	
					25	0.3751	
					30	0.3083	
					35	0.2534	
						3.4460	3 046
Internal Walls – Finishes							
Redecoration	–	£2 652 total	7	100%	7	0.7599	
					14	0.5775	
					21	0.4388	
					28	0.3335	
					35	0.2534	
						2.3631	6 267
Cleaning Lining	932 m²	£2.31/m²	7	100%		2.3631	5 088
Floor Finishes							
Vinyl tiling	280 m²	£13.21	30	75%	30	0.3083	855
Carpet	596 m²	£8.44	10	50%	10	0.6756	
					20	0.4564	
					30	0.3083	
						1.4403	3 623
Plumbing							
Sanitary Fittings	–	£5 023 total	20	25%	20	0.4564	573
					Carried forward		£26 081
Heating							
Pumps	–	£3 119 total	30	100%		0.3083	962
Boiler	–	£14 227 total	30	100%		0.3083	4 386
Stoker	–	£11 889 total	20	100%	20	0.4564	5 426
Electric							
Rewiring		£28 869 total	25	100%		0.3751	10 829
Furniture/Equipment							
Furniture		£26 836 total	10	33%	10	0.6756	
					20	0.4564	
					30	0.3083	
						1.4403	12 755
Kitchen Equipment		£9 665 total	10	40%		1.4403	5 568
External Works							
Resurfacing tarmac paved areas	786 m²	£264/m²	10	100%	10	0.6756	
					20	0.4564	
						1.1320	2 349
Resurfacing footpaths and hard play	853 m²	£1.49/m²	10	100%		1.1320	1 439

OAKTOWN SOUTH PRIMARY SCHOOL, KENSHIRE – SCHEDULE OF REPLACEMENT AND INTERMITTENT MAINTENANCE

Element	*Quantity*	*Cost*	*Replacement Interval years*	*Proportion Replaced*	*Yr*	*Present Value Factor*	*Present Value £*
Replacement of tarmac	786 m²	£13.11/m²	30	100%		0.3083	3 176
Replacement of footpath/ hardplay	853 m²	£9.12/m²	30	100%		0.3083	2 398
Fencing	174 m	£18.73/m	30	100%		0.3083	1 005
						Total	£76 374

Note: no replacements are carried out in the last year of the building's life.

OAKTOWN SOUTH PRIMARY SCHOOL, KENSHIRE – SCHEDULE OF ANNUAL MAINTENANCE

GROUND MAINTENANCE

	hours	£
Labour		
Groundsman		
Summer 26 weeks @ 6 hours	= 156.00	
Winter 26 weeks @ 1 hour	= 26.00	
	182.00 @ £8.80	1601.60
Plant		
Summer 26 weeks @ 5 hours	= 130.00	
Winter 26 weeks @ 1 hour	= 26.00	
	156.00 × £4.30	670.80
Materials		135.00
	Total ground maintenance	£2407.40

ROUTINE SERVICING

Building Repairs	£680 per annum
Services	£950 per annum
Lighting	£400 per annum

Note: To convert to present value (PV), the multiplier for PV of £1 per annum at compound interest of 4% for 40 years (throughout the life of the building) is 19.7927.
Hence the PV of ground maintenance would be £2407.40 × 19.7927 = £47 648.95

REFERENCES

Applied Property Research (1990). *Living with BI.*

Aqua Group (1990a). *Tenders and Contracts for Building*. BSP.

Aqua Group (1990b). *Contract Administration for the Building Team*. BSP.

Artingstall B. and Jeffreys M. (1992). The right questions. *Housing, June, 45.*

Ashworth A. (1986). Cost models – their history, development and appraisal, *CIOB Technical Information Service nr 64.*

Ashworth A. (1994). *Cost Studies of Buildings*. Longman.

Ashworth A. and Skitmore M. (1982). Accuracy in estimating. *CIOB Occasional Paper 27.*

Atkin B. L. (1987). A time/cost planning technique for early design evaluation. *Building Cost Modelling and Computers*. Spon.

Balchin P. N., Bull G. H. and Kieve J. L. (1994). *Urban Land Economics*. Macmillan.

Baron T. (1980). Using derelict land for housing. *Chartered Surveyor, August.*

Barry P. (1992). Marketplace *Chartered Quantity Surveyor, December/January*, 34–5.

BCIS (1986). *Design/performance data – Building owner's reports: 2. Flat roofs.*

BCIS (1992a). *The Building Cost Information Service.*

BCIS (1992b). *Average element prices (on line).*

BCIS (1993). *Quarterly Review of Building Prices, December.*

Beard Dove (1990). *Value Engineering pamphlet.*

Bennett J. and Ferry D. (1987). Towards a simulated model of the total construction process. *Building Cost Modelling and Computers*. Spon.

Benson J., Evans B., Colomb P. and Jones G. (1980). *The Housing Rehabilitation Handbook*. Architectural Press.

Berryman A. (1971). Controlling the cost of engineering services in buildings. *Chartered Surveyor, August.*

Bishop D. (1975). Productivity in the construction industry. *Aspects of the Economics of Construction.* Godwin.

Bishop D. (1981). If he build what is good. *Building Technology & Management, December.*

BMCIS (1985). *Study of Average Occupancy Costs.*

BMI (1989). *Special Report 176/177: Building Maintenance – Investment for the Future.*

BMI (1990). *Special Report 190: Occupancy Cost Indices.*

BMI (1991). *Special Report 199: University Occupancy Expenditure.*

BMI (1992). *Building Maintenance Information.*

BMI (1993). *Special Report 220: Rehabilitation Costs.*

Bowen P. *et al.* (1987). Cost modelling: a process modelling approach. *Building Cost Modelling and Computers*. Spon.

Brandon P., Stafford B. and Atkin B. (1990). *The potential for expert systems within quantity surveying.* RICS.

BRE (1969). *Coding and data co-ordination for the construction industry.*

BRE (1972). *Digest 64: Soils and foundations 2.* HMSO.

BRE (1979a). *Digest 223: Wall claddings: designing to minimise defects due to inaccuracies and movement.* HMSO.

BRE (1979b). Why drains get blocked. *BRE News, Spring.*

BRE (1980). *Digest 67: Soils and foundations 3.* HMSO.

BRE (1982). *Quality in Traditional Housing, Vol 1: An investigation into their faults and their avoidance.*

BRE (1985). *Quality in Timber Frame Housing.*

BRE (1987). *Digest 260: Smoke control in buildings: design principles.* HMSO.

BRE (1988). *Digest 338: Insulation against external noise.* HMSO.

BRE (1989). *Information paper IP 23/89: CFCs and the building industry.*

BRE (1990a). *Digest 350: Climate and soil development, part 3: Improving microclimate through design.* HMSO.

BRE (1990b). *Good Building Guide 6 (GRB6): Outline guide to assessment of traditional housing for rehabilitation.*

BRE (1990c). *Rehabilitation – a review of quality in traditional housing.*

BRE (1990d). *BREEAM/An environmental assessment for new office designs 1/90.*

BRE (1991). *Digest 367: Fire modelling*. HMSO.

BRE (1992). *Information paper IP20/92: Energy use in office buildings.*

BRE (1993a). *Digest 379: Double glazing for heat and*

sound insulation. HMSO.
BRE (1993b). *Information paper IP2/93: Industrial building refurbishments: opportunities for energy efficiency*.
British Council for Offices (1994). *Specification for Urban Offices*.
British Steel (1992). *The Market for Structures in Multi Storey Buildings*.
Brown R. P. (1992). Value management: a new service for the client. *RICS QS Bulletin, April, iii*.
Bryant A. (1970). The best use of our land. *Chartered Surveyor, September*.
BS 5427: 1976. *Code of practice for performance and loading criteria for profiled sheeting in building*.
BS 5750: 1987/BS EN ISO 9000: 1994. *Quality Systems*.
BS 5930: 1981. *Code of practice for site investigations*.
BS 6262: 1982. *Code of practice for glazing of buildings*.
BS 6399: 1984. *Design loading for buildings; Part 1: Code of practice for dead and imposed loads*.
BS 7750: 1992. *Specification for environmental management systems*.
BS 8004: 1986. *Code of practice for foundations*.
BS 8110: 1985. *Structural use of concrete; Part 1: Code of practice for design and construction*.
Business Round Table (1994). *Controlling the Upwards Spiral*.
Cadman D. (1990) The environment and the property market. *Town and Country Planning, October, 267–70*.
Cadman D. and Austin-Crowe L. (1990). *Property Development*. Spon.
Cambridge Economics Group (1981). *Policies for Recovery – An Evaluation of Alternatives*.
Cantaouzino S. and Brandt S. (1980). *Saving Old Buildings*. Architectural Press.
Carter T. G. (1991/92). *Value Management Selected Papers*. Davis Langdon Management.
Carter T. G. (1992a). Value engineering: a comparison between the 40 hour workshop and a one/two day study. *RICS QS Bulletin, April, v*.
Carter T. G. (1992b). Case study: computer center – northern England. *RICS QS Bulletin, April, v–vi*.
Catt R. (1991a). Upgrading – but at a price. *Chartered Quantity Surveyor, October, 30*.
Catt R. (1991b). Servicing the built environment. *Chartered Quantity Surveyor, September, 30*.
Catt R. (1992). Model fire-fighting. *Chartered Quantity Surveyor, March, 26*.
Catterick P. (1992). *Total Quality: An introduction to quality management in social housing*. CIH.
CEEC/UNTEC (1993). *Conference in Paris: Introduction to CEEC*.
Centre for Strategic Studies in Construction, University of Reading (1991). *Construction Management Forum Report*.
CIBSE (1991). *Energy Audits and Surveys: AM5*.
CIOB (1983). *Code of Estimating Practice*.
CIOB (1987). *Code of Estimating Practice: Supplement No 1: Refurbishment and Modernisation*.
CIOB (1989). *Quality Assurance in the Building Process*.
CIOB (1992). *Construction Computing, No 37, Spring*.
CIRIA (1983). *Buildability: An Assessment*.
CIRIA (1989). *Special Publication 69, The engineering implications of rising groundwater levels in the deep aquifer beneath London*.
Civic Trust (1988). *Urban Wasteland Now*.
Civic Trust for Wales/Civic Trust Regeneration Unit (1992). *Llanidloes: A Town Study*.
CLASP (1990). *CLASP: Introduction to the System*.
Colquhon I. and Fauset P. G. (1991). *Housing Design in Practice*. Longman.
Committee chaired by HRH the Duke of Edinburgh. *Report of the Inquiry into British Housing* (1985) and *Second Report* (1991).
Committee commissioned by the Archbishop of Canterbury (1985). *Faith in the City*.
Cox W. P. (1971). Cost information service for engineering systems for buildings. *Journal of Institution of Heating and Ventilating Engineers (CIBS), February*.
CP 153 (1969). *Windows and rooflights, Part 1: Cleaning and safety*.
CPI (1987a). *Production Drawings: A code of procedure for building works*.
CPI (1987b). *Project Specification: A code of procedure for building works*.
CPI (1988). *SMM7: A code of procedure for measurement of building works*.
Curtis A. (1993). Greenhouse programme. *Inside Housing, 12 March, 9–10*.
Davis Langdon & Everest (1994). *A Survey of Building Contracts in Use in 1993*. RICS.
Debenham Tewson & Chinnocks (1990). *Business Parks – Out of town or out of touch*.
DES (1989). Technical co-ordination working party. *Building Industry Code*.
DoE (1971). Whitehall Development Group. *Planned Open Offices: CBA*.
DoE (1987). *Re-using Redundant Buildings: Good Practice in Regeneration*. HMSO.
DoE (1990). *House Renovation Grants*.
DoE (1991). *Environmental Action Guide for Building Purchasing Managers*. HMSO.
DoE (1992). *Handbook of Estate Improvement 3: Dwellings*.
DoE (1994a). Consultation Paper. *Access to local authority and housing association tenancies, January*.
DoE (1994b). *Vital and Viable Town Centres*.
DoE/DTp (1992). *Residential Roads and Footpaths: Layout Considerations: Design Bulletin 32*. HMSO.
Doyle N. (1993). Straight talking. *New Builder, 22 January, 12–13*.
Drake M. (1992). *Europe and 1992: A handbook for local housing authorities*. CIH.
Duell J. (1991). *Windows: Specification 91, technical*.

Dyas D. (1992). What do tenants want? *Housing, June, 42–3.*

Economic Development Committee for Building (1967). *Action on the Banwell Report.* HMSO.

EEO (Energy Efficiency Office) (1987a). *Degree Days, Fuel Efficiency Booklet 7.*

EEO (1987b). *Energy Efficiency in Buildings: How to bring down energy costs in schools.*

EEO (1991). *Energy Efficiency in Offices: Good practice case study 16, Heslington Hall.*

EEO (1992). *Saving Energy in Schools.*

EEO (1993a). *Good Practice Guide 61: Design Manual – Energy efficiency in advance factory units.*

EEO (1993b). *Good Practice Case Study 141: Energy efficiency in advance factory units.*

English Estates *et al.* (1986). *Industrial and Commercial Estates: Planning and Site Development.* Telford.

English House Condition Survey 1991. HMSO.

Estates Times (1993a). *Review: Offices, 24 September.*

Estates Times (1993b). *Supplement: Business Parks, 25 June.*

European Liaison Committee for Social Housing (CECODHAS) (1993). *A roof over the head of every European: 5 years of involvement and action.*

Farr J. and Nevin B. (1994). Strategy lost in the rubble. *Housing, February, 19–20.*

Ferry D. J. and Brandon P. S. (1991). *Cost Planning of Buildings.* BSP.

Field A. J. (1992). *BRE Information Paper IP12/92: Energy Audits and Surveys, May.*

Flanagan R. *et al.* (1979). *UK and US construction industries: a comparison of design and contract procedures.* RICS.

Flanagan R. and Norman G. (1982). Making good use of low bids. *Chartered Quantity Surveyor, March.*

Flanagan R. and Norman G. (1983). *Life Cycle Costing for Construction.* RICS.

Flanagan R. and Norman G. (1989). *Life Cycle Costing.* BSP.

Flanagan R. and Norman G. (1993). *Risk Management and Construction.* BSP.

Fraser W. D. (1993). *Principles of Property Investment and Pricing.* Macmillan.

Freeman I. L. (1975). *BRE Current Paper 30/75: Building failure patterns and their implications.*

Galbraith A. (1991). *Building Law for Students.* Heinemann.

Ganesan S., Yip S. L. and So P. H. (1980). *Resource and cost analyses of building projects on a digital computer.* University of Hong Kong, School of Architecture.

Gardiner & Theobold (1992). Onward and upward: lifts and escalators. *Chartered Quantity Surveyor, June 11–17.*

Golland A. and Oxley M. (1993). Learning the lessons on students. *Housing, October, 13*

Greater London Council (1978). *An Introduction to Housing Layout.* Architectural Press.

Green S. and Moss G. (1993). Value for money from SMART management. *Chartered Builder, October, 5–7.*

Griffith A. (1990). *Quality Assurance in Building.* Macmillan.

Griffith A. (1994). *Environmental Management in Construction.* Macmillan.

Griffith A. and Sidwell A. C. (1995). *Constructability in Building and Engineering Projects.* Macmillan.

Gunn D. (1992). Maintain the advantage. *Housing, June, 33–4.*

Harvey R. C. and Ashworth A. (1993). *The Construction Industry of Great Britain.* Butterworth – Heinemann.

Hassan G. (1995). *Advanced Building Services.* Macmillan.

Hayward D. (1993). Smooth operation. *New Builder, 12 March, 12–13.*

HDB, Singapore (1985). *Designed for Living: Public housing architecture in Singapore.*

HDB, Singapore (1989). *Public Housing in Singapore.*

Heap D. (1991). *An Outline of Planning Law.* Sweet & Maxwell.

Herbert J. (1991). Construction management: What's in it for us. *Chartered Quantity Surveyor, October, 20.*

Highfield D. (1987). *Rehabilitation and Re-use of Old Buildings.* Spon.

Hillier Parker (1993). *Hillier Parker Rent Index Digest, August.*

Hong Kong Government, Lands and Works Branch (1989). *Hong Kong: The Development Challenge.*

Housing Corporation (1992). *A review of housing needs assessment.*

HRH The Duke of Edinburgh (1985 and 1991). *Reports of Committee of Inquiry into British Housing.*

HRH The Prince of Wales (1989). *A Vision of Britain: A personal view of architecture.* Doubleday.

ICE (1969). *An Introduction to Engineering Economics.*

ICE, Infrastructure Policy Group (1988). *Urban Regeneration.* Telford.

Imaginor Systems (1991). *KBS Knowledge Based Software.*

Institute of Housing (1992). *Housing – The First Priority.*

Institute of Housing/RIBA (1983). *Homes for the Future – Standards for New Housing Development.*

Isaac D. (1994). *Property Finance.* Macmillan.

Jackson A. E. (1981). Controlling costs in industry: scope in cost management. *Chartered Quantity Surveyor, January.*

James W. (1954). A new approach to single price-rate approximate estimating. *The Chartered Surveyor, May.*

Janssens D. E. L. (1991). *Design and Build Explained.* Macmillan.

Johnson S. (1993). *Greener Buildings: Environmental impact of property.* Macmillan.

Jones Lang Wootton (1994). *50 Centres: Office, industrial and retail rents, January*.
Joseph Rowntree Trust (1993). *Innovations in Social Housing, No 4, March*.
Josey B. (1991a). Profiled metal sheeting. *Specification 91, technical*.
Josey B. (1991b). GRP cladding. *Specification 91, technical*.
Josey B. (1991c). GRC cladding. *Specification 91, technical*.
Josey B. (1991d). Curtain walling. *Specification 91, technical*.
Just R. and Williams D. (1996). *Urban Regeneration: A practical guide*, Macmillan.
Kay J. (1991). Controlling the cost. *Chartered Quantity Surveyor, September, 32, 34*.
Kelly J. and Male S. (1991). *The Practice of Value Management: Enhancing Value or Cutting Cost?* Heriot-Watt University/RICS QS Division.
Kelly J. and Male S. (1992). Functional analysis method. *RICS QS Bulletin, April, iv*.
Kelly J. and Male S. (1993). *Value Management in Design and Construction*. Spon.
Knowles P. (1991). Structural steelwork. *Specification 91, technical*.
Koh S. (1979). Contract cleaning services in Singapore today. *Maintenance Man (Journal of Building Society, Ngee Ann Polytechnic, Singapore), June*.
Latham Report (1994). *Constructing the Team*. HMSO.
Lee How Son and Yuen G. C. S. (1993). *Building Maintenance Technology*. Macmillan.
Levinson, DSSR and Gleeds (1990). *A Guide to Environmental Friendly Buildings*.
Maver T. (1979). Cost performance modelling. *Chartered Quantity Surveyor. December*.
MDA (1992). In the frame. *Chartered Quantity Surveyor, May, 22–4*.
Melia C. (1992). Public procurement in the European Community. *Chartered Surveyor Monthly, January, 9*.
Metra Martech (1991). *European Environmental Legislation and its Application*.
MHLG (1952). *The cost of house-building. Third Report of Committee of Inquiry*. HMSO.
MHLG (1961). *Parker Morris Report: Homes for today and tomorrow*.
MHLG (1962). *Planning Bulletin No 2: Residential areas – higher densities*. HMSO.
Miles L. D. (1972). *Techniques for Value Analysis and Engineering*. McGraw-Hill.
Miller S. (1992). Going for the green option. *Chartered Quantity Surveyor, March, 9–11*.
Monk Dunstone Associates (1992). Estimating and economics: structural frames. *Chartered Quantity Surveyor, May, 25–7*.
Moorhead M. (1992/93). More than meets the eye. *Chartered Quantity Surveyor, December/January, 10–14*.
MPBW (1965). *The placing and management of contracts for building and civil engineering work (Banwell Report)*. HMSO.
MPBW (1966). *Early selection of contractors and serial tendering*. HMSO.
National Association of Lift Manufacturers (NALM) (1992a). *Principles of planning and programming a lift installation*.
NALM (1992b). *Guidance on a range of car finishes related to mass*.
NALM (1992c). *Some examples of specifier's requirements which can have a significant effect upon cost and delivery*.
Nationwide Building Society (1993). *House prices in 1993*.
NEDO (1990). *Study of the UK lift and escalator industry*.
Newark & Sherwood (1992/93). *Care and Repair: Annual Report*.
NFHA (National Federation of Housing Associations) (1980). *Minimum standards for housing rehabilitation*.
NFHA (1989). *Maintenance Planning: basic methods in assessing future requirements: Research report 7*.
NFHA (1990). *Committee Members Handbook*.
NFHA (1992). *Affordable new low energy housing for housing associations*. DoE Energy Efficiency Office.
NJCC for Building (1994). *Code of procedure for single-stage selective tendering*.
NJCC for Building (1994). *Code of procedure for two-stage selective tendering*.
NJCC for Building (1985). *Code of procedure for selective tendering for design and build*.
Norton B. (1992a). Value added. *Chartered Quantity Surveyor, June, 21–3*.
Norton B. (1992b). A value engineering case study. *Chartered Quantity Surveyor, September, 7*.
Norton B. and McElligott W. (1995). *Value Management in the Construction Industry: A practical guide*. Macmillan.
Nottinghamshire County Council (1993). *Listed Buildings at Risk in Nottinghamshire*.
Ong S. B. (1981). A study of residential land values in Singapore. *Unibeam: Annual Magazine of Building & Estate Management Society, National University of Singapore*.
Park J. A. (1992). *Facilities for growth. Chartered Quantity Surveyor, February, 17–19*.
Park J. A. (1994). *Facilities Management: An explanation*. Macmillan.
Parker D. (1993). Back to reality. *New Builder, 16 April, 22–3*.
Parker Morris Report (1961). *Homes for Today and Tomorrow*. HMSO.
Patchell B. R. T. (1987). The implementation of cost modelling theory. *Building Cost Modelling and Computers*. Spon.
Paterson M. (1992). Keeping it in-house. *Housing, June, 47*.

Pell-Hiley P. (1974). Integration in design. *Chartered Surveyor, Building & Surveying Quarterly, March.*

Pickles D. (1982). *Cost in use: a reappraisal.* Unpublished dissertation, Honours degree in quantity surveying, Trent Polytechnic (Nottingham Trent University).

Property Helpline (1994). The Guide to Better Decisions in Facilities Management. *The Facilities Handbook Vol. 1.* CML Data Ltd.

PSA. (1988). *The Conservation Handbook.*

Raftery J. (1987). The state of cost/price modelling in the UK construction industry: a multicriteria approach. *Building Cost Modelling and Computers.* Spon.

Raftery J. (1993). *Risk Analysis in Project Management.* Spon.

Raftery J. and Newton S. (1995) *Building Cost Modelling.* BSP.

RIBA (1954). *RIBA rules for cubing buildings for approximate estimates: D/1156/54.*

RIBA (1973). *Handbook of Architectural Practice and Management.*

RIBA (1985). *Decaying Britain.*

Richmond D. (1994). *Introduction to Valuation.* Macmillan.

RICS (1980). Planning and Development Division. Seminar on shopping development partnerships. *Chartered Surveyor, March.*

RICS (1986). *Housing: The Next Decade.*

RICS (1989). *Quality Assurance: Introductory Guidance.*

RICS (1991). *Market requirements for the profession (Lay Committee).*

RICS (1992). *Living Cities.*

RICS (1993a). *Energy Efficiency in Buildings: Energy Appraisal of Existing Buildings; A Handbook for Surveyors.*

RICS (1993b). *Environmental Management and the Chartered Surveyor.*

RICS (1993/94). *European Alert, December/January.*

RICS (1994). *Economic Cycles and Property Cycles.*

RICS Facilities Management Skills Panel (1993). Facilities management: what you should know. *Chartered Surveyor Monthly, July/August, 4.*

RICS Planning & Development Division (1980). Seminar on shopping development partnerships. *Chartered Surveyor, March.*

RICS QS Division, Essex Branch (1982). *Pre-contract cost control and cost planning.*

RICS QS Division (1990). *Report of the review carried out by the cost information and data services working party.*

RICS QS Division (1991). *QS 2000: The Future Role of the Chartered Quantity Surveyor.*

RICS/English Heritage (1993). *Investment performance of listed buildings.*

Robertson J. D. M. (1983). *Maintenance Audit.* BMCIS.

RTPI (1993). *The Character of Conservation Areas.*

Seeley I. H. (1974). *Planned Expansion of Country Towns.* Godwin.

Seeley I. H. (1984a). *Blight on Britain's Buildings: A Survey of Paint and Maintenance Practice.* Paintmakers Association.

Seeley I. H. (1984b). *Quantity Surveying Practice.* Macmillan.

Seeley I. H. (1987). *Building Maintenance.* Macmillan.

Seeley I. H. (1992). *Public Works Engineering.* Macmillan.

Seeley I. H. (1993). *Civil Engineering Contract Administration and Control.* Macmillan.

Seeley I. H. (1995). *Building Technology.* Macmillan.

Sharp E. (1978). Inner cities in decay: problems, priorities and possibilities, *Chartered Surveyor, November.*

Silk & Frazier (1992). Blowing hot and cold: air conditioning. *Chartered Quantity Surveyor, September, 10–13.*

Simon Committee (1944). *The placing and management of building contracts.* HMSO.

Singh S. (1995). *Cost Estimation of Structures in Commercial Buildings.* Macmillan.

Sisk J. (1993). *Prefabrication: four star quality in half the time. Chartered Quantity Surveyor, April, 18–19.*

Skitmore M. (1989). *Contract Bidding in Construction.* Longman.

Smith A. J. (1995). *Estimating, Tendering and Bidding for Construction. Theory and Practice.* Macmillan.

Smith G. (1980). Cost planning the design process. *Chartered Quantity Surveyor, August.*

Smith M. E. H. (1989). *Guide to Housing.* Housing Centre Trust.

Smith M. (1993). VE: is it withering on the vine? *Chartered Quantity Surveyor, February, 18–19.*

Smith T. (1992). Managing cost for long term competitiveness. *RICS General Practice Members Information Service, July.*

Snedecor G. W. and Cochrane W. G. (1976). *Statistical Methods.* Iowa State University Press.

Society of Chief Quantity Surveyors in Local Government (SCQSLG) (1984). *Life Cycle Cost Planning.*

Southwell J. (1971). *Building Cost Forecasting.* RICS QS Research and Information Committee.

Speight B. A. (1994). Maintenance policy, programming and maintenance feedback. *Building Maintenance and Preservation.* (Ed. Mills E. D.). Butterworth-Heinemann.

Spon (1994). *Architects' and Builders' Price Book.*

Stephens F. R. (1992). Legionella in domestic hot water systems. *Building Research and Information (20)2, 96–101.*

Stewart A. (1993). Impossible challenge. *Building, 16 April.*

Stone P. A. (1975). *Building Design Evaluation: Costs in Use.* Spon.

Taylor L. (1993). *Energy Efficient Homes: A guide for housing professionals. Association for the Conser-*

vation of Energy/IOH.

Taylor N. P. (1991). *Development Site Evaluation*. Macmillan.

Telling and Duxbury (1993). *Planning Law and Procedure*. Butterworths.

TRADA (1988). *Timber Frame Construction*.

Treasury (1994). The UK recession 1990–92. *Economic Briefing, No. 6, February*.

Trench D. (1991). *On Target: A design and manage target cost procurement system*. Telford.

Turner A. (1990). *Building Procurement*. Macmillan.

Tysoe B. (1991). Making the right choice. *Chartered Quantity Surveyor, November, 12–13*.

Uthwatt Committee (1942). Expert committee on compensation and betterment. *Cmnd 6386*.

Vann R. (1993). Contaminated land: digging up the dirt. *Chartered Quantity Surveyor, October, 13–14*.

Wager D. (1992). The Construction Industry Computing Association. *RICS QS Bulletin, March, iii*.

Watson B. (1981). Controlling costs in industry: setting the scene. *Chartered Quantity Surveyor, January*.

Watson B. (1990). Cost planning engineering services contracts. *The Cost Engineer, July*.

Watson B. (1992). Cost planning techniques for engineering services. *CIBS Seminar*.

Watts T. (1992). Facilities management typifies shift to demand led market. *RICS keynote address, 21 February*.

Whittaker R. S. (1975). Services. *Chartered Surveyor, Urban Quarterly, Autumn*.

Wilderness Cost of Building Study Group (1964). *An investigation into building cost relationships of the following design variables: storey heights, floor loadings, column spacings, number of storeys*. RICS.

Wilson A. J. (1984). Introductory report on cost modelling. *CIB Ottawa Vol. 1, 61–2*.

Wong A. K. and Yeh S. H. K. (1985). *Housing a Nation: 25 years of public housing in Singapore*. HDB, Singapore.

Woodward C. and Campbell K. (1982). Two perspectives on Odham's Walk, Covent Garden. *Architects' Journal, 3 February*.

Wordsworth P. (1992). The greening of building maintenance. *The Building Surveyor, June*.

W. T. Partnership (1993). The big heat. *Chartered Quantity Surveyor, May, 12–14*.

INDEX

Agrément Board 14, 79
Agricultural land and property values 401–2
Air conditioning 61, 82–4, 272–5, 323, 347–9
All-in items 171–2
American building procedures and costs 94
Amount of £1 pa table 302–3, 480–1
Amount of £1 table 301–2, 476–7
Annual equivalent 327–9, 343, 346, 350
Annual sinking fund table 303–4, 482–3
Approximate estimating 21, 22, 154–73
  approximate quantities 164–8
  classification of procedures 155–6
  comparative costs 171–2
  comparative estimates 173
  cube method 156–8
  elemental cost analyses 168–70, 172
  floor area method 159–60
  interpolation method 173
  purpose 154–5
  storey-enclosure method 160–4
  unit method 156
Approximate quantities 164–8
Automation 87–8

Banwell committee 15
Basements 64–5
BCIS 23–4, 238
  on-line approximate estimating package 204
Brick walls 65–6, 320
Briefing 182–5
Bucknall Austin building cost model 204–5
Budgeting
  hospitals 417–18
  housing association housing 413–15
  local authority housing 413
  non-commercial properties 417
  student halls of residence 418
Building
  cost indices 224–5, 226–8, 231, 232–3
  cost information service 23–4, 238
  costs 8–10, 242–3
  development 403–33
  economics 14
  firms 442–5
  importance of 7–8
  industry code 191, 193
  leases 433
  listed 467
  maintenance 311–12, 371–2
  maintenance information 24, 238, 371–2
  nature and value 7–9
  of special interest 390
  output 10–12, 441
  price books 239
  prices 242–3, 270
  procurement 15–21
  productivity 12–14
  regulations 91, 392–3
  workload 445–7
Business parks 410–11
Business Round Table Report 85, 93

C1/SfB classification system 193–4
CAD 203, 246, 377–8
Capital allowances 370–1
Car parking 62, 141–2
Cashflow 199–200, 415–16
CCT 62, 110
CDM regulations 393
CEEC 3, 24–5
Churches 417
CIRIA 25
City challenge 455
Civic Trust 467–9
Civil engineering 442
Cladding 71–3, 320
CLASP 69, 111–12
  project costs 112
Cleaning 323–4, 325, 337
Client needs 181–3
Coefficient of variation 333
Column spacings 50–1
Commercial properties 52, 334–5
Comparative cost 171–2
  estimates 173
  planning system 190–1, 192, 193
Components 340–2, 351–5
Computers 26–7, 204, 213
Concise cost plan 214–16, 249
Concrete 320
Conservation 466–7, 468–9
  areas 391, 466
Constructability 52
Construction industry
  council 449
  effect of government action 448–52
  manpower 442–5
  needs 445–8
  output 441, 445–7, 448
  resources 448
  structure 442–5
  training board 8
  workload 445–8
Consultants for inner city developments 461
Contaminated land 387–8, 435–6
Contingencies 250, 254, 262, 263, 267
Contours 90–1, 387
Contractor's cost control 199–201
Corporate finance 423–4
Cost
  adjustment opportunities 6
  analysis 22, 211–24, 272–4
  benefit analysis 470, 472–5
  checks 21–2, 189, 267–9
  control 22
  data 236–9
  definition 6–7
  indices 224–35
  information 237–41
  limits 222–4
  management 22
  modelling 202–10
  plan 21, 22–3, 189, 267
  plus contracts 16
  prediction 241–2
  records 371
  research 22, 244–7
  study 22
  unit 23
  value reconciliation 201
  yardsticks 120
Cost analyses 22, 211–24, 272–4
  application 235–6
  concise 214–16, 249
  detailed 216–22, 254–6
  educational buildings 213–14
  engineering services 272–4
  factories 217–20, 249
  flats 215, 254–6
  nature and purpose 211–13
  office block 188
  property occupancy 373–5
  social club 267
  standard form 214–22
  warehouse 216
Cost control 22
  application of techniques 248–75
  contractor's 199–201
  engineering services 270–5
  main aims of 6
  need for 4–6
  on site 200–1
  post-contract 197–9
  procedure 179–81, 271
  techniques 185–201
Cost modelling
  expert systems 27–9, 209–10
  historical development 203–5
  introduction 202–3
  other applications 208–10
  purpose 205–6
  types 206–8

Cost planning
  application of techniques 248–75
  computer aids 26–7
  mechanical and electrical services 194–7, 243–4, 270–5
  methods 185–93
  theories and techniques 174–201
CPI 15, 111
Culs de sac 129
Curtain walling 73

Decoration 336, 338
  costs 172
Demolition 343
Department of Education and Science 237–8
Design
  cost criteria 52, 53
  criteria 184
  detail 177, 179
  guidelines 469
  scheme 177, 179
  sequence of work 174–9
  stage 174–5, 177–9
  variables 31–44
Design and build contracts 17–18
Design and manage contracts 18
Develop and construct contracts 18–19
Developer's return or budget 405–6, 427–33
  causes of loss 422
  needs 420–1
  selection process 461–2
  worked examples 429–33
Development 389–90
  companies 5–6
  costs 421–2, 465–6
  decisions 473–4
  essence of 404–5
  factors influencing 387–8
  financial considerations 420–3
  properties 405
  public and private 405–19
  quantity surveyor involvement 422–3
  sources of finance 423–7
  town 464–6
  urban renewal 452–63
  working arrangements 462–3
Dimensional co-ordination 25–6
Direct labour organisations 326
Disabled persons' dwellings 122
Discounting future costs 327–9
Doors 78, 256
  costs 78, 171, 260
Drainage 172
Drawings and specification contracts 17

Easements 395
Educational buildings 223–4
  analysis of elements 213–14, 233
  cost analysis 213–14
Elderly persons' dwellings 122
Electrical installations 46, 61
Elemental
  cost analysis 168–70, 172, 255
  cost planning 186–90, 256–62
  costs 337–9
  unit rate 258–62
Elements 22
Encumbrances 394–6
Energy conservation 356–67, 368
  audits 356–7
  efficiency in new buildings 150–2
  ratings 357–61
  usage 358
English Partnerships 455
Enterprise zones 455
Environmental
  assessment 436
  economics 439–40
  friendly buildings 150
  impact assessment 438–9
  issues 435–8
  labelling 436–7
  management 434–5
Escalators 85
Eurogrants 456
European funding 152–3, 452
European procedures 3–4, 20–1, 452
Examiners' reports 29–30
Expanding towns 464–6
Expert systems 27–9, 209–10
External works 89, 350
Externalities 472

Facilities management 376–8
Factories 51–2, 57–8, 216–20, 331, 333
  cost analysis 217–20, 249
  development 409–11, 430–1, 432–3
  energy use 365–6
  site requirements 389
Feasibility 178, 182–5, 427–33
  studies 427–33
Features 23
Finance
  cost of 429, 430–1
  private 413–18, 423–7
  projects 424–7
  public 405–13
  reporting 201
  sources of 423–7
Finishes 62, 77–8, 79–80, 174, 177, 178
  costs 172
Fire protection 61, 88
Flats 39–41, 44–50, 56–7, 120–2, 148–9, 250–62
  concise cost analysis 215
  initial cost plan 263
Floor
  area estimates 159–60
  construction 77–8
  costs 78, 259, 354–5
  finishes 77–8, 260, 325, 354–5
  loadings 51–2, 60–1
  space index 390
  spans 51
Formwork 67–8
Foundations
  economics of 64
  piled and ground beam 63
  raft 63
  strip 63
  vibrocompaction 63–4
Friendly buildings 150
Future payments 327–9

Gas barriers 65
General building indices 226–7
Glazing 347
Government
  action on construction industry 448–52
  as client 450
  control 450–1
Greener buildings 372, 375–6
Gross development value 427–31
Ground conditions 90–1, 387–8
Groundwater 437
Group of eight 449
Grouping of buildings 44

Health and safety 393, 437
Heating 46, 81–2, 261, 338–9, 345–7, 356–66
Heights
  storey 41, 43, 60
  total 43–4
Highway requirements 391–2
Hospitals 223, 417–18
Hotels 41–2
House prices 121
Housing
  and construction statistics 238–9
  assessment of need 117–18
  association developments 5, 114–15, 124, 145–6, 361, 413–15, 416–17
  background to provision 113–15
  budgeting 413–15
  care and repair schemes 468–9
  condition 434
  condition survey 114, 319
  cost analysis 215
  cost control procedures 222–3
  cost indices 231
  cost limits 222–3
  cost yardsticks 120
  costs 36, 56–7
  costs of layouts 135–41
  densities 125–6
  design 119
  development 127–42
  dwelling types 118–23
  energy saving 150–2, 357–61
  financial aspects 118
  improvement grants 412–13
  inner city 454
  land use 115–17
  latest trends 114–15, 150–3
  layouts 128–41
  local authority 5, 413, 465
  maintenance costs 329–30
  modernisation 142–50
  overseas developments 132, 134–5
  Parker Morris standards 120
  prices 412
  private 411–12
  quality management 152
  refurbishment costs 147–9
  rehabilitation 143–50
  renovation grants 144–5
  reports 115
  requirements of occupants 123–5
  standard house types 122–3
  tenants' surveys 125
  tenures 115
  unfit 114, 319, 453, 455
  weighted analysis 233

Improvement grants 143–4, 412–13
Inception 174, 177, 178
Indices
  application 232–3, 235–6
  BCIS 226–7
  Davis Belfield and Everest 228
  PUBSEC 229
  purpose 224

sources 225
validity 232
Industrial development 409–11
land values 400–1
rental values 400–1
site requirements 389
Industrial engineering 276
Industrialised building 57–8
advantages 100–1
classifying systems 99–100
comparison with traditional buildings 106–7
contractual arrangements 101
costs 104, 105
defects 108–9
economics of production 105–6
future 107–8
historical background 99
housing 101–2, 103–4
labour implications 104–5
nature 99
problems 102–3
Information technology 2–3, 61, 379
Inner Urban Areas Act 1978 455
Insulation 74, 359, 361
Insurance 369–70
Intelligent buildings 378–9
Interest 328–9
Internal finishes 45
Interpolation estimates 173
Investment 297, 440, 442
Investment grants 455–7
Investor's needs 421, 424–5

Joinery fittings 45

Labour
availability 91–2
costs 10
Land
acquisition 419–20
ownership 387
use determinants 396
use patterns 380–1
use planning 381–4
values 384–7, 396–7, 400–2
Latham report 55
Leasehold Reform, Housing and Urban Development Act 1993 455
Life cycle costing
air conditioning 347–9
analysis 310–11
building designs 342–5
components 351–5
cost plan 342, 488–92
definition 22
discounting 327–9
examples 342–55
external works 350
heating systems 345–7, 356–66
lifts 349–50
management 310–11
nature 308–9
planning 310–11, 312, 488–92
problems 314–16
schools 361–3, 364, 388–9, 488–92
terminology 316–17
Lifts 46, 61, 84–6, 322–3, 349–50
Listed buildings 467
Lives of buildings 339–40
Lives of components 340–2
Loadbearing brickwork 65–6

Local authority developments 5
Local Government Planning and Land Act 1980 326
Location factors (BCIS) 234–5
Location of site 90
Locational factors 248, 257–8
Lump 442

M & E indices 228
Maintenance
audit 372
cost records 371
costs 329–32
execution of work 324–6
feedback 326–7
management 324–7
manuals 324–5
origins of problems 318–24
technology 317–24
types 318
volume of work 311–13
Maisonettes 121–2, 148–9
Management contracting 19–20
Market considerations 96–7
Masonry 320
Materials
analysis 233
availability 91–2
wastage 92, 95–6
Mean 234, 332
Measure and value contracts 17
Metals 321
Metric conversion table 486–7
Mode 234
Monte Carlo simulation 207

National Building Agency 102, 103
National Building Specification 111
Natural rights 393–4
NEDO price adjustment formula 231–2
Network cost modelling system 209
New towns 463–5
NHER ratings 110, 150–2, 357–61

Occupancy costs 322–39
Offices
air conditioned open plan 474–5
application of cost benefit analysis 474–5
cost plan 188
costs 406–8
design criteria 59–62
development 406–8, 429
energy costs 363, 365
maintenance costs 333, 335
rental values 397, 398, 408
requirements 59–62
types 363, 365
Outline building proposals 177, 178–9

Painting 321–2
Parker Morris standards 120
Partitions 73–4, 256
costs 74, 171, 260
Partnership (local authority and developers) 458–63
Pavings costs 172, 350
Performance expectations 152
Perimeter/floor area ratios 31–4, 37–9, 249
Plan of work 174–8
Plan shape 31–4, 60
Planning
balance sheet 473

controls 389–91
gain 394–5
regulations 91, 389–90
section 106 agreements 394–5
Plant
availability 91–2
usage 92–3
Plastics 79, 110, 321, 351, 352–4
Plumbing 46, 80–1
Pollution 437
Precast concrete structures 68
Prediction errors 367, 369
Prefabrication 97–9
Preliminaries 254, 262, 267
Premiums 306–7
Present value 328, 342–5, 348–9
Present value of £1 pa table 304–5, 484–5
Present value of £1 table 302, 478–9
Price 6–7
Price indices 224–35
Private finance projects 469–71
Profiled metal sheeting 72
Project finance 424–7
Property characteristics 403–4
Property occupancy cost analyses 373–5
PUBSEC index 228–9
Purchasing power relatives (EU) 3–4

QS 2000 1
Quality assurance and management 52–5, 152
Quantity factors 258
Quantity surveyor, role of 14–15, 422–3

Radburn layouts 129–31
Rainwater goods 351
Range 234
Ransom strips 395–6
Rationalisation 98–9
Real property 403–4
Recession problems 449–50, 452
Redundant buildings, re-use 466–7
Refurbishment of buildings 109–11, 146–50, 418–19
Refuse disposal 87, 322, 339
Regression analysis 207–8
Rehabilitation of buildings 146–50, 466–7
Rehabilitation of twilight areas 143–50
Reinforced concrete frames 66–8
Renovation grants 144–5
Rental value 305–6, 397–401, 408, 415
Residential development 127–42, 432
Residential land values 396–7
Restrictive covenants 394
Retail warehousing 410
RIBA plan of work 174–8
RICS cost planning system 190–1, 192, 193
Risk analysis 207
Rolling programmes 449–50
Roofs 45, 75–7, 320
costs 76–7, 171, 259, 337, 351–2
finishes 75–6, 171, 337, 351–2
types 75
Running costs 329–37

Sale and leaseback 426–7
Sanitary appliances 261
Schools 59, 188–92, 223–4, 361–3, 364, 388–9
energy conservation 361–3, 364
maintenance costs 331, 333
site requirements 388–9

weighted elements 233
Science parks 411–12
Security 61–2
Service charges 307
Services 80–8, 251, 256, 322, 338
cost control 194–7, 270–5
cost information 83, 243–4, 255
mechanical and electrical 243–4
planning 183
Shops 58–9, 408–9
development 431
environmental impact assessment 438–9
rental values 397, 399, 400
Simon committee 15
Simulated modelling 209
Sinking fund 300
Site
access 388
area 388
aspect 388
boundaries 388
characteristics 387–8
clearance 90
conditions 90–1
contamination 387–8
contours 90–1, 387
groundwater 387, 437
location 90
obstructions 388
operations 176
productivity 93–5
requirements 388–9
services 91, 388
soil conditions 387
works 46–7
Size of buildings 34–6
Sketch plans 174, 175
Social club 264–9
Soil conditions 387
Sound insulation 74–5
Specialist firms 442–3
Specification and design notes 251, 256, 275
Stairs 259
Standard deviation 234, 333
Standardisation 98
Steel frames 68–70
Storey enclosure estimates 160–4
Storey heights 41, 43, 60
Structural components 65–71
Student halls of residence 334–5, 418
Substructures 44
Superstructures 45
Sustainability 437
System building
defects 102–3, 108–9
methods 99–108

Target cost contracts 16–17
Taxation 369–71
Team approach 181
Tender price indices 225–6, 228–31
Tendering arrangements 15, 47
Terotechnology 311
Thermal insulation 74, 359, 361
Timber 320–1
Timber frames 70–1
Time factors 248, 257–8
Total costs 308–10
Tower block refurbishment 109–11
Town
centre redevelopment 457–63
conservation 467–9
planning 381–4, 385–6
Tree preservation orders 390–1

Unit estimating 156
University buildings 224, 330–2, 334–7, 373–5
Urban development areas 454
Urban regeneration 455–7, 467–9
Urban renewal 452–7
funding 455–7

Valuation
comparison method 297–8
contractor's method 298
investment method 299
of work in progress 198–9
profits method 298–9
reinstatement method 298
residual method 298
tables 300–5, 476–85
Value 6–7, 296–7
for money 313–14
Value management
case studies 287–93
charette 278
comparison with cost management 281–2
conclusions 293–5
criteria scoring/alternative analysis matrix 287, 288
definitions 277–8
fast diagrams 284–7, 289–91
functional analysis 282–4
general principles 277
introduction 26
strategy 281
techniques 282–7
workshops 279–80
Variations 199
Ventilation 61

Wall to floor ratio 31–4, 37–9, 249
Walling 71–3, 337
costs 171, 259
Warehousing 410
Wastage of materials on site 92, 95–6
Waste disposal 80–1, 261
Wilderness study group 50–1, 238
Windows 78–9, 256, 259, 338, 352–4
Working drawings 174–5, 177

Yardsticks 120
Years' purchase 299–300
table 304–5, 484–5